The Spectrum of Atomic Hydrogen
ADVANCES

The Spectrum of Atomic Hydrogen
ADVANCES

a collection of progress reports by experts

edited by **G W Series**

and incorporating

The Spectrum of Atomic Hydrogen

G W Series
Oxford University Press ... 1957

World Scientific
Singapore • New Jersey • Hong Kong

.03108338

Published by

World Scientific Publishing Co. Pte. Ltd.
P. O. Box 128, Farrer Road, Singapore 9128

U. S. A. office: World Scientific Publishing Co., Inc.
687 Hartwell Street, Teaneck NJ 07666, USA

Part I of this book was reprinted from *Spectrum of Atomic Hydrogen*
(Oxford University Press, 1957)

Library of Congress Cataloging-in-Publication data is available.

THE SPECTRUM OF ATOMIC HYDROGEN: ADVANCES

ISBN 9971-50-261-5
9971-50-287-9 (pbk)

Printed in Singapore by Utopia Press.

EDITOR'S INTRODUCTION

Hydrogen was news. Hydrogen is news. *The Spectrum of Atomic Hydrogen*, here reprinted as Part I, was written only ten years after Lamb and Retherford, in 1947, had so convincingly demonstrated that even the best contemporary theory of an atom in an empty universe fell short of what their apparatus told them was true; and in those ten years the foundations of quantum electrodynamics had been consolidated and experiments had been refined to the point where — so it appeared — only the *t*'s remained to be crossed and the *i*'s to be dotted. In Part I, I tried to tell the story up to that time.

And yet, thirty more years on, hydrogen is still news, as Part II of this book testifies. The *t*'s indeed have been crossed and the *i*'s dotted, and QED has been taken as exemplar for QCD, but work on hydrogen continues to flourish. The reason, surely, is that the apparent simplicity of the hydrogen atom invites meticulous scrutiny of its very depths, yet those depths are sufficiently impenetrable to present a challenge both for the exercise of the unbelievably sharp experimental tools which are now in use and in contemplation, and for the exercise of the enormous computing power now available to theorists.

Part II tells that more recent story. The field has widened. 'Hydrogen and hydrogen-like systems' would be more accurate than 'Hydrogen' in the title, but atomic hydrogen remains the prototype of the systems we describe, and that is what is reflected in the title we use.

Authors active in their respective fields have written the chapters, which are different in style and in scope from the chapters of Part I. The reason is that, to exhibit the detail, there is so much more to be said. Yet we have tried to present a text which will serve several purposes: to give physicists at large some idea of the tremendous effort that has gone into these studies and of the conclusions that have been reached; to give newcomers to the field an understanding of where the barriers to progress are now set; to give those who labour in one corner the opportunity to survey the achievements of their colleagues labouring in others.

Part I has been reprinted because, for some readers, it may serve a purpose as an introduction to Part II and because, even in elementary physics, the time comes when it needs to be revealed to the serious student that the spectroscopy of hydrogen did not come to an end in 1913. Relativistic fine structure, the Lamb shift, hyperfine structure, positronium — these already find a place in Part I.

No book on hydrogen can claim completeness of survey, nor is that claim made for this. In Part II we review experimental spectroscopy in widely different regions, the optical and the radiofrequency, and we review quantum electrodynamics calculations. We give an account of transition theory and of the electroweak interaction. We review the simpler coulombic systems muonium and positronium and the more complicated systems, hydrogen-like ions of higher Z. We study the symmetries of the coulomb field itself by examining the response to external fields of hydrogen and alkali atoms in Rydberg states. We look at hydrogen as a system which has been mined deeply to provide material for the evaluation of the fundamental atomic constants. We give a theoretical treatment of the minute, yet perceptible changes brought about by black body radiation above the zero-point. The chapters differ in approach, in comprehensiveness, even in the mundane detail of style of referencing. These differences reflect the individuality of the several authors. There has been some cross-referencing but very little emendation: even some overlap of material has been permitted to remain in order to preserve the unity of presentation within each chapter.

Some current investigations we are aware of but do not include in this book are: level shifts for atoms confined in cavities, level shifts in a 'squeezed' vacuum, level shifts in non-Euclidean spaces. These topics are referenced in a supplementary list at the end of the book. Also referenced are some items of a different character illustrating some of the widespread applications of hydrogen spectroscopy: in laboratory and astrophysical plasmas — as diagnostic of their physical condition; in astrophysical plasmas additionally — as indicating masers in the sky; in quasars — by observation of Lyman-α Doppler-shifted into the visible spectrum, as evidence of extraordinary speeds of recession. The list of applications is endless. But in this book we are mainly concerned with exploring the extent of agreement between experimental observations of the spectrum of atoms — unperturbed as far as is practicable — and the predictions of theory. As to which theory, there is no alternative in point of detail to quantum electrodynamics (embedded in contemporary physics within the electroweak theory), but will experimental refinement ever push it beyond its limitations? Will optical activity before long be demonstrated in a gas of hydrogen atoms? Will some successor to the electroweak theory ultimately claim a signature in the spectrum? And for every experimental technique that is yet to be discovered: is there anything new it might tell us about hydrogen? Some of these questions will surely be answered and others raised within the next ten years.

In a work of this kind the authors lean heavily on their friends and colleagues for up-to-date, often unpublished material, and on owners of copyright for permission to reproduce figures that have been published. The authors and Publisher here express their gratitude for information that has been generously given and for permissions that have been granted. Where figures have not been specially drawn for the present work the sources are acknowledged in the text.

The Editor wishes especially to thank Ms Chong and her colleagues in the publishing office for their patience and willingness to accept corrections and supplementary material to a very late stage. And finally, with pleasure I record my thanks to Heini Kuhn, my former supervisor and esteemed colleague, who first suggested to me that the spectrum of atomic hydrogen might be worth a crack.

Oxford
January 1987

LIST OF AUTHORS AND THEIR AFFILIATIONS

Dr. G. Barton: *School of Mathematics and Physical Sciences, University of Sussex, England.*

Prof. G. W. F. Drake: *Department of Physics, University of Windsor, Ontario, Canada.*

Dr. J. C. Gay: *Laboratoire de Spectroscopie Hertzienne del'E.N.S. Université Pierre et Marie Curie, Paris.*

Prof. T. W. Hänsch: *Max-Planck Institut fur Quantenoptik, Garching, West Germany, formerly of Department of Physics, Stanford University, U. S. A.*

Prof. E. A. Hinds: *Department of Physics, Yale University, U. S. A.*

Dr. A. P. Mills, Jr.:   *AT&T Bell Laboratories, Murray Hill, U. S. A.*

Dr. P. J. Mohr: *National Science Foundation, Washington, and National Bureau of Standards, Gaithersburg, U.S.A., formerly of Yale University.*

Prof. G. W. Series: *Clarendon Laboratory, Oxford, England.*

Dr. E. Träbert: *Institut für Experimentalphysik, Bochum, West Germany.*

CONTENTS

PART II

The Spectrum of Atomic Hydrogen
ADVANCES

CHAPTER I

INTRODUCTION

ATOMIC spectra furnish material for testing theories of atomic structure. Since hydrogen is the simplest kind of atom, the interpretation of its spectrum has been of the greatest interest to theorists, particularly since all but the most recent theories can be applied with mathematical rigour to the two-body problem which the hydrogen atom presents.

The spectrum of atomic hydrogen is simple, but its details are difficult to measure. The imperfections of successive theories have been revealed only through measurements of increasing precision. Current interest springs from very small differences between the observed spectrum and the predictions of the most complete and logically satisfying of the quantum mechanical theories: that of Dirac. These small differences are of profound theoretical significance.

The inadequacy of Dirac's theory was reported in the nineteen thirties by Houston, Gibbs and others working on the fine structure of the red line of hydrogen. More refined measurements by R. C. Williams in 1938 [139] (see also Chapter VII) allowed a quantitiative estimate of a discrepancy. But in the opinion of other spectroscopists there were no such discrepancies: Dirac's theory gave a perfectly satisfactory account of the spectrum. The situation was clarified beyond doubt in 1945 by the success of a remarkable microwave experiment by Lamb and Retherford (Chapter VIII). Displacements of energy levels were accurately measured, and have since been confirmed by optical spectroscopy (Chapter X). They are known as Lamb shifts.

Vigorous theoretical activity followed Lamb's experiment. Ways have been found of treating the atom, not in isolation, but in interaction with the electromagnetic field (Chapter IX). These developments have been remarkably successful in explaining the Lamb shifts. At the same time they predict a small change in the formerly accepted value for the magnetic moment of the electron, in agreement with recent measurements of the magnetic moments of free atoms. The hyperfine structure of the hydrogen spectrum is also affected, since this depends on the interaction between the magnetic moments of the electron and the nucleus (Chapter XI).

For so long as it has been studied, the hydrogen spectrum has excited interest as great as that which prevails today. It so happens that successive theories, though sometimes differing widely in their approach, have led to remarkably similar conclusions. The distinctions between their predictions have lain, not in gross variations in the positions of the components of a line structure, but in their relative intensities, or in the presence or absence of weak satellites, or in very small differences of position. Since these features are greatly influenced by the action of electric fields on the emitting atoms, false conclusions as to the validity of the several theories have often been drawn. The subject has therefore been the occasion of frequent discussion and controversy, and advance in it has waited upon the more careful control of experimental conditions.

The history of the spectrum of atomic hydrogen—with which we include that of ionized helium and other one-electron atoms—is therefore closely linked with the history of the quantum theory. The study of the spectrum at the present time rests upon half a century of detailed experimental and theoretical work which illustrates in a remarkable way the interdependence of theory and experiment in physics. Interest in the contemporary work is enhanced when it is seen in its historical context. The following chapters will give an outline of the successive theories, the experimental findings which they could interpret, and the points of detail at which they proved deficient. The work on the Lamb shift itself, and related topics, will be described in considerably greater detail.

CHAPTER II

THE EXCITATION OF THE SPECTRUM

2.1. The atomic and the molecular spectrum

AN electric discharge in hydrogen gas excites both the atomic and the molecular spectrum. The latter is frequently called the 'many-lined' spectrum, since, although it is a band spectrum for purposes of analysis, its appearance is quite unlike that of a typical band spectrum. The low atomic weight gives rise to a low moment of inertia of the molecule, with correspondingly widely spaced rotational lines which fall irregularly throughout the spectrum. A well-known feature of the molecular spectrum is the continuum which extends from 1600 to 5000 Å. It is frequently used as a continuous background for absorption measurements in the ultra-violet.

The main feature of the atomic spectrum in the visible region is a very strong red line known as H_α (6563 Å) which is the same as the C line of Fraunhofer in the solar absorption spectrum. A strong blue line, H_β, and a violet line, H_γ (the Fraunhofer F and f lines) at 4861 and 4340 Å respectively, are easily visible, and H_δ at 4101 Å may also be seen. These four lines are the first members of the celebrated Balmer series which is prominent in the spectra of stars and nebulae as well as in the spectrum of hydrogen excited in the laboratory. A stronger series in the far ultra-violet was found by Lyman [89] in a laboratory source, although air is not transparent at these wavelengths. Its first member has recently been detected in the solar absorption spectrum from a high-altitude rocket [23]. Other series are described in section 3.3.

The relative intensities of the atomic and molecular spectra in laboratory sources depend strongly on the conditions of the discharge. Although it is generally true that high gas pressure favours the molecular spectrum and violent discharge conditions favour the atomic, it is also found that a great deal depends on whether there is in the discharge tube a catalyst for the recombination of atoms. Most metals and extremely clean, dry glass surfaces act as catalysts. Ground glass or the rough edges of broken glass also act in this way.

The early experimenters relied on the violent excitation produced by condensed spark discharges to obtain a strong atomic spectrum, but Wood [144] developed a type of discharge tube which enabled him not only to observe an almost pure atomic spectrum, but also to draw off in a side tube a stream of atomic hydrogen and to investigate its properties.

2.2. Atomic hydrogen

Atomic hydrogen has a lifetime of about $\frac{1}{3}$ second, and can be made to travel distances of the order of 10 cm at pressures of 0·1 mm mercury, if no catalyst is present. Collisions between two atoms in the gas do not result in recombination, and collisions between two atoms at the wall do not always do so. If recombination does take place, a large amount of energy is liberated, so that recombination on a fine tungsten wire, for example, causes it to become red hot.

2.3. Wood's discharge tubes

The principal feature of Wood's discharge tubes was that the electrode regions were well separated from one another by a very long, clean glass tube. Thus, he used a tube of 7 mm bore, and 140 cm long, with platinum electrodes at the end carrying either DC or AC. The tube could be bent if desired. At a pressure of about $\frac{1}{2}$ mm mercury, a current of 100 mA in such a tube will excite the molecular spectrum at the electrodes and the atomic spectrum in the region well away from the electrodes. The atomic spectrum may not appear at first if the tube is not clean, for then the discharge will show the molecular spectrum at those parts of the tube which are dirty. After a while, the discharge itself cleans the tube and the atomic spectrum can be obtained practically pure. It is necessary to evacuate the tube and refill it with clean hydrogen. After prolonged running, the glass may become extremely clean and dry, and under these conditions the discharge in pure, dry hydrogen will exhibit almost exclusively the molecular spectrum.

This behaviour can be understood in terms of the catalytic activity of the glass. Very clean, dry glass assists the recombination of atomic hydrogen, but this activity is easily 'poisoned' by the presence of water vapour. Thus, unless the glass has been brought to the extremely clean state by prolonged running of the discharge, the surface will not favour recombination. Gross particles of dirt and rough edges of glass on the other hand, may encourage recombination, so that the glass has to be reasonably clean and smooth if the atomic spectrum is desired. Cooling the outer surface of the glass with liquid air encourages the formation of molecules, but it is still possible to observe a strong atomic spectrum in a cooled tube. As one would expect, metal surfaces encourage recombination even if they are not scrupulously clean. Aluminium does not favour recombination as strongly as most metals.

Wood's tubes have been used by most investigators of the atomic spectrum of hydrogen, but not in some of the most recent work [76]. It has become a standard practice deliberately to introduce water vapour into the hydrogen, and in later work, when the desirability of mild excitation was realized, to add helium also. This allows a stable discharge at lower voltage, and has been shown to have no influence on the position of the hydrogen lines [139]. In sealed hydrogen discharge tubes it is customary to use a large volume, since the gas is driven into the walls by the discharge, with consequent loss of pressure.

Wood's tubes have been used as a source of hydrogen atoms in atomic beam experiments [71]. A small slit in the middle of the long, glass section allows the atoms to diffuse into the beam chamber. A concentration of hydrogen atoms of 0·7 to 0·9 was reported. In recent microwave work [143], the use of a mixture of dimethyldichlorosilane and methyltrichlorosilane proved very effective as a wall-poisoner.

2.4. The 'ring' discharge

An alternative, very efficient source of atomic hydrogen is the so-called 'ring' discharge. This relies on the inductive excitation due to a coil carrying high frequency oscillations which surrounds a glass tube containing hydrogen gas at low pressure. This was first described by Masson [91], and used notably by Herzberg [59] in his investigation of the Balmer series continuum (sections 3.5 and 6.3). The appearance of the ring discharge depends on the gas pressure and on the current in the coil, there being very little wall effect since the ions in the discharge are constrained to move in circles concentric with the coil. The degree of dissociation and excitation of the hydrogen depends on the voltage drop per mean free path, so that reducing the pressure has the same effect as increasing the current in the coil. It is possible to observe concentric rings showing different degrees of excitation, the greater excitation being at the centre. Thus, at pressures below 1 mm mercury, Herzberg observed a white outer ring showing mainly the molecular spectrum, then a red ring in which the molecular spectrum was weak and the Balmer spectrum strong, and a blue inner ring with no molecular spectrum, and H_β stronger than H_α.

Ring discharges have found a recent application in providing sources of protons or deuterons in high energy accelerators. In one such source, for example [130], clean, dry hydrogen is admitted through a palladium tube into a glass vessel about 5 cm in diameter where the pressure is maintained at about $\frac{1}{100}$ mm mercury. A few turns of heavy gauge copper wire surrounding the glass form part of a tuned circuit which carries a current of 10–100 amp at 20 Mc/s. The bright red colour of the discharge is due to H_α, and signifies a high concentration of atomic hydrogen. In fact, the yield of protons in this source is about 90 per cent.

2.5. Electron bombardment

The conditions in a gas discharge are clearly more complicated than those which would obtain if the gas were excited by a controlled beam of electrons. Nevertheless, investigators of the optical spectrum of hydrogen have always used the gas discharge method of excitation for reasons of simplicity, and because intense spectra are readily obtained in this way. The influence of discharge conditions on the fine structure has been studied by the more recent investigators, who find that the factor which limits the precision of their measurements, if the discharge is sufficiently mild, is the Doppler effect rather than the uncertainties of Stark and pressure shifts.

For a number of elements it has proved feasible to reduce by a large factor the Doppler width of spectrum lines in the radiation produced by electron or photon bombardment of a collimated atomic beam (the former process for emission, the latter generally for absorption spectra). The possibility of applying this technique to hydrogen has been discussed in a number of laboratories, but the difficulties of producing an atomic beam of sufficient intensity for optical spectroscopy have proved too great. Excitation by gas discharge was the technique employed in all the optical investigations described in this book.

For radiofrequency measurements of the fine structure of hydrogen, on the other hand, the Doppler effect is completely unimportant, since it is proportional to the frequency which is actually measured. The fine structure intervals are given by frequency differences in optical spectroscopy: in radiofrequency spectroscopy they are measured directly. The greater precision of radiofrequency measurements would compel careful investigation of the conditions in a gas discharge if this method were chosen for the excitation of the atoms. In fact, in the radiofrequency experiments which have so far been performed on hydrogen and ionized helium, the method of excitation by electron bombardment has been used.

Excited atomic states in hydrogen can be produced by electron bombardment of molecular or of atomic hydrogen. Lamb and Retherford used the latter method (see Chapter VIII), since their method of observing radiofrequency resonances relied on the detection of the atoms themselves, which were produced as a beam. The relative merits of various ways of breaking down the molecule are described in one of their papers [82]. Lamb and Sanders, who investigated a different excited state (section 10.2.2), detected radiofrequency resonances in a way which did not require the use of an atomic beam. Consequently they bombarded the molecule directly.

Pulsed electron bombardment was used by Novick, Lipworth and Yergin in experiments on the metastable level $2^2S_{\frac{1}{2}}$ of ionized helium (section 10.2.1).

CHAPTER III

THE GROSS STRUCTURE

3.1. Balmer's formula

BALMER'S celebrated representation in a single empirical formula of the wavelengths of the principal lines in the visible and near ultra-violet spectrum of hydrogen was not the first attempt in this direction, but was indisputably the most successful. Rydberg [115] records that he heard of Balmer's formula as he was trying out various forms of the function $f(n)$ in his expression

$$\nu_n = C - f(n), \tag{3.1}$$

n being a positive integer, for the series relationship between the wavenumbers ν_n of the lines in the alkali spectra. Balmer's formula, $\lambda_n = An^2/(n^2-4)$, which is equivalent to

$$\nu_n = 1/\lambda_n = R(1/2^2 - 1/n^2), \tag{3.2}$$

is a special case of Rydberg's formula when, in the function $f(n) = R/(n+\phi)^2$, one takes $\phi = 0$. Since the function $R/(n+\phi)^2$ fitted the alkali spectra, with ϕ taking different values for the different series, the close relation between the spectrum of hydrogen and those of the alkalis was known from the earliest period of spectrum analysis.

3.2. Bohr's analysis

Bohr's theory of the dynamical behaviour of an electron under the influence of a massive electrically charged atomic nucleus is well known. It is of interest to recall the way in which he was led [14] not only to provide a theoretical basis for

Balmer's formula, but also to establish one of the most important results of quantum mechanics: the quantization of angular momentum in units of $h/2\pi$. This result arises from the analysis in his paper; it is not his starting point. Bohr's quantum postulate was based on Planck's assumption of the quantization of the energy of harmonic oscillators. By analogy with this, and arguing from a correspondence with classical physics, he set a restrictive condition on the mechanically possible electron orbits, and postulated that this limited set of orbits should be non-radiating.

Let W be the energy necessary to remove to infinity an electron moving without radiation in a circle of radius a under the Coulomb attractive force Ze^2/a^2. (W is the negative of what is now known as the energy of the state, E.) Then, if $f = \omega/2\pi$ is the frequency of revolution of the electron in its orbit, we have

$$-W = \tfrac{1}{2}ma^2\omega^2 - Ze^2/a \qquad (3.3)$$

and $\qquad m\omega^2 a = Ze^2/a^2, \qquad\qquad\qquad (3.4)$

so that $\qquad 2a = Ze^2/W \qquad\qquad\qquad\quad (3.5)$

and $\qquad f = \omega/2\pi = (1/\pi Ze^2) . (2/m)^{\frac{1}{2}} . W^{\frac{3}{2}}. \qquad (3.6)$

Thus, an electron at infinity has $W = 0$ and therefore $f = 0$. At finite distances its frequency of revolution is finite. If we assume that the electron, in coming from infinity to a non-radiating orbit where its frequency is f emits light whose frequency is the mean of the orbital frequencies f_∞ and f (this is the correspondence with classical physics) we have

$$W_n - W_\infty = nh\left(\frac{f+f_\infty}{2}\right) \text{ (from Planck's assumption),}$$

i.e. $\qquad\qquad\qquad W_n = nhf/2. \qquad\qquad\qquad (3.7)$

This equation, which is the restrictive condition, may be interpreted as representing the emission either of n quanta of the basic frequency $f/2$, or of one quantum of the nth harmonic, $\dfrac{nf}{2}$.

Eliminating f, we find

$$W_n = 2\pi^2 Z^2 e^4 m/n^2 h^2 = RZ^2/n^2 \qquad (3.8)$$

2

The next step is to appeal again to Planck's assumption in equating the frequency of the light radiated when the state of the atom changes from orbit n to orbit m, to $(W_m - W_n)/h$:

$$\nu_{nm} = (W_m - W_n)/h \qquad (3.9)$$

Balmer's formula follows when we set $m = 2$.

Bohr points out in a later paper [16] that for any system containing one electron rotating in a closed orbit, the replacement of W by the average kinetic energy (which is equal to W), secures an identity between the frequency of radiation calculated by equation 3.9 and that to be expected from ordinary electrodynamics in the limit when the difference between the frequencies of rotation of the electron in successive stationary states is much less than the absolute value of the frequency. The requirement that these two frequencies be identical is a partial statement of the Correspondence Principle.

We may go on to work out the angular momentum p_n of the electron in state n,

$$p_n = ma^2\omega_n = nh/2\pi, \text{ from equations 3.4, 3.5 and 3.7.}$$

This is the condition which is usually taken in conjunction with equations 3.3 and 3.4, to determine the stationary state n.

3.3. Experimental tests

It is not profitable to make a precise comparison between the most recent experimental value of the constant R in equation 3.8 and its value in terms of atomic constants since as we shall see in later chapters, we must recognize several small corrections to Bohr's expression. The agreement at the time was well within the limits of error in the measurement of the atomic constants, and was very powerful evidence in support of Bohr's theory.

The validity of Balmer's formula itself, however, was called in question by Curtis [27], who undertook precise measurements of the wavelengths of the first six members of the series. Using Balmer's formula, he calculated the value of R (equation 3.2) from the measured wavelengths. The results showed a definite trend in R with the running integer n. Although it was possible to meet this situation by using a constant R and non-integral n, Bohr showed later that at least part of the discrepancy could be taken up by considering the relativistic change of mass with velocity. This will be discussed more fully in the next chapter.

Further support for Bohr's theory came from the discovery of line series in the hydrogen spectrum for which the other integer m took values other than 2. The far ultra-violet series for which $m = 1$ has already been mentioned. Lyman announced the discovery of the first two members a year after Bohr's paper, recognizing the close connection between the wavelengths of these lines and Balmer's formula. Balmer himself had asked whether there might not exist a series for which $m = 3$, and Paschen had observed its first two members in the infra-red in 1908 [101]. The series with $m = 4$ was found by Brackett in 1922 [18], with $m = 5$ by Pfund in 1924 [108], and with $m = 6$ by Humphreys in 1953 [66]. It is to be understood that in all these series, the 'running' integer n takes $(m+1)$ as its first value.

3.4. Hydrogen-like spectra

It is further to be deduced from Bohr's theory that an atom other than hydrogen, ionized so that only one electron remains bound to the nucleus, will also emit hydrogen-like spectra. This has been verified for elements up to oxygen by the work of Edlén and Tyrén [93].

The spectrum of singly ionized helium has been as important in the history of spectroscopy as that of hydrogen itself. From the time of its discovery in a stellar spectrum [109] until Bohr's theory, it was believed to be part of the hydrogen spectrum, since the series discovered by Pickering converged to the same limit as the Balmer series. The Pickering and the Balmer series were interpreted as the sharp and diffuse series of hydrogen, respectively. On this interpretation it was to be expected that the first line of the principal series would be found at 4686 Å, and such a line was indeed observed in the spectrum of the sun's chromosphere [48] and in a laboratory source containing hydrogen and helium [49]. Bohr interpreted this line as an ionized helium line, with m and n equal to 3 and 4 respectively, and similarly accounted for the Pickering series by setting $Z/m = \frac{2}{4}$, which would give it the same convergence limit as the Balmer series with $Z/m = \frac{1}{2}$.

Bohr's interpretation was still disputed [13], [50] since there was a discrepancy of about 2 Å between the observed wavelength of the line at 4686 Å and the calculated wavelengths using the value of the Rydberg constant derived from the Balmer series, but Evans [43] showed convincingly that the 4686 line belonged to helium by observing it in a discharge which showed no trace of the Balmer lines. The wavelength discrepancy is accounted for by a slight refinement of Bohr's theory, namely, by allowing for the finite mass of the nucleus. The motion of electron (mass m) and nucleus (mass M) then takes place about their common centre of mass, and the formulae derived from the same equations of motion as before are changed only to the extent that m is replaced by the reduced mass, $mM/(m+M)$. One must therefore use slightly different Rydberg's constants for nuclei of different masses. This aspect of Bohr's theory is completely vindicated by observations and has been used as a basis for measuring the atomic mass of the electron [140].

3.5. High series members and the continuum

A difficulty noticed by the early observers was that more members of the Balmer series had been observed in celestial spectra than in laboratory sources. Bohr suggested that, since high series members represented transitions from orbits of large radius ($a \propto n^2$), conditions in laboratory light sources might be such that large orbits could not be realized. Possibly the lines might appear at lower pressures. Experiments in this direction were completely unsuccessful, and further progress was not made until Wood developed his discharge tubes. The condition for high series members is the suppression of the molecular spectrum, for against this background weak atomic lines cannot be seen. The limit of the Balmer series and the continuum beyond were photographed by Herzberg [59] in the light of his ring discharge (section 2.4).

CHAPTER IV

THE FINE STRUCTURE—OLD QUANTUM THEORY

IT had been known for over twenty-five years before Bohr's theory that the Balmer lines were not single, but appeared to be close doublets under instruments of high resolving power. There was, however, considerable disagreement as to the separation between the doublet peaks. It was particularly important to the early spectroscopists to know whether the doublet separation remained constant throughout the Balmer series, or whether it gradually decreased, since this is an important distinction in alkali spectra between the sharp and diffuse series on the one hand, and the principal series on the other. In fact, each Balmer line is a blend of sharp, principal and diffuse doublets. Complete resolution even in the most favourable case of H_α has not yet been achieved.

4.1. Relativistic effects

Bohr himself outlined a possible theoretical interpretation of the fine structure [15]. In noticing that the relativistic change of mass with velocity would slightly alter the energy of the orbit, he remarked that if the velocity of the electron were small compared with the velocity of light, the energy would be the same whether the orbits were circular or not, but that for higher velocities, orbits which were not circular would rotate in their own planes. Bohr did not calculate the energy of these rotating orbits, but conceived that this extra frequency of revolution of the electron would appear in a 'combination tone' in the light radiated by the atom. This is the kernel of the correspondence principle which was subsequently highly developed by Bohr and Kramers.

An important generalization of the quantum theory by Sommerfeld [125] and independently by Wilson [142] allowed a detailed study of the non-radiating non-circular orbits, and led to Sommerfeld's celebrated fine structure formula which represents the energy levels of hydrogen-like atoms to a precision which was substantiated by the most refined experiments over the twenty years following its derivation. Comparison with experiment, however, implies a consideration, not only of energy levels, but also of the relative intensities of spectral lines. We shall see that on this point the theory failed.

4.2. Quantum conditions

The Sommerfeld–Wilson postulate was that non-radiating orbits are specified by the condition that the phase integrals $J_i = \oint p_i dq_i$, which are constants of periodic motions in classical mechanics, should be set equal to integral multiples of Planck's action constant h. This condition is equivalent to the postulate of Ehrenfest [42] that those quantities should be quantized which are adiabatic invariants of the motion (see, for example, [126]). The phase integrals are adiabatic invariants, that is to say, they remain unchanged during the variation of any of the parameters of the motion—for example, the field of force under which the body moves—provided that this variation is gradual and not correlated with the phase of the motion. A number of integers (quantum numbers) equal to the number of phase integrals, i.e. to the number of degrees of freedom of the atomic system is introduced in this way. In the case of a hydrogen-like atom, the electron moving relative to a fixed nucleus has three degrees of freedom, and therefore there are three quantum numbers.

4.3. Energy levels

Sommerfeld's formula for the energy of hydrogen-like atoms is derived under the assumptions of relativistic motion under an inverse square attractive force, subject to the above quantum postulates. It exhibits only two quantum numbers, n and k. The third, m, appears explicitly if the atom is situated in a magnetic field, in which case orbits inclined to the field at different angles have different m and different energies. The formula may be derived rigorously, and expressed in closed form as follows:

$$E_{\text{rel}} = m_0 c^2 \left\{ 1 + \frac{\alpha^2 Z^2}{(n' + \sqrt{k^2 - \alpha^2 Z^2})^2} \right\}^{-\frac{1}{2}}, \quad \begin{array}{l} n' = 0,1... \\ k = 1,2... \end{array} \quad (4.1)$$

where E_{rel} includes the rest mass $m_0 c^2$, and α is the dimensionless fine structure constant $2\pi e^2/hc \approx \frac{1}{137}$. It is convenient to expand equation 4.1 in powers of $(\alpha Z)^2$. Subtracting $m_0 c^2$ we are left with

$$E = -\frac{RZ^2}{n^2} \left[1 + \frac{\alpha^2 Z^2}{n} \left(\frac{1}{k} - \frac{3}{4n} \right) + ... \right], \quad \begin{array}{l} n = n'+k = 1,2... \\ k = 1,2 ... n. \end{array} \quad (4.2)$$

which shows the relativistic correction (the second and subsequent terms within the bracket) as a small addition to the 'Bohr' term $-RZ^2/n^2$.

4.4. Intensities and the Correspondence Principle

Although Sommerfeld made certain proposals concerning the relative intensities of the fine structure components, they were rather arbitrary and were not supported by experiment, in particular, by observations on the line 4686 Å of He^+. The development of the Correspondence Principle by Bohr and Kramers was more successful, and more convincing in its logical basis.

The Correspondence Theorem is a formal expression of the requirement that the radiation calculated by quantum theory must agree in frequency, intensity and polarization with that calculated by classical theory when the difference between the initial and final quantum states tends asymptotically to zero. This condition is met when the difference between the two values of a particular quantum number is small compared with the absolute value.

In its application to hydrogen-like spectra the most important result of the Correspondence Principle was the prediction of a 'selection rule' for the quantum number k. When the atom undergoes a change from one stationary state to another, k must change by ± 1, although no restriction is placed on the change of n. These rules are consequences of the fact that k is the quantum number belonging to the slow precessional motion of the orbital ellipses (corresponding to the relativistic correction term in equation 4.2), and this is a uniform precessional motion without harmonics. The number n', on the other hand, belongs to the radial motion, whose Fourier representation, since the motion in the ellipse is not uniform, contains harmonics as well as the basic frequency (see, for example, [106]).

While this severe restriction on all conceivable transitions was an indisputable consequence of the Correspondence Principle, the prediction of the relative intensities of the allowed transitions was less certain since this implied some sort of averaging between the classical radiations from the initial and final states. The principle was applied to predict the intensities and polarization of the Stark effect components of the Balmer lines with considerable success for those electric fields at which the Stark effect is much greater than the fine structure [74]. The selection rule $\Delta k = \pm 1$ is no longer valid when the atom is in an electric field. Thus 'forbidden' lines were predicted even for relatively weak fields where the theory could not be tested, but the intensity of such lines would decrease to zero with the field.

CHAPTER V

OLD QUANTUM THEORY—COMPARISON WITH EXPERIMENT

PASCHEN's study [102] of the lines of ionized helium, and, in particular, of the line 4686 Å ($n = 4 \rightarrow n = 3$) was claimed to provide striking experimental support for Sommerfeld's theory. The fine structure of the Balmer series, as far as it was resolved at that time, also supported the theory, and further weighty evidence came from measurements of X-ray fine structure. In the following sections we shall briefly review these lines of argument.

5.1. The ionized helium lines

These were produced by Paschen in DC and in condensed spark discharges, and resolved by a powerful grating. More components appeared in the condensed spark than in the DC discharges, but in neither pattern was the structure completely resolved. The condensed spark spectrograms showed the closest agreement with the predicted patterns, and measurements of these allowed an experimental determination of the fine structure constant α. In this comparison between theory and experiment, Sommerfeld's original intensity rules, later abandoned, were used.

The interpretation of the condensed spark patterns was later challenged by Leo [87] who showed that an important component labelled IIIa by Paschen belonged to the spectrum of the helium molecule He_2 (see Fig. 8, page 58). Paschen's interpretation of the structure, and his value of α, were consequently invalid. Leo's evidence was accepted by Paschen [103], who re-interpreted his old observations, now making use of the DC spectrograms. At this time, new intensity formulae had been derived by Sommerfeld and Unsöld [127], and, making use of these, Paschen derived a new value for α differing slightly from his old value, but agreeing well with the theoretical value based on accepted values of e, h and c.

The formulae of Sommerfeld and Unsöld, however, made use of the concept of electron spin. The quantum numbers describing the energy levels were changed and transitions were allowed between levels which originally had the same value of k (compare Figs. 1, page 18, and 2, page 33). Paschen's new interpretation therefore violated the selection rule $\Delta k = \pm 1$. Seen in retrospect, therefore, the agreement between Paschen's observations and Sommerfeld's theory is much less convincing than it appeared to be at the time.

5.2. The Balmer lines

Paschen's 1916 paper also records observations on the fine structure of the Balmer lines, which appeared double to him as to earlier observers. Recognizing in these doublets a blend of the components predicted by Sommerfeld's theory, Paschen

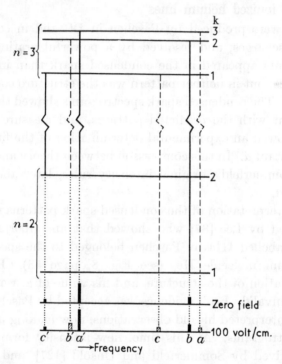

FIG. 1. Fine structure of H_α showing the relative intensities of the allowed transitions according to the old quantum theory. Forbidden transitions, which may appear under an electric field, are drawn with broken lines.

showed that the doublet intervals were consistent with the value of α derived from the ionized helium line. The lack of resolution did not allow a precise test of the theory.

The experimental situation was improved in 1925 by Hansen [55] who made new, very careful observations of the fine structure of the Balmer lines. His use of a Lummer plate allowed greater resolving power than Paschen's grating, but the Doppler broadening of the lines still limited the resolution.

Hansen compares his observations with Sommerfeld's term values and the transition probabilities calculated by Kramers from the Correspondence Principle (Fig. 1). The predicted pattern in zero field is very simple since the selection rule $\Delta k = \pm 1$ prohibits the transitions a, c and b'. These components should appear weakly, however, in consequence of the Stark effect, in an electric field of 100 volt/cm. Hansen brings evidence to show that his fields were certainly less than this.

He finds that the doublet interval is rather smaller than the interval between components b and a', and that the intensity ratio of the members of the doublet is approximately one, in disagreement with the theory. Further, he finds a definite asymmetry in the short wave member which implies a weak component roughly in the position of c. The presence of such a component, which might arise through the Stark effect, would possibly account for the slight discrepancy in the separation of the main peaks. But the appearance of c would be accompanied by that of b' with even greater intensity, and this would increase the peak separation. Under more violent discharge conditions, i.e. under a greater Stark effect, Hansen found that the short wave component increased in intensity relative to the long wave component. The reverse would be expected on the theory because of b'. Hansen's improved observations thus revealed further weakness in the theory. Especially noteworthy in view of later developments is his discovery of the component c.

5.3. X-ray spectra

In his paper in 1916 [125], Sommerfeld applied his formula to interpret X-ray spectra and their fine structure. The doublet interval which is now known as L_{II}—L_{III} had been measured for a number of elements. Sommerfeld interpreted this as the difference in energy of an electron between the states $n = 2$, $k = 1$ and $n = 2$, $k = 2$, the electron in question moving under the influence of the nucleus but being partially screened from it by the other electrons in the atom. The screening effect he took into account by reducing the nuclear charge from Z to $(Z-s)$. His fine structure formula (equation 4.2) then predicts for the energy difference $L_{II}-L_{III}$:

$$\Delta \nu_L = \frac{R\alpha^2}{16}(Z-s)^4 + \text{terms in higher powers of } \alpha. \quad (5.1)$$

The agreement between this formula and the experimental material was very good, and the value of α derived from it was in excellent agreement with that obtained from the line 4686 Å of ionized helium.

5.4. The dilemma

The evidence in support of Sommerfeld's formula was too strong to be set aside: its essential correctness could hardly be doubted. Yet in order to interpret the observations, both on hydrogen and on ionized helium, it was necessary to violate the selection rule $\Delta k = \pm 1$, a rule inescapable on the Correspondence Principle, which itself was so strongly supported both by its appeal to classical physics and by its success in predicting the Stark effect in hydrogen.

The resolution of the dilemma is now well known and will be treated later: the property of electron spin brings about a re-labelling of the energy levels, but does not displace them. Sommerfeld's formula remains valid, but not the assumptions on which it was derived. The Correspondence Principle is not violated, but in its application it must now be realized that the electron possesses an extra degree of freedom.

CHAPTER VI

THE FINE STRUCTURE—NEW QUANTUM THEORY*

6.1. Wave and matrix mechanics

WAVE mechanics and matrix mechanics do not represent a natural development from the theory of Sommerfeld and Wilson:

---

* The well-known article by Bethe in the 1933 edition of the *Handbuch der Physik* has been re-written for the 1957 edition [9] in the light of recent developments.

they have their origins in new ideas (see, for example [138]). The relation between these two formulations of the new quantum theory was established by Schrödinger [120]. Their connection with classical mechanics and with the old quantum theory can be seen in the Wentzel–Kramers–Brillouin method for the approximate solution of problems in wave mechanics. It is assumed, in this method, that the wave function is proportional to an exponential function whose argument is a power series in Planck's constant h. Correspondence with classical mechanics is obtained by setting h equal to 0, under which condition the equations obtained are identical with those appearing in Hamiltonian mechanics, and represent a wave whose surfaces of constant phase are orthogonal to the particle trajectories.* The approximation obtained by taking the term linear in h leads to the quantization formula

$$\oint p_x dx = \oint \{2m(E-V)\}^{\frac{1}{2}} dx = (n+\tfrac{1}{2})h \qquad (6.1)$$

in the case of the harmonic oscillator, and this differs from the rule of the old theory only by the addition of $\frac{1}{2}$ to the quantum integer.

6.2. Quantum mechanics

Emphasis on the necessity of correspondence with classical mechanics led Dirac [34] to a general formulation of quantum mechanics. He showed that if two dynamical quantities ξ, η, functions of the generalized co-ordinates and momenta q_r, p_r, were to be regarded as non-commuting quantities, then their Poisson Bracket $[\xi,\eta] \equiv \sum\limits_{r} \left(\dfrac{\partial \xi}{\partial q_r}\dfrac{\partial \eta}{\partial p_r} - \dfrac{\partial \eta}{\partial q_r}\dfrac{\partial \xi}{\partial p_r}\right)$ must be related to their commutator $\xi\eta - \eta\xi$ as follows

$$\xi\eta - \eta\xi = \lambda[\xi, \eta], \qquad (6.2)$$

where λ is a pure imaginary constant with the dimensions of action. To secure agreement with experiment $\lambda = i\hbar = ih/2\pi$. Classical mechanics, in so far as it can be expressed in terms of

---

* The correspondence between Hamiltonian mechanics and wave mechanics is rigorous, not requiring the neglect of powers of h, if the symbols are understood in a certain way [137].

Poisson Brackets, may be immediately taken over into quantum mechanics by this relation.

Since, in particular,

$$[q_r, q_s] = 0; \quad [p_r, p_s] = 0; \quad [q_r, p_r] = 1, \; r = s, \qquad (6.3)$$
$$= 0, \; r \neq s,$$

it follows that

$$p_r q_r - q_r p_r = -i\hbar. \qquad (6.4)$$

In matrix mechanics, the dynamical variables p_r and q_r are associated with matrices. Equation 6.4, interpreted as a relation between matrices, expresses the basic law of matrix mechanics.

The idea that physical quantities may be represented by operators was suggested by Born and Wiener [17]. Equation 6.4 is satisfied by the choice of the multiplication operator q_r to represent the dynamical variable q_r, and the differential operator $-i\hbar \partial/\partial q_r$ to represent the dynamical variable p_r, conjugate to q_r. A particular 'representation' in quantum mechanics is made if, for the q_r, we take Cartesian co-ordinates. The Hamiltonian function of classical mechanics $H(p_r, q_r)$, the energy of the system, then becomes

$$H(p_r, q_r) = \sum_{r=x,y,z} (p_r^2/2m) + V(x, y, z), \qquad (6.5)$$

which, regarded as an operator equation with the interpretation of p_r and q_r given above, leads to Schrödinger's equation

$$\frac{\hbar^2}{2m} \nabla^2 \psi + (E - V)\psi = 0. \qquad (6.6)$$

6.3. Relativity

Schrödinger's wave equation is not relativistic. Its application to the hydrogen atom is made by setting $V = -Ze^2/r$. The result is obtained that stationary states exist having energies given by the simple formula of Bohr, $E_n = -RZ^2/n^2$. Unquantized states exist for which E is positive. This form of the wave equation, therefore, predicts a spectrum of simple lines without fine structure, having continua beyond the series limits. The energy levels are degenerate. As in Sommerfeld's theory, three quantum numbers (since there are three degrees of freedom) are necessary and sufficient for the description of

each state. Again, as in Sommerfeld's theory, the degeneracy
in one of the remaining two quantum numbers is lifted if a
relativity correction is applied. A possible (but unsatisfactory)
way of doing this is to introduce the relativistic relation

$$\frac{1}{c^2}(E-V)^2 = \mathbf{p}^2 + m^2 c^2 \qquad (6.7)$$

between energy E and momentum $\mathbf{p}$ (see, for example, [114]).
The substitution of the operator $-i\hbar$ **grad** for the dynamical
variable $\mathbf{p}$ allows the construction of a relativistic wave
equation from which proper values of the energy may be found.
The solution for the hydrogen atom is an expression similar to
Sommerfeld's (equation 4.2) in which $(l+\frac{1}{2})$ takes the place of k.
Since l, the quantum number associated with the orbital
angular momentum in the new theory, is integral, the new
formula implies different energy levels from the old, in definite
disagreement with experiment.

6.4. Electron spin

In 1926 it was realized that there is a property of the electron
other than the charge which must be taken into account,
namely, the magnetic moment associated with intrinsic spin.
It was shown by Goudsmit and Uhlenbeck [53] that this
property, which represents an extra degree of freedom and
therefore demands a fourth quantum number, could account
for the doublet structure of the alkali spectra and the anomalous
Zeeman effect. It was necessary and sufficient that the extra
quantum number be two-valued.

The motion of the magnetic dipole through the electric
field of the nucleus implies a magnetic force, and an extra
contribution to the energy. If this spin-orbit energy is regarded
as a small perturbation, it is legitimate to take the average over
the unperturbed orbit, that is to say, Schrödinger functions
may be used to evaluate the correction to the energy levels.
The postulate of two-valuedness for the spin orientation
implies two possible values for the spin-orbit energy to be
associated with a particular state specified by (n, l).

6.5. Spin and relativity corrections to the energy levels

The result obtained by Thomas [129] for the spin-orbit energy (calculated relativistically) was added by Heisenberg and Jordan [58] to their calculation of the relativistic correction to the Bohr energy levels. The most remarkable result was obtained that Sommerfeld's formula and energy levels were regained, but with a different interpretation of the quantum numbers. The k in Sommerfeld's formula (equation 4.2) is replaced by $(j+\frac{1}{2})$, where j is the quantum number belonging to the total angular momentum, the vector sum of orbital and spin angular momenta. $(j+\frac{1}{2})$ is integral since j itself is half-integral. Moreover, since l may take any value between 0 and $(n-1)$, the values of $(j+\frac{1}{2})$ run from 1 to n. But these are precisely the values allowed to k on Sommerfeld's theory. Thus the new theory predicts the same energy levels as before, although these are now labelled by (n,j) instead of (n,k). The fact that the orbital quantum number l does not appear in the result means that the state $[l, j = (l+\frac{1}{2})]$ has exactly the same energy as $[(l+1), j = (l+1)-\frac{1}{2})]$.

While the new theory predicts the same energy levels as that of Sommerfeld, it does not predict the same fine structure since the different labelling of the energy levels permits the appearance of components which on the old theory were forbidden. In particular, the component c observed by Hansen (section 5.2) now corresponds to an allowed transition (cf. Figs. 1, 2).

6.6. Pauli spin matrices—two-valued wave functions

The degeneracy in l which is a feature of the Schrödinger equation for the hydrogen atom is lifted by the relativity correction: levels with different l, which are described by different wave functions, are now separated. With the introduction of the extra degree of freedom of electron spin, we find a change of a different kind. The wave functions of the different spin states associated with a given l (e.g. $P_{\frac{1}{2}}$ and $P_{\frac{3}{2}}$) are described by combinations of functions of different m but the same l. A method developed by Pauli for finding the correct linear combinations (see, for example, [114]) will now be outlined.

Pauli [105] introduced the spin operators s_x, s_y, s_z, to represent the components of spin angular momenta. Each operator could have the eigenvalues $\pm\frac{1}{2}\hbar$. Since the components of ordinary angular momenta obey the relations

$$[L_x, L_y] = L_z; \quad [L_y, L_z] = L_x; \quad [L_z, L_x] = L_y; \qquad (6.8)$$

and hence, in quantum mechanics, the relations

$$L_x L_y - L_y L_x = i\hbar L_z; \quad L_y L_z - L_z L_y = i\hbar L_x; \quad L_z L_x - L_x L_z = i\hbar L_y; \qquad (6.9)$$

it was reasonable to postulate that

$$s_x s_y - s_y s_x = i\hbar s_z; \quad s_y s_z - s_z s_y = i\hbar s_x; \quad s_z s_x - s_x s_z = i\hbar s_y. \qquad (6.10)$$

These conditions on the s_x, s_y, s_z are satisfied by a particular choice of two-by-two matrices which are known as the Pauli spin matrices:

$$\tfrac{1}{2}\hbar \begin{pmatrix} 0 & 1 \\ 1 & 0 \end{pmatrix}, \qquad \tfrac{1}{2}\hbar \begin{pmatrix} 0 & -i \\ i & 0 \end{pmatrix}, \qquad \tfrac{1}{2}\hbar \begin{pmatrix} 1 & 0 \\ 0 & -1 \end{pmatrix}. \qquad (6.11)$$

In order to manipulate these matrices, it was necessary to introduce two-component wave functions on which they could operate. (Such functions had previously been used by Darwin [29].) It is then possible to set up eigenvalue equations of the type

$$\mathbf{J}^2 \begin{pmatrix} \psi_1 \\ \psi_2 \end{pmatrix} = j(j+1)\hbar^2 \begin{pmatrix} \psi_1 \\ \psi_2 \end{pmatrix}, \qquad (6.12)$$

where the eigenvalues $j(j+1)\hbar^2$ of the angular momentum-operator $\mathbf{J}^2$ are a consequence of the fundamental commutation relations (6.9, 6.10), and to solve such equations for ψ_1 and ψ_2. Under the approximation that the spin-orbit energy is small, these wave functions turn out to be particular combinations of the m-dependent Schrödinger functions.

6.7. The Dirac theory of the electron

The method outlined on p. 23 for constructing a relativistic wave equation is not the only possible approach. It suffers from the disadvantage that the Hamiltonian function obtained from it (the expression for the energy in terms of the space co-ordinates and the conjugate momenta),

$$E = H = V \pm c(\mathbf{p}^2 + m^2 c^2)^{\frac{1}{2}}, \tag{6.13}$$

is not linear in the momenta, and hence does not yield a linear differential equation when the substitution $\mathbf{p} \rightarrow (-i\hbar)$ **grad** is made. Dirac [36] proposed to construct a linear equation by means of the substitution

$$(\mathbf{p}^2 + m^2 c^2) = (\boldsymbol{\alpha} \cdot \mathbf{p} + \beta mc)^2, \tag{6.14}$$

leading to

$$H = V + c(\boldsymbol{\alpha} \cdot \mathbf{p} + \beta mc), \tag{6.15}$$

where the four quantities, α_x, α_y, α_z and β are not ordinary numbers but four-by-four matrices which are regarded as linear operators. The wave function ψ on which they operate must therefore have four components which are taken to be the elements of a single column matrix.

Dirac's procedure may be applied to a particle of charge $-e$ in an electromagnetic field by making the substitutions $c\mathbf{p} \rightarrow (c\mathbf{p} + e\mathbf{A})$ and $V = -e\phi$ in equation 6.15, where $\mathbf{A}$ is the vector and ϕ the scalar potential of the field. After reduction of the equation the non-relativistic limiting value of the energy may be found by using the condition $E - mc^2 \ll mc^2$. Comparison with the non-relativistic expression for the energy now reveals an extra term in Dirac's expression, of the form which would describe a magnetic dipole of moment $\mu_0 = e\hbar/2mc$ in the specified field* [37].

Applied to a central field, the theory may be tested for the constancy of the orbital angular momentum of the charged particle. It is found that the operator representing this quantity does not commute with the Hamiltonian operator, i.e. that the orbital angular momentum is not a constant of the motion. A quantity may be constructed which will commute with the Hamiltonian operator by adding to the orbital

---

* The exact value for the magnetic moment of the Dirac electron in a central field is not μ_0 but $\mu_0(1 + 2\sqrt{1 - \alpha^2 Z^2})/3$ [19]. The small correction is a consequence of the use of retarded potentials which is necessary when the particle velocity is comparable with that of light. Likewise a small correction must be made to the Landé g-factor.

angular momentum a further angular momentum of magnitude $\pm\frac{1}{2}\hbar$.

These results show that the spinning electron with its associated magnetic moment is already incorporated into the Dirac theory by the choice of Hamiltonian. This stands in contrast with the case we discussed in section 6.4 where spin and magnetic moments of certain particular magnitudes were added arbitrarily after the Hamiltonian operator had been constructed. The magnitudes of spin and magnetic moment which follow from Dirac's theory are the same as those previously assumed in order to bring agreement (as it was thought at the time) between theory and experiment.

6.8. The Dirac theory of the hydrogen atom

Dirac's theory may be applied to the hydrogen atom in the usual way by choosing the form $-Ze^2/r$ for the potential in which the electron moves. The nucleus exists merely as the source of the field. Properly speaking, therefore, the theory is still that of the electron. With this reservation (which we shall follow up in the next section), the equations allow rigorous solution for the energy levels, which are found to be exactly as given by Sommerfeld's formula (equation 4.2), with k replaced by $(j+\frac{1}{2})$. As in the last section, j is the quantum number associated with the total angular momentum, and is half-integral. This celebrated formula, which was regained on the new quantum theory by adding spin and relativity corrections to the non-relativistic result of Schrödinger, now appears for the third time as a direct consequence of requiring the wave equation to be linear and Lorentz-invariant.

Although we have remarked that we have been considering a theory of a single particle, we are, nevertheless, forced to contemplate many particles by a feature of the theory to which we have not yet drawn attention, the ambiguity of sign in equation 6.13. The negative sign implies negative energy. There is no mathematical reason why this choice should be discarded since the wave equation may be solved and the proper values of the energy found for this case as for the case of positive energy. Dirac therefore allowed that the negative energy states were meaningful, but was faced with the difficulty of explaining why transitions to these states should not always take place spontaneously by electrons in states of positive energy. To meet this situation he postulated that space was filled by electrons in states of negative energy, so that electrons of positive energy could not normally make transitions to these states. The electrons in states of negative energy could, however, be excited to states of positive energy. One would then observe, together with the electron itself, an abnormality in its place of origin. For suppose an electric force to act on an electron in a state of negative enery. The effect of this force we regard as the normal state of the vacuum. If, then, the electron were to be removed, the force would have a negative effect in comparison with the normal state of affairs. Hence the absence of an electron from a state of negative energy would simulate the presence of a particle of the same mass but of opposite charge. This is Dirac's theory of positrons.

6.9. The Breit interaction

Dirac's equation for an electron moving under a central force described by the potential $V = -Ze^2/r$ can be solved rigorously. We have remarked that this, strictly speaking, does not represent the problem of the hydrogen atom since the nucleus does not appear explicitly in the equations. It is not known how to formulate exactly the relativistic expression for the interaction of two charged particles. Notable attempts in this direction were made by Darwin [28], in classical theory, who took account of retarded potentials, and Breit [20], who attempted quantum formulations of Darwin's expressions. Breit's Hamiltonian operator, which is formally expressed in terms of the Dirac α's, can be reduced to show the spin interactions explicitly. It is recognized as being inadequate when the distance between the particles is too small, or their velocities too great. Its application is limited to first-order perturbation theory, under which approximation it gives a good account of the fine structure in the two-electron spectrum of helium.* The solution which has been obtained when this operator is used for hydrogen [80] must therefore be regarded as approximate, but valid to a stated, high order of accuracy.

6.10. Corrections to the Dirac theory

The relativistic two-body problem is formulated by writing down the combined relativistic Hamiltonian

$$H = H_e + H_n - \frac{Ze^2}{r} + U, \qquad (6.16)$$

where
$$H_e = -e\,\phi_e + \boldsymbol{\alpha}_e \cdot (\mathbf{p}_e + e\mathbf{A}_e) + \beta_e mc,$$
$$H_n = Ze\phi_n + \boldsymbol{\alpha}_n \cdot (\mathbf{p}_n - Ze\mathbf{A}_n) + \beta_n Mc.$$

ϕ, $\mathbf{A}$ are the scalar and vector potentials of an external field, the suffixes $_e$ and $_n$ refer to the electron and nucleus respectively, and U is the Breit interaction. The spin-orbit interaction of the electron is included in H_e, that of the nucleus, which is smaller by the factor $(m/M)^2$, in H_n. In the term U are included the magnetic orbit-orbit and spin-spin interactions, and additional spin-orbit terms, all proportional to $(mM)^{-1}$.

This formulation takes no account of the size of the particles, and assumes that the expression for the Coulomb interaction is valid at all distances. It does not take account of the zero-point energy of the electro-magnetic field (section 9.2). It implies that the nucleus, since it is a relativistic charged particle obeying Dirac's equation, possesses an intrinsic spin and magnetic moment. In fact the proton is known to possess a magnetic moment in excess of this intrinsic moment. The interaction between the nuclear magnetic moment and the magnetic field due to the electron is conveniently treated separately (Chapter XI).

Calculations based on the Hamiltonian (6.16) yield, for the case of zero external field, the same solution for the energy levels as the one-particle equation, except for the expected reduced mass correction $m \to mM/(m+M)$, (section 3.4), and

---

* A recent discussion of its validity is given by Salpeter [118], who finds on the basis of the new quantum electrodynamics, the need for a very small extra term for S levels in hydrogen, $-(4R\alpha^3 m/3\pi n^3 M)$.

a small shift $-R\alpha^2 m/4n^4 M$, which is the same for all the fine structure levels of given n [80]. The small orbit-orbit interaction is included in this shift.

While the reduced mass effect is easily detectable spectroscopically, the small extra shift is well beyond the limits of accuracy of any spectroscopic work which has yet been done. (It might just be detectable in an absolute measurement of the wavelength of the Lyman α line of Li^{++}.) We shall consider in the next chapter the experimental test of the Dirac theory.

6.11. Intensities : new quantum theory

Guided by the result in classical mechanics, that the amount of energy radiated by an oscillating charge is related to the dipole moment, the new quantum theory attempts to construct in terms of wave functions the quantity which will represent a fluctuating electric dipole moment. The function

$$\mathbf{M}(p, q) = \int \Psi_p^* e\mathbf{r}\Psi_q \, d\tau,$$

where $\Psi_p = \psi_p \exp\,(iE_p/\hbar)t$ is the time-dependent wave function, is a quantity of the dimensions dipole moment multiplied by a harmonically varying function of time $\exp\,[i(E_q - E_p)/\hbar]t$ depending in the required way on the energies E_p, E_q of the states p and q, and therefore possesses the required properties, except that it is not symmetrical between p and q. Replacement of the square of the dipole moment in the classical formula by the symmetrical product $\mathbf{M}(p,q)\mathbf{M}(q,p)$ yields a formula for transition probabilities which, as we shall see in the case of H_α, is in good agreement with experiment.

The appeal to classical theory is avoided in a theory of radiation due to Dirac [35] in which the same formula for intensities is derived by the application of perturbation theory to a weakly interacting system of atoms and radiation. A serious objection to the theory, however, is that for higher approximations to the perturbation interaction entirely meaningless results— diverging integrals—are obtained. A fuller discussion of these difficulties is given in Chapter IX.

The calculation of transition probabilities on the new quantum theory is thus essentially an evaluation of integrals of the form $\int \psi_p^* \mathbf{r} \psi_q \, d\tau$. The result is often identically zero. Thus, the selection rule $\Delta l = \pm 1$ follows from the properties of spherical harmonics, which represent the angular part of the wave function for central systems. Further, since one quantum of radiation cannot carry away more than one unit of angular momentum, the quantum number j, which measures the total angular momentum of the atom, cannot change by more than one unit in the emission of one quantum. The change in j may be zero. This is not incompatible with a change in l of one unit since the electron spin may change in direction (though not in magnitude) in such a way as to preserve j constant.

If Schrödinger wave functions, which are simpler than those of Dirac, are used in the evaluation of the non-vanishing integrals, the results are insufficiently detailed, for Schrödinger functions do not describe the electron spin, and no distinction is drawn, for example, between $P_{\frac{1}{2}}$ and $P_{\frac{3}{2}}$. Since, however, there is no direct interaction between the radiation and the spin, one would expect the same probability for transition to a D level (say) from each of the possible states of combination of the P level with the spin. Of these, two are associated with $P_{\frac{1}{2}}$ and four with $P_{\frac{3}{2}}$. The total intensity of the P-D transition should therefore be divided between $P_{\frac{1}{2}}$ and $P_{\frac{3}{2}}$ in the ratio two to four. This is a particular example of the sum rule of Ornstein, Burger and Dorgelo for intensities in multiplets. Transition probabilities in hydrogen calculated from Schrödinger wave functions and the multiplet sum rules (derived quantum mechanically) may be found in Condon and Shortley ([26]. p. 133; see also [9]).

The distribution of intensity between the different multiplet levels is given directly if Dirac functions are used to evaluate the transition probabilities, since the electron spin is incorporated in the equations. The total intensity of all the transitions within a multiplet, calculated from Dirac functions, is inappreciably different from that obtained with Schrödinger functions [116]. This may be understood in that the radial factor of the electron probability distribution does not differ greatly between the Dirac and the Schrödinger theories.

We have been speaking of calculations of probabilities for spontaneous transition, whereas measurements are of relative intensities. Calculated intensities are obtained by multiplying the transition probabilities by the populations of the upper states. In gas discharges it is usually the case that all excited states of approximately the same excitation potential are equally populated—this is known as statistical population—but special conditions may sometimes lead to over- or under-population of some particular state. Thus abnormal intensities may reflect non-statistical populations.

Abnormalities in intensity are also to be expected when the atoms are perturbed by external influences, which is exceedingly likely in gas discharges. On the new quantum theory, as on the old, transition probabilities may be altered and 'forbidden' transitions may occur with appreciable intensity in the presence of an electric field.

CHAPTER VII

NEW QUANTUM THEORY—COMPARISON WITH EXPERIMENT

7.1. The H_α line

SINCE most spectroscopic investigations have been concerned with the H_α line, we shall examine in detail the structure to be expected for this line on the basis of the new theory, and review the work done after Heisenberg and Jordan had given the new interpretation to Sommerfeld's formula.

Fig. 2 shows the fine structure levels belonging to $n = 2$ and $n = 3$, labelled according to modern spectroscopic terminology. The capital letters signify the value of l, according to the convention that $l = 0, 1, 2 \ldots$ is denoted by $S, P, D \ldots$. The numerical subscript, e.g. $P_{\frac{3}{2}}$, denotes the value of j and the superscript on the left, e.g. ^2P, indicates the doublet nature of the spectrum—that there are two possible values of j for each value of l (except where $l = 0$).

It will be noticed that all the levels except the uppermost levels in a set are double since every j smaller than $(n-\frac{1}{2})$ occurs in combination with two different values of l. Some of the predicted transitions are, in consequence, also double.

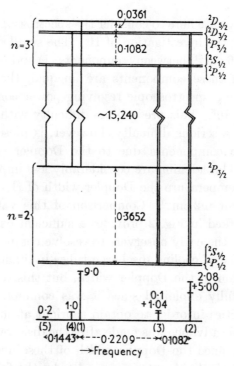

FIG. 2. Fine structure of $H_\alpha(D_\alpha)$ according to the Dirac theory. Intervals in cm⁻¹.

The predicted relative intensities indicate that the fine structure should be dominated by two components, in agreement with the old theory and with experiment. The components labelled (3) and (4), however, which formerly were not allowed, are now expected to appear with significant intensity.

Particular interest attaches to component (3). We have mentioned that it was observed by Hansen, but could be explained at the time neither as an allowed nor as a Stark-induced transition. Kent, Taylor, and Pearson [72] confirmed Hansen's discovery just as Heisenberg and Jordan's paper was published, and regarded the existence of (3) as establishing the superiority of the new theory. Measurements of the position of (3) were, however, quite unreliable since it was observed not as a resolved line, but as a blend with the short wave component of the doublet.

It is appropriate here to consider the spectroscopic conditions necessary for the investigation of H_α. The smallest interval in Fig. 2 is 0.036 cm^{-1}, between components (1) and (4). Since the intensities of these components are unequal, their resolution would require a spectroscopic resolving power considerably in excess of 5×10^5. This presents no difficulty with interference techniques; a serious difficulty, however, is presented by the width of the components due to the Doppler effect. Other causes of line broadening are considerably less important.

At room temperature the Doppler width of the components of H_α is about 0.2 cm^{-1}. Comparison of this value with the intervals marked in Fig. 2 provides a sufficient explanation of the failure of the early observers to resolve component (3).

The practice of cooling discharge tubes in liquid air allowed some reduction in the Doppler width, but this technique was not always fully exploited since it was common to use high current densities in order to obtain a bright atomic spectrum. A considerable advance was made after the discovery of deuterium in 1932, since the Doppler width of these lines compared with those of the light isotope is smaller by the factor $\sqrt{2}$ (see Fig. 3).

Measurements made on H_α were usually directed towards one of the following:

(a) Measurement of the absolute wavelength.
(b) Determination of the interval ν_{12}.
(c) (In the later work.) Determination of the interval ν_{23}.

7.1.1. *The wavelength*

The absolute wavelength, together with the fine structure formula, can be used to determine the Rydberg constant. The

Rydberg constant so deduced for hydrogen was then compared with that for deuterium or ionized helium in order to obtain a value for the atomic mass of the electron (section 3.4). This result, divided into the Faraday, gave a 'spectroscopic' value of e/m for comparison with the 'deflection' value, which at one time appeared to differ significantly [10]. Recent assessments of the best values of the atomic constants [39] do not maintain this distinction, which has mainly been resolved by the use of more reliable observational material. The agreement of spectroscopic and deflection values of e/m therefore supports the

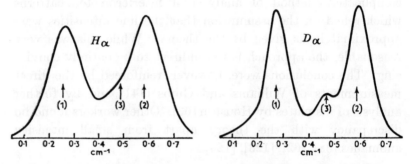

FIG. 3. Fine structure of H_α and D_α as resolved by R. C. Williams [139]. The reduced Doppler width in D_α allowed the resolution of (3).

fine structure formula, but the test is not so sensitive as examination of the fine structure itself. The adjustments to the formula which are detailed in Chapter IX in fact necessitate a change in the value of the Rydberg of about one part in five million [25].

7.1.2. The interval ν_{12}

The large line width which prevented the resolution of (3) also prevented a definite check of the theory by measurement of the interval ν_{12}. Differences between the measurements of different observers were of the same order as deviations from the predicted value [79]. In retrospect it is significant that the measured intervals were nearly always too small. Some workers felt sufficiently confident of this difference (which they detected also in other members of the Balmer series) to record definite disagreement with the theory. Thus Houston and

Hsieh [64] are 'forced to the conclusion that the theory is inadequate to explain the observations'. They consider possible reasons for the discrepancy, and point out the likelihood of its originating in the omission from the theory of terms which represent the interaction between the atom and radiation (section 6.10). The magnitude of this effect would be of order α times the fine structure separation, or again, of the order of the natural (radiation) width of the lines, which is roughly the magnitude of the discrepancy.

These conclusions of Houston and Hsieh were based on a complicated method of analysis of interferometer patterns which relied on the assumption that the line intensities were approximately as given by the theory. While this was very reasonable, the approach is too indirect to be entirely convincing. The conclusions were, however, reinforced by the direct measurements of Williams and Gibbs [141] and by further analysis of his plates by Houston [63]. Other workers found no discrepancy with the theory apart from small intensity anomalies [38], [62], [123], [128].

7.1.3. *The interval ν_{23}*

Measurements of the interval ν_{23} showed even greater differences between different observers. The best resolution of (3) was obtained by Williams [139] (see Fig. 3) who concluded that it was not in the predicted position. Pasternack [104] pointed out that Williams's observations could be explained on the assumption of an upward shift of the $2^2S_{\frac{1}{2}}$ level by $0 \cdot 03$ cm^{-1}. On the other hand, Drinkwater, Richardson and W. E. Williams [38] and Heyden [62] found no discrepancy with theory.

7.1.4. *Relative intensities of the components*

Intensity anomalies were found by nearly all observers. Component (2) nearly always appeared stronger than (1), contrary to the theory. These anomalies were ascribed to the conditions prevailing inside the discharge tube, and were not looked upon as providing a test of the theory.

7.2. The ionized helium line 4686 Å

An interferometric examination of this line was undertaken by Chu [24a] in Houston's laboratory. The gas discharge was mild, but, owing to the large instrumental width which is unavoidable when a single Fabry-Perot etalon is used to analyse a wide structure, the resolution was no better than Paschen's. Chu concluded that, within the accuracy of the experiment, the positions of the components agreed generally with the Dirac theory.

7.3. Conclusion

Dirac's formula still awaited its decisive test. Suspicion had been cast on it through the spectroscopic observations and suggestions for its amendment had been explored (an account is given by Lamb [79]), but none of these gave results of the required magnitude. The microwave experiment of Lamb, which we shall describe in the next chapter, confirmed the discrepancy, and developments in the theory of quantum electrodynamics explained it. In Chapter X we shall review the quantitative tests of the new theory.

CHAPTER VIII

THE EXPERIMENT OF LAMB AND RETHERFORD

THE selection rules of the new quantum theory allow electric dipole transitions between levels of the same n, but since the probability of spontaneous transition depends on the cube of the frequency, such transitions in hydrogen are exceedingly improbable. Stimulated transitions, on the other hand, may take place under quite small alternating electric fields of the appropriate frequency. Absorption and emission of energy by the atom are equally probable, so that a change in an assembly of atoms may only be detected if the two states between which transitions are taking place are unequally populated at the outset.

Attempts in the nineteen thirties to detect the absorption of energy from high frequency spark gap oscillators by excited atoms in a hydrogen discharge met with very uncertain results. The development of microwave techniques during the war encouraged Lamb and Retherford to re-examine the possibility of stimulating transitions between fine structure levels of the same n, but they no longer attempted to detect the transitions by looking for an absorption of energy: rather, they looked for changes in the atoms themselves.

Their method depends on the exceptionally long life of the $2^2S_{\frac{1}{2}}$ level. There were differences of opinion on what the lifetime was likely to be. On the one hand, it was argued, since decay to the ground level $1^2S_{\frac{1}{2}}$ was forbidden by the selection rules, the lifetime of $2^2S_{\frac{1}{2}}$ must be very great. On the other hand, the smallest electric field would invalidate the description of the level as $2^2S_{\frac{1}{2}}$ because of its degeneracy with $2^2P_{\frac{1}{2}}$. The correct wave functions for atoms in these levels would be $\sqrt{\frac{1}{2}}[\psi(2^2S_{\frac{1}{2}})\pm\psi(2^2P_{\frac{1}{2}})]$, for which the lifetime would be twice that of the pure P level. Although objections could be made against the latter argument, it seemed clear that the $2^2S_{\frac{1}{2}}$ level would only be metastable if the atoms experienced electric fields no greater than a fraction of a volt per cm.

Lamb and Retherford [81] proposed to make a beam of atomic hydrogen and to excite it by electron bombardment. Most of the atoms in the $2^2S_{\frac{1}{2}}$ level, but not those in the $2^2P_{\frac{1}{2},\frac{3}{2}}$ levels, would live long enough to reach a detector. A radiofrequency field of the correct frequency in the path of the atoms would induce transitions from $2^2S_{\frac{1}{2}}$ to one of the 2^2P levels. Decay of the latter would result in a reduction in the number of atoms detected, which would indicate radiofrequency resonance. We shall give here a brief account of the work and quote the results. Details of the experiments are given in a series of six papers [82], [83], [80], [84], [131], [30] and a review [79].

The hydrogen is introduced into the highly evacuated atomic beam chamber through a tungsten oven heated to 2500°K, where the molecules are dissociated. Cross bombardment of the beam by electrons of 10·8 volt energy excites the atoms, and (incidentally) deviates them. The excited atoms now pass into the radiofrequency interaction space—a region between the inner and outer conductors of a transmission line—which they enter through slots in the outer cylinder. They are detected by the ejection of electrons from a tungsten plate on which they fall. The current constituted by the ejected electrons is amplified by an electrometer valve and measured on a galvanometer.

The beam, along most of its path, traverses a homogeneous magnetic field provided by an external magnet. In this way Lamb and Retherford planned to separate some of the Zeeman components of the $2^2S_{\frac{1}{2}}$ and 2^2P levels and thus to decrease the likelihood of decay from the former states through mixing with the latter. An important experimental advantage is also gained by the use of a magnetic field, for then the radiofrequency oscillator can be used at fixed frequency and detailed resonance curves of the Zeeman components can be explored by varying the magnetic field. This avoids the difficulty of producing a radiofrequency field of constant intensity but variable frequency.

In Fig. 4 are to be seen the Zeeman component energy levels which belong to $n = 2$ in hydrogen calculated on the Dirac theory. The allowed transitions are represented by the full lines in Fig. 5. The experimental points of Lamb and Retherford do not lie on these lines, but are displaced by about 1000 Mc/s. Fig. 4 therefore gives a wrong representation of the energy levels. Agreement can be secured by an upward shift of the components of $2^2S_{\frac{1}{2}}$ by about 1000 Mc/s (0·033 cm^{-1}). Refinement of the apparatus and a very detailed study by Lamb of the corrections led to the ultimate determination of the $2^2S_{\frac{1}{2}}-2^2P_{\frac{1}{2}}$ and $2^2S_{\frac{1}{2}}-2^2P_{\frac{3}{2}}$ intervals in hydrogen and deuterium to a precision of 0·1 Mc/s—about one part in 10^4 of the line width [131], [30].

The discrepancy in the components of H_α affirmed by Williams, but contradicted by others, is thus entirely confirmed by Lamb and Retherford. An account of further spectroscopic work on H_α and observations of 'Lamb shifts' in levels other

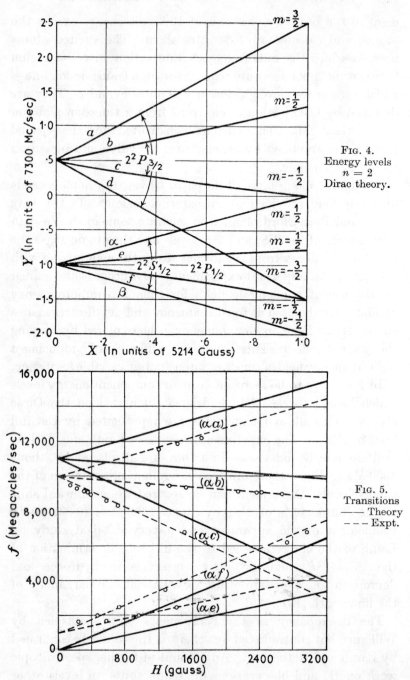

FIG. 4.
Energy levels
$n = 2$
Dirac theory.

FIG. 5.
Transitions
—— Theory
- - - Expt.

than $n = 2$ in hydrogen will be given after a chapter on developments in the theory. At the end of Chapter X is to be found a comparison between observed and calculated Lamb shifts. It will be noticed that there is a small difference, both experimental and theoretical, between the shifts in hydrogen and in deuterium.

CHAPTER IX

THE NEW QUANTUM ELECTRODYNAMICS

THE confirmation by Lamb and Retherford of the inadequacy of the Dirac theory stimulated a re-examination of a theoretical problem to which only a very incomplete solution had so far been found: the problem of the interaction between charged particles and the electromagnetic field. We shall briefly refer to the problem as it presented itself in classical physics, and then (following Weisskopf [135]) notice the further difficulties which the quantum theory introduces. Finally we shall see how these difficulties have been circumvented by the new quantum electrodynamics, and how a small correction is thereby introduced to the energy levels predicted by Dirac's theory. The new theory, however, is not a complete and logically satisfactory solution to the problems we shall state: a difficulty of principle remains now, as formerly.

9.1. The problem in classical physics

In the classical theory of the electron, the energy of the electrostatic field generated by an electron at rest is $\int (E^2/8\pi) \, d\tau$, where the field E is given by e/r^2. If we assume that the charge of the electron resides on its surface, the integral is to be taken over the whole of space outside it, and has the value $e^2/2a$ for a spherical electron of radius a. This value for the self-energy, which would become divergent for a point electron, cannot exceed mc^2, and if we suppose that it accounts for the whole mass of the electron, we must ascribe to the latter the finite radius $e^2/2mc^2$. But this procedure does not allow a consistent explanation of the mass of a moving electron, since there is momentum in the electromagnetic field which the moving charge generates, and the energy and momentum of a particle are not related in the same way as the energy and momentum of an electromagnetic field. The structure of the electron therefore remains an outstanding problem in the classical theory.

9.2. The problem in quantum physics

The same difficulty is present also in quantum theory, since the idea of a charge located at a point is preserved, as is shown explicitly by calculating the probability that part of the charge on one electron be present simultaneously at two points r and r' [134]. The probability is zero unless $r = r'$.

Further difficulties arise when the electromagnetic field is described in terms of the quantum theory. This is accomplished by treating the potentials of the electromagnetic field as operators, subject to the quantum commutation rules. It is convenient first to expand the vector potential $\mathbf{A}$ (r, t) into harmonic waves of all frequencies, $\omega = 2\pi\nu$. The attempt to express the energy H of the field in terms of the number of photons n_ω of angular frequency ω (see, for example, [117]) then results in the expression

$$H = \sum_\omega (n_\omega + \tfrac{1}{2})\hbar\omega, \quad \left(\hbar = \frac{h}{2\pi}\right). \tag{9.1}$$

Thus, even in a space devoid of photons there must still be present 'zero-point' electromagnetic energy of amount $\sum_\omega \tfrac{1}{2}\hbar\omega$.

This implies oscillating electric and magnetic fields to which charged particles are subject. The following non-relativistic calculation shows how the resultant motion is of infinitely great amplitude and infinitely great energy.

The number of modes of standing waves in an enclosure, having angular frequencies in the range ω, $(\omega+d\omega)$, is well known to be $\left(\dfrac{1}{\pi^2 c^3}\right) \omega^2\, d\omega$ per unit volume. Thus, the total electromagnetic energy due to the zero-point energy of the radiation in this frequency interval enclosed in volume V is

$$\left[\left(\frac{1}{\pi^2 c^3} \right) \omega^2 \, d\omega \right] V(\tfrac{1}{2}\hbar\omega). \tag{9.2}$$

This energy may be equated to the integral over V of the energy density in an electromagnetic field $\mathbf{E}$, $\mathbf{H}$,

$$\frac{1}{8\pi} \int (\overline{\mathbf{E}_\omega^2} + \overline{\mathbf{H}_\omega^2}) \, \underline{d\omega} \, dV = \frac{1}{8\pi} \overline{\mathbf{E}_\omega^2} \, \underline{d\omega} \, V, \tag{9.3}$$

where $\mathbf{E}_\omega$ is the amplitude of the field component of frequency ω and $\overline{\mathbf{E}_\omega^2}$ its mean square value. Hence,

$$\overline{\mathbf{E}_\omega^2} = \frac{4\hbar}{\pi c^3} \omega^3 . \tag{9.4}$$

Electrons in the volume V will oscillate under these electric fields, the equation of motion at the frequency ω being

$$\ddot{\mathbf{r}}_\omega = \frac{e\mathbf{E}_\omega}{m} \cos \omega t . \tag{9.5}$$

The mean square displacement for the field component of frequency ω will be

$$\overline{\mathbf{r}_\omega^2} = \frac{e^2 \overline{\mathbf{E}_\omega^2}}{2m^2 \omega^4} = \frac{2e^2 \hbar}{\pi m^2 c^3 \omega} . \tag{9.6}$$

Since the oscillators are independent, the mean square displacement due to all frequencies between ω_1 and ω_2 will be the integral of equation 9.6, namely

$$\overline{\mathbf{r}^2} = \frac{2e^2 \hbar}{\pi m^2 c^3} \ln \left(\frac{\omega_2}{\omega_1} \right). \tag{9.7}$$

In this expression, the lower limiting frequency will be determined by the state of binding of the electron: if the average binding energy $\approx \hbar\omega_1$, then the electron will not respond freely to frequencies lower than ω_1. The upper limit is closely connected with the size of the electron, for if the electron radius is a, then frequencies greater than $\omega_2 \approx c/a$ would not affect the electron as a whole. For a point electron the mean square displacement would diverge logarithmically unless one arbitrarily placed some restriction on ω_2.

The energy possessed by the electron in virtue of these oscillations is also a divergent integral, since, as may easily be shown, it is proportional to $\int_{\omega_1}^{\omega_2} \omega \,.\, d\omega$. This type of divergent energy, due to the zero-point energy of the radiation field, is called transverse, in contrast with the longitudinal self-energy of the static field.

A relativistic calculation leads to the results that the amplitude of the zero-point oscillations is finite, but that the energy remains divergent, although less severely so. This result is completely at variance with the experimental findings, since a treatment of atomic systems which ignores the radiation interaction yields formulae for the energy which agree very closely with experiment. It is, however, significant that the theoretical expressions remain finite if one takes a finite value for the electron radius. One is again led to seek a solution of the difficulty in a theory of the structure of the electron.

9.3. Re-normalization of mass

The new theory of quantum electrodynamics does not solve this problem. It recognizes that the mass of the electron is finite, and that this finite quantity must embrace the apparently infinite self-energy. It is therefore meaningful to obtain the difference between the self-energies of two electrons in different states of binding by subtracting the divergent integrals which represent these self-energies. The mathematical difficulty is avoided with the argument that, although the self-energies are infinite in form, they are physically represented by finite quantities. If we recognize this, we may calculate the contribution of the self-energy to the binding energy of a bound electron by subtracting from it the self-energy of a free electron. For all high frequencies the radiation interactions are identical: by subtraction, we avoid difficulties connected with the size and structure of the electron. The lower limiting frequency is different in the two cases since it depends on the strength of the binding: a finite residue is left after the subtraction. This is the measurable part of the self-energy.

9.4. Bethe's calculation

The first successful attempt[*] to evaluate these differences was made by Bethe [7] whose procedure is outlined below.

The interaction energy between an electron moving with velocity $\mathbf{v}$ and an electromagnetic radiation field described by the vector potential $\mathbf{A}(r, t)$ is (non-relativistically) $-e\mathbf{v} \cdot \mathbf{A}(r, t)$. In first-order perturbation theory this interaction energy vanishes for a stationary state since it is periodic in the time, and the diagonal matrix elements vanish. The non-diagonal matrix elements, however, do not vanish, so that there is a non-vanishing second-order perturbation for the state m whose value is

$$W_m = -\frac{2e^2}{3\pi\hbar c^3} \int_0^K \frac{\sum_n |\mathbf{v}_{mn}|^2}{k(E_n - E_m + k)} k^2 \, dk. \tag{9.8}$$

This is Bethe's starting point. In this expression the part $e^2|\mathbf{v}_{mn}|^2/k$ comes from the matrix element squared between states m and n of the operator $-e\mathbf{v} \cdot \mathbf{A}_k$, where $\mathbf{A}_k$ is the component of the field whose quanta have energy k when the field is expanded in plane waves. $(E_n - E_m + k)$ is the energy denominator. The factor $k^2 dk$ comes from the well-known expression used in equation 9.2 for the number of waves per unit volume which have energies (frequencies) in the interval dk. W_m diverges if no bound is put on the upper limit of integration, K.

For a free electron $\mathbf{v}$ has only diagonal elements, and equation 9.8 is replaced by

$$W_0 = -\frac{2e^2}{3\pi\hbar c^3} \int \frac{\mathbf{v}^2}{k^2} k^2 \, dk. \tag{9.9}$$

The difference $W_m - W_0$ will yield the significant part of the radiation interaction for the state m. In taking the difference we use the expectation value $\langle \mathbf{v} \rangle^2_{mm}$, which, by the sum rule, is equal to $\sum_n |\mathbf{v}_{mn}|^2$.

Then

$$W_m - W_0 = \frac{2e^2}{3\pi\hbar c^3} \int dk \sum_n \frac{|\mathbf{v}_{mn}|^2(E_n - E_m)}{(E_n - E_m + k)}. \tag{9.10}$$

[*] For an earlier attempt, see H. A. Kramers, Collected Papers, page 845.

Integrating over k between the limits 0 and K, where K is very large compared with $(E_n - E_m)$, we obtain

$$W_m - W_0 = \frac{2e^2}{3\pi\hbar c^3} \left[\sum_n |\mathbf{v}_{mn}|^2 (E_n - E_m)\right] \ln \frac{K}{|E_n - E_m|}, \quad (9.11)$$

where $|E_n - E_m|$ is a weighted average value. We observe that the divergence is now less severe, being only logarithmic in K. Proper account has now been taken of the low frequency fluctuations. High frequency fluctuations have been eliminated.

The evaluation of the term in brackets is achieved by using the relation $\mathbf{v} = \mathbf{p}/m = (\hbar/im)\,\mathbf{grad}$. The term can then be reduced to $\frac{\hbar^2}{2m^2} \int \nabla^2 V \psi_m^2 \, d\tau$, and we have

$$\delta E_m = W_m - W_0 = \frac{2e^2}{3\pi\hbar c^3} \frac{\hbar^2}{2m^2} \left[\int \nabla^2 V \psi_m^2 \, d\tau\right] \ln \frac{K}{|E_n - E_m|}. \quad (9.12)$$

But $\int \nabla^2 V \, d\tau = 0$ except at the origin where its value is $4\pi Ze^2$ if the nuclear charge is Z. Hence

$$\int \nabla^2 V \psi_m^2 \, d\tau = 4\pi Ze^2 \psi_m^2(0) = 4\pi Ze^2 \frac{Z^3}{\pi n^3 a_0^3} \quad (9.13)$$

for an S state of principal quantum number n, where a_0 is the Bohr radius. $\psi(0)$ is zero for states other than S states.

Thus

$$\delta E_{nS} = \frac{e^2 \hbar}{3\pi m^2 c^3} \frac{4Z^4 e^2}{n^3 a_0^3} \ln \frac{K,}{|E_n - E_m|},$$

$$= \frac{8\alpha^3}{3\pi} R \frac{Z^4}{n^3} \ln \frac{K}{k_0}, \quad (9.14)$$

where α is the fine structure constant $e^2/\hbar c$,
$\quad R$ is the Rydberg constant $me^4/2\hbar^2$,
$\quad k_0$ is the average excitation energy for the state nS,
and we have used $a_0 = \hbar^2/me^2$.

The value of K, the upper limit in the integration over photon energies, was taken by Bethe to be $\sim mc^2$, on the grounds that a limit of this order would appear in a relativistic calculation. The justification for this belief lay in a comparison with the old relativistic and non-relativistic calculations of the radiation divergences. Whereas the latter give expressions for the energy which diverge as $\int \omega \, d\omega$ (section 9.2), the former give $\int d\omega/\omega$. The severity of the divergence is reduced in the relativistic calculation through the effect of positron-electron pairs which appear at high energies. Bethe argued, therefore, that one would expect the logarithmic divergence of his non-relativistic calculation to become a convergence in a relativistic calculation.

The value of k_0, the lower limit in the integration, was found by Bethe to be $17 \cdot 8R$ for the $2S$ state in hydrogen, an unexpectedly high value. The logarithm term in equation 9.14 then becomes $7 \cdot 63$, and $\delta E_{2S} = 1040$ Mc/s.

The large value of K compared with k_0 implies that the logarithm is very insensitive to n, so that the n-dependence of δE is approximately proportional to $1/n^3$. The Z dependence is given partly by the Z^4 term and partly through the relation $(k_0)_Z = Z^2(k_0)_H$.

9.5. Welton's calculation

We have reproduced Bethe's calculation since the concept 're-normalization of mass' on which it is based was used in the subsequent development of the theory. It is interesting also to follow an argument given by Welton [136] which develops the calculation in section 9.2 of the oscillations which the electrons perform under the fluctuating electromagnetic fields, and which leads to Bethe's result. If an electron is situated in an external, non-uniform electric field (for example, the field of the proton) the average potential which it experiences will, by virtue of its oscillations, be different from the potential at its mean position. The latter potential is taken into account in the conventional theory. The difference in potential arising from the oscillations may be calculated by using the expansion:

$$V(\mathbf{r}+\boldsymbol{\delta r}) = V(\mathbf{r})+(\mathbf{grad}\ V) \cdot \boldsymbol{\delta r}+\tfrac{1}{2}(\nabla^2 V)\,\frac{(\boldsymbol{\delta r})^2}{3} + \dots \quad (9.15)$$

so that $\qquad \delta V = \overline{V(\mathbf{r}+\boldsymbol{\delta r})} - V(\mathbf{r}) = \frac{1}{6}\nabla^2 V(\overline{\boldsymbol{\delta r}})^2$ $\qquad$ (9.16)

to the first significant term, since the time average of $\boldsymbol{\delta r}$, but not of $(\boldsymbol{\delta r})^2$, is zero. Making use of the result (equation 9.7)

$$(\overline{\boldsymbol{\delta r}})^2 = \frac{2e^2\hbar}{\pi m^2 c^3} \ln \frac{\omega_2}{\omega_1},$$

we have

$$\delta E_m = \frac{e^2\hbar}{3\pi m^2 c^3} \left[\int \nabla^2 V \psi_m^2 \, d\tau \right] \ln \frac{\omega_2}{\omega_1}, \qquad (9.17)$$

which is of exactly the same form as Bethe's expression, equation 9.12.

9.6. Re-normalization of charge

We have used the term 're-normalization of mass' to describe Bethe's method of eliminating the self-energy of the electron. There is another effect described as 're-normalization of charge' which gives a contribution to the energy of bound states. This effect, which had been treated earlier [132], is a consequence of the presence of electrons which, according to Dirac's interpretation, occupy negative energy states. The electric field which accompanies a charge distorts the wave functions of the electrons in negative energy states which fill the vacuum. The charge distribution in the vacuum is thereby altered, and the vacuum appears to be polarized. This effect is proportional to the field, and therefore to the charge which produces it, and is always and unavoidably present. The charge which one observes experimentally is not the same as the 'bare' charge—the latter can never be measured. The objectionable feature of this theory is that the constant factor by which the 'bare' charge must be multiplied in order to obtain the observed charge turns out to be a divergent integral. Here again, the difficulty is met by recognizing that the observed charge contains this divergence. A subtraction procedure is again justified on the grounds that the observed charge is finite.

9.7. Relativistic calculations

The relativistic theory which was in due course worked out [75], [51], [46] enabled a calculation of the radiation interaction of a bound electron to be made by constructing, as Bethe had done, the operator representing the total self-energy of a bound electron and subtracting from this an operator representing the self-energy of a free electron. In these expressions the terms corresponding to the infinite self-charge could easily be recognized, but the term corresponding to the infinite self-mass presented a little more difficulty. The condition is imposed, after the charge and mass operators have been subtracted, that the residue should be invariant under a Lorentz transformation. The operator so constructed is used in a second-order perturbation calculation. Bethe's argument that a relativistic calculation would give a finite result without arbitrarily imposing a limiting frequency is justified.

These relativistic calculations disclose a term in the residual self-energy which can be interpreted as the interaction with electric and magnetic fields of an additional magnetic moment of the electron, of magnitude $\left(\dfrac{\alpha}{2\pi}\right)$ Bohr magnetons. Thus, while Bethe's calculation gave a level shift proportional to $\nabla^2 V$ which vanished for all terms other than S terms, the relativistic calculations imply a small shift also for other terms, proportional to the fine structure interactions. The magnitude of the total shift of the $2S$ level in hydrogen is 1034 Mc/s, to which charge re-normalization contributes about -27 Mc/s, and the extra magnetic moment about 68 Mc/s. The shift in the $2P_{\frac{1}{2}}$ level, arising from the magnetic moment only, is -17 Mc/s, while the $2P_{\frac{3}{2}}$ level is shifted half as much, and in the opposite direction.

9.8. Results of the calculations—formulae

The formulae for the level shifts derived from the relativistic calculations are given on p. 50. They are taken from Bethe, Brown and Stehn [8] where the computation of the average excitation energy for the $2^2S_{\frac{1}{2}}$ level in hydrogen is to be found. Calculations of the average excitation energy k_0, for the levels $1s$ to $4p$ inclusive have been made by Harriman [56]. As anticipated (p. 47), these quantities are not very sensitive to n.

$$\Delta E(n, 0, \tfrac{1}{2}) = \frac{8}{3\pi} \alpha^3 R_\infty \frac{Z^4}{n^3} \left[\ln \frac{mc^2}{k_0(n, 0)} - \ln 2 + \frac{5}{6} - \frac{1}{5} \right], \quad (9.18)$$

$$\Delta E(n, l, j) = \frac{8}{3\pi} \alpha^3 R_\infty \frac{Z^4}{n^3} \left[\ln \frac{R}{k_0(n, l)} \pm \frac{3}{8(j+\tfrac{1}{2})(2l+1)} \right], \quad (9.19)$$

where the $\pm$ sign corresponds with $j = l \pm \tfrac{1}{2}$.

The quantity $\left(\dfrac{8}{3\pi} \alpha^3 R_\infty \right)$ is known as the Lamb constant L.
We shall use $L' = L Z^4 / n^3$. In that R_∞ is measured in cm^{-1}, the conversion factor c is needed if L is required in Mc/s. The relation $R_\infty = mc^2 \alpha^2 / 2$ is useful in comparing the results of different writers.

Harriman's values for k_0 are, in units of R_∞,

$1s$:	19·76967(6)	$2p$:	0·97042964(1)
$2s$:	16·6398 (2)	$3p$:	0·96253147(1)
$3s$:	15·9214 (2)	$4p$:	0·9589139 (3)
$4s$:	15·6404 (1)	$3d$:	0·9947815 (2)

where the figures in brackets are the estimated errors in the last digit.

Values of k_0 for ionized helium etc. are obtained by multiplying Harriman's values by Z^2. In $\Delta E(n, l, j)$ the numerator of the logarithm term should also be multiplied by Z^2.

In these formulae, the anomalous magnetic moment of the electron contributes $\tfrac{3}{2} L'$ to the shift of S levels (equation 9.18) and is responsible for the last term inside the bracket in equation 9.19. It thus contributes $\tfrac{1}{2} L'$ to the $S_{\frac{1}{2}} - P_{\frac{1}{2}}$ interval. The vacuum polarization term included in equation 9.18 is $-L'/5$. The other terms, which comprise the greater part of the Lamb shift, we refer to as radiative shifts.

It is to be emphasized that these formulae have been obtained by the use of second-order perturbation theory taken to the first significant terms. The Lamb shifts obtained from them have been refined by the evaluation of smaller terms in the second order, and by the calculation of higher orders in the perturbation expansion; also by corrections for the finite mass and volume of the nucleus. The papers [119], [1a], [52] (see also [9]), summarize the position at the time of writing.

The most important of these further terms, [69]*, [2], is a second-order relativistic shift $3\pi Z\alpha L'\left(1+\dfrac{11}{128}-\dfrac{1}{2}\ln 2+\dfrac{5}{192}\right)$ applicable to S states only. Next in importance are the fourth-order corrections. The fourth-order radiative shift is $L'\left(\dfrac{3\alpha}{2\pi}\right)(0\cdot 52)$, [6], and affects only S levels. The fourth-order magnetic moment of the electron [70] contributes $-\dfrac{3}{4}L'\left(2\cdot 973\dfrac{\alpha}{\pi}\right)^\dagger$ to S levels and $\mp\dfrac{3}{4}L'\left(2\cdot 973\dfrac{\alpha}{\pi}\right)\dfrac{1}{(j+\frac{1}{2})(2l+1)}{}^\dagger$, $(j=l\pm\frac{1}{2})$, to other levels. The fourth-order vacuum polarization term, which affects only S levels, is $-L'\left(\dfrac{41\alpha}{54\pi}\right)$ [3]. Higher order corrections of these types have not yet been evaluated.

The corrections for the finite mass include, first of all, small terms in the fine structure found by Salpeter [118] who evaluates them only for the $2^2S_{\frac{1}{2}}-2^2P_{\frac{1}{2}}$ interval in hydrogen and deuterium. In a later article [119] he shows the need for the correction factor $\left(1+\dfrac{m}{M}\right)^{-3}\approx\left(1-3\dfrac{m}{M}\right)$ in the formulae 9.18, 9.19, for the Lamb shift. Further, R and k_0 inside the brackets in these formulae are to be multiplied by $(1+m/M)^{-1}$. For the terms in these formulae which arise from the anomalous magnetic moment of the electron, however, the factor $(1+m/M)^{-3}$ is inapplicable. The correct factor has not yet been calculated, but is of the order of magnitude $0\cdot 1$ per cent. of $L'/2$, so that only a small error is introduced if the whole bracket is multiplied by $(1+m/M)^{-3}$.

The finite sizes of the proton and deuteron lead to corrections for S levels similar to the small energy differences which give rise to spectroscopic isotope shifts in heavy elements. If a_n is the nuclear radius, then the correction is

$$\Delta E = -\int_0^{a_n}\left(\frac{Ze^2}{a_n}-\frac{Ze^2}{r}\right)\psi^2(0)4\pi r^2\,dr,$$

---

* A factor 8 has been omitted from the formula quoted in the abstract of [69]. It may be found in the body of the paper.
† An independent calculation [124a] yields $0\cdot 328$ instead of $2\cdot 973$.

where we have taken that average value of a_n which will allow us to use the constant potential $- Ze^2/a_n$ within the nucleus, and assumed that the wave function does not vary appreciably over the nuclear volume. We then find

$$\Delta E \text{ (in frequency units)} = \frac{Ze^2}{3\hbar} \, \psi^2(0)(a_n^2)_{av} = \frac{1}{3\pi} \frac{e^2}{a_0\hbar} \left(\frac{a_n}{a_0}\right)^2 \frac{Z^4}{n^3}.$$

This correction is less important for hydrogen than for deuterium by a factor of about six [119], [1a].

A very small correction was estimated by Salpeter on account of the internal structure of the nucleon: the possession of a magnetic moment implies a small change in the electrical potential. This correction is significant for hydrogen, but not for deuterium, since the non-Dirac parts of the magnetic moments of the proton and the neutron in the deuteron are approximately equal and opposite. The Dirac part of the proton moment contributes a negligible amount.

9.9. Numerical results for hydrogen and deuterium

The relative importance of the correction terms may be judged from the following numerical values of their contributions to the $2^2S_{\frac{1}{2}} - 2^2P_{\frac{1}{2}}$ interval in hydrogen and deuterium.

	Mc/s	
All second-order effects, for infinite nuclear mass	$1052 \cdot 14 \pm 0 \cdot 08$	
(the estimated error in α accounts for	$\pm 0 \cdot 05$)	
All fourth-order corrections	$6 \cdot 20 \pm 0 \cdot 10$	
Nuclear mass correction	$- 1 \cdot 295 \pm 0 \cdot 08$	(H)
	$- 0 \cdot 643 \pm 0 \cdot 08$	(D)
Uncertainties which arise in the nuclear mass correction on account of the anomalous magnetic moment of the electron	$\pm 0 \cdot 08$	(H)
	$\pm 0 \cdot 04$	(D)
Nuclear volume correction	$0 \cdot 733 \pm 0 \cdot 025$	(D)
	$0 \cdot 118 \pm 0 \cdot 03$	(H)
Nucleon structure correction	$0 \cdot 025 \pm 0 \cdot 005$	(H)

The corrections for He$^+$ are tabulated by Novick, Lipworth and Yergin [99].

Comparison with experiment is shown in Table 1, page 64.

9.10. Development of the radiation theory

Although the calculations leading to the formulae 9.18 and 9.19 were relativistic and took account of the negative energy states of the Dirac theory, the need was felt for a new formulation of the problem in which the relativistic invariance of the radiation interaction would be apparent from the outset. Such a theory was developed by Schwinger and independently by Tomanaga, while at the same time Feynman developed a technique for depicting graphically the interactions between electromagnetic fields and particles in such a way as to enable the interaction operator to be formulated to any required order in a perturbation expansion. These two treatments were unified and extended by Dyson [40] who showed how the interaction divergences were removed in every term of the expansion by charge and mass re-normalization. Many of the correction terms of the last section were obtained by this technique.

The language used in this formulation of the theory is that of the quantum theory of fields, to which reference was made in section 9.2. Interactions between charged particles are represented by the emission and subsequent absorption of photons. It is not necessary that energy be conserved in these processes, since if the excited particles live only a time Δt, their energy is imprecise to the extent $\Delta E \sim \hbar/\Delta t$. The exciting photons may therefore possess energy much greater than that which would be available in classical physics. Such photons are termed 'virtual.' Similarly, the theory contemplates 'virtual' electron-positron pairs, whose momentary existence implies a temporary deficit of energy. The divergences of quantum electrodynamics arise through the virtual photons and virtual pairs of very high energy: these are eliminated by mass and charge re-normalization.

The picture presented by Dyson is that of a fluid in violent quantum-mechanical fluctuation, the fluctuations becoming more violent as the region of observation is made smaller. At any point in space-time the properties of the fluid define the electromagnetic and the particle fields. It is, however, meaningless to make statements concerning the behaviour of the fluid at a point—an experiment to measure the force on a test-object measures the average force over a finite volume of space-time. If, therefore, one expresses the properties of the fluid in terms of averages over regions of space-time, the fluctuations are smoothed out and one obtains a divergence-free description of the fluid. The new theories of quantum electrodynamics are therefore expressed in terms of operators which are time-integrals of the operators used in the earlier theories. The results are given in the form of a power series in e^2, the successive terms being successive orders in a perturbation expansion. The individual coefficients of these terms are finite, so that it is sometimes asserted that the technique gives a finite result to any required order of perturbation. On the other hand, the series itself is believed to be divergent [41].

A recent account of the theory has been given by Gunn [54]. In the same volume, Peierls and others [107] give a non-specialist review of field theory.

The authors of the theories do not claim to have developed a theory of the interaction between radiation and matter which is completely free of divergences, but a method of transformation which, although not mathematically rigorous, allows calculations to be made to any desired degree of approximation. What is still lacking is a theory of the structure of the electron. In the absence of a satisfactory theory, the new quantum electrodynamics provides a technique for side-stepping the difficulties It is to be regarded as justified by its success in applications rather than by its logical foundation.

In the next chapter we shall compare the predictions of the theory with experimental observations on hydrogen-like spectra (see Table 1, page 64).

CHAPTER X

FURTHER MEASUREMENTS OF LAMB SHIFTS

DEVIATIONS from the Dirac theory have now been found in the levels $n = 1$, 2 and 3 of deuterium, $n = 2$ of tritium, and $n = 2$, 3 and 4 of He$^+$ by optical spectroscopy. Shifts in the level $n = 3$ of hydrogen and $n = 2$ of He$^+$ have been measured by radio-frequency spectroscopy.

10.1. Optical spectroscopy

10.1.1. *Deuterium: $n = 2$ and 3*

After the publication of Lamb and Retherford's first paper, Kuhn and Series made a further attempt to resolve the components of D_α [76], [121]. Doppler broadening was reduced by the use of liquid hydrogen as a cooling agent. Temperatures of

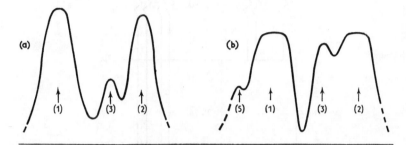

FIG. 6(a) and 6(b). Fine structure of D_α as resolved by Kuhn and Series [76],[121]. The curves are smoothed microphotometer tracings of the photographs. The exposure time was chosen suitably for component (3) in (a) and for (5) in (b). Components (1) and (2) appear broadened through over-exposure.

the excited atoms as low as 50°K were realized in a feeble discharge in deuterium cooled in this way. Metal sections of the U-shaped discharge tube served as electrodes and allowed efficient cooling. Light was taken from the positive column since the discharge is less violent there than at the cathode.

Component (3) was easily resolved (Fig. 6a), and by development of the interferometric technique, so also was the very weak component (5) (Fig. 6b). Under certain conditions component (4) could be seen as an asymmetry on the long-wave side of (1). Moreover, component (2) was wider than (1), showing that it is not simple. Fig. 7 (which should be compared with Fig. 2, p. 33, the Dirac term scheme), shows the terms and transitions for H$_\alpha$ drawn under the assumption that the $2^2S_{\frac{1}{2}}$ and $3^2S_{\frac{1}{2}}$ levels are shifted. The measurements of Kuhn and Series allowed the magnitude of these shifts to be ascertained, and showed that there were no comparable shifts in the other levels. The intensities were in good agreement with those calculated from the Dirac theory, with the exception that (3) was slightly

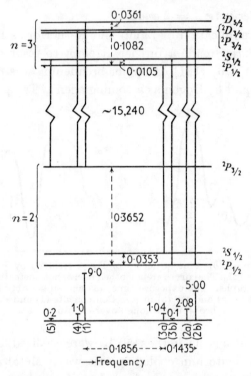

FIG. 7. Fine structure of H_α (D_α) showing calculated Lamb shifts. (Compare Fig. 2, p. 33.) Intervals in cm^{-1}.

stronger than expected, although hardly in excess of the (rather wide) limits of error. The measured shifts

$$\delta(2^2S_{\frac{1}{2}}) = 0 \cdot 0369 \pm 0 \cdot 0016 \text{ cm}^{-1},$$
$$\delta(3^2S_{\frac{1}{2}}) = 0 \cdot 0083^{+0 \cdot 002}_{-0 \cdot 003} \text{ cm}^{-1}$$

are not in disagreement with the values $0 \cdot 0353$, $0 \cdot 0105$ cm^{-1} respectively which the new theory predicts.

10.1.2. *Tritium: n = 2*

Kireyev [72a] has recently excited the tritium α-line in a high frequency discharge tube cooled by liquid nitrogen. The factor

$\sqrt{\frac{2}{3}}$ in the Doppler width allowed rather better resolution than that obtained by R. C. Williams for deuterium. The mean value 0.0355 cm^{-1} for the shift of the $2^2S_{\frac{1}{2}}$ level was obtained by treating components (1), (2), and (3) as blends with the theoretical intensity ratios, but the subsequent application of corrections for the overlap of (3) by the wings of (1) and (2) led to the final value 0.037 cm^{-1}. (This has been incorrectly reported in *Physics Abstract* 4138 of May, 1957.) Estimation of errors led to the conclusion $0.033 < \delta(2^2S_{\frac{1}{2}}) < 0.039$ cm^{-1}.

10.1.3. Helium$^+$: $n = 3$ and 4

The fine structure of the ionized helium lines, and in particular of the line 4686 Å ($n = 4-3$) is sixteen times as wide as the corresponding structure in hydrogen. $\lambda 4686$ Å therefore offers to optical spectroscopists a better opportunity than H_α for the resolution of its components. For this reason, and because of its greater complexity, Paschen's observations provided Sommerfeld with more material than did the Balmer lines for testing his fine structure formula (section 5.1). In so far as 4686 Å is a spark line, its excitation requires a relatively violent gas discharge, which favours neither a small Doppler width nor freedom from the influence of electric fields. Its attraction as a test of the radiation theory lies particularly in the possibility of checking the predicted dependence of term shifts on n and on Z.

Of a number of studies of this line since the war, the best resolution was obtained by Series [122]. Reduction of the Doppler width was again achieved by liquid hydrogen cooling of a discharge as gentle as possible. Spectroscopic analysis was by means of multiple Fabry-Perot interferometers, which allow an extension of spectral range without sacrifice of resolving power.

Fig. 8 shows a term diagram for this line, together with the relative probabilities of the allowed transitions calculated from the Dirac theory. Beneath the term diagram is a micro-photometer tracing obtained by Series. The reality of the

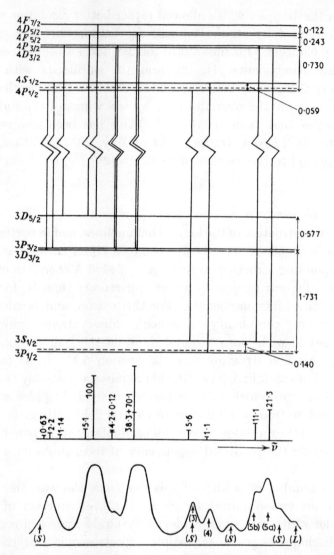

FIG. 8. Fine structure of the He$^+$ line 4686 Å, showing calculated Lamb shifts. Beneath is a photometer tracing of the structure resolved by Series [122]. The line marked (L) is a band line due to molecular helium. Lines marked (S) are false lines which arise with the use of double interferometers, and which may readily be moved to different places within the pattern.

Lamb shifts in the S levels is immediately apparent in that components (5a) and (5b), and (3) and (4) are separated. On the Dirac theory, these pairs would coincide exactly.

The intervals $(3^2S_{\frac{1}{2}} - 3^2P_{\frac{1}{2}}) = 0 \cdot 140 \pm 0 \cdot 005$ cm^{-1},

$$(3^2P_{\frac{3}{2}} - 3^2P_{\frac{1}{2}}) = 1 \cdot 735 \pm 0 \cdot 003 \text{ cm}^{-1},$$

and the shift $\delta(4^2S_{\frac{1}{2}}) = 0 \cdot 056 \pm 0 \cdot 003$ cm^{-1} measured by Series are in good agreement with the theoretical values $0 \cdot 140$ $(1 \cdot 731 + 0 \cdot 003)$, and $0 \cdot 059$ cm^{-1} respectively (see ‖, ¶, Table 1). The position of component (3), however, could not be explained without postulating a displacement of the level $4^2P_{\frac{1}{2}}$ downward from the Dirac position by $0 \cdot 011 \pm 0 \cdot 003$ cm^{-1}, in disagreement with the radiation theory. It was not possible to explain the discrepancy on the basis of a Stark effect.*

This anomalous shift of the $4^2P_{\frac{1}{2}}$ term was not confirmed by later work of Herzberg [61]. In these experiments the line was excited in a liquid-nitrogen cooled discharge, and analysed spectroscopically with a powerful grating. The resolution was slightly inferior to that of Series, but the interval (3)—(4) could be measured with comparable precision. No disagreement with the radiation theory was found.

An experiment is now being undertaken at Oxford to measure the interval $4^2P_{\frac{1}{2}} - 4^2S_{\frac{1}{2}}$ by radiofrequency methods.

10.1.4. Helium$^+$: $n = 2$ and 3

Herzberg's paper [61] also reports measurements on the He$^+$ line 1640 Å ($n = 3-2$), excited and analysed in the same way as 4686 Å. The fine structure is the same as that of H$_\alpha$ enlarged by the factor 16, except that the Lamb shifts are increased by a slightly smaller factor. All seven predicted components can be seen (one with difficulty), in disagreement with the exact coincidences of some of them which one would expect on the Dirac theory. The measured intervals agree with the radiation theory to within the accuracy of measurement ($\pm 0 \cdot 02$ cm^{-1}).

---

* The discrepancy is not due to the coincidence with (3) of the false line (S) shown in Fig. 8. By a different choice of interferometer spacer, the false lines can be moved, and were moved from (3) when this component was to be measured.

10.1.5. *An intensity anomaly*

An intensity anomaly noticed by Series in 4686 Å was found also in both lines by Herzberg, who was able to account for it as an effect depending on the relative magnitudes of the collision lifetime and the natural lifetime of the excited atom. If the former is shorter than the latter, one can expect a statistical distribution of atoms between the excited states, and the relative intensities of the components will be proportional to the transition probabilities. This is the assumption usually made in calculating relative intensities. If, on the other hand, the atoms generally decay before they collide, then (since the photons escape and the Principle of Detailed Balancing does not apply) the statistical distribution is distorted and intensity anomalies arise. In this way Herzberg explained his observations that the relative intensities of the components approached the theoretical values at higher pressures in the discharge tube. Since hydrogen lifetimes are sixteen times greater than He$^+$ lifetimes, hydrogen is less liable to these anomalies than ionized helium.

10.1.6. *Deuterium:* $n = 1$

Herzberg [60] has measured the absolute wavelength of the Lyman α-line ($n = 2-1$) in deuterium with a view to finding the shift of the $1^2S_{\frac{1}{2}}$ term, (see Fig. 9). The line (1215 Å) lies outside the spectral region accessible by interferometry: Herzberg photographed it in the fifth order of a large grating. It was produced by absorption in a stream of atomic deuterium produced in a Wood's tube and subsequently cooled by liquid air. Deuterium rather than hydrogen was investigated principally because residual hydrogen in the apparatus gave rise to an unnecessarily broad absorption line. Since this lies 22·4 cm^{-1} from the deuterium line, there is no danger of overlap. Lyman α, however, has a structure (shown in Fig. 9) consisting of two components of unequal intensity which lie 0·365 cm^{-1} apart. (The radiation theory does not alter the Dirac value, to this precision.) Since the Doppler width at 80°K is almost the same as the doublet interval, the components were not resolved. Herzberg's problem was to locate the peak of an asymmetrical blend of overall width about 0.7 cm^{-1} and to measure its wavelength to a precision which would be significant in comparison with the expected shift of the $1^2S_{\frac{1}{2}}$ term, 0.273 cm^{-1}. His final result was $\delta(1^2S_{\frac{1}{2}}) = 0.26_2 \pm 0.03_8$ cm^{-1}.

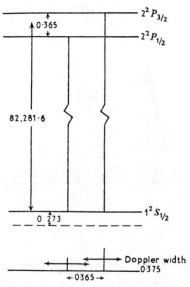

Fig. 9. Fine structure of the Lyman α-line (deuterium) showing the calculated Lamb shift in the ground level. The Doppler width corresponds to a temperature of $80°$K. Intervals in cm^{-1}.

10.2. Radiofrequency spectroscopy

10.2.1. Helium$^+$: $n = 2$

Lamb and Skinner [86] were able to measure the interval $2^2S_{\frac{1}{2}} - 2^2P_{\frac{1}{2}}$ in He$^+$ by a method which, as in the experiment on hydrogen, depended on the metastability of the $2^2S_{\frac{1}{2}}$ level, but in which the method of detection of radiofrequency resonance was different. In the case of hydrogen, the method of detection depended on the possibility of distinguishing metastable atoms from atoms in the ground state through their power of liberating electrons from tungsten. Helium ions can also liberate electrons from metals, but the relative ejection efficiency of metastable and non-metastable ions is not known. Lamb and Skinner therefore proposed to detect the 41 eV photons emitted in the decay of the level $2^2P_{\frac{1}{2}}$ (see Fig. 10). An increase in the population of this level (induced by radiofrequency resonance from $2^2S_{\frac{1}{2}}$) would be reflected in an increased photoelectric current.

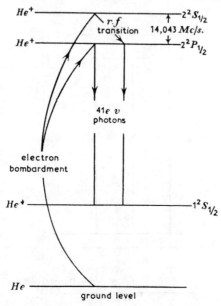

FIG. 10. The detection of radio-frequency resonance in the $n = 2$ levels of He⁺.

Helium gas at a pressure of a few microns inside a waveguide is subjected to electron bombardment. Of the small number of ions excited to the $2^2P_{\frac{1}{2}}$ and $2^2S_{\frac{1}{2}}$ levels, those in the former decay immediately. An excess accumulates in the $2^2S_{\frac{1}{2}}$ level because of its metastability. Transitions from $2^2S_{\frac{1}{2}}$ to $2^2P_{\frac{1}{2}}$ are induced by the radiofrequency fields in the waveguide, and are detected by observation of an increase in the 41 eV photoelectric current. As in the hydrogen experiment, the interaction region is situated in a magnetic field which can be varied through a resonance line, while the microwave frequency remains constant.

The value $14{,}020 \pm 100$ Mc/s obtained by Lamb and Skinner for the $2^2S_{\frac{1}{2}} - 2^2P_{\frac{1}{2}}$ interval in He⁺ has been superseded by the more accurate determination of Novick, Lipworth and Yergin [99]. The method used by these authors differs from that of Lamb and Skinner in that the radiofrequency field and the detecting equipment are applied in a short pulse about a quarter microsecond later than a pulse of exciting electrons. In the short interval between the exciting and detecting pulses, the ions and atoms in short-lived states decay, and no longer contribute to the background signal as they do when the exciting and detecting processes take place simultaneously. The consequent improvement in signal-to-noise ratio, together with other advances in technique, allowed a detailed study of the line shape and a considerable increase in the precision of the result. The experimental value, 14,043±13 Mc/s is consistent with the most recent theoretical value, 14,043±3·0 Mc/s.

10.2.2. *Hydrogen: $n = 3$*

The state $n = 3$ of hydrogen has been studied by Lamb and Sanders [85] by a related method, although here there are no metastable states. The method is the following. Suppose that initially $3s$ and $3p$ states are unequally populated (this can be achieved by electron bombardment of hydrogen gas at low pressure). Decay from $3s$ to $2p$ is more probable than from $3p$ to $2s$, although each transition constitutes part of H_α (6563 Å). If, then, transitions between $3s$ and $3p$ are induced by a radiofrequency field, the intensity of the H_α emitted in the subsequent decay will change. This change, detected by a photomultiplier, will indicate radiofrequency resonance.

The interaction region is placed in a uniform magnetic field, which, as in the other experiments, is varied through resonance. Of the six Zeeman components which are to be expected by the selection rule $\Delta m = \pm 1$, five have been observed by Lamb and Sanders at various frequencies. The measurements are in agreement with the predictions of the radiation theory to within about one Mc/s.*

---

* Private communication from Professor Lamb.

TABLE 1

Lamb shifts. Comparison between theory and experiment*††

	H	D	He⁺
$n = 1$		(a) $\delta(S_{1/2})$ $0.26_2 \pm 0.03_8$ cm^{-1} (0.273)	
$n = 2$	(b) $S_{1/2} - P_{1/2}$ 1057.77 ± 0.10 Mc/s (1057.19 ± 0.16)*	(b) $S_{1/2} - P_{1/2}$ 1059.00 ± 0.10 Mc/s (1058.43 ± 0.16)* (d) $\delta(S_{1/2})$ 0.0369 ± 0.0016 cm^{-1} (0.0353)	(e) $S_{1/2} - P_{1/2}$ 14,043 ± 13 Mc/s (14,043 ± 3)* (i) $S_{1/2} - P_{1/2}$ 0.48$_0$ cm^{-1}†† (0.468)
§§		(c) $P_{3/2} - S_{1/2}$ 9912.59 ± 0.10 Mc/s§§	
$n = 3$	(f) $\left.\begin{array}{l} S_{1/2} - P_{1/2} \\ P_{3/2} - P_{1/2} \end{array}\right\}$ Agreement with theory to ± 1 Mc/s (315)	(g) $\delta(S_{1/2})$ $0.0083 \begin{array}{l}+\,0.002 \\ -\,0.003\end{array}$ cm^{-1} (0.0105)	(h) $S_{1/2} - P_{1/2}$ 0.140 ± 0.005 cm^{-1} (0.140) (i) $S_{1/2} - P_{1/2}$ 0.12, 0.14$_7$ cm^{-1**},†† (0.140) (h) $P_{3/2} - P_{1/2}$ 1.735 ± 0.003 cm^{-1}‖ (1.731 ± 0.003)¶‖
$n = 4$			(h) $\delta(S_{1/2})$ 0.056 ± 0.003 cm^{-1} (0.058) (h) $\delta(P_{1/2})$ 0.011 ± 0.003 cm^{-1} (0.001) (i) $S_{1/2} - P_{1/2}$ 0.05$_9$ cm^{-1}†† (0.059)

(a) Herzberg [60].
(d) Kuhn and Series [76].
(g) Series [121].

(b) Triebwasser, Dayhoff, and Lamb [131].
(e) Novick, Lipworth, and Yergin [99].
(h) Series [122].

(c) Dayhoff, Triebwasser, and Lamb [30].
(f) Lamb and Sanders [85].
(i) Herzburg [61].

10.3. Comments on Table 1

The broad agreement between experiment and theory is most satisfactory. The dependence on Z is particularly well verified by the radiofrequency measurement on He+, $n = 2$. The dependence on n over the wide range from 1 to 4 is verified to within the precision of optical measurements.

There is a small discrepancy in the $n = 2$ level, both for hydrogen and deuterium, which is outside the range of experimental error, and of theoretical calculations so far as the latter have been taken. A suggestion [2] that the term of order $Z^2\alpha^2 \ln (Z\alpha)L'$ might be large enough to account for the discrepancy cannot be upheld in the light of the measurements on $n = 2$ in He+ [99].

Some information on the size of the correction for nucleon structure can be gained from a comparison of the difference between the experimental values for hydrogen and deuterium, $1\cdot23\pm0\cdot15$ Mc/s with the difference between the theoretical values, $1\cdot24\pm0\cdot035-e_{st}$. This shows that e_{st}, the correction for nucleon structure (section 9.8) cannot be very large.

A discrepancy of a different order of magnitude appears in $4^2P_{\frac{1}{2}}$ of He+. It remains to be seen whether more refined experiments will remove this discrepancy.

---

* Theoretical values are given in brackets. They are taken principally from Salpeter [119]. The proton volume correction [1a] and certain mass corrections [52] have been added. Karplus and Kroll's value of μ_e has been used [70], see p. 68. Sommerfield's value [124a] leads to the intervals 1057·99, 1059·23, 14055·9 Mc/s for $2(S_{\frac{1}{2}}-P_{\frac{1}{2}})$ in H, D and He+ respectively.

† Radiofrequency and optical measurements may be distinguished by the units employed.

‡ It has sometimes been possible to refer a shift to a particular term, in which case the symbol $\delta(\ )$ is used.

§ This measurement, together with that of the $S_{1/2}-P_{1/2}$ interval, is at present used to calculate the most reliable experimental value for α.

‖ These results are taken from the body of the paper.

¶ 1·731 cm⁻¹ is the Dirac value for this interval.

** Calculated on the assumption of the theoretical value for $2^2S_{1/2}-2^2P_{1/2}$.

†† Calculated on the assumption of the theoretical value for $3^2S_{1/2}-3^2P_{1/2}$.

‡‡ These values are given with reserve.

§§ The value $\delta(2S_{1/2}) = 0\cdot037^{+0\cdot002}_{-0\cdot004}$ cm⁻¹ has recently been reported for tritium [72a].

CHAPTER XI

HYPERFINE STRUCTURE

It is well known that hyperfine structure in spectra arises from the interaction of nuclear magnetic dipoles and the magnetic field due to the electron cloud. In the case of hydrogen the interaction energy is appreciably greater than the natural width of the energy levels only in the case of the ground level and the abnormally narrow (metastable) $2^2S_{\frac{1}{2}}$ level. The hyperfine structure of both these levels has been measured with great precision by atomic beam magnetic resonance methods. The ground level in particular has been repeatedly examined by this technique, and also by the method of paramagnetic absorption of microwaves.

Calculation of the hyperfine structure requires knowledge of the magnetic moment of the proton. If this were a Dirac particle, its moment would be $eh/4\pi Mc$ (one nuclear magneton), which stands in conflict with the value 2·79275 nuclear magnetons measured by a variety of methods. This experimental value, used in the simplest hyperfine structure formula, leads to substantial agreement between experiment and theory. Even so, there remains a discrepancy of about 0·1 per cent, which is greatly in excess of the experimental errors. This discrepancy is closely related to the interactions which give rise to the Lamb shift, as we shall now see.

11.1. The Fermi formula

For s-states the hyperfine energy is the product of the magnetic induction at the origin which arises from the density of the electron-spin magnet there, multiplied by the nuclear magnetic moment. Thus

$$E_{\text{h.f.s.}} = \frac{8\pi}{3}\,\mu_e \psi^2(0) \cdot \mu_n, \qquad (11.1)$$

where μ_e and μ_n are the electron and nuclear magnetic moments respectively. The numerical factor arises from the relation between the magnetic induction and the magnetic moment of a uniformly magnetized sphere. Substitution of the Schrö-dinger function $\psi^2(0) = Z^3/\pi a_0^3 n^3$, where a_0 is the Bohr radius, leads to the well-known Fermi equation [44] for the hyperfine structure a-factor

$$a = \frac{16\pi}{3}\,\mu_e\,\frac{\mu_n}{I}\,\frac{Z^3}{\pi a_0^3 n^3} = \frac{8}{3}\,\frac{hcR_\infty\alpha^2 Z^3}{n^3}\,g_I\,\frac{m}{M}. \qquad (11.2)$$

a is defined by the equation

$$E_{\text{h.f.s.}} = a\mathbf{I}.\mathbf{J}$$

which we are here applying to a one-electron system in which $\mathbf{J} = \mathbf{S}$. The magnitude of $\mathbf{S}$ is $\frac{1}{2}$. For a given nuclear spin $\mathbf{I}$ there exist two hyperfine structure levels separated in energy by $\Delta\nu = a(I+\frac{1}{2})$. The value of I is $\frac{1}{2}$ for the proton, triton and helium^3 nucleus; for the deuteron $I = 1$.

The occurrence of the term $1/a_0^3$ in $\psi^2(0)$ indicates that a reduced-mass correction factor $(1+m/M)^{-3}$ is necessary. A small relativistic correction also arises when Dirac functions are used to evaluate $\psi^2(0)$ [21]. Its value in the $1^2S_{\frac{1}{2}}$ level is

$$\left[1+\frac{3}{2}\,Z^2\alpha^2+0(\alpha^4)\right], \text{ and in the } 2^2S_{\frac{1}{2}}, \left[1+\frac{17}{8}\,Z^2\alpha^2+0(\alpha^4)\right].$$

11.2. The Fermi formula—comparison with experiment

The Fermi formula (equation 11.2) together with the reduced-mass and relativistic correction terms, evaluated with Dumond and Cohen's values of the atomic constants [39], leads to the hyperfine structure interval

$$(\Delta\nu)_{\text{H}} = 1418\cdot90\pm0\cdot03 \text{ Mc/s}.$$

The most recent experimental values are:

$$1420\cdot40573\pm0\cdot00005 \text{ Mc/s [77]}$$
$$1420\cdot40580\pm0\cdot00006 \text{ Mcs [143]}.$$

The discrepancy between theory and experiment first became apparent with the measurement of $(\Delta\nu)_{\text{H}}$ by Nafe, Nelson and Rabi [94], [95]. At the time the discrepancy appeared to be worse than is shown by the figures here, since the method of magnetic field calibration, upon which the value of the proton magnetic moment depended, was based on the assumption that the magnetic moment of the alkali metals in the ground state was exactly one Bohr magnetron. We shall see in the next section that correction of this false assumption leads, on the one hand, to self-consistent measurements of magnetic fields, and, on the other hand, to a great improvement between the experimental value of $(\Delta\nu)_{\rm H}$ and the theoretical value quoted above. There remains a much smaller discrepancy which will be discussed in sections 11.4 and 11.5.

11.3. The anomalous magnetic moment of the electron

The suggestion of Breit [22] that the magnetic moment of the electron differs from one Bohr magneton substantially resolved the major hyperfine structure discrepancy. In section 9.7 we mentioned that the new quantum electrodynamics, through the relativistic treatment of renormalization of mass, predicts a correction to the Dirac value of the electron magnetic moment. To second order in the perturbation approach, the correction factor is $(1+\alpha/2\pi)$. A higher order calculation [70] yields $\left(1+\dfrac{\alpha}{2\pi}-2\cdot973\dfrac{\alpha^2}{\pi^2}\right)$, i.e., $1\cdot00114536*$, using the atomic constants of 1955. This suggestion affects $(\Delta\nu)_{\rm H}$ directly through μ_e, and indirectly through μ_n, if the latter has been determined with reference to an atomic g_J value based on the uncorrected electron moment.

$(\Delta\nu)_{\rm H}$, corrected for the anomalous magnetic moment of the electron, is $1420\cdot53\pm0\cdot03$ Mc/s, a value much closer to the experimental value, and capable, as we shall see, of further refinement by the new theory.

Experimental support for Breit's suggestion comes not only from hydrogen hyperfine structure, but also from experiments in which atomic magnetic moments are compared directly through their precessional frequencies in the same field. Kusch and Foley [78] compared the moments of gallium in the

---

* Sommerfield [124a] obtains $\left(1+\dfrac{\alpha}{2\pi}-0\cdot328\dfrac{\alpha^2}{\pi^2}\right)=1\cdot0011596$.

states $P_{\frac{1}{2}}$ and $P_{\frac{3}{2}}$, in which spin and orbital contributions combine differently. The ratio was not integral, as it should have been if the electron spin moment had been one Bohr magneton. From the measurements they deduced that, if the discrepancy lay with the electron spin moment, its value must be

$$\mu_e = \mu_o(1 \cdot 00114 \pm 0 \cdot 00004).$$

Similar results were obtained in comparisons between sodium and gallium, and between sodium and indium.

While these conclusions need defence against the objection that the states of these atoms may not be quite pure, similar results have now been obtained for many other cases, and in particular for the very pure ground state of hydrogen. The magnetic moment of the atom in the $1^2S_{\frac{1}{2}}$ state is identical with that of the bound electron (to the approximation that the hyperfine structure is negligible compared with the binding energy of the electron). A small correction (see p. 26) is necessary to obtain the magnetic moment of the free electron. The value obtained by Koenig, Prodell and Kusch [73] was

$$\mu_e = 1 \cdot 001146 \pm 0 \cdot 000012 \text{ Bohr magnetons.}$$

Beringer and Heald [5] obtained

$$\mu_e = 1 \cdot 001148 \pm 0 \cdot 000006.$$

The magnetic moment of a free electron has not yet been measured directly.

11.4. Further electrodynamic corrections

There are consequences of the new quantum electrodynamics other than the correction to the electron magnetic moment which arise in the theory of hyperfine structure. The most important of these is due to the spatially distributed nature of entities in the new theory (section 9.10). Thus, magnetic dipole interactions are not local, but are to be averaged over small regions. The calculation of Karplus and Klein [67] leads to the result that the whole effect, for S states, is equivalent to increasing the magnetic moment of the electron by

$$-\tfrac{1}{2}Z\alpha^2(5-2\ln 2) \text{ Bohr magnetons,}$$

i.e., a correction of about one part in 10^{-4} for hydrogen.

A further correction, a type of reduced mass effect, arises from the recoil of the nucleus in the emission of virtual photons which, in the theory, describe the attraction between the nucleus and the electron. The effect is not independent of the structure of the nucleus. It is, however, possible to separate the contributions due to mass and structure, so that the correction appears as the product of two terms, C_r and C_s. The magnitude of C_r, the contribution from a point proton (equation 11.6), has been estimated as parts in 10^5 [97]. C_s, the structure correction [146], is found to be $(1-2\bar{r}/a_0)$, where $\bar{r}$ is an average electromagnetic radius, and a_0 is the Bohr radius. C_s is independent of n([146], p. 1773).

$\bar{r}$, and hence C_s, cannot be evaluated without assuming a particular structure for the proton. On the other hand, by comparison of the complete hyperfine structure formula with experiment (section 11.5), limits can be set to the value of C_s, from which an upper bound $\sim 5\times10^{-14}$ cm can be set to $\bar{r}$. This is rather smaller than the radius $(7\cdot7\pm1\cdot0)\times10^{-14}$ cm obtained by Chambers and Hofstadter [24] from electron scattering experiments.

11.5. The corrected formula—comparison with experiment

The complete formula for the hyperfine structure in the $S_{\frac{1}{2}}$ levels of one-electron spectra is thus

$$(\Delta\nu)_{\text{H}} = (I+\tfrac{1}{2})\times \left[\frac{8}{3}\frac{hcR_\infty\alpha^2Z^3}{n^3}g_I\cdot\frac{m}{M}\right] \text{(Fermi formula)}$$

$$\times\left[1+\frac{m}{M}\right]^{-3} \qquad \text{(reduced-mass correction)}$$

$$\times C_{\text{rel}} \qquad\qquad \text{(relativistic correction)}$$

$$\times\left[1+\frac{\alpha}{2\pi}-2\cdot973\frac{\alpha^2}{\pi^2}\right] \quad \text{(anomalous electron moment) (but see p. 68)}$$

$$\times C_q \qquad\qquad \text{(quantum electrodynamic correction)}$$

$$\times C_r \qquad\qquad \text{(proton recoil correction)}$$

$$\times C_s \qquad\qquad \text{(proton structure correction)} \qquad (11.3)$$

where $C_{\rm rel} = [1+\tfrac{3}{2}Z^2\alpha^2+0(\alpha^4)]$ for $n = 1$ (11.4a)

$$[1+\tfrac{17}{8}Z^2\alpha^2+0(\alpha^4)] \text{ for } n = 2 \qquad (11.4b)$$

$$C_q = [1-\tfrac{1}{2}Z\alpha^2(5-2\ln 2)] \qquad (11.5)$$

$$C_r = \left\{1+\frac{\alpha}{\pi\mu_p}\cdot\frac{m}{M}\left[\frac{3}{4}(\mu_p-3)(\mu_p+1)\ln\frac{M}{m}\right.\right.$$

$$\left.\left.+\frac{1}{8}(\mu_p-1)^2-\frac{9}{4}(\mu_p-1)^2\ln\frac{2k_0}{M}\right]\right\}$$

$$= \left[1-(1\cdot86\times10^{-6})-(3\cdot26\times10^{-6})\ln\frac{2k_0}{M}\right], \qquad (11.6)$$

where k_0 is a 'cut-off' energy which may reasonably be taken $\sim M$, and

$$C_s = (1-2\bar{r}/a_0).$$

Of the terms in this formula, $C_{\rm rel}$ and C_r depend on n . C_r and C_s have been calculated only for the case of hydrogen, as distinct from deuterium, helium$^+$ etc. The complications introduced by compound nuclei will be treated in the next section. For comparison with experiment, we restrict ourselves at present to the ground state of hydrogen.

The atomic constants which arise in the formula, and in particular the fine structure constant, are known far less accurately than $(\Delta\nu)_{\rm H}$. The latter has been measured to a precision of four parts in 10^8 [77], [143]. The fine structure constant is now known (from the measurements of Dayhoff, Triebwasser and Lamb on the $P_{\frac{3}{2}}$—$P_{\frac{1}{2}}$ interval in hydrogen) to one part in 10^5, and to this precision the formula is in exact agreement with experiment. Since, formerly, the value of α was even less accurately known, it was at one time computed from the hyperfine structure formula itself. The uncertainty in the correction factors C_r, C_s leaves this determination open to objection, but a comparison of the values of α obtained in the two different ways sheds light on C_r and C_s. Thus, Dayhoff et al. [30] find

$$(\alpha^{-1})_{\rm f.s.} = 137\cdot0365\pm0\cdot0012, \qquad (11.7)$$

while

$$(\alpha^{-1})_{\rm h.f.s.} = (137\cdot0365\pm0\cdot0006)\,(C_rC_s)^{\frac{1}{2}} \qquad (11.8)$$

We have, therefore, $C_r C_s = 1 \pm (2 \times 10^{-5})$. Now, for $k_0 = M$, $C_r \approx 1 + (2 \times 10^{-5})$. An upper limit can therefore be set to $(1 - C_s)$ if we assume the correctness of the hyperfine structure formula. This, as we mentioned in the last section, allows inferences to be drawn concerning the finite extension of the proton.

11.6. Compound nuclei

We have been concerned with the absolute value of hydrogen hyperfine structure. Comparison of $\Delta \nu$ between hydrogen and deuterium in the ground state eliminates uncertainty in the values of μ_e, α, C_{rel} and C_q, and discloses a discrepancy of a kind different from any we have yet considered [111].

From equation 11.3 (ignoring C_r and C_s), we find

$$R_{H,D} = (\Delta \nu_H)/(\Delta \nu_D) = (\tfrac{4}{3})(\mu_{proton}/\mu_{deuteron})(m_H/m_D)^3, \quad (11.9)$$

where the last factor is the correction for the reduced mass of the electron in hydrogen, m_H, and in deuterium, m_D.

Using the experimental value [124]

$$(\mu_p/\mu_d) = 3 \cdot 2571995 \pm 0 \cdot 000001,$$

we find

$$(R_{H,D})_{th} = 4 \cdot 3393876 \pm 0 \cdot 000001,$$

whereas the value of R measured directly [77] is

$$(R_{H,D})_{exp} = 4 \cdot 33864947 \pm 0 \cdot 00000043.$$

There is a discrepancy between experiment and theory of about two parts in 10^4 which can be conveniently expressed:

$$\Delta \equiv 1 - (R_{exp}/R_{th}) = (1 \cdot 703 \pm 0 \cdot 005) \times 10^{-4}. \quad (11.10)$$

The hydrogen/deuterium discrepancy is taken up to the extent $\Delta = (0 \cdot 3 \pm 0 \cdot 1) \times 10^{-4}$ by corrections of the type C_r [88], but effects due to the structure of the deuteron play a more important part. Thus, A. Bohr [12] pointed out that the electron in the deuterium atom is bound to the proton as its centre of force, and that the neutron is in motion relative to this centre. The effective field due to the neutron is therefore reduced to zero over the volume of the deuteron. Since the magnetic moment of the neutron in the deuteron is opposed to that of the proton, the result will be an increase in the hyperfine structure, of the order of magnitude

$$(\mu_n/\mu_d) \ (\text{deuteron radius/atom radius}) \sim 1{\cdot}8 \times 10^{-4}.$$

More refined models of the deuteron will give slightly different values; nevertheless, Bohr's suggestion explains the anomaly both as to sign and as to order of magnitude. The most recent discussion is that of Low and Salpeter [88].

In the case of tritium and helium[3], but not in the case of deuterium, there is evidence from the magnitude of the nuclear moments that exchange currents within these nuclei modify the magnetic moments of the individual nucleons [11]. The calculated hyperfine structure must therefore include an 'exchange current' term in addition to the structure term mentioned in the last paragraph. The net effect for tritium [1] is a correction of a few parts in 10^6, which is smaller than the uncertainty of the experimental value of the ratio of the magnetic moments of the triton and proton $(\mu_t/\mu_p) = 1{\cdot}066636 \pm 0{\cdot}000010$ [11a]. The structure correction cannot, therefore, be tested experimentally although the hyperfine structure of tritium in the ground state, $1516{\cdot}70170 \pm 0{\cdot}00007$ Mc/s, has been measured with the necessary precision [96], [111a].

For ionized helium[3], on the other hand, the experimental value of the hyperfine structure of the metastable $2^2S_{\frac{1}{2}}$ level, $1083{\cdot}360 \pm 0{\cdot}020$ Mc/s [98] differs from the 'uncorrected' calculated value by 182 ± 22 parts in 10^6. This calculated value includes all the terms included in equation 11.3 except C_r and C_s. The values of the nuclear structure and exchange current corrections depend on the choice of the nuclear wave functions and exchange current operators. Foley and Nessler [47] find a structure correction of 138 parts in 10^6, and exchange current corrections of 2 parts in 10^6 for one type of interaction current, and 230 parts in 10^6 for another. The measured hyperfine structure therefore discriminates against the latter.

11.7. Excited levels in hydrogen and deuterium

The measurements of Heberle, Reich and Kusch [57] of the hyperfine structure of the metastable level $2^2S_{\frac{1}{2}}$ in hydrogen allow a further check on equation 11.3. In taking the ratio $\rho = (\Delta\nu)_{n=1}/(\Delta\nu)_{n=2}$ the leading terms are

$$\rho_{\text{th}} = \tfrac{1}{8}[1+(\tfrac{5}{8})\alpha^2] = \tfrac{1}{8}(1\cdot 0000333), \tag{11.11}$$

where $\tfrac{1}{8}$ expresses the n-dependence, and the second factor is the relativistic correction. The reduced-mass factor cancels, as do all other factors with the possible exception of C_r.

The experimental value $(\Delta\nu_{\text{H}})_{n=2} = 177\cdot 55686 \pm 0\cdot 00005$ Mc/s leads to the ratio

$$\rho^{(\text{H})}_{\text{exp}} = \tfrac{1}{8}(1\cdot 0000346 \pm 0\cdot 0000003). \tag{11.12}$$

We notice first that the relativistic correction (~ 3 parts in 10^5)—and hence the use of Dirac rather than Schrödinger wave functions in the calculation of $\psi^2(0)$—is fully justified, since the ratio $(C_r)_{n=1}/(C_r)_{n=2}$ can hardly exceed parts in 10^6. It is, in fact, considered unlikely that such a differential effect in C_r could account even for the very small discrepancy $\tfrac{1}{8}(13\pm 3)\times 10^{-7}$ between ρ_{th} and $\rho^{(\text{H})}_{\text{exp}}$. The reason for the discrepancy lies more probably in the neglect in ρ_{th} of quantum electrodynamic terms of order α^3 [92].

Equation 11.3 applies only to hydrogen. In deuterium the factor $(1+\Delta)$, which replaces C_r and C_s, allows for the compound nucleus. Comparison between ρ_{th} and $\rho^{(\text{D})}_{\text{exp}}$ tests whether Δ is the same for different states.

Reich, Heberle and Kusch [113] have measured the hyperfine structure of deuterium in the level $n = 2$. Their value $40\cdot 924439 \pm 0\cdot 000020$ Mc/s leads to

$$\rho^{(\text{D})}_{\text{exp}} = \tfrac{1}{8}(1\cdot 0000342 \pm 0\cdot 0000006). \tag{11.13}$$

Comparison with ρ_{th} reveals the very small discrepancy $\tfrac{1}{8}(9\pm 6)\times 10^{-7}$. In view of the closeness of this value to the corresponding figure for hydrogen we may draw the conclusions:

(i) That there is no evidence for a differential effect in Δ for hydrogen and deuterium between the $1S$ and $2S$ levels, and

(ii) that the discrepancy between ρ_{th} and ρ_{exp} is real.

CHAPTER XII

POSITRONIUM

THE system comprised of one electron and one positron can exist in bound states for times sufficiently long to allow the measurement of energy differences between them. Positronium*, as this substance is called, thus allows a further test of the theory of the two-particle atom. Details of the theory were worked out before there was experimental proof of the existence of positronium.

12.1. The gross structure

The most important interaction between the two particles is the Coulomb attraction, which leads to a gross structure of the energy levels identical in form with that of hydrogen. The scale is smaller by the factor two because of the reduced-mass correction, so that the wavelengths are doubled. The matrix elements of the dipole moment are doubled compared with hydrogen, since the electron-positron distance is twice as great as the electron-proton distance. The transition probabilities, which are proportional to $v^3\langle er\rangle^2$, are therefore halved. These predictions are confidently affirmed, although experimental confirmation has not yet been reported.

12.2. The fine and hyperfine structure

The fine structure was calculated by Pirenne [110] and independently by Berestetski [4]. Minor errors are corrected, and numerical results are given by Ferrell [45]. The approach used by these authors is to write down the Dirac equations for the two particles, and the interaction terms as they are expressed in quantum field theory. The equations can be transformed so that the particle spins appear explicitly. The interaction terms are found to comprise the Coulomb energy, the Breit interaction, and a term analogous to the Fermi expression for

* Excellent review articles have been written by Deutsch [32], and by De-Benedetti and Corben [31].

the hyperfine structure (p. 67). Just as the Breit interaction is an approximation, so also in Pirenne's formulae the restriction to non-relativistic velocities is imposed, although the particles themselves are relativistic in the sense of possessing intrinsic spin. Divergent terms characteristic of the radiation interaction are omitted.

The equations for positronium are now formally identical with those for hydrogen discussed on p. 29. All interaction terms other than $e_1 e_2 / r_{12}$ are treated as perturbations. The eigenvalues for the energy in zero order then give simply the gross structure. The relative magnitudes of the various terms which contribute to the fine and hyperfine structure differ profoundly from the corresponding terms in hydrogen because of the different masses of the proton and the positron. Thus the spin-orbit, spin-spin and orbit-orbit terms are all of comparable magnitude, whereas for hydrogen the first is by far the most important. For positronium, therefore, the natural classification of the terms is into singlets and triplets. The levels $^1P_1$, $^3P_{0,1,2}$ of positronium, for example, correspond to the hyperfine structure levels $P_{\frac{1}{2}}$, $F = 0, 1$; and $P_{\frac{3}{2}}$, $F = 1, 2$ of hydrogen.

12.3. The annihilation force

A further term, which has no analogue in hydrogen, arises in the fine structure of positronium. This comes from the possibility of virtual annihilation and re-creation of the electron-positron pair. A virtual process is one in which energy is not conserved. Real annihilation limits the lifetimes of the bound states and broadens the energy levels (section 12.6). Virtual annihilation and re-creation shift the levels. It is essentially a quantum-electrodynamic interaction. The energy operator for the double process of annihilation and re-creation is different from zero only if the particles coincide, and have their spins parallel. There exists, therefore, in the triplet states, a term proportional to $\psi^2(0)$. It is important only in $^3S_1$ states, and is of the same order of magnitude as the Fermi spin-spin interaction. Humbach [65] has given an interpretation of this annihilation term shift in a manner analogous to Welton's interpretation of the Lamb shift (section 9.5). The uncertainty relation implies a mean distance of closest approach of the electron and positron in the region of their annihilation. This distance is of the order $\hbar/mc$. The mean potential experienced by the electron at position $\mathbf{r}$ is then, not $V(\mathbf{r})$, but $V(\mathbf{r}+\hbar/mc)$. The change in energy due to this change in potential is

$$\delta E = -e(\overline{\delta V})_{\mathbf{r}=0}\,\psi^2(0),\tag{12.1}$$

since the positron is at $\mathbf{r} = 0$. Using for $\overline{\delta V}$ the expression $\frac{1}{6}\nabla^2 V(\hbar/mc)^2$, (compare equation 9.16), where $\nabla^2 V = -4\pi$ (charge on positron), and recalling that $(e\hbar/2mc) = \mu_0$, we obtain

$$\delta E = \frac{8\pi\mu_0^2}{3}\,\psi^2(0).\tag{12.2}$$

Pirenne and Berestetski, by a more formal calculation, obtain three times this amount.

12.4. The energy levels

Combination of all these terms results in the following formulae [45]:

(a) Gross structure

$$E_n = -\frac{R}{2n^2} = -\frac{mc^2\alpha^2}{4n^2}\tag{12.3}$$

(b) Fine and hyperfine structure

$$E_{n,l,j} = \frac{11mc^2\alpha^4}{64n^4} + \frac{mc^2\alpha^4}{n^3}\left[-\frac{1}{2(2l+1)}+\varepsilon\right],\tag{12.4}$$

where
$$\varepsilon = 0 \text{ for singlet states,}\tag{12.4a}$$
$$= \tfrac{7}{12}* \text{ for } {}^3S_1 \text{ states,}\tag{12.4b}$$

$$= \frac{1}{8(l+\frac{1}{2})} \times \begin{cases} \dfrac{(3l+4)}{(l+1)(2l+3)} & \text{for } j = l+1 \\[2mm] -\dfrac{1}{l(l+1)} & \text{for } j = l \\[2mm] -\dfrac{(3l-1)}{l(2l-1)} & \text{for } j = l-1 \end{cases} \begin{array}{l} \text{for} \\ \text{triplet} \\ \text{states,} \\ l > 0. \end{array}\tag{12.4c}$$

* The factor $\tfrac{7}{12}$ has been omitted from Ferrell's article.

The fine structure terms are of order α^2 times the gross structure, as in hydrogen. The term in n^{-4} displaces all the levels of a given n by the same amount, and is reminiscent of a similar term in hydrogen (section 6.10). The fine structure splitting appears in the term in n^{-3}. There is no degeneracy between states of different orbital angular momentum.

The interval which has been determined experimentally, ΔW, is that between the $^1S_0$ and $^3S_1$ levels of the ground state, $n = 1$. From the formulae above,

$$\Delta W = (\tfrac{7}{12})mc^2\alpha^4 = (\tfrac{7}{3})\mu_0^2/a_0^3 = 2 \cdot 044 \times 10^5 \text{ Mc/s.} \quad (12.5)$$

The interactions which contribute to this are the Fermi spin-spin terms $[(8\pi/3)\mu_0^2\psi^2(0)$ for $^3S_1$; $-8\pi\mu_0^2\psi^2(0)$ for $^1S_0$;] and the annihilation term $8\pi\mu_0^2\psi^2(0)$ for $^3S_1$. Insertion of $\psi^2(0) = 1/\pi(2a_0)^3$ leads to the result quoted.

Corrections to the above formulae are introduced by the new quantum electrodynamics. There is no spectacular splitting of levels since there is no degeneracy, but in so far as ΔW depends on μ_0^2, it is clear that the anomalous magnetic moment of the electron will influence the result. Calculations to order α^3, relative to the gross structure, have been made by Karplus and Klein [68]. These authors obtain the result

$$\Delta W = (\mu_0^2/a_0^3)[\tfrac{7}{3} - (\tfrac{32}{9} + 2\ln 2)\alpha/\pi] = 2 \cdot 0337 \times 10^5 \text{ Mc/s} \quad (12.6)$$

More complete calculations, with particular reference to the $n = 2$ levels, have been made by Fulton and Martin [52].

12.5. The Zeeman effect

The determination of ΔW depends on the behaviour of positronium in magnetic fields, which we must now consider.

The atom is diamagnetic: the orbital contribution to the magnetic moment is always zero because the charges are equal and opposite while the masses are identical; the spin contribution is also zero in zero field since, if the spin momenta are parallel, the spin magnetic moments are opposed, while if the spin momenta are antiparallel there is no preferred direction and the time average of the magnetic moment in any particular direction is zero.

In a magnetic field the spin-spin coupling is partially broken down, so that the effective magnetic moment becomes a function of the field. The $M_J = 0$ states of the $^3S_1$ and $^1S_0$ levels begin to lose their individuality, and their energies, measured from the unperturbed singlet state, are given by

$$\delta W = (\Delta W/2)[1 \pm (1 + x^2)^{\frac{1}{2}}], \tag{12.7}$$

where $\qquad x = 4\mu_0 H / \Delta W.$

(This is the well-known Breit-Rabi formula: see, for example, [112].) Since there are no $M_J = \pm 1$ states in $^1S_0$, the states $M_J = \pm 1$ of $^3S_1$ are unaffected. Since, further, the net magnetic moment of these states is zero, they are completely unaffected by the field.

A plot of the energy levels as a function of field is given in Fig. 11.

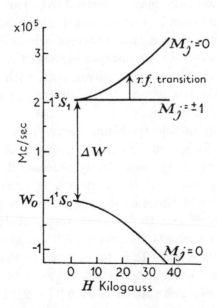

Fig. 11. Zeeman splitting of the lowest triplet and singlet levels of positronium. (The r.f. transition has been drawn, for clarity of illustration, at a higher field than that used by Deutsch and Brown [33].)

12.6. Lifetimes

The energy of annihilation of positronium may be emitted as two, three, or more photons, according to the symmetry of the state which decays. Single photon emission can take place only in the presence of fields strong enough to absorb the momentum: we shall not be concerned with this case.

Conservation of linear momentum requires that, if two photons are emitted, their directions be opposite. A consequence of this is, as we shall show, that states with $J = 1$ cannot decay by double quantum emission. For, the angular momentum carried by the photons is $2\hbar$ or zero; the former is unacceptable if, before annihilation, the total angular momentum was $\hbar$ $(J = 1)$; the latter is also unacceptable on grounds of symmetry, since states with $J = 1$, $M_J = 0$ change sign under a rotation of $180°$ about the x-axis, but this is not true of the two-quantum state with $M = 0$, i.e., that which has angular momentum zero.

Other selection rules may be derived from parity arguments* [145]. The conclusion is reached that double quantum emission is allowed from all even states (except where $J = 1$) and all odd states of even J. In interpreting these rules it is to be understood that an electron-positron pair, with both particles at rest, has odd parity. S, D . . . states are therefore odd, while P, F . . . are even.

In particular, double quantum decay from $^1S_0$ is allowed; from $^3S_1$ it is forbidden. From the latter state, annihilation into three quanta is the most likely process. The ratio of the decay rates by two and three quanta respectively is about 1100 [100]. The lifetimes of the singlet and triplet states are of the order 10^{-10} and 10^{-7} seconds respectively. This is the basis of the experiments from which ΔW is determined, which we shall now describe. It is to be noticed that the natural widths of these energy levels ($\sim 10^4/2\pi$ and $\sim 10/2\pi$ Mc/s respectively) are much smaller than $\Delta W (\sim 2 \times 10^5$ Mc/s).

---

* Although it is now known that parity is not always conserved in 'weak' interactions, we are here dealing with electromagnetic interactions for which no cases are known in which parity is not conserved.

12.7. Experimental test

Positrons emitted from radioactive sources (e.g. Na^{22} or Cu^{64}) into gases at ordinary pressures generally form positronium before they annihilate, since the rate of energy loss by ionization is fast compared with the decay rate of $^3S_1$. Positronium was formed in this way in a resonant cavity excited by a magnetron oscillator working at about 3000 Mc/s [33]. A steady magnetic field of about 9000 gauss was established perpendicular to the magnetic component of the radiofrequency field. These are the conditions necessary for stimulating magnetic dipole transitions between the magnetic states of a given level, e.g., between the $M_J = 0$ and ± 1 states of $^3S_1$ (see Fig. 11). Absorption and emission of radiofrequency quanta by the positronium are equally likely: a net effect can only be observed if the states are unequally populated.

We have explained (section 12.5) how the steady magnetic field mixes the $M_J = 0$ states of $^1S_0$ and $^3S_1$. By virtue of the fast decay rate of $^1S_0$, the state $^3S_1$, $M_J = 0$ will therefore be less populated than $^3S_1$, $M_J = \pm 1$. Under these conditions, there will be a net transfer of positronium atoms from $M_J = \pm 1$ to $M_J = 0$ at radiofrequency resonance.

In the experiment of Deutsch and Brown, the positronium was detected through the annihilation γ-rays. Radiofrequency resonance was indicated by an increase in the ratio of two-photon to three-photon processes. The value of ΔW calculated by means of equation 12.7 was $(2 \cdot 032 \pm 0 \cdot 003) \times 10^5$ Mc/s. More recent experiments [133], [90], give

$$(2 \cdot 0338 \pm 0 \cdot 0004) \times 10^5 \text{ Mc/s and } (2 \cdot 0333 \pm 0 \cdot 0004) \times 10^5 \text{ Mc/s}$$

These agree very satisfactorily with the theoretical value $2 \cdot 0337 \times 10^5$ Mc/s (equation 12.6) which includes the quantum electrodynamic corrections to order $R\alpha^3$. The uncorrected value, $2 \cdot 044 \times 10^5$ Mc/s, is clearly inadequate. The importance of the annihilation term is demonstrated beyond doubt.

APPENDIX

Useful relations

(In the original text these relations were given in the cgs electrical system of units. They are here given in the International System (SI), the conversion factors, where appropriate, being shown explicitly. Numerical values are taken from CODATA 1986. Figures in brackets are the estimated standard deviations of the last digits quoted.)

Fine structure constant

$$\alpha = \left(\frac{e^2}{\hbar c}\right)\left(\frac{\mu_0 c^2}{4\pi}\right)$$

$$= \frac{1}{137.035\,989\,5\,(61)} \ .$$

Rydberg constant
(expressed as energy)

$$R_\infty^{(E)} = R_\infty hc = \left(\frac{e^2}{2a_0}\right)\left(\frac{\mu_0 c^2}{4\pi}\right)$$

$$= \frac{\alpha^2 mc^2}{2} \ .$$

Rydberg constant
(expressed as wavenumber)

$$R_\infty = \left(\frac{me^4}{4\pi\hbar^3 c}\right)\left(\frac{\mu_0 c^2}{4\pi}\right)^2$$

$$= 1.097\,373\,153\,4\,(13) \times 10^7\,\text{m}^{-1} \ .$$

First Bohr radius

$$a_0 = \left(\frac{\hbar^2}{me^2}\right)\left(\frac{4\pi}{\mu_0 c^2}\right)$$

$$= 5.291\,772\,49\,(24) \times 10^{-11}\,\text{m} \ .$$

Bohr magneton

$$\mu_B = \left(\frac{e\hbar}{2m}\right)$$

$$= 9.274\,015\,4\,(31) \times 10^{-24}\,\text{JT}^{-1} \ .$$

Bohr energy levels

$$E_n = -\frac{R_\infty^{(E)} Z^2}{n^2} \ , \quad n = 1, 2, \ldots$$

Reduced mass

$$\mu = \frac{mM}{m+M} \ .$$

Dirac energy levels

$$E_{n,j} = mc^2 \left\{ 1 + \frac{\alpha^2 Z^2}{(n' + \sqrt{(j+\frac{1}{2})^2 - \alpha^2 Z^2})^2} \right\}^{-\frac{1}{2}} - mc^2,$$

$$n' = 0, 1 \ldots; \quad j = \tfrac{1}{2}, \tfrac{3}{2} \ldots$$

$$= -\frac{RZ^2}{n^2} \left[1 + \frac{\alpha^2 Z^2}{n} \left(\frac{1}{j+\frac{1}{2}} - \frac{3}{4n} \right) + \ldots \right],$$

$$n = n' + j + \tfrac{1}{2} = 1, 2 \ldots$$

Value at the origin of Schrödinger wave function for nS states

$$\psi^2(0) = \frac{Z^3}{\pi a_0^3 n^3}$$

REFERENCES AND AUTHOR INDEX

1. Adams, E. N. (1951) *Phys. Rev.*, **81**, 1 . . . 73
1a. Aron, W., and Zuchelli, A. J. (1957) *Phys. Rev.*, **105**, 1681 65, 52
2. Baranger, M., Bethe, H. A., and Feynman, R. P. (1953) *Phys. Rev.*, **92**, 482 . . . . . 51, 65
3. ——, Dyson, F. J., and Salpeter, E. E. (1952) *Phys. Rev.*, **88**, 680 51
4. Berestetski, V. B. (1949) *Zh. eksp. teor. Fiz.*, **19**, 673, 1130 75
5. Beringer, R., and Heald, M. A. (1954) *Phys. Rev.*, **95**, 1474 69
6. Bersohn, R., Weneser, J., and Kroll, N. M. (1953) *Phys. Rev.*, **91**, 1257 51
7. Bethe, H. A. (1947) *Phys. Rev.*, 72, 339 45
8. ——, Brown, L. M., and Stehn, J. R. (1950) *Phys. Rev.*, **77**, 370 49
9. Bethe, H. A., and Salpeter, E. E. (1957) *Handbuch der Physik*, **35**, 88, Berlin: Springer . . 20, 31, 50
10. Birge, R. T. (1942) *Rep. Progr. Phys.*, **8**, 120 . . 35
11. Blin-Stoyle, R. J. (1957) *Theories of Nuclear Moments*, Oxford University Press 73
11a. Bloch, F., Graves, A. C., Packard, M., and Spence, R. W. (1947) *Phys. Rev.*, **71**, 551 73
12. Bohr, A. (1948) *Phys. Rev.*, **73**, 1109 72
13. Bohr, N. (1913) *Nature*, **92**, 231 12
14. —— (1913) *Phil. Mag.*, **26**, 1. 8
15. —— (1915) *Phil. Mag.*, **29**, 332 13
16. —— (1915) *Phil. Mag.*, **30**, 394 10
17. Born, M., and Wiener, N. (1926) *J. Math. Phys.*, **5**, 84; *Z. Phys.*, **36**, 174 22

18. Brackett, F. S. (1922) *Nature*, **109**, 209 . . . 11
19. Breit, G. (1928) *Nature*, **122**, 649 26
20. —— (1929) *Phys. Rev.*, **34**, 553; (1930) *Phys. Rev.*, **36**, 383; (1932) *Phys. Rev.*, **39**, 616 28
21. —— (1930) *Phys. Rev.*, **35**, 1447 67
22. —— (1947) *Phys. Rev.*, **72**, 984 68
23. Byram, E. T., Chubb, T., Friedman, H., and Gailar, N. (1953) *Phys. Rev.*, **91**, 1278 3
24. Chambers, E. E., and Hofstadter, R. (1956) *Phys. Rev.*, **103**, 1454 70
24a. Chu, D. Y. (1939) *Phys. Rev.*, **55**, 175 . . . 37
25. Cohen, E. R. (1952) *Phys. Rev.*, **88**, 353 . . . 35
26. Condon, E. U., and Shortley, G. H. (1951) *The Theory of Atomic Spectra*, Cambridge University Press 31
27. Curtis, W. E. (1914) *Proc. Roy. Soc.* A, **90**, 605 . . 10
28. Darwin, C. G. (1920) *Phil. Mag.*, **39**, 537 . . . 28
29. —— (1927) *Nature*, **119**, 282 25
30. Dayhoff, E. S., Triebwasser, S., and Lamb, W. E. (1953) *Phys. Rev.*, **89**, 106. . . . . 38, 39, 71
31. De-Benedetti, S., and Corben, H. C. (1954) *Ann. Rev. Nucl. Sci.*, **4**, 191 75
32. Deutsch, M. (1953) *Prog. Nucl. Phys.*, **3**, 131 . . 75
33. ——, and Brown, S. C. (1952) *Phys. Rev.*, **85**, 1047 79, 81
34. Dirac, P. A. M. (1925) *Proc. Roy. Soc.* A, **109**, 642 . 21
35. —— (1927) *Proc. Roy. Soc.* A, **114**, 243 . . . 30
36. —— (1928) *Proc. Roy. Soc.* A, **117**, 610 . . . 26
37. —— (1947) *The Principles of Quantum Mechanics*, 3rd edition, p. 263. Oxford University Press . . 26
38. Drinkwater, J. W., Richardson, Sir O., and Williams, W. E. (1940) *Proc. Roy. Soc.* A, **174**, 164 . . . 36
39. Dumond, J. W. M., and Cohen, E. R. (1955) *Rev. Mod. Phys.*, **27**, 363 . . . . . 35, 67, 82
40. Dyson, F. J. (1951) *Proc. Roy. Soc.* A, **207**, 395 . . 53
41. —— (1952) *Phys. Rev.*, **85**, 631 54
42. Ehrenfest, P. (1911) *Ann. Phys., Lpz.*, **36**, 91; (1916) *Ann. Phys., Lpz.*, **51**, 327 14
43. Evans, E. J. (1913) *Nature*, **92**, 5 12
44. Fermi, E. (1930) *Z. Phys.*, **60**, 320 67
45. Ferrell, R. A. (1951) *Phys. Rev.*, **84**, 858 . . 75, 77
46. Feynman, R. P. (1949) *Phys. Rev.*, **76**, 769 . . . 48
47. Foley, H. M., and Nessler, A. M. (1955) *Phys. Rev.*, **98**, 6 73
48. Fowler, A. (1901) *Phil. Trans.*, **197**, 202 . . . 11
49. —— (1912) *Mon. Not. R. Astr. Soc.*, **73**, 62 . . . 12
50. —— (1913) *Nature*, **92**, 231 12
51. French, J. B., and Weisskopf, V. F. (1949) *Phys. Rev.*, **75**, 1240 48
52. Fulton, T., and Martin, P. C. (1954) *Phys. Rev.*, **95**, 811 65, 78

53. Goudsmit, S., and Uhlenbeck, G. E. (1926) *Nature*, **117**, 264 23
54. Gunn, J. C. (1955) *Rep. Progr. Phys.*, **18**, 127 . . 54
55. Hansen, G. (1925) *Ann. Phys.*, *Lpz.*, **78**, 558 . . 18
56. Harriman, J. M. (1956) *Phys. Rev.*, **101**, 594 . . 49
57. Heberle, J. W., Reich, H. A., and Kusch, P. (1956) *Phys. Rev.*, **101**, 612 74
58. Heisenberg, W., and Jordan, P. (1926) *Z. Phys.*, **37**, 263 24
59. Herzberg, G. (1927) *Ann. Phys.*, *Lpz.*, **84**, 553 . 6, 13
60. —— (1956) *Proc. Roy. Soc.* A, **234**, 516 . . . 60
61. —— (1956) *Z. Phys.*, **146**, 269 59
62. Heyden, M. (1937) *Z. Phys.*, **106**, 499 36
63. Houston, W. V. (1937) *Phys. Rev.*, **51**, 446 . . . 36
64. ——, and Hsieh, Y. M. (1934) *Phys. Rev.*, **45**, 263 . 36
65. Humbach, W. (1955) *Z. Naturforsch.*, **10a**, 347 . . 76
66. Humphreys, C. J. (1953) *J. Res. Nat. Bur. Stand.*, **50**, 1 11
67. Karplus, R., and Klein, A. (1952) *Phys. Rev.*, **85**, 972 . 69
68. ——, —— (1952) *Phys. Rev.*, **87**, 848 78
69. ——, ——, and Schwinger, J. (1952) *Phys. Rev.*, **86**, 288 51
70. Karplus, R., and Kroll, N. M. (1950) *Phys. Rev.*, **77**, 536 51, 68
71. Kellog, J. M. B., Rabi, I. I., and Zacharias, J. R. (1936) *Phys. Rev.*, **50**, 472 5
72. Kent, N. A., Taylor, L. B., and Pearson, H. (1927) *Phys. Rev.*, **30**, 266 34
72a. Kireyev, P. S. (1957) *Dokl. Akad. Nauk SSSR*, **112**, 41 . 56
73. Koenig, S. H., Prodell, A. G., and Kusch, P. (1952) *Phys. Rev.*, **88**, 191 69
74. Kramers, H. A. (1919) *K. danske vidensk. Selsk. Skr.*, (8), **3**, 3, 287 16
75. Kroll, N. M., and Lamb, W. E. (1949) *Phys. Rev.*, **75**, 388 48
76. Kuhn, H., and Series, G. W. (1950) *Proc. Roy. Soc.* A, **202**, 127 55
77. Kusch, P. (1955) *Phys. Rev.*, **100**, 1188 . . 67, 71, 72
78. ——, and Foley, H. M. (1948) *Phys. Rev.*, **74**, 250 . 68
79. Lamb, W. E. (1951) *Rep. Progr. Phys.*, **14**, 23 . 35, 37, 38
80. —— (1952) *Phys. Rev.*, **85**, 259 . . . 29, 30, 38
81. ——, and Retherford, R. C. (1947) *Phys. Rev.*, **72**, 241 . 38
82. ——, —— (1950) *Phys. Rev.*, **79**, 549 . . . 7, 38
83. ——, —— (1951) *Phys. Rev.*, **81**, 222 38
84. ——, —— (1952) *Phys. Rev.*, **86**, 1014 . . . 38
85. ——, and Sanders, T. M. (1956) *Phys. Rev.*, **103**, 313 . 63
86. ——, and Skinner, M. (1950) *Phys. Rev.*, **78**, 539.. . 61
87. Leo, W. (1926) *Ann. Phys.*, *Lpz.*, **81**, 757 . . . 17
88. Low, F., and Salpeter, E. E. (1951) *Phys. Rev.*, **83**, 478 72, 73
89. Lyman, T. (1914) *Phys. Rev.*, **3**, 504 3
90. Marder, S., Wu., C. S., and Hughes, V. W. (1957) *Phys. Rev.*, **106**, 934 81
91. Masson, I. (1914) *Nature*, **92**, 503 6

92. Mittleman, M. (1956) *Bull. Am. Phys. Soc. Ser. II*, **1**, 46   74
93. Moore, Charlotte, E. (1949) *Atomic Energy Levels*, Washington, D.C.: National Bureau of Standards: and references there given   11
94. Nafe, J. E., and Nelson, E. B. (1948) *Phys. Rev.*, **73**, 718   67
95. ——, ——, and Rabi, I. I. (1947) *Phys. Rev.*, **71**, 914 .   67
96. Nelson, E. B., and Nafe, J. E. (1949) *Phys. Rev.*, **75**, 1194   73
97. Newcomb, W. A., and Salpeter, E. E. (1955) *Phys. Rev.*, **97**, 1146   70
98. Novick, R., and Commins, E. (1956) *Phys. Rev.*, **103**, 1897   73
99. ——, Lipworth, E., and Yergin, P. F. (1955) *Phys. Rev.*, **100**, 1170 . . . . . 8, 52, 63, 65
100. Ore, A., and Powell, L. J. (1949) *Phys. Rev.*, **75**, 1696 .   80
101. Paschen, F. (1908) *Ann. Phys., Lpz.*, **27**, 565 . .   11
102. —— (1916) *Ann. Phys., Lpz.*, **50**, 901 . . .   16
103. Paschen, F. (1927) *Ann. Phys., Lpz.*, **82**, 689 . .   17
104. Pasternack, S. (1938) *Phys. Rev.*, **54**, 1113 . . .   36
105. Pauli, W. (1927) *Z. Phys.*, **43**, 601 . . .   25
106. Pauling, L., and Goudsmit, S. (1930) *Structure of Line Spectra*, p. 134. New York: McGraw-Hill .   16
107. Peierls, R. E., Salam, A., Matthews, P. T., and Feldman, G. (1955) *Rep. Progr. Phys.*, **18**, 424 . .   54
108. Pfund, A. H. (1924) *J. Opt. Soc. Amer.*, **16**, 193 . .   11
109. Pickering, E. C. (1896) *Astrophys. J.*, **4**, 369 . .   11
110. Pirenne, J. (1947) *Arch. Sci. phys. nat.*, **29**, 121, 207, 265   75
111. Prodell, A. G., and Kusch, P. (1950) *Phys. Rev.*, **79**, 1009; (1952) *Phys. Rev.*, **88**, 184   72
111a. ——, —— (1957) *Phys. Rev.*, **106**, 87 . . .   73
112. Ramsey, N. (1956) *Molecular Beams*, p. 80. Oxford University Press   79
113. Reich, H. A., Heberle, J. W., and Kusch, P. (1956) *Phys. Rev.*, **104**, 1585   74
114. Rojansky, V. (1939) *Introductory Quantum Mechanics*, London: Blackie. (See p. 472 for section 6.3; Chapter 13 for section 6.6) 23, 24
115. Rydberg, J. R. (1889) *K. svenska Vetensk Akad. Handl.*, **23**   8
116. Saha. M., and Banerji, A. C. (1931) *Z. Phys.*, **68**, 704 .   32
117. Salam, A. (1955) *Rep. Progr. Phys.*, **18**, 433 . .   42
118. Salpeter, E. E. (1952) *Phys. Rev.*, **87**, 328 .   29, 51
119. —— (1953) *Phys. Rev.*, **89**, 92 . . 50, 51, 52, 65
120. Schrödinger, E. (1926) *Ann. Phys., Lpz.*, **79**, 734 . .   21
121. Series, G. W. (1951) *Proc. Roy. Soc.* A, **208**, 277 .   55
122. —— (1954) *Proc. Roy. Soc.* A, **226**, 377 .   57, 58
123. Shane, C. D., and Spedding, F. H. (1935) *Phys. Rev.*, **47**, 33   36
124. Smaller, B., Yasaitis, E., and Anderson, H. L. (1951) *Phys. Rev.*, **81**, 896; **83**, 812   72
124a. Sommerfield, C. A. (1957) *Phys. Rev.* **107**, 328 . 51, 65, 68
125. Sommerfeld, A. (1916) *Ann. Phys., Lpz.*, **51**, 1   14, 19

126. —— (1934) *Atomic Structure and Spectral Lines*, vol. I.
       Translated by Brose, H. L. 3rd edition, p. 342.
       London: Methuen   14
127. ——, and Unsöld, A. (1926) *Z. Phys.*, **36**, 259 . .   17
128. Spedding, F. H., Shane, C. D., and Grace, N. S. (1935)
       Phys. Rev., **47**, 38   36
129. Thomas, L. H. (1926) *Nature*, **117**, 514; (1927) *Phil.
       Mag.*, **3**, 1   24
130. Thonemann, P. C., Moffatt, J., Roaf, D., and Sanders,
       J. H. (1948) *Proc. Phys. Soc.*, **61**, 483 . . .   6
131. Triebwasser, S., Dayhoff, E. S., and Lamb, W. E. (1953)
       Phys. Rev., **89**, 98 . . . . . .   38, 39
132. Uehling, E. A. (1935) *Phys. Rev.*, **48**, 55 . . .   48
133. Weinstein, R., Deutsch, M., and Brown, S. (1955) *Phys.
       Rev.*, **98**, 223 (A)   81
134. Weisskopf, V. F. (1939) *Phys. Rev.*, **56**, 72 . . .   42
135. —— (1949) *Rev. Mod. Phys.*, **21**, 305   41
136. Welton, T. A. (1948) *Phys. Rev.*, **74**, 1157 . . .   47
137. Whittaker, Sir E. T. (1940) *Proc. Roy. Soc., Edin.*, **61**, 1   21
138. —— (1953) *A History of the Theories of Aether and Elec-
       tricity*, 1900–1926, Edinburgh: Nelson . . .   21
139. Williams, R. C. (1938) *Phys. Rev.*, **54**, 558 . . 5, 35, 36
140. —— (1938) *Phys. Rev.*, **54**, 568   12
141. ——, and Gibbs, R. C. (1934) *Phys. Rev.*, **45**, 475; (1936)
       Phys. Rev., **49**, 416   36
142. Wilson, W. (1915) *Phil. Mag.*, **29**, 795; (1916) *Phil. Mag.*,
       31, 156   14
143. Wittke, J. P., and Dicke, R. H. (1956) *Phys. Rev.*, **103**,
       625 . . . .   5, 67, 71
144. Wood, R. W. (1922) *Proc. Roy. Soc. A*, **102**, 1; *Phil.
       Mag.*, **44**, 538   4
145. Yang, C. N. (1950) *Phys. Rev.*, **77**, 242 . . .   80
146. Zemach, A. C. (1956) *Phys. Rev.*, **104**, 1771 . . .   70

SUBJECT INDEX

CHAPTER 1

ADVANCES IN EXPERIMENTAL METHODS

E. A. Hinds and G. W. Series

1.1   Conventional Laboratory Spectroscopy: Gas Discharges: Spontaneous Emission

Spectra of gases are conventionally excited by passing an electrical discharge through a gas at low pressure — typically, a few torr. In the case of hydrogen the discharge dissociates the molecules and the atoms are excited, for the most part, by electron impact. When the material is in short supply or needs to be totally contained (tritium, for example) the glass container may be sealed off and the gas — typically now at a considerably lower pressure — may be excited inductively by placing it inside a coil, or capacitatively by placing electrodes outside the container.

It is the spontaneously emitted light from an assembly of atoms in random motion that is focussed into a spectrograph and analysed by the conventional tools of prisms, gratings, or interferometers; or, exceptionally for hydrogen, the spectra are studied in absorption. In either case the spectra are subject to Doppler broadening and the manifold perturbations experienced by excited atoms under bombardment by electrons, ions, other atoms and molecules in the discharge. It is indeed a dirty environment, and it is remarkable that almost our whole understanding of atomic and molecular structure has been achieved through quantitative interpretation of spectra emitted by such sources. They are not, in fact, so perturbing as might have been supposed.

For hydrogen it was not the perturbations within gas discharges that limited progress, but the Doppler broadening. Chapter X of *Spectrum of Atomic Hydrogen* (which, throughout these Advances, we shall refer to as SAH) records the progress that was made in reducing this broadening by cooling the gas discharge first with liquid oxygen, and then with liquid hydrogen, while Chapter 5 of these Advances records the use of liquid helium. But the great advance in optical spectroscopy came with the invention of the tunable laser towards the end of the nineteen sixties. An outline of the methods by which laser spectroscopy overcame the problem of Doppler broadening is given in section 4 of this chapter; the application of the methods is described in Chapter 5.

Already before the invention of lasers revolutionary changes in optical methods applied to spectroscopic problems had left their mark on hydrogen studies: optical pumping, fluorescence spectroscopy and (for ions) beam foil spectroscopy. We describe these in sections 5, 6 and 7 of this chapter. A number of these methods called for radiofrequency resonance techniques to be combined with the optical methods.

Radiofrequency resonance methods themselves had been successfully applied to hydrogen in Lamb's experiment, on the one hand, and in studies of hyperfine structure in the ground state, on the other. This work is briefly recalled in sections 2 and 3 below. The great advantage of these radiofrequency methods is that they measure small intervals directly rather than as the difference between independent transitions, as is the case in optical spectroscopy, where each transition may suffer from some broadening or perturbing influence.

Lamb's experiment was carried out on an atomic beam of hydrogen. The use of atomic beams eliminates perturbations due to ions and electrons and vastly reduces, but does not entirely eliminate, atomic collisions. There are other problems such as the development of parasitic charges on the walls, and stray fields that penetrate the interaction region. Atomic beams are chosen nowadays as the most desirable environment for studies of atomic hydrogen. But it is a mistake to suppose that the use of gas discharges is entirely outmoded. For optical studies up to the level of about 1 part in 10^8 in accuracy, the environment of a mild gas discharge at low pressure perturbs only minimally the atoms within it — apart from the highly excited states — and these small perturbations can be investigated experimentally by varying the conditions of the discharge and corrected for. It is only at the highest level of accuracy in contemporary studies that gas discharges are considered unacceptably turbulent.

Moreover, gas discharges are often used as a source of hydrogen atoms. Hydrogen survives in atomic form long after charged particles have recombined, and the atoms can be led away from the discharge in teflon tubes as long as several metres to an apparatus where the experiment proper is to be performed.

1.2   Radiofrequency Spectroscopy: Atomic Beams: Lamb's Experiment

Lamb's experiment is described in SAH, p. 38. The novel feature, as compared with conventional spectroscopy — apart from the use of a beam of atoms rather than a gas discharge — was the reliance on stimulated rather than spontaneous transitions to drive a resonance. The tremendous advance over conventional spectroscopy, arising, in the main, from the direct measurement of a small spectroscopic interval, can be appreciated by comparing the precision with which he reported his 'Lamb shift', about 0.1 MHz, with the precision reported a few years later for the same interval determined by fine-structure analysis of optical spectra, 50 MHz, the source in this case being a liquid-helium-cooled discharge (Chapter 5). A full account of later measurements of the Lamb shift by radiofrequency methods is given in section 4.2.

1.3   Radiofrequency Spectroscopy: Hyperfine Structure of the Ground State

Work on this most important spectroscopic interval was initiated by Nafe, Nelson and Rabi (SAH, Chapter XI) who used the atomic beam technique, the source of atomic hydrogen being a gas discharge. The atoms subjected to magnetic resonance in the beam are those which have issued from the discharge and not those within it. It was this work which first disclosed that the magnetic moment of the electron was not exactly one Bohr magneton and which, twenty years later, brought forth the hydrogen maser.

But before the maser, two other techniques had successfully been applied to the ground state hyperfine structure: Wittke and Dicke [WIT 56] had applied the method of paramagnetic absorption of microwaves to a sample of hydrogen atoms in a container into which they were introduced from a gas discharge. And Pipkin and his colleagues had applied to hydrogen and its isotopes the method of optical pumping, which we will describe in section 1.5. A requirement of all these techniques where the observed resonances are stimulated, not spontaneous, is that a difference of population be established between the two levels of the spectroscopic interval. The method of paramagnetic absorption relies on thermal equilibrium processes for this, the dominating relaxation process in the experiment of Wittke and Dicke being spin-exchange collisions between hydrogen atoms; optical pumping relies on differences in optical transition probabilities; the maser on state selection by magnetic focussing.

In section 4.1 we give a detailed account of the hyperfine structure in hydrogen, muonium and positronium, paying particular attention to the hydrogen maser. Table 4.1 lists the results that have been obtained by the various techniques.

1.4   Optical Spectroscopy: the Laser Era

With the advent of tunable lasers (towards the end of the nineteen sixties) some of the characteristics of radiofrequency spectroscopy were brought into the optical range: atoms could be regarded as passive systems to be probed by sharp-edged tools; quasi-monochromatic, unidirectional beams of radiation.

Three properties of laser light are especially to be noticed: the spectral purity, the directional property, and the brightness. Spectral purity and directionality can be achieved — up to a point — with conventional sources using filters and collimators but at the cost of intensity. Thus, experiments become possible with laser light that were considered, and might have been attempted with conventional sources, but would have failed for lack of intensity. An example is quoted in section 5.3.4.2 where extremely high-resolution spectra of Balmer-α are obtained in absorption from atoms in a collimated beam. The experiment carried out with a tunable, cw laser operating in the red permits such high collimation of the beam that the first-order Doppler effect is reduced below other line-broadening factors.

Conventional absorption spectra are classified as linear in the intensity of the light. Most of the new effects in laser spectroscopy depend on non-linear interactions. Three such effects, whose special virtue was the elimination of Doppler broadening from samples containing atoms in random motion, were soon discovered and applied to hydrogen. Their application is described in Chapter 5; here, we describe the principles. A convenient textbook reference to laser spectroscopic techniques is [DEM 81]. An excellent survey of laser and other recently-developed spectroscopic techniques is to be found in the Proceedings of a Discussion Meeting [SER 82].

1.4.1 *Saturated absorption spectroscopy*

The technique requires that the assembly of atoms to be investigated be irradiated by a relatively strong laser beam (the saturating beam) and a weaker, counter-propagating beam (the probe beam). Consider an atom — one of the assembly — moving with velocity **v** obliquely to a beam of laser light whose frequency measured in the laboratory frame, is ω. (Fig. 1.1). In the frame of the atom the frequency appears to be $\omega(1 - v_z/c)$ (first order Doppler effect), where v_z is the component of **v** in the direction of the laser beam. Notice that, for light travelling in the opposite sense, the frequency appears to be $\omega(1 + v_z/c)$.

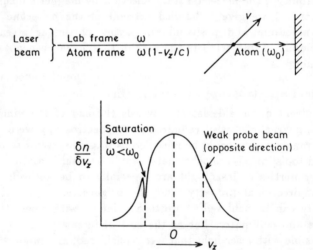

Fig. 1.1. (*a*) First Doppler effect for an atom moving obliquely to a laser beam. (*b*) Distribution of atoms by v_z, and showing the 'hole' burnt in the ground state distribution by a laser beam tuned to resonance with obliquely-moving atoms, and sufficiently strong to excite a significant number of them.

Now consider the one-dimensional velocity distribution of atoms (supposed to be in thermal equilibrium) represented in the lower part of the figure. If the laser be tuned to an absorption resonance for obliquely moving atoms, and if it is sufficiently strong to excite a substantial number of them (partial or complete saturation in absorption), the distribution will be depleted, as shown in the figure, for a range of velocities depending on the homogeneous width of the transition — the radiative width, possibly increased by some pressure broadening or laser power broadening. But this width is typically about 100 times smaller than the width of the velocity distribution. One commonly speaks of 'burning a hole' in the velocity distribution.

The counter-propagating beam interacts, of course, with a different set of atoms, *unless the laser is tuned to that particular set which is moving transversely to the beam* $(v_z \approx 0)$, in which case both beams interact with the same atoms. Now, regard the counter-propagating beam as a probe of the velocity distribution by measuring its intensity after it has traversed the gas. When the laser is tuned to the peak of the distribution the detector of the probe beam will register *increased* intensity. The spectral width over which this occurs will be the width of the 'hole' (approximately), many times narrower than the spectral width of the distribution, which is the Doppler width. We speak of a 'velocity-selective' technique by which the Doppler broadening has been eliminated.

1.4.2   Two-photon spectroscopy

Superficially, the experimental arrangement for this technique is similar to that used in saturated absorption spectroscopy, but the Doppler effect is eliminated in a way which allows *all* atoms in the sample to participate, irrespective of their velocity.

Consider atoms in a gas in random motion irradiated, as before, by opposing beams of laser light of the same frequency (in the laboratory frame). In the atom frame the frequency of the light appears to be shifted, as before, by the factor $(1 + v_z/c)$ (see Fig. 1.2). But now, suppose the atoms absorb one photon from each beam. The increase in energy is $\hbar\omega[(1 - v_z/c) + (1 + v_z/c)] = 2\hbar\omega$, independent of v_z, to first order in v_z/c. The possibility exists, therefore, for resonant absorption of light by all atoms in the assembly to a state energetically $2\hbar\omega$ above the ground state. The selection rules governing this process are, of course, different from those governing a one-photon transition: in particular, the initial and final states must be of the same parity. The resonant absorption is detected by observing some convenient fluorescent radiation from the atoms, indicating spontaneous decay of the excited state.

In practice the laser light is usually introduced into the sample in a converging beam and focussed down to a 'waist' only a few microns in diameter. From this waist it emerges to encounter a con-focal, concave mirror from which the beam is reflected to converge again into the waist. The high intensity gained by this focussing is required on account of the typically small values of two-photon transition probabilities.

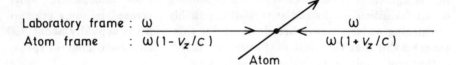

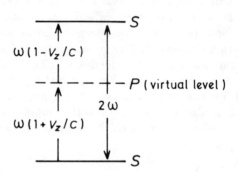

Fig. 1.2. Two-photon absorption: elimination of Doppler effect in first order. (*a*) Doppler-shifted frequencies seen by an atom moving obliquely to a laser beam. (*b*) Absorption of one photon from each beam results in an increase of energy which is the same (to first order in v_z/c) for all atoms, independently of v_z

1.4.3 Polarization spectroscopy

This is a variant of saturation spectroscopy (velocity-selective, therefore, and useful for one-photon transitions) which has the advantage that much weaker laser beams can be used, with greatly reduced power broadening of the resonances. The primary beam, now called the 'pumping beam' rather than the 'saturating beam', is polarized, often by circular polarization, though not necessarily so. The atoms interacting with such a beam themselves become polarized and present to the probe beam at the resonant frequency an anisotropic sample. The probe beam is polarized (neither identically nor orthogonal to the pumping beam) and is intercepted by an orthogonal polarizer before it reaches the photodetector. No light therefore would reach the photodetector if the sample had remained isotropic or if the probe beam were far off resonance for the polarized atoms. But if the probe beam is near resonance and can interact with these atoms its state of polarization changes through differential dispersion (birefringence) and differential absorption (dichroism). In such a case some portion of the light (one may regard it as that light which has been forward-scattered by the atoms) reaches the detector. The Doppler-free signal therefore appears against a null background — another important advantage, for it reduces the photon shot noise. In practice, the second polarizer is often rotated slightly to allow a small fraction of the primary light to reach the detector. This, being coherent with the signal light, serves to homodyne it which is a form of amplification.

We shall not discuss other refinements of polarization-spectroscopy here. There is, however, an artefact of these velocity-selective techniques, false signals called 'cross-overs', which we will describe in the next section.

1.4.4   Cross-over resonances

Figure 1.3(i) represents the situation we have described: pump and probe both acting resonantly at a transition frequency associated with a pair of levels a and b for atoms moving transversely to the laser beams. This situation gives rise to the principal resonances. But if, as in (ii) a transition is allowed between a and a third level c, then a signal will arise from atoms moving obliquely to the laser beams such that the pump beam is Doppler-shifted to the frequency ab and the probe beam to the frequency ac, and vice versa. This signal will be recorded when the laser is tuned halfway between the frequencies ab and ac as measured for atoms moving transversely to the beam — Doppler shift zero. Such a signal is called a 'cross-over resonance'.

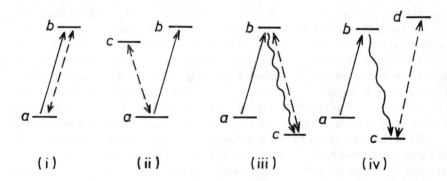

(i)   (ii)   (iii)   (iv)

Fig. 1.3. Cross-over resonances. The solid lines with single-ended arrows represent pump transitions. The wavy lines represent spontaneous decay. The broken lines with double-ended arrows represent probe transitions (i) illustrates a direct resonance, (ii) , (iii) and (iv), cross-over resonances. (iii) and (iv) depend on the existence of structure in the ground state and redistribution of population through optical pumping.

Cross-over resonances may also occur through the re-distribution of population arising through the spontaneous decay of atoms which have been excited by the pump beam (optical pumping — see section 1.5), illustrated in (iii) and

(iv) of the figure. In (iii) the cross-over resonance occurs halfway between ab and bc; in (iv) it occurs halfway between ab and cd. The strength of these contributions will depend, among other things, on the lifetime of the lower state c in relation to the time for the pumping cycle $a - b - c$. An example of case (iii) is seen in Fig. 2 of section 5.2.1. A detailed analysis of these cross-over effects has been given by Nakayama [NAK 81].

1.4.5 *Interaction times: the Ramsey technique: cooling*

Among the causes of spectral line broadening that are encountered in high-resolution spectroscopy is that arising from the finite interaction time, T, between the atoms and the radiation field. The 'natural', that is, the radiation width Γ of spectral lines is an example of this: Γ (in radians) $= 1/T_e$, where T_e is the mean lifetime of the excited state. For atoms in the ground state T is, in principle, under the control of the experimenter. The longer the interaction time can be made, the sharper the line (provided other causes of line broadening do not dominate).

For example, if atoms are confined by some technique to the locality of a radiation field, T will be the inverse of the probability of excitation; dependent, therefore, on the power density of the field. More commonly, if atoms are moving through a finite region of radiation field, and if the probability of excitation is small enough, T will vary inversely with the velocity of the atoms. Thus, for the two-photon Doppler-free spectroscopy of atoms moving in a beam with a distribution of velocities, there will be a velocity-dependent spectral profile.

In the radio-frequency spectroscopy of atomic beams there is a well-known technique invented by N. F. Ramsey to prolong the interaction time by allowing the atoms to traverse, in turn, two spatially separate regions of coherent field: the interaction with the field need not be continuous. The line width of the fringes of the interference pattern so created is determined by the overall time.

While the methods of radiofrequency spectroscopy cannot always be applied in laser spectroscopy, the Ramsey two-field technique can be so applied and is attractive for two-photon processes in particular in that the strong laser fields required for two-photon excitation can be secured only over small regions of space (an exception is the arrangement used by the Paris group, section 5.3.4.3, where the atoms were sent along the laser beams).

The arrangement contemplated, therefore (Fig. 1.4) allows an atomic beam to be intercepted in two regions by forward and counter-propagating beams, each of which will stimulate two-photon transitions. There must be stable coherence between the separated regions and standing waves along the beams in each region, otherwise the rapidly-changing phase of the light along each beam would destroy its coherence with respect to the other. The method has been applied successfully to the $5S - 32S$ transitions in Rb [LEE 79], where the subtleties of the technique were explored, and is being attempted for the $1S - 2S$ transition in H (section 5.3.5). Under discussion also is a 'fountain' variant of the double-interaction principle. Atoms would be permitted to escape vertically from a nozzle, would intersect a standing wave laser beam and — being decelerated by the earth's gravitational field — would intersect it again on their way down. To make this feasible it would be necessary to reduce the initial velocity to manageable size: hence a technique of preliminary cooling is called for. Conventional cryogenics, and cooling by laser irradiation [WIN 79] are among the alternatives envisaged.

But cooling is not necessarily to be associated with 'fountain' experiments. Of itself it is valuable as a means of lengthening interaction times; also as a means of reducing shifts arising from the Doppler effect in second order. We give some details in section 1.8.2 of the preparation of a beam of cold hydrogen.

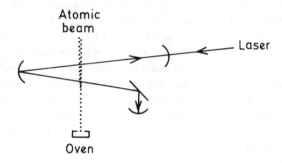

Fig. 1.4. The Ramsey two-field technique in laser spectroscopy. Special care is needed to ensure coherence between the fields in the interaction regions (see text).

1.5   Optical Pumping: Ground States

Optical pumping is a technique whereby an assembly of atoms in equilibrium under some relaxation processes (as, for example, an atomic vapour in a sealed bulb whereby the atoms are thermally distributed over the hyperfine states of the ground electronic level) can be perturbed by irradiation with polarized resonance radiation. The cycle: absorption of resonance radiation followed by fluorescence, with return of the atoms to a lower state, leaves the assembly in a non-equilibrium distribution, and therefore susceptible to some probe radiation, as, for example, to a radiofrequency field in a resonance experiment. Thus, hyperfine and Zeeman structures in ground states can be measured with an accuracy limited by ground state lifetimes and virtually free of Doppler broadening, although the light sources used for the irradiation themselves emit Doppler and pressure-broadened spectral lines. The favourite subjects to study by this technique have been the alkali atoms and the isotopes of mercury; for these atoms the resonance radiation is easy to generate and manipulate with lenses and polarizers made of glass, silica, or plastic materials.

But for hydrogen the case is different because those optical elements which serve in the visible and near ultraviolet will not serve for Lyman-α. Optical pumping was applied to hydrogen by a development of the technique known as 'spin-exchange'. In this process the species under study (hydrogen atoms formed by a pulse of rf radiation applied to the molecular gas in a bulb) were mixed with an atomic vapour which could be easily polarized by optical pumping (rubidium, whose resonance lines are in the deep red). The polarized rubidium atoms, by collisions with unpolarized hydrogen atoms, polarized the latter and thereby prepared the assembly for an rf resonance experiment at the frequency of the ground state hyperfine interval in hydrogen (or deuterium, or tritium). The resonance in hydrogen alters the population difference between the levels which, again by virtue of collisions, changes the polarization of the rubidium atoms. Such changes were monitored by measuring the changes of the absorption of resonance radiation of the rubidium vapour. The assembly of hydrogen atoms, therefore, was both polarized and monitored by rubidium resonance radiation: the problems associated with Lyman-α were avoided.

The idea of optical pumping is due to Kastler [KAS 50]; many reviews have been written [SER 81, for example]. The applications to hydrogen and its isotopes was due to Pipkin and his colleagues [PIP 62]: the results are included in Chapter 4 of this book.

1.6 Fluorescence Spectroscopy: Excited States

Fluorescence spectroscopy offers a variety of techniques for studying the properties of excited, as distinct from ground state atoms. The essence of the matter is that one measures the *intensity* of the fluorescent light, not its spectral distribution. Changes in the intensity are brought about in various ways: the object of all of them is to induce a change of state of the atom by some combination of static or oscillating fields that will lead to a resonant response determined by the Bohr interval between the states. The point is that, by disregarding the spectral distribution of the fluorescent radiation, one determines the Bohr frequency directly, and not as the difference between two much larger frequencies that are subjected to Doppler broadening.

We shall cite some specific examples but readers may care to contrast the distinctive feature of fluorescence spectroscopy, the monitoring of a change of state by observation of the *intensity of fluorescent light*, with the distinctive feature of atomic beam resonance experiments, the monitoring of a change of state by observation of the *intensity of a beam of particles*. It is generally agreed that the latter is potentially the more sensitive technique, though it may not always be practicable. The former is often more convenient.

Experiments of this type performed by Lamb and his colleagues [LAM 50a; LAM 56, LAM 60a, LAM 60b] were described as the 'Lamb shift in a bottle'. Many other studies of the same type were made [BEY 78] in which the transitions investigated were between states of opposite parity (mainly $S - P$ transitions in H, D and He$^+$), and therefore required an oscillating *electric* field to induce them. It is not at all surprising that a change in the intensity of fluorescent light should result from perturbing the *electric* dipole properties of atoms. Less obvious is that a change in the *magnetic* properties should result in a change in fluorescent intensity. That it may do so is a characteristic of the next class of experiment.

1.6.1 *Optical-radiofrequency double resonance*

This kind of experiment, first applied to excited states of mercury by Brossel and Bitter [BRO 52], relies on changing the spatial distribution of fluorescent light, and not its total intensity. The spatial distribution is intimately related to changes in the direction of the angular momentum of the atoms. A detector in a fixed position is therefore sensitive to transitions between neighboring Zeeman levels, or between different fine structure levels of the same family, as, for example, between $P_{1/2}$ and $P_{3/2}$ levels. The optical resonance implied by the title describes the method of preparation of the state — excitation by absorption of resonance radiation — coupled with the method of monitoring by detection of resonance radiation in emission.

Optical-radiofrequency double resonance was found unattractive for studies on hydrogen, partly because of the problems with Lyman-α, partly because a technique superficially similar though conceptually different was discovered: the technique of spectroscopy by level-crossing. And the advantage of level-crossing over optical-radiofrequency spectroscopy was a gain in simplicity and in line-profile analysis.

1.6.2 *Level-crossing spectroscopy*

This again relies on the dependence of the angular distribution of fluorescent radiation on the angular momentum of the radiating atoms: in addition, it relies on the *breaking of a symmetry* in the environment of atoms by the imposition of a magnetic field.

Consider an atom in free space excited by polarized resonance radiation (or, indeed, by any other non-isotropic perturbation). The electronic configuration of the atoms will take on the symmetry of the electric vectors of the exciting light, and this symmetry will again appear in the fluorescent light unless it is broken before the radiation takes place. Breaking of the symmetry may occur by perturbation of the orbital motion by coupling to the electronic or nuclear spin (depolarization of resonance radiation) or by the action of an external magnetic field in a skew direction. It is this latter technique which has proved most attractive and which has been applied to hydrogen.

The phenomenon is more familiar when described as an interference effect in the radiation from coherently-excited states. Consider two or more degenerate states of the same family; a Zeeman multiplet in zero magnetic field, or Zeeman components of different members of a spin multiplet e.g., $P_{1/2}, m = -1/2; P_{3/2}, m = +3/2$. These latter can be brought to degeneracy by the application of a magnetic field whose magnitude depends on the $P_{1/2} - P_{3/2}$ interval. A coherent superposition of these states is created if the atom is excited from resonance radiation linearly polarized at right angles to the magnetic field. But if the excited states are not degenerate, the excitation is *incoherent* and the radiation pattern is that from incoherent states. Radiation from the coherent superposition has a different spatial distribution by virtue of interference.

This depolarization of resonance radiation by a magnetic field was discovered by Hanle in 1924, but it was not until 1959, when the phenomenon was re-discovered by Colegrove *et al.* as an effect associated with the crossing of energy levels in non-zero fields, that its spectroscopic potentialities began to be realized. For, by using the effect to locate those magnetic fields at which the crossings take place, and with the help of the (separately determined) Landé *g*-values of the states concerned, it is possible to calculate the zero-field splittings of the structures under investigation. The experimental curves — obtained by scanning a magnetic field through the crossing region — are of Lorentzian shape with characteristic denominator $\omega_{ab}^2 + \Gamma^2$. The width parameter, Γ, is the natural (radiation) width of the crossing levels. There is no Doppler broadening and no power broadening.

Hydrogen was an obvious challenge for the new technique. The $P_{1/2} - P_{3/2}$ crossings (with hyperfine structure taken into account) would yield the zero field doublet interval and a value for the fine-structure constant, calculable in the most direct and unequivocal way. The problem over the use of Lyman-α remained, but was addressed directly by Fontana and Himmel [FON 67] and by Wing [WIN 67, WIN 68] at Ann Arbor, where the level-crossing technique was discovered and at Brown University, Rhode Island, by Baird *et al.* [BAI 72]. Details of their work are described in Chapter 4 of this book.

1.6.3 *Anticrossing spectroscopy*

Levels cross only if the states concerned are truly orthogonal. The crossing becomes a repulsion if a perturbation linking the states exists or is imposed —

but the interaction must be strong enough to take effect within the lifetime of the states; formally, $\delta_{ab}\tau > \hbar$ where δ_{ab} is the interaction matrix element and τ is the lifetime of the states. This is rather a simplistic view, because the states in which one is interested do not always have the same lifetime, but it serves to bring out the point that the perturbation must have some minimum value before the crossing becomes a repulsion, or, to use the term which has come into general use, an 'anticrossing'.

The perturbation may be internal, such as a hyperfine interaction linking fine structure states; the implication here is that the hyperfine interaction was omitted from the Hamiltonian whose diagonalization yielded the fine structure 'crossing'. Indeed, it was this type of interaction through which the phenomenon was first discovered [ECK 63]. Or the perturbation may be external, in which case, of course, it is controllable. An example would be the application of a static electric field to crossing levels of opposite parity.

The spectroscopic use of anticrossings will be understood when it is remembered that the degree of mixing of states depends not only on the perturbation linking them, but also on their separation. Thus, as a magnetic field is scanned through an anticrossing the degree of mixing changes. Provided that the method of excitation populates the (unmixed) states at different rates, and provided the method of detection records different fluorescent rates for the two states, then a change in the degree of mixing results in a change in the total light detected, and so the position of nearest approach of the two levels, the 'anticrossing', can be determined. Coherence and interference are not essential here as they are for level crossing spectroscopy, though the signals may contain coherent as well as incoherent contributions.

Experimental line contours for anticrossings are broader than those for the corresponding crossings would be: the mixing perturbation makes a contribution to the line-width. Indeed, if one regards an anticrossing resonance as a resonance driven by a 'radio-frequency' field at zero frequency, one can immediately understand the characteristic resonance denominator: $\omega_{ab}^2 + \Gamma_a\Gamma_b + |\delta_{ab}/\hbar|^2$, where a, b label the states, ω_{ab} is the interval between the levels; Γ_a, Γ_b are the natural widths and δ_{ab} is the matrix element of the perturbation. Anticrossing spectroscopy has found very extensive application to H and He$^+$, particularly at the hands of Kleinpoppen and his colleagues [BEY 78], and by Glass-Maujean [GLA 74]. Fine structure intervals have been determined to $n = 4$ in H and to $n = 5$ in He$^+$. The method has also allowed the study of the Stark effect in weak fields, but the precision attained in these experiments has not been sufficient to provide a check on theory at the level of the best level-crossing or rf resonance type of experiment; it falls below contemporary theory by one or two orders of magnitude. Nevertheless, the measurements have provided useful checks against theory of the position of levels which had not been determined by other methods and the technique deserves a place in any account of the spectroscopy of hydrogen.

1.7   Beam Foil Spectroscopy

This technique, introduced in the early nineteen sixties [SIL 84], has proved exceptionally important for the systematic production of highly stripped ions and for the study of their spectra. The method of production, outlined in section 1.8.4 and given in more detail in Chapter 6, results in a well-collimated beam of ions in a mixture of charge states and, for each ion, a mixture of excited states. Light, gradually decreasing in intensity, is emitted from the beam as it travels downstream. The primary spectroscopic problem is to identify the ion, and the particular transition within each ion, corresponding to each line. Having done this, and by concentrating on particular spectral lines, one may determine lifetimes of states by studying the decrease of intensity of particular lines, spatially, along the beam; one may study the effect of applying electric and magnetic fields to the ions as they travel; and one may induce transitions by irradiating the ions with laser beams. Figure 1 of Chapter 6 illustrates the general arrangement and indicates some typical experiments that can be carried out.

In the spectroscopy of hydrogenic ions the following are the principal kinds of investigation that have been made:

(*a*) straightforward measurement of the frequency of ionic Lyman-α radiations against convenient standards, the result being used to give values for the ground state Lamb shift (see section 5.3.3);

(*b*) direct determination of $2S - 2P$ intervals by — essentially — Lamb's method, except that laser rather than radiofrequency radiation is used to induce the resonances;

(*c*) 'quench' and 'anisotropy' studies, based on the mixing of $2S$ and $2P$ states by electric fields, or by motion through magnetic fields. The point here is that the radiative properties of the $2S$ level are profoundly affected by admixture of the P-state. Now, the amount of mixing in a given field depends on the $2S - 2P$ intervals. Observation of the lifetime of $2S$ as a function of the field therefore allows a determination of these intervals. Further, the application of an electric field results in asymmetry in the polarization of the fluorescent radiation, and the $2S - 2P$ intervals can be inferred from measurements of this asymmetry. The theory is presented in Chapter 3.

Details of these techniques, the particular ions to which they have been applied, and the results obtained, are presented in Chapter 6.

1.8 Production of Hydrogenic Atoms

The production of atomic hydrogen is not always simple. It involves the dissociation of molecules such as H_2 or H_2O because the monatomic gas is far too reactive to last for long. The dissociation can be accomplished thermally or by means of an electrical discharge. Recombination of the H atoms so formed occurs principally at the wall of the container, which is able to take up the binding energy as well as the linear momentum; as two-body process in the volume it is extremely improbable. The main techniques vary according to the temperature and density of hydrogen required, as we shall see below. The exotic atoms muonium and positronium present their own problems which are discussed in sections 1.8.5 and 1.8.6 respectively.

1.8.1 Traditional methods

We have already mentioned one standard technique for hydrogen, that of allowing atoms to escape from a gas discharge. The best conditions for a direct current discharge are obtained by admitting a small amount of molecular hydrogen into a discharge in helium. Another traditional method is that used by Lamb [LAM 50b], in which the molecules were dissociated at 2850 K in a tungsten oven from which the atoms escaped to form an atomic beam.

For the more recently-developed techniques described below there is some overlap between these subsections and later chapters devoted to specific systems, but readers may find an advantage in having an overall summary here.

1.8.2 Cold hydrogen beams

Cold hydrogen beams are a by-product of attempts to observe the Bose-Einstein condensation of cold, dense atomic hydrogen gas [WAL 84]. The cooling is relatively simple to do provided it does not go below 4 K, and the techniques are described, for example, by Silvera and Walraven [SIL 79]. A room temperature rf discharge (typically tens of watts at 50 MHz) dissociates hydrogen in a source bulb. A teflon tube connects the source to a liquid helium-cooled accomodator in which the atoms are cooled by collisions with walls and the molecular H_2 freezes out. The cold atoms then effuse from an orifice to form a beam. With such a source one can obtain a flux of 2×10^{16} atoms/s from an accomodator at 8 K [SIL 79]. Below 4 K the atomic hydrogen is lost through recombination at the walls and more complicated methods using helium wall coatings are required.

A slow hydrogen beam is more efficiently excited by electron impact, because the atoms spend a longer time in the electron gun, however the recoil angles are very large [LAM 50b] and the metastable beam brightness is correspondingly small. Harvey [HAR 82] has pointed out that slow hydrogen beams can be optically pumped into the $2S$ state by exciting the $1S - 3P$ (Lyman-β) transition using a high power ultraviolet discharge lamp. The $3P - 2S$ decay has a 12% branching ratio and thus a large fraction of the beam can be pumped into the metastable state. This scheme has yet to be fully demonstrated but seems promising.

1.8.3 Fast hydrogen beams

The production of fast hydrogen beams starts with the electrostatic extraction of a proton beam from a plasma. After being accelerated to the required energy the beam passes through a target where electrons may be captured to produce neutral hydrogen. Generally all states of excitation are formed, the high lying states being populated roughly in proportion to $1/n^3$ (n is the principal quantum number). (Since the energy spacing between the excited states of hydrogen is proportional to $1/n^3$, this population distribution may be understood as the classical equipartition of energy.) The production of highly excited states is discussed more fully by Berry [BER 77]. Population of the lower lying states depends in a more specific way on the target material, the ion beam energy and the state of hydrogen produced.

Some broad distinctions can be drawn between solid and gaseous targets. Solid targets, usually foils of carbon between 1 and 100 μg/cm^2 thick are generally more suitable for stripping heavy atoms to produce highly ionized states. Being thicker than a gas target, they produce greater beam divergence and energy spread but are useful for the study of transient phenomena such as radiative decay and quantum beats because the interaction time with the foil is very short. Gabrielse [GAB 81] has measured the density matrix of the $n = 2$ hydrogen beam formed by passing protons through a foil.

For the production of a beam of fast metastable hydrogen atoms the most efficient technique is to pass the protons through a Cs vapour target. The process $H^+ + Cs \rightarrow H(2S) + Cs^+$ has a very small energy defect (0.49 eV) which causes a large peak in the cross section at unusually low energy (6×10^{-15} cm^2 at 500 eV) where competing processes have relatively small cross sections [PRA 74]. This reaction provides the most intense metastable hydrogen beams available today. Other more convenient gas targets such as H_2 and N_2 can also be used to produce fast metastable hydrogen beams.

1.8.4 Hydrogenic ions

Helium is readily ionized by electron impact, a technique that has been widely used to produce beams of both ground state and excited He$^+$. A suitable electron gun is described, for example in [DWO 68]. The extension of this technique to heavier atoms requires larger fluxes and higher energies in the electron beam because a larger number of electrons must be removed from the atom. A better method is to ionize the atom to a modest charge state and then to accelerate that ion to an energy such that collision with a foil or gas target will yield the hydrogenic state. The cross section for production of the hydrogenic state peaks when the ion velocity is comparable with that of the K-shell electrons. Thus the required energy for an atom having Z protons is of order $25Z^2$ keV/amu, corresponding for example, to about 32 MeV/amu for the production of hydrogenic Kr^{35+}. Such energies are available at heavy ion accelerators.

A different, more recent technique is that of recoil ion production in which a heavy, high energy ion passes through a gaseous target, leaving a track of highly ionized atoms behind. These ions recoil radially outward from the path of the projectile with typically a few eV of kinetic energy and can be focussed into a beam or held in an ion trap.

Methods of producing hydrogenic ions are discussed in Chapter 6. A recent conference publication [SIL 84] is a rich source of information on highly ionized atoms in general.

1.8.5 Positronium (e^+e^-)

Positrons are emitted from radioactive sources (e.g. ^{22}Na) typically with several hundred keV of kinetic energy. Originally positronium was formed in targets of high pressure ($\sim$ 1 atm) gas or fine grained powder where the positrons were slowed down by multiple interactions eventually capturing an electron to form positronium (e^+e^-). In both gas and powder targets the measured spectroscopic intervals were perturbed by the presence of the target and an empirical extrapolation to zero target density was necessary [HUG 84a]. It has now become possible to form beams of e^+e^- at low energy in a vacuum [CAN 74], a development which led immediately to the discovery of the $n = 2$ states [CAN 75] and which is obviously desirable for the spectroscopy of e^+e^- in general. The positrons from a radioactive source impinge on a metal surface from which a small fraction $(10^{-3} - 10^{-7})$ are re-emitted in a very low energy peak at about 1 eV [MIL 79]. The slow positrons are then accelerated to form a beam whose energy is in the range of a few eV to a few hundred eV. This beam impinges on a second metal target where more than half of the positrons capture electrons and emerge from the target into the vacuum as free e^+e^- atoms.

Recently it has been found that pulses of high energy electrons (100 MeV) striking a tantalum target produce intense pulses of slow positrons (1-1000 eV) [HOW 82]. This type of positron source may eventually lead to much higher intensity positronium beams. Further references, and a description of work done with positronium beams are to be found in Chapter 8.

1.8.6 Muonium $(\mu^+ e^-)$

When high energy protons strike a target, pion bremstrahlungen are produced. Most of the negative pions are captured by the target nuclei but the positive pions escape and decay in flight to produce polarized muons through the weak decay $\pi^+ \rightarrow \mu^+ + \nu_\mu$. The earliest method of muonium production was to send those muons into a high pressure gas target (~ 1 atm) where they would slow down and eventually capture an electron to form muonium. Some of the muon polarization is retained in this process and can be detected subsequently through the angular distribution of the decay positron $(\mu^+ \rightarrow e^+ + \nu_e)$. Hyperfine transitions in the muonium atom are observed as a change in the positron angular distribution [HUG 77] (see sections 4.1.6 and 4.1.7).

The desire to eliminate pressure shifts and to study excited states has stimulated the development of muonium beams in a vacuum. The new technique is based on the electron capture by low energy muons (< 20 keV) passing through a thin foil [BOL 81]. This approach was made possible by the development of a "surface muon" source in which the pions are produced and stopped near the surface of a carbon target. The decay muons penetrate to the surface and provide a source of much higher intensity at low energy than does the pion decay in flight [REI 78]. The new method led immediately to measurements of the $2S_{1/2} - 2P_{1/2}$ Lamb shift interval in $\mu^+ e^-$. It has also made possible attempts to observe spontaneous conversion of $\mu^+ e^-$ to $\mu^- e^+$ [HUG 84b].

REFERENCES

BAI 72   J. C. Baird, J. Brandenberger, K. -I. Gondaira and H. Metcalf, *Phys. Rev.* **A5** (1972) 564.

BER 77   H. G. Berry, *Rep. Prog. Phys.* **40** (1977) 155.

BEY 78   H. -J. Beyer, in *Progress in Atomic Spectroscopy* B, eds. W. Hanle and H. Kleinpoppen (Plenum, 1978) p. 538.

BOL 81   P. R. Bolton and 9 others, *Phys. Rev. Lett.* **47** (1981) 1441.

BRO 52   J. Brossel and F. Bitter, *Phys. Rev.* **86** (1952) 308.

CAN 74   K. F. Canter, A. P. Mills and S. Berko, *Phys. Rev. Lett.* **33** (1974) 7.

CAN 75   K. F. Canter, A. P. Mills and S. Berko, *Phys. Rev. Lett.* **34** (1975) 177.

DEM 81   W. Demtröder, *Laser Spectroscopy* (Springer, 1981).

DWO 68   S. Dworetsky, R. Novick, W. W. Smith and N. Tolk, *Rev. Sci. Instrum.* **39** (1968) 1721.

ECK 63   T. G. Eck, L. L. Foldy and H. Wieder, *Phys. Rev. Lett.* **10** (1963) 239.

FON 67   P. R. Fontana and L. C. Himmel, *Phys. Rev.* **162** (1967) 23.

GAB 81   G. Gabrielse, *Phys. Rev.* **A23** (1981) 775.

GLA 74   M. Glass-Maujean, *Opt. Commun.* **8** (1973) 260 and thesis, University of Paris (1974).

HAR 82   K. C. Harvey, *J. Appl. Phys.* **53** (1982) 3383.

HOW 82   R. H. Howell, R. A. Alvarez and M. Stanak, *Appl. Phys. Lett.* **40** (1982) 751.

HUG 77   *Muon Physics*, eds. V. W. Hughes and C. S. Wu. (Academic Press, 1977).

HUG 84a   V. W. Hughes, in *Precision Measurements and Fundamental Constants II*, eds. B. N. Taylor and W. D. Phillips., *Nat. Bur. Stand. Spec. Publ.* **617** (1984) p. 237.

HUG 84b   V. W. Hughes and G. zuPutlitz, *Comments Nucl. Part. Phys.* **12** (1984) 259.

KAS 50   A. Kastler, *J. Phys. Rad.* **11** (1950) 255.

LAM 50a   W. E. Lamb and M. Skinner, *Phys. Rev.* **78** (1950) 539.

LAM 50b   W. E. Lamb Jr. and R. C. Retherford, *Phys. Rev.* **79** (1950) 549.

LAM 56   W. E. Lamb and T. M. Sanders, *Phys. Rev.* **103** (1956) 313.

LAM 60a   W. E. Lamb and T. M. Sanders, *Phys. Rev.* **119** (1960) 1901.

LAM 60b   W. E. Lamb and L. R. Wilcox, *Phys. Rev.* **119** (1960) 1915.

LEE 79   S. A. Lee, J. Helmcke and J. L. Hall, in *Laser Spectroscopy IV*, eds. H. Walther and K. W. Rothe (Springer, 1979) p. 130.

MIL 79   A. P. Mills Jr. and L. Pfeifer, *Phys. Rev. Lett.* **43** (1979) 1961.

NAK 81   S. Nakayama, *J. Phys. Soc. Japan* **50** (1981) 609 and *J. Opt. Soc. Am.*, **B2** (1985) 1431.

PIP 62   F. M. Pipkin and R. H. Lambert, *Phys. Rev.* **127** (1962) 787 and references therein.

PRA 74   P. Pradel, F. Rousell, A. S. Schlachter, G. Spiess and A. Valance, *Phys. Rev.* **A10** (1974) 797.

REI 78   H. W. Reist and 9 others, *Nucl. Instrum. Methods* **153** (1978) 61.

SER 81   G. W. Series, *Contemp. Phys.* **22** (1981) 487.

SER 82   G. W. Series and B. A. Thrush, eds. *New Techniques in Optical and Infrared Spectroscopy*, *Phil. Trans. Roy. Soc.* **307**, (1882) 465-687. Also published as a separate volume.

SIL 79   I. F. Silvera and J. T. M. Walraven, *Phys. Lett.* **74A** (1979) 193.

SIL 84   *The Physics of Highly Ionized Atoms*, eds. J. D. Silver and N. J. Peacock (North Holland, Amsterdam, 1984).

WAL 84   J. T. M. Walraven, in *Atomic Physics 9*, eds. R. S. Van Dyck Jr. and E. N. Fortson (World Scientific, Singapore, 1984) p. 187.

WIN 67   W. H. Wing and P. R. Fontana, *Bull. Am. Phys. Soc.* **12** (1967) 95.
WIN 68   W. H. Wing, thesis, University of Michigan (1968).
WIN 79   D. J. Wineland and W. M. Itano, *Phys. Rev.* **A20** (1979) 1521.
WIT 56   J. P. Wittke and R. H. Dicke, *Phys. Rev.* **103** (1956) 625.

CHAPTER 2

QUANTUM ELECTRODYNAMICS CALCULATIONS

P. J. Mohr

2.1 Introduction

This chapter is meant to provide a brief introduction to the basic formulation of quantum electrodynamics as it relates to calculations of magnetic moments and atomic level shifts. Readers who are interested in a historical perspective of the interpretation of quantum electrodynamics effects will find the early ideas concerning the interaction of the atom with the zero-point electromagnetic field of the vacuum and their formulation summarized in SAH, Chapter IX.

Contemporary work on the interpretation of quantum electrodynamic effects in terms of vacuum fluctuations, radiative reaction or classical electrodynamics is reviewed elsewhere. A model calculation of radiative level shifts and damping has been given, for example, by Cohen-Tannoudji for a small system coupled to a reservoir [Cohen-Tannoudji 1986]. Cohen-Tannoudji and his colleagues [Dupont-Roc 1978] have also given a physical interpretation of the magnetic moment anomaly as have Grotch and Kazes [Grotch 1977]. Radiation reaction interpreted in the Absorber Theory of Radiation [Wheeler 1945, 1949], has been described in a recent account by Pegg [Pegg 1986]. Series [Series 1986] has utilized a direct action formulation of Absorber Theory to derive the conventional expression for the level shift and decay constants. A more descriptive account of Lamb shifts and radiative damping, in the context of a discussion of the general physical properties of the vacuum in quantum field theory, has been written by Aitchison [Aitchison 1985].

In this chapter, we shall take the view that renormalized quantum electrodynamics with the concepts of virtual photon interactions and pair creation effects provides a complete description of the interaction of radiation and atoms. This description involves the well-known phenomenon that infinite quantities appear in intermediate stages of computation, but on a practical level, there is a well-defined way to set up the calculations so that the physical predictions are unique and finite. The remarkable agreement between theory and experiment based on such calculations is evidence for the credibility of this approach.

One example where the effectiveness of the contemporary formulation of quantum electrodynamics is evident is in the unified description of radiative damping and level shifts. The connection between radiative decay and spectral line shape was already understood in the work of Weisskopf and Wigner [Weisskopf 1930]. However, in that work the real part of the energy level shift, now known as the Lamb shift, was not calculated. In present day quantum electrodynamics, a finite unified treatment of the radiative level shift and decay rate, in terms of the real and imaginary part of the correction as indicated in equations (2.28) and (2.29) can be made in a simple way. Within this same framework, the connection between the radiative level shift and natural line shape has been described, for example, by Low [Low 1952]. More recently,

Low's approach has been extended to the case of an atom in a weak electric field [Hillery 1980].

2.2 Quantum Electrodynamics

In the years since the formulation of the modern covariant form of quantum electrodynamics (QED), with charge and mass renormalization, done in the late 1940's and early 1950's, there has been steady progress in both experimental and computational techniques leading to increasingly stringent tests of the basic theory. Throughout this period, QED has been successful in explaining experiments of increasing precision over a widening range of phenomena. Over the decade ending in the early 1970's, the electroweak theory was formulated as a unified description of QED and weak interaction effects, based on spontaneous symmetry breaking. This unified theory has so far been successful in describing weak interaction effects in atoms, as well as in high energy phenomena.

In the description of atomic structure, the electroweak theory and quantum electrodynamics differ only by extremely small corrections, so that it is sufficient to take QED alone as a basis for atomic theory and treat weak interaction effects, particularly parity violating interactions, as small corrections. The theory of weak interaction effects will be found in section 3.6 of this book: experimental tests are treated in section 4.3. In fact, most of the properties of atoms are described by the Dirac or Schrödinger equation for electrons coupled to the external electromagnetic field of the nucleus, omitting QED effects that require renormalization methods. However, it is necessary to apply a complete description, including QED effects, in order to understand some of the more detailed features of atomic structure.

QED can be formulated in a number of ways. For an electron in a weak magnetic field, QED effects give rise to the well-known anomalous magnetic moment. In this case, calculations are based on free-particle QED in which the unperturbed electron is described by a free-particle solution of the Dirac equation. Highly-sophisticated calculations have been done with corrections up to eighth order in perturbation theory for the electron anomaly. For many bound state problems, such as the hydrogen atom, the nucleus is at rest to a first approximation and acts as the source of a static field. In this case, the most natural formulation of QED is the bound-interaction or Furry picture in which the zeroth order of states of the electron are solutions of the Dirac equation for an electron in the external field. In this approach, the binding of the electron, in the absence of interactions with radiation, is taken into account exactly, and the effects of interaction of the electrons with the quantized electromagnetic field are calculated as perturbations. For systems such as positronium, where recoil effects of both particles are important in lowest order, the external field approach is inappropriate, and calculations are based on a more general formulation of QED such as the Bethe-Salpeter equation or one of a number of approximations to it.

2.3 Magnetic Moment Anomaly

One of the most basic predictions of QED is the radiative correction to the magnetic moment of Dirac particles such as the electron. In 1947, the effect of the radiative correction on atomic magnetic moments was observed by Foley and Kusch [Foley 1948], and shortly thereafter the leading radiative correction was calculated by Schwinger [Schwinger 1948]. The leading correction, corresponding to the Feynman diagram in Fig. 2.1, is second order in the

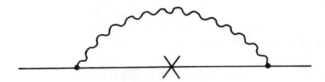

Fig. 2.1. Feynman diagram for the lowest order radiative correction to the electron magnetic moment. The straight line represents the free electron, the wavy line represents the virtual photon, and the X represents interaction of the electron with the external magnetic field.

coupling of the electron to the electromagnetic field. It arises from the virtual emission and re-absorption of a photon as the free electron interacts with the external magnetic field. Only the term linear in the magnetic field is necessary for typical field strengths. The calculation of the lowest order diagram, described in most introductory field theory texts, gives the well-known correction term $\alpha/2\pi$ for the electron magnetic moment. Over the years, both theory and experiment for the anomalous magnetic moment of the electron have been extended to high precision. The calculations have reached the level of eighth order corrections, i.e., corrections corresponding to Feynman diagrams with four virtual photons in all possible configurations. Extremely large scale numerical computation is required to deal with the 891 Feynman diagrams in this order.

Current results of calculations are summarized as follows. The g-factor anomaly a_e, defined by the equation

$$g_e = 2(1 + a_e) \tag{2.1}$$

is expressed as a power series in the fine structure constant α as

$$\begin{aligned} a_e = a_e^{(2)}(\alpha/\pi) + a_e^{(4)}(\alpha/\pi)^2 + a_e^{(6)}(\alpha/\pi)^3 \\ + a_e^{(8)}(\alpha/\pi)^4 + \dots, \end{aligned} \tag{2.2}$$

where [Schwinger 1948]

$$a_e^{(2)} = 1/2 , \qquad (2.3)$$

[Karplus 1950, Sommerfield 1957, Petermann 1957a, b]

$$a_e^{(4)} = 197/144 + \pi^2/12 - (\pi^2/2)\ln 2 + (3/4)\varsigma(3)$$
$$= -0.328478966 , \qquad (2.4)$$

[Kinoshita 1981]

$$a_e^{(6)} = 1.1765(13) , \qquad (2.5)$$

[Kinoshita 1981, 1984a]

$$a_e^{(8)} = -0.8(1.4) . \qquad (2.6)$$

Summation of these terms yields

$$a_e = 0.001159652307(41)(102) \qquad (2.7)$$

where the first number in parentheses is the error due to uncertainty in the theoretical evaluation of the eighth order terms and the second number in parentheses is the uncertainty associated with the fine structure constant $\alpha^{-1} = 137.035$ $981(12)$ [Taylor 1985]. This number is in agreement with the experimental value [Van Dyck 1984]

$$a_e = 0.001159652193(4) . \qquad (2.8)$$

If the theory is assumed to be correct, then comparison of theory and experiment leads to a determination of the fine structure constant that gives the value $\alpha^{-1} = 137.035994(5)$.

The theoretical radiative correction to the magnetic moment of the muon is given by [Kinoshita 1984b]

$$a_\mu = 0.0011659200(20) \qquad (2.9)$$

which differs from the electron value, because virtual particle pair corrections are more important for the heavier muon. The uncertainty in the muon anomaly is due to uncertainties in the contribution of virtual pairs of strongly interacting particles. Comparison of the theoretical value for the anomaly with the experimental value [Bailey 1977, Farley 1979]

$$a_\mu = 0.001165911(11) \qquad (2.10)$$

shows good agreement.

2.4   Bound Interaction Picture QED

Quantum electrodynamics is the field theory of charged particles interacting via the exchange of photons, the quanta of the electromagnetic field. For an electron bound in an atom, the most natural theoretical approach is QED in the bound-interaction picture. The generalization to other particles such as muons or taus is straightforward. In the bound-interaction or Furry picture [Furry 1951, Schweber 1961] the electron-positron field operator $\psi(x)$ is expanded as a sum over solutions of the Dirac equation in the external field with annihilation operators for electrons a_n multiplying the positive energy solutions and creation operators for positrons $b_n^\dagger$ multiplying negative energy solutions (natural units in which $\hbar = c = m = 1$ are employed in this chapter)

$$\psi(x) = \sum_{E_n > 0} a_n \phi_n(x) + \sum_{E_n < 0} b_n^\dagger \phi_n(x) \tag{2.11}$$

where

$$\phi_n(x) = \phi_n(\mathbf{x}) \exp\left(-i E_n t\right) \tag{2.12}$$

and $\phi_n(\mathbf{x})$ is a solution of the coordinate-space Dirac equation

$$[-i\boldsymbol{\alpha} \cdot \nabla + V(x) + \beta - E_n] \phi_n(\mathbf{x}) = 0 . \tag{2.13}$$

The interaction Hamiltonian is

$$H_I(x) = j_\mu(x) A^\mu(x) - \delta M(x) \tag{2.14}$$

where the electron-positron current $j_\mu(x)$ is

$$j_\mu(x) = -\frac{1}{2} e[\overline{\psi}(x)\gamma_\mu, \psi(x)] \tag{2.15}$$

$A^\mu(x)$ is the radiation field operator, and the mass renormalization term $\delta M(x)$ is

$$\delta M(x) = \frac{1}{2}\delta m[\overline{\psi}(x), \psi(x)] . \tag{2.16}$$

In (2.15) and (2.16), the commutator refers only to the creation and annihilation operators; the Dirac spinors and matrices are in the order shown. Explicit expressions for the radiative level shifts can be obtained by applying the prescription of Gell-Mann and Low [Gell-Mann 1951] and Sucher [Sucher 1957] that relates energy-level shifts ΔE to S-matrix elements or Feynman diagrams

$$\Delta E = \lim_{\substack{\varepsilon \to 0 \\ \lambda \to 1}} \frac{1}{2} i\varepsilon \frac{\frac{d}{d\lambda}\langle S_{\varepsilon,\lambda}\rangle_c}{\langle S_{\varepsilon,\lambda}\rangle_c} . \tag{2.17}$$

The subscript c denotes the fact that only connected Feynman diagrams are included in the S-matrix in (2.17). State vectors in (2.17) are given by an electron creation operator acting on the Furry picture vacuum

$$|n\rangle = a_n^\dagger |0\rangle \tag{2.18}$$

for one-electron states, and by k creation operators acting on the vacuum for a k-electron state. The S-matrix is written in a perturbation expansion as

$$S_{\epsilon,\lambda} = \sum_{j=0}^{\infty} S_{\epsilon,\lambda}^{(j)} \tag{2.19}$$

where

$$
\begin{aligned}
S_{\epsilon,\lambda}^{(j)} =& \frac{(-i\lambda)^j}{j!} \int d^4 x_j \ldots d^4 x_1 \exp\left(-\varepsilon|t_j|\right) \ldots \\
& \times \exp(-\varepsilon|t_1|) T[H_I(x_j) \ldots H_I(x_1)]
\end{aligned}
\tag{2.20}
$$

and T denotes the time-ordering operator. S-matrix elements in (2.17) can be evaluated order-by-order with the aid of Wick's theorem, to obtain expressions for the energy level shifts corresponding to the appropriate Feynman diagrams. Contractions of the electron-positron field operators are given by the bound state propagation function

$$
\begin{aligned}
S_F(x_2, x_1) &= \langle 0|T[\psi(x_2)\overline{\psi}(x_1)]|0\rangle \\
&= \begin{cases}
\displaystyle\sum_{E_n > 0} \phi_n(x_2)\overline{\phi}_n(x_1), & t_2 > t_1 \\
-\displaystyle\sum_{E_n < 0} \phi_n(x_2)\overline{\phi}_n(x_1), & t_2 < t_1 .
\end{cases}
\end{aligned}
\tag{2.21}
$$

Contractions of the photon field operators give the standard photon propagator

$$D_F(x_2 - x_1) = -\frac{i}{(2\pi)^4} \int d^4 k \frac{e^{-ik\cdot(x_2 - x_1)}}{k^2 + i\delta} . \tag{2.22}$$

From this formalism, the leading (second order) corrections for a one-electron atom in state n are

$$
\begin{aligned}
E_{SE}^{(2)} =& -4\pi i\alpha \int d(t_2 - t_1) \int d\mathbf{x}_2 \int d\mathbf{x}_1 D_F(x_2 - x_1) \\
& \times \overline{\phi}_n(x_2)\gamma_\mu S_F(x_2, x_1)\gamma^\mu \phi_n(x_1) - \delta m \int d\mathbf{x}\overline{\phi}_n(\mathbf{x})\phi_n(\mathbf{x})
\end{aligned}
\tag{2.23}
$$

and

$$E_{VP}^{(2)} = 4\pi i \alpha \int d(t_2 - t_1) \int d\mathbf{x}_2 \int d\mathbf{x}_1 \, D_F(x_2 - x_1)$$
$$\times \, \mathrm{Tr}\left[\gamma_\mu S_F(x_2, x_2)\right]\overline{\phi}_n(x_1)\gamma^\mu \phi_n(x_1) \tag{2.24}$$

corresponding to the self-energy and vacuum-polarization diagrams shown in Figs. 2.2(a) and 2.2(b) respectively. These corrections are examined in the next two sections.

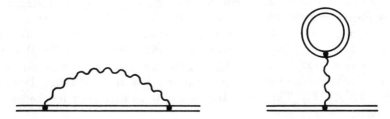

Fig. 2.2. Feynman diagrams for the (a) self-energy and (b) vacuum polarization in one-electron atoms. The double lines represent the electron in the static nuclear field.

2.5   Self-Energy

Virtual emission and reabsorption of a photon by a bound electron gives rise to the self-energy level shift corresponding to the Feynman diagram in Fig 2.2(a). This is the main contribution to the Lamb shift, i.e., the splitting of the $2S_{1/2}$ and $2P_{1/2}$ levels in hydrogen. Bethe's original calculation of this effect [Bethe 1947], to interpret the experimental result of Lamb and Retherford [Lamb 1947], was based on non-relativistic quantum electrodynamics. Bethe introduced a procedure for mass renormalization to isolate the physical level shift from the infinite mass shift of a free electron.

In the relativistic calculation, the basis for mass renormalization is the fact that the hypothetical bare mass in the Dirac equation (2.13) with no radiative corrections, is not the physical measured mass. (The mass appears as the coefficient of β when $m \neq 1$.) The physical mass m of a free electron is the bare mass m_0 plus the radiative corrections δm to the mass of a free electron: $m = m_0 + \delta m$. In order to express the results of calculations in terms of the measured mass rather than the unknown bare mass, we effectively replace the bare mass m_0 in the Dirac equation by m, and pick up the correction term proportional to $-\delta m$ as the mass renormalization counter term $-\delta M(x)$. This procedure leads to a finite result for the energy level shift in each order of perturbation theory even though δm and the calculated radiative corrections are separately infinite. In fact, for a free electron, these corrections exactly cancel to any order to give zero net mass shift. In the case where the electron is bound in an atom, the radiative corrections are modified and there is a non-zero finite result for the energy shift that gives the main contribution to the Lamb shift.

Relativistic calculations of the self-energy correction have been carried out by Feynman [Feynman 1948, 1949], by Fukuda, Miyamoto, and Tomonaga [Fukuda 1949], by Kroll and Lamb [Kroll 1949], by Schwinger [Schwinger 1949], and by French and Weisskopf [French 1949]. These calculations gave the leading order in powers of $Z\alpha$ for the Lamb shift. Subsequent calculations by Baranger, Bethe, and Feynman [Baranger 1951, 1953] and by Karplus, Klein, and Schwinger [Karplus 1951, 1952a] gave the next order in powers of $Z\alpha$. Higher order terms have also been calculated [Layzer 1960, 1961a, Fried 1960, Erickson 1965, 1971]. These calculations lead to an expression for the level shift of the form

$$E_{SE}^{(2)} = \frac{\alpha}{\pi} \frac{(Z\alpha)^4}{n^3} F(Z\alpha) mc^2 \qquad (2.25)$$

where

$$F(Z\alpha) = A_{40} + A_{41} \ln(Z\alpha)^{-2} + A_{50}(Z\alpha) + A_{60}(Z\alpha)^2$$
$$+ A_{61}(Z\alpha)^2 \ln (Z\alpha)^{-2} + A_{62}(Z\alpha)^2 \ln^2(Z\alpha)^{-2} + \dots . \qquad (2.26)$$

This power series converges slowly, so that for high accuracy at low Z or an evaluation at high Z the truncated series is a poor representation for the exact function in equation (2.25). An alternative method of calculating the level shift has been to carry out a non-perturbative numerical evaluation of the complete expression in (2.23) to all orders in $Z\alpha$ [Brown 1959, Desiderio 1971, Mohr 1974, Cheng 1976, Soff 1982, Mohr 1982]. Calculations have been done for both the binding field of a finite size nucleus and a pure Coulomb binding field. Results of numerical calculations for the self-energy of an electron in a Coulomb field are shown in Fig. 2.3. The consistency of the numerical calculations and power-series results has been confirmed for the $1S$ state [Sapirstein 1981].

The preceding discussion is concerned with the real part of the self-energy level shift. For excited states the level shift in equation (2.23) is complex, and the imaginary part is simply related to the decay rate. For photons with momenta less than K, where K is of order mc^2, the expression in (2.23) is equal (up to an additive constant that cancels other terms in the complete

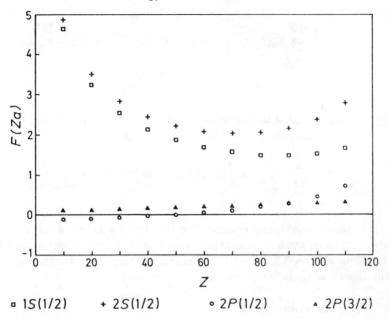

Fig. 2.3. Results of numerical calculations of the self-energy for an electron bound in a Coulomb field [Mohr 1974, 1982]

calculation) to

$$E_{SE}^{(2)}(K) = -\frac{\alpha}{4\pi^2} \int_{k<K} d\mathbf{k} \frac{1}{k}\left(\delta_{jl} - \frac{k_j k_l}{k^2}\right)$$

$$\times \left\langle \alpha^j e^{i\mathbf{k}\cdot\mathbf{x}} \frac{1}{H - E_n + k - i\delta} \alpha^l e^{-i\mathbf{k}\cdot\mathbf{x}} \right\rangle$$

$$= -\frac{\alpha}{4\pi^2} \int_{k<K} d\mathbf{k} \frac{1}{k} \sum_{\lambda,m} \frac{|\langle m|\hat{\varepsilon}_\lambda \cdot \boldsymbol{\alpha} e^{-i\mathbf{k}\cdot\mathbf{x}}|n\rangle|^2}{E_m - E_n + k - i\delta} . \tag{2.27}$$

This expression reduces to the original non-relativistic formula of Bethe for the level shift in the non-relativistic limit, (SAH, section IX.4) in which $\boldsymbol{\alpha}$ is replaced by p, and the dipole approximation, in which $e^{-i\mathbf{k}\cdot\mathbf{x}}$ is replaced by 1.

To identify the decay rate in (2.27), we employ the identity

$$\int_0^K dk \sum_m \frac{f_m(k)}{E_m - E_n + k - i\delta} = P \int_0^K dk \sum_m \frac{f_m(k)}{E_m - E_n + k}$$
$$+ \pi i \sum_{\substack{m \\ E_m < E_n}} f_m(E_n - E_m) \qquad (2.28)$$

which divides the level shift into its real and imaginary parts, since f_m is real. Then for the decay rate $A = -2 \, \text{Im}(\Delta E_n)$, we have

$$A = \frac{\alpha}{2\pi} \sum_{\substack{\lambda, m \\ E_m < E_n}} k \int d\Omega_k |\langle m|\widehat{\varepsilon}_\lambda \cdot \boldsymbol{\alpha} e^{-i\mathbf{k}\cdot\mathbf{x}}|n\rangle|^2 \Big|_{k = E_n - E_m} \qquad (2.29)$$

which is the relativistic expression for the decay rate of an excited state in hydrogen. An equivalent expression, with e, m, h, c written out explicitly, is to be found in section 3.2, where it forms the starting point for the development of the theory of radiative processes.

2.6   Vacuum Polarization

Early discussions of vacuum polarization predate modern quantum electrodynamics and renormalization. In 1935, Serber [Serber 1935] and Uehling [Uehling 1935] considered effects of vacuum polarization from the point of view of Dirac hole theory in which negative energy states are occupied by electrons. In this picture, there is an accumulation of the electrons that fill the negative energy sea near a positive nucleus. The effective charge of the nucleus for a bound electron far from the nucleus is the net charge $+Ze$ of the nucleus together with accumulated vacuum electrons. Inside the characteristic distance associated with this screening, i.e., the Compton wavelength of the electron $\lambda_e = 4 \times 10^{-11}$ cm, the net effective charge is larger and approaches the (infinite) bare charge $+Ze_0$ of the nucleus. Hence, S electrons are more tightly bound when vacuum polarization is taken into account; the higher l states are affected less, because the electron is less likely to be near the nucleus. From the point of view of field theory, the vacuum polarization correction is the result of the process in which the bound electron exchanges a photon with a virtual electron-positron pair, which in turn interacts with the Coulomb field of the nucleus.

A field theoretical formulation of vacuum polarization was given by Schwinger [Schwinger 1951a], and a thorough study was made by Wichmann and Kroll [Wichmann 1954] of the higher-order vacuum polarization effects in which the virtual electron-positron pair interacts with nuclear Coulomb field more than once. The complete second order vacuum polarization level shift, given by equation (2.24), can be written as

$$E_{VP}^{(2)} = \frac{\alpha}{\pi} \frac{(Z\alpha)^4}{n^3} H(Z\alpha) mc^2 \qquad (2.30)$$

where

$$H(Z\alpha) = H_1(Z\alpha) + H_3(Z\alpha) + H_5(Z\alpha) + \dots . \qquad (2.31)$$

The term $H_1(Z\alpha)$ arises from the expectation value of the Uehling potential, the lowest order vacuum polarization potential $V_1(r)$ [Serber 1935, Uehling 1935]

$$V_1(r) = -\frac{\alpha}{\pi} \frac{Z\alpha}{3r} \int_1^\infty dt (t^2 - 1)^{1/2} (2t^{-2} + t^{-4}) \exp(-2tr) \qquad (2.32)$$

in which the virtual electron-positron pair interacts once with the nuclear Coulomb field, and the higher order terms $H_n(Z\alpha)$ arise from expectation values of vacuum-polarization operators $V_n(r)$ that are of order n in the external potential. The Uehling term follows from equation (2.24) with the approximation that S_F is expanded in powers of the external potential, and only the term of first order is retained. Charge renormalization is implemented by dropping (infinite) terms from the potential that vary as $1/r$ at large distances, because such terms are already taken into account by using the physical charge for the nucleus rather than the hypothetical bare charge. We obtain the well-known leading contribution to the level shift for S states by evaluating $\langle V_1 \rangle$ to lowest order in $Z\alpha$

$$H_l(Z\alpha) = -(4/15)\delta_{l0} + \dots \qquad (2.33)$$

where l is the orbital angular momentum quantum number. The complete function $H_1(Z\alpha)$ is readily evaluated exactly numerically [see for example Mohr 1982]. Higher order terms have been calculated in various approximations for electrons or muons in a strong Coulomb field [Wichmann 1954, Blomqvist 1972, Brown 1974, 1975, Gyulassy 1974, Mohr 1983, Johnson 1985].

2.7   Higher Order Radiative Corrections

Higher order quantum electrodynamics effects arise from Feynman diagrams with more than one virtual photon. These effects are smaller by a factor α than the one-loop self-energy and vacuum-polarization corrections. Two-photon (fourth order) diagrams have been calculated for hydrogenlike atoms to lowest order in $Z\alpha$.

The complete fourth order correction is given by the formalism described in section 2.4, however to obtain just the lowest order term in $Z\alpha$, calculations have been carried out in the scattering approximation which takes into account radiative corrections to the free-electron form factors and vacuum polarization corrections.

The calculated terms are classified as vacuum-polarization corrections [Baranger 1952, Källén 1955], magnetic-moment corrections [Karplus 1950, Sommerfield 1957, Petermann 1957a, b], and self-energy corrections [Weneser 1953, Mills 1955, Soto 1966, 1970, Appelquist 1970a, b, Lautrup 1970, Barbieri 1970, Peterman 1971, Fox 1973]. The lowest order total shift for a level with angular momentum-parity number k and orbital angular momentum l is

$$E^{(4)} = \left(\frac{\alpha}{\pi}\right)^2 \frac{(Z\alpha)^4}{n^3} mc^2$$

$$\times \left\{ \delta_{l0} \left[2\pi^2 \ln 2 - \frac{49\pi^2}{108} - \frac{6131}{1296} - 3\varsigma(3) \right] \right.$$

$$\left. + \frac{1}{k(2l+1)} \left[\frac{\pi^2}{2} \ln 2 - \frac{\pi^2}{12} - \frac{197}{144} - \frac{3}{4}\varsigma(3) \right] \right\} . \tag{2.34}$$

There is some uncertainty in the magnitude of the complete fourth order level shift because higher order terms in $Z\alpha$ have not been calculated.

2.8 Finite Nuclear Size Effects

The main effect of the finite size of the nucleus on energy levels in hydrogen is readily calculated by perturbation theory. A model-independent evaluation was done by Karplus, Klein and Schwinger [Karplus 1952a]. For a spherically symmetric nuclear charge distribution $\rho(r)$, normalized so that $\int dr \rho(r) = 1$, the correction to the Coulomb potential is

$$\delta V(r) = -Z\alpha \int d\mathbf{r'} \frac{1}{|\mathbf{r} - \mathbf{r'}|} [\rho(\mathbf{r'}) - \delta(\mathbf{r'})] \tag{2.35}$$

with the corresponding level shift

$$E_{NS} = \int d\mathbf{r} |\phi_n(r)|^2 \delta V(r) . \tag{2.36}$$

Since the wave function varies slowly near the nucleus where $\delta V(r)$ is nonzero, we have

$$E_{NS} \approx -Z\alpha |\phi_n(0)|^2 \int d\mathbf{r} \int d\mathbf{r'} \frac{1}{|\mathbf{r} - \mathbf{r'}|} [\rho(\mathbf{r'}) - \delta(\mathbf{r'})]$$

$$= -Z\alpha |\phi_n(0)|^2 \int d\mathbf{r'} 4\pi \int_0^{r_m} dr r^2 \frac{1}{r_>} [\rho(\mathbf{r'}) - \delta(\mathbf{r'})]$$

$$= Z\alpha |\phi_n(0)|^2 \frac{2\pi}{3} \int d\mathbf{r'} r'^2 \rho(r')$$

$$= \delta_{l0} \frac{2(Z\alpha)^4}{3n^3} (R/\lambda_e)^2 mc^2 \tag{2.37}$$

for the level shift, where it is assumed that $\delta V(r)$ vanishes for r greater than some large radius r_M, and R is the rms radius of the nuclear charge distribution.

If Z is large, the non-relativistic treatment described above is no longer valid. In this case, relativistic corrections are important, and owing to the singular nature of the perturbation, first order perturbation theory does not give the entire relativistic correction. A complete calculation of the effect of the nuclear size is made by solving the Dirac equation for the electron in the finite nucleus potential. For intermediate Z, a non-perturbative solution of the Dirac equation yields [Friar 1979, Mohr 1983]

$$E_{NS} = [1 + 1.19(Z\alpha)^2]\frac{1}{12}(Z\alpha)^2(Z\alpha R/\lambda_e)^{2s}mc^2 \qquad (2.38)$$

for the Lamb shift with $s = [1-(Z\alpha)^2]^{1/2}$. This expression is based on a nuclear model with a uniform spherical distribution of charge, and terms of relative order $(Z\alpha)^4$ and $Z\alpha R/\lambda_e$ are neglected. A tabulation of size corrections that extends to high Z, based on numerical eigenvalues of the Dirac equation with a finite nuclear size potential has been given by Johnson and Soff [Johnson 1985].

Finite nuclear size corrections to the radiative corrections themselves grow rapidly with Z for Z of order 100 or larger, but these corrections are not presently accessible experimentally [Cheng 1976, Soff 1982]. For $Z = 90$, the size correction to the $1S$ self-energy is less than 2% [Johnson 1985]. At low Z, the size correction to the self-energy is negligible compared to other uncertainties [Borie 1981, Lepage 1981, Hylton 1984].

For the vacuum polarization, the nuclear size correction to the Uehling potential has been calculated numerically over a wide range of Z [Johnson 1985]. For small Z, the nuclear size correction to the vacuum polarization can be evaluated in the non-relativistic approximation [Hylton 1985]

$$E_{VP,NS} = \delta_{l0}\frac{\alpha}{2}\frac{(Z\alpha)^5}{n^3}(R/\lambda_e)^2mc^2 \qquad (2.39)$$

and is negligible compared to other uncertainties. Finite nuclear size corrections to the higher order vacuum polarization level shifts have been included in various calculations [Gyulassy 1974, 1975, Borie 1982, Neghabian 1983].

2.9 Nuclear Recoil Corrections

Motion of the atomic nucleus is neglected in the external field approximation. Corrections for nuclear motion, which are higher order in m/M, the ratio of the electron mass to the nuclear mass, are calculated in the more general framework of relativistic two-body theory. The Bethe-Salpeter equation [Salpeter 1951,

Schwinger 1951b] provides the basic theory for two-body effects, and approximation methods have been developed over the years to apply the theory in practical calculations.

The leading nuclear motion correction is the ordinary non-relativistic reduced-mass effect which is taken into account by replacing the mass of the electron by the electron-nucleus reduced mass $m_R = mM/(m + M)$ in the expressions for the energy levels. Early studies of nuclear motion effects by Breit and Meyerott [Breit 1947] for hyperfine structure and by Breit and Brown [Breit 1948] for the fine structure were based on the Breit equation [Breit 1929]. The latter study led to a relativistic correction to energy levels of order (m/M) $(Z\alpha)^4 mc^2$.

By developing a systematic approximation procedure for solving the Bethe-Salpeter equation, Salpeter calculated a relativistic recoil correction of order $(m/M)(Z\alpha)^5 mc^2$ to the Lamb shift in hydrogen [Salpeter 1952]. These results were confirmed by Grotch and Yennie who used a formalism based on effective potential that produces the correct field-theoretical scattering amplitude to an appropriate level of accuracy [Grotch 1967, 1969]. In this approach, recoil effects are partially accounted for within the external field approximation by means of the effective interaction potential. This approach has been extended to evaluate higher order finite nuclear mass corrections [Bhatt 1985, 1987].

A good deal of theoretical effort has gone into studies of recoil effects in positronium and muonium. In these systems, the recoil effects are larger than in hydrogenlike atoms and require extensive calculations to include higher order corrections. A review of recoil calculations that emphasizes this work is given by Bodwin, Yennie, and Gregorio [Bodwin 1985].

2.10   Lamb Shift in Hydrogenlike Atoms

A classic test of bound state quantum electrodynamics is the comparison of theory and experiment for the Lamb shift $S = E(2S_{1/2}) - E(2P_{1/2})$ in hydrogen. The dominant theoretical contributions arise from the self-energy and vacuum polarization. The self-energy can be written as

$$\begin{aligned}
S_{SE}^{(2)} = {} & (\alpha/\pi)[(Z\alpha)^4/6]mc^2\{\ln(Z\alpha)^{-2} + B(2,0) - B(2,1) \\
& + 11/24 + 1/2 + 3\pi[1 + 11/128 - (1/2)\ln 2](Z\alpha) \\
& - (3/4)(Z\alpha)^2\ln^2(Z\alpha)^{-2} \\
& + (299/240 + 4\ln 2)(Z\alpha)^2\ln(Z\alpha)^{-2} + (Z\alpha)^2 G_{SE}(Z\alpha)\}
\end{aligned}$$

$$(2.40)$$

and the vacuum polarization is

$$\begin{aligned}
S_{VP}^{(2)} = {} & (\alpha/\pi)[(Z\alpha)^4/6]mc^2\{-1/5 + (5/64)\pi(Z\alpha) \\
& - (1/10)(Z\alpha)^2\ln(Z\alpha)^{-2} + (Z\alpha)^2 G_{VP}(Z\alpha)\}
\end{aligned}$$

$$(2.41)$$

where $B(2,0) = -2.811\,770$ and $B(2,1) = 0.030\,017$. In the above expressions, the energy shift is divided into calculated lower order parts (see section 2.5) and a function that contains the complete higher order remainder. An estimate of the function G_{SE} is obtained from the complete numerical calculations by subtracting the lower order terms in (2.40). Extrapolation of the values for G_{SE} from $Z = 10, 20$ and 30 to $Z = 1$ yields the value $G_{SE}(\alpha) = -23.4 \pm 1.2$ [Mohr 1975]. The value of G_{SE} obtained by this approach has been confirmed in the case of the $1S$ state by an independent calculation of Sapirstein [Sapirstein 1981]. To evaluate the vacuum polarization contribution to the Lamb shift in hydrogen, it is sufficient to consider the Uehling potential contribution. This gives the lower order terms in $S_{VP}^{(2)}$ and the dominant part $G_U(Z\alpha)$ of $G_{VP}(Z\alpha)$ [Wichmann 1954, Mohr 1975]

$$G_{VP}(Z\alpha) \approx G_U(Z\alpha) = - 1199/2100 + (5/128)\pi(Z\alpha)\ln(Z\alpha)^{-2}$$
$$+ 0.5(Z\alpha) + \dots . \qquad (2.42)$$

There are additional leading order corrections to the Lamb shift that are discussed in previous sections of this chapter. They are the fourth order radiative correction

$$S^{(4)} = (\alpha/\pi)^2[(Z\alpha)^4/6]mc^2\{\pi^2\ln 2 - 37\pi^2/144$$
$$- 3767/1728 - (3/2)\varsigma(3)\} , \qquad (2.43)$$

the reduced-mass correction

$$S_{RM} = (\alpha/\pi)[(Z\alpha)^4/6]mc^2(-3m/M)\{\ln(Z\alpha)^{-2}$$
$$+ B(2,0) - B(2,1) + 23/60$$
$$+ 3\pi[1 + 11/128 - (1/2)\ln 2](Z\alpha)\} , \qquad (2.44)$$

the relativistic recoil correction

$$S_{RR} = (\alpha/\pi)[(Z\alpha)^4/6]mc^2(Zm/M)\{(1/4)\ln(Z\alpha)^{-2}$$
$$+ 2[B(2,0) - B(2,1)] + 97/12\} \qquad (2.45)$$

and the finite nuclear size correction

$$S_{NS} = [(Z\alpha)^2/12](Z\alpha R/\bar{\lambda}_e)^2 mc^2 . \qquad (2.46)$$

The nuclear size correction depends on the rms nuclear charge radius R which is determined by experiments on muonic atoms and by electron scattering experiments. In the case of hydrogen, only the latter information is presently available, and the nuclear size effect is based here on the value $R = 0.862(12)$ fm determined by Simon *et al.* [Simon 1980]. Theoretical contributions and the total are listed in Table 1. The total value $S = 1057.873(20)$ MHz is consistent with measurements of the Lamb shift. In $^4He^+$ the corresponding theoretical value is $S = 14042.05(55)$ MHz [Mohr 1976]. Experiments on the Lamb shift are discussed in section 4.2.

Table 1. Theoretical contributions to the Lamb shift in hydrogen.

Contribution	Order[mc^2]	Value[MHz]
Self-Energy	$\alpha(Z\alpha)^4 \ldots$	1085.812
Vacuum Polarization	$\alpha(Z\alpha)^4 \ldots$	−26.897
Fourth Order	$\alpha^2(Z\alpha)^4$	0.101
Reduced Mass[1]	$(m/M)\alpha(Z\alpha)^4 \ldots$	−1.647
Relativistic Recoil	$(m/M)(Z\alpha)^5$	0.359
Nuclear Size	$(Z\alpha)^4(R/\lambdabar_e)^2$	0.145
Total		1057.873(20)

[1] A further small contribution, -0.53kHz, arising from terms of order $(m/M)\alpha(Z\alpha)^5$, has been evaluated by Bhatt and Grotch [Bhatt 1987].

The Z-dependence and strong binding effects in radiative corrections are tested in high-Z hydrogenlike atoms. The development of ion sources has made it possible to study experimentally both the Lamb shift and QED effects in the $1S - 2P$ separations in highly-ionized atoms. (The work is described in Chapter 6 of this book.) The dominant contributions to the Lamb shift at high Z are the self-energy, the vacuum polarization, and the effect of the finite charge radius of the nucleus. Theoretical values for the self-energy and vacuum polarization follow from the numerical calculations described in sections 2.5 and 2.6. At intermediate and high Z, the simple non-relativistic expression for the finite nuclear size effect is not valid, but for intermediate Z values, the more general expression in equation (2.38) accurately represents the nuclear size effect. A summary of the theoretical contributions for hydrogenlike argon is given in Table 2. Figure 2.4 shows a comparison of theory and experiment for the Lamb shift over a range of Z.

The dominant contribution to the energy difference between states of different n is given by the Dirac hydrogenic formula. There is, in addition, the ordinary reduced-mass correction

$$E_M = [M/(M + m) - 1](E_n - mc^2) , \tag{2.47}$$

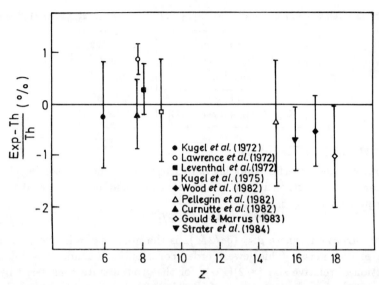

Fig. 2.4. Comparison of theory and experiment for the Lamb shift in one-electron ions. See Chapter 6 (E. Träbert) for the experimental references.

Table 2. Theoretical contributions to the Lamb shift for $Z=18$

Contribution	Order[mc^2]	Value[THz]
Self-Energy	$\alpha(Z\alpha)^4 \ldots$	40.546
Vacuum Polarization	$\alpha(Z\alpha)^4 \ldots$	−2.597
Fourth Order	$\alpha^2(Z\alpha)^4$	0.011
Relativistic Recoil	$(m/M)(Z\alpha)^5$	0.012
Nuclear Size	$(Z\alpha)^4(R/\lambdabar_e)^2$	0.276
Total		38.247(19)
Experimental [Gould 1983]		37.89(38)

the relativistic reduced-mass correction [Breit 1948, Bechert 1935]

$$E'_M = -[(Z\alpha)^4/(8n^4)](m/M)mc^2 , \qquad (2.48)$$

and the Lamb shift corrections. A summary of the theoretical contributions and a comparison with experiment for $Z = 18$ is shown in Table 3.

Theoretical energy levels in hydrogenlike atoms have been tabulated by Erickson [Erickson 1977], by Mohr [Mohr 1976, 1983], and by Johnson and Soff [Johnson 1985].

Table 3. Theoretical contributions to the $1S_{1/2}$ - $2P_{1/2}$ transition at $Z = 18$ [10^3 cm^{-1}]

Dirac Energy	26 772.502
Reduced Mass	-0.366
Lamb Shift	-9.240
Total Energy	26 762.896(5)
Experimental [Beyer 1985]	26 762.86(14)

2.11 Hyperfine Structure

The lowest order hyperfine splitting in hydrogen is obtained by solving the Dirac equation for an electron in a Coulomb field and adding an external magnetic dipole potential

$$\mathbf{A} = \boldsymbol{\mu} \times \mathbf{r}/r^3 \tag{2.49}$$

as a perturbation, where $\boldsymbol{\mu} = (ge/2M)\mathbf{I}$ is the nuclear dipole operator and M is the proton mass. This lowest order contribution includes the relativistic correction of relative size $(3/2)(Z\alpha)^2$ for the ground state considered by Breit [Breit 1930]. A correction for the finite size of the proton has been calculated by Zemach [Zemach 1956].

The leading radiative effects correspond to the lowest order self-energy and vacuum-polarization corrections. It is convenient to formulate the radiative corrections by considering hypothetical expressions, analogous to equations (2.23) and (2.24), in which the zeroth order Dirac equation contains both the Coulomb field and the hyperfine interaction field. In an expansion of such expressions in powers of the hyperfine field, the zeroth order term is independent of the hyperfine interaction and gives the ordinary Lamb shift correction. The term of first order in the hyperfine interaction strength is the dominant radiative correction to the hyperfine interaction. It consists of three parts, corresponding to first order corrections to the wave functions, to the Green's function, and to the bound state energy level. These expressions have been evaluated by systematically expanding in powers of $Z\alpha$. The leading term is just the radiative correction to the static magnetic moment of the electron. The next correction, of relative order $\alpha(Z\alpha)$ was calculated by Kroll and Pollock [Kroll 1952], by Karplus and Klein [Karplus 1952b], and by Sapirstein, Terray, and Yennie [Sapirstein 1984]. Higher order terms have been calculated by Layzer [Layzer 1961b, 1964], by Zwanziger [Zwanziger 1961, 1964], by Brodsky and Erickson [Brodsky 1966], and Sapirstein [Sapirstein 1983].

Leading recoil corrections to the ground state hyperfine splitting in hydrogen beyond the ordinary reduced mass correction, of relative order $(m/M)\alpha$, were calculated by Arnowitt [Arnowitt 1953], by Newcomb and Salpeter [Newcomb 1955], and by Grotch and Yennie [Grotch 1967]. In these calculations, it is

Table 4   Theoretical contributions to the hyperfine splitting in hydrogen. (The terms are written more explicitly in Chapter 4, equation (4.3)).

Contribution	Value[MHz]
E_F	1 418.840 8
$a_e E_F$	1.645 4
$(3/2)\alpha^2 E_F$	0.113 3
$O(\alpha^2)E_F$	-0.136 5
$O(\alpha^3)E_F$	-0.010 5(1)
$O(\alpha m/M)E_F$	-0.049 1(13)
δ_p	?
Total	1 420.403 4 MHz
Experimental	1 420.405 75 . . . MHz

necessary to take into account the finite size of the proton to avoid divergences in the result.

The experimental value of the hyperfine splitting is known accurately as [Hellwig 1970, Essen 1971]

$$E_{HF} = 1420.4057517667(10)\text{MHz} . \qquad (2.50)$$

A summary of the theoretical contributions is listed in Table 4. In that table, E_F is the Fermi contact splitting that includes a reduced-mass factor $[M/(m+M)]^3$. The term of order $\alpha(Z\alpha)^2 E_F$ has recently been calculated by Sapirstein with an improvement in accuracy over the original work of Brodsky and Erickson. The term δ_p represents the unknown proton polarizibility correction that depends on the internal structure of the proton. If the theoretical value of the hyperfine splitting with δ_p taken as a free parameter is set equal to the experimental value, then the solution for the fractional level shift due to the proton magnetic polarizability is $\delta_p = 1.6(9)$ ppm. An independent limit imposed by polarized electron-proton scattering data is $|\delta_p| < 4$ ppm [Hughes 1983].

2.12   Helium Energy Levels

Energy levels in heliumlike atoms have been measured and calculated with sufficiently high precision to be sensitive to radiative corrections. Despite much progress in understanding the theory and in the development of numerical methods, the comparison of theory and experiment in low-lying levels of helium is often limited by uncertainties in the theoretical results.

In neutral helium, comparison of theory and experiment for the fine structure of the $2P$ levels of helium (section 10.3.2) provides a test of atomic structure theory, or assuming that the theory is correct, provides a determination of the value of the fine structure constant α to a precision of about 1 ppm. The fine structure splittings are not sensitive to the radiative corrections. A theoretical program for a calculation of the splittings was outlined by Schwartz in 1964 [Schwartz 1964]. Derivation of the formalism from the Bethe-Salpeter equation and calculation of the fine structure splittings was carried out by Daley, Douglas, Hambro, and Kroll [Daley 1972]. The splittings have been calculated through order $\alpha^6 mc^2$ and $(m/M)\alpha^4 mc^2$. Additional numerical work has been done by Lewis and Serafino [Lewis 1978].

Theory and experiment for S states in helium are sufficiently accurate that radiative corrections are significant. For example the ground state $1s^2 \; ^1S_0$ radiative correction, derived from the experimental transition energy by calculating and subtracting off the non-radiative contributions, is [Martin 1984] $\Delta_L = -1.26(15)$ cm^{-1}, which can be compared to the calculated value [Ermolaev 1985] $\Delta_L = -1.381(23)$cm^{-1}.

Studies of high-Z two-electron ions provide an important test of relativistic bound state quantum electrodynamics. In neutral helium, the relative size of the radiative corrections to the total energy of the $2^3S_1 - 2^3P_0$ separation is 0.002%, while at higher Z the relative size of the radiative correction grows rapidly with Z, and at $Z = 20$ the ratio is about 1%. Hence a relatively imprecise determination of the energy splitting is sensitive to radiative corrections. The theory simplifies to some extent at high Z. The electrons are well approximated as non-interacting Dirac electrons in lowest order. The electron-electron interaction and radiative corrections can be calculated in perturbation theory by applying the formalism described in section 2.4. The uncertainties in the theory presently limit the precision of the comparison of theory and experiment, and corrections that involve relativistic electron interactions and radiative corrections simultaneously need to be calculated in order to improve the theory [Mohr 1985a, b].

REFERENCES

Aitchison 1985: I. J. R. Aitchison, *Contemp. Phys.* **26** (1985) 333.

Appelquist 1970a: T. Appelquist and S. J. Brodsky, *Phys. Rev. Lett.* **24** (1970) 562.

Appelquist 1970b: T. Appelquist and S. J. Brodsky, *Phys. Rev. A* **2** (1970) 2293.

Arnowitt 1953: R. Arnowitt, *Phys. Rev.* **92** (1953) 1002.

Bailey 1977: J. Bailey, K. Borer, F. Combley, H. Drumm, F. J. M. Farley, J. H. Field, W. Flegel, P. M. Hattersley, F. Krienen, F. Lange, E. Picasso and W. von Rüden, *Phys. Lett.* **68B** (1977) 191.

Baranger 1951: M. Baranger, *Phys. Rev.* **84** (1951) 866.

Baranger 1952: M. Baranger, F. J. Dyson, and E. E. Salpeter, *Phys. Rev.* **88** (1952) 680.

Baranger 1953: M. Baranger, H. A. Bethe, and R. P. Feynman, *Phys. Rev.* **92** (1953) 482.

Barbieri 1970: R. Barbieri, J. A. Mignaco, and E. Remiddi, *Lett. Nuovo Cimento* **3** (1970) 588.

Bechert 1935: K. Bechert and J. Meixner, *Ann. Phys.* **22** (1935) 525.

Bethe 1947: H. A. Bethe, *Phys. Rev.* **72** (1947) 339.

Beyer 1985: H. F. Beyer, R. D. Deslattes, F. Folkmann, and R. E. LaVilla, *J. Phys. B* **18** (1985) 207.

Bhatt 1985: G. Bhatt and H. Grotch, *Phys. Rev. A* **31** (1985) 2794.

Bhatt 1987: G. Bhatt and H. Grotch, *Phys. Rev. Lett.* (1987) 2 Feb.

Blomqvist 1972: J. Blomqvist, *Nucl. Phys.* **B48** (1972) 95.

Bodwin 1985: G. T. Bodwin, D. R. Yennie, and M. A. Gregorio, *Rev. Mod. Phys.* **57** (1985) 723.

Borie 1981: E. Borie, *Phys. Rev. Lett.* **47** (1981) 568.

Borie 1982: E. Borie and G. A. Rinker, *Rev. Mod. Phys.* **54** (1982) 67.

Breit 1929: G. Breit, *Phys. Rev.* **34** (1929) 553.

Breit 1930: G. Breit, *Phys. Rev.* **35** (1930) 1447.

Breit 1947: G. Breit and R. E. Meyerott, *Phys. Rev.* **72** (1947) 1023.

Breit 1948: G. Breit and G. E. Brown, *Phys. Rev.* **74** (1948) 1278.

Brodsky 1966: S. J. Brodsky and G. W. Erickson, *Phys. Rev.* **148** (1966) 26.

Brown 1959: G. E. Brown, J. S. Langer, and G. W. Schaefer, *Proc. Roy. Soc. (London)* **A251** (1959) 92.

Brown 1974: L. S. Brown, R. N. Cahn, and L. D. McLerran, *Phys. Rev. Lett.* **33** (1974) 1591.

Brown 1975: L. S. Brown, R. N. Cahn, and L. D. McLerran, *Phys. Rev.* **D12** (1975) 581, 596.

Cheng 1976: K. T. Cheng and W. R. Johnson, *Phys. Rev.* **A14** (1976) 1943.

Cohen-Tannoudji 1986: C. Cohen-Tannoudji, *Physica Scripta* **T12** (1986) 19.

Daley 1972: J. Daley, M. Douglas, L. Hambro, and N. M. Kroll, *Phys. Rev. Lett.* **29** (1972) 12.

Desiderio 1971: A. M. Desiderio and W. R. Johnson, *Phys. Rev.* **A3** (1971) 1267.

Dupont-Roc 1978: J. Dupont-Roc, C. Fabre and C. Cohen-Tannoudji, *J. Phys. B* **11** (1978) 563.

Erickson 1965: G. W. Erickson and D. R. Yennie, *Ann. Phys.* (N. Y.) **35** (1965) 271, 447.

Erickson 1971: G. W. Erickson, *Phys. Rev. Lett.* **27** (1971) 780.

Erickson 1977: G. W. Erickson, *J. Phys. Chem. Ref. Data* **6** (1977) 831.

Ermolaev 1985: A. M. Ermolaev, in *Atomic Theory Workshop on Relativistic and QED Effects in Heavy Atoms,* eds. H. P. Kelly and Y. -K Kim, AIP Conference Proceedings (AIP, New York, 1985).

Essen 1971: L. Essen, R. W. Donaldson, M. J. Bangham, and E. G. Hope, *Nature* **229** (1971) 110.

Farley 1979: F. J. M. Farley and E. Picasso, *Ann. Rev. Nucl. Part. Sci.* **29** (1979) 243.

Feynman 1948: R. P. Feynman, *Phys. Rev.* **74** (1948) 1430.

Feynman 1949: R. P. Feynman, *Phys. Rev.* **76** (1949) 769.

Foley 1948: H. M. Foley and P. Kusch, *Phys. Rev.* **73** (1948) 412.

Fox 1973: J. A. Fox and D. R. Yennie, *Ann. Phys. (N.Y.)* **81** (1973) 438.

French 1949: J. B. French and V. F. Weisskopf, *Phys. Rev.* **75** (1949) 1240.

Friar 1979: J. L. Friar, *Ann. Phys. (N.Y.)* **122** (1979) 151.

Fried 1960: H. M. Fried and D. R. Yennie, *Phys. Rev. Lett.* **4** (1960) 583.

Fukuda 1949: H. Fukuda, Y. Miyamoto and S. Tomonaga *Prog. Theor. Phys.* (Kyoto) **4** (1949) 47, 121.

Furry 1951: W. H. Furry, *Phys. Rev.* **81** (1951) 115.

Gell-Mann 1951: M. Gell-Mann and F. Low, *Phys. Rev.* **84** (1951) 350.

Gould 1983: H. Gould and R. Marrus, *Phys. Rev.* **A28** (1983) 2001.

Grotch 1967: H. Grotch and D. R. Yennie, *Z. Phys.* **202** (1967) 425.

Grotch 1969: H. Grotch and D. R. Yennie, *Rev. Mod. Phys.* **41** (1969) 350.

Grotch 1977: H. Grotch and E. Kazes, *Am. J. Phys.* **45** (1977) 618.

Gyulassy 1974: M. Gyulassy, *Phys. Rev. Lett.* **33** (1974) 921.

Gyulassy 1975: M. Gyulassy, *Nucl. Phys.* **A244** (1975) 497.

Hellwig 1970: H. Hellwig, R. F. C. Vessot, M. W. Levin, P. W. Zitzewitz, D. W. Allan and D. J. Glaze, *IEEE Trans. Instrum. Meas.* **IM19** (1970) 200.

Hillery 1980: M. Hillery and P. J. Mohr, *Phys. Rev* **A21** (1980) 24.

Hughes 1983: V. W. Hughes and J. Kuti, *Ann. Rev. Nucl. Part. Sci.* **33** (1983) 611.

Hylton 1984: D. J. Hylton and P. J. Mohr, *Satellite Workshop and Conference Abstracts,* Ninth International Conference on Atomic Physics, Seattle, 1984.

Hylton 1985: D. J. Hylton, *Phys. Rev.* **A32** (1985) 1303.

Johnson 1985: W. R. Johnson and G. Soff, *At. Data and Nucl. Data Tables* **33** (1985) 405.

Källén 1955: G. Källén and A. Sabry, *Dan. Mat. Fys. Medd.* **29** No. 17 (1955).

Karplus 1950: R. Karplus and N. M. Kroll, *Phys. Rev.* **77** (1950) 536.

Karplus 1951: R. Karplus, A. Klein, and J. Schwinger, *Phys. Rev.* **84** (1951) 597.

Karplus 1952a: R. Karplus, A. Klein, and J. Schwinger, *Phys. Rev.* **86** (1952) 288.

Karplus 1952b: R. Karplus and A. Klein, *Phys. Rev.* **85** (1952) 972.

Kinoshita 1981: T. Kinoshita and W. B. Lindquist, *Phys. Rev. Lett.* **47** (1981) 1573.

Kinoshita 1984a: T. Kinoshita and J. Sapirstein, in *Atomic Physics 9*, eds. R. S. Van Dyck, Jr. and E. N. Fortson (World Scientific, Singapore, 1984).

Kinoshita 1984b: T. Kinoshita, B. Nižić, and Y. Okamoto , *Phys. Rev. Lett.* **52** (1984) 717.

Kroll 1949: N. M. Kroll and W. E. Lamb, Jr., *Phys. Rev.* **75** (1949) 388.

Kroll 1952: N. M. Kroll and F. Pollock, *Phys. Rev.* **86** (1952) 876.

Lamb 1947: W. E. Lamb, Jr. and R. C. Retherford, *Phys. Rev.* **72** (1947) 241.

Lautrup 1970: B. E. Lautrup, A. Peterman and E. de Rafael, *Phys. Lett.* **B31** (1970) 577.

Layzer 1960: A. J. Layzer, *Phys. Rev. Lett.* **4** (1960) 580.

Layzer 1961a: A. J. Layzer, *J. Math. Phys.* **2** (1961) 292, 308.

Layzer 1961b: A. J. Layzer, *Bull. Am. Phys. Soc.* **6** (1961) 514.

Layzer 1964: A. J. Layzer, *Nuovo Cimento* **33** (1964) 1538.

Lepage 1981: G. P. Lepage, D. R. Yennie and G. W. Erickson, *Phys. Rev. Lett.* **47** (1981) 1640.

Lewis 1978: M. L. Lewis and P. H. Serafino, *Phys. Rev.* **A18** (1978) 867.

Low 1952: F. Low, *Phys. Rev.* **88** (1952) 53.

Martin 1984: W. C. Martin, *Phys. Rev.* **A29** (1984) 1883; **A30** (1984) 651.

Mills 1955: R. L. Mills and N. M. Kroll, *Phys. Rev.* **98** (1955) 1489.

Mohr 1974: P. J. Mohr, *Ann. Phys.* (N.Y.) **88** (1974) 26, 52.

Mohr 1975: P. J. Mohr, *Phys. Rev. Lett.* **34** (1975) 1050.

Mohr 1976: P. J. Mohr, in *Beam-Foil Spectroscopy*, eds I. A. Sellin and D. J. Pegg (Plenum, New York, 1976), Vol. 1, p. 89.

Mohr 1982: P. J. Mohr, *Phys. Rev.* **A26** (1982) 2338.

Mohr 1983: P. J. Mohr, *At. Data Nucl. Data Tables* **29** (1983) 453.

Mohr 1985a: P J. Mohr, *Nucl. Instrum. Methods Phys. Res.* **B9** (1985) 459.

Mohr 1985b: P. J. Mohr. *Phys. Rev.* **A32** (1985) 1949.

Neghabian 1983: A. R. Neghabian, *Phys. Rev.* **A27** (1983) 2311.

Newcomb 1955: W. A. Newcomb and E. E. Salpeter, *Phys. Rev.* **97** (1955) 1146.

Pegg 1986: D. T. Pegg, *Physica Scripta* **T12** (1986)14.

Peterman 1957a: A. Petermann, *Helv. Phys. Acta* **30** (1957) 407.

Peterman 1957b: A. Petermann, *Nucl. Phys.* **3** (1957) 689.

Peterman 1971: A. Petermann, *Phys. Lett.* **B35** (1971) 325.

Salpeter 1951: E. E. Salpeter and H. A. Bethe, *Phys. Rev.* **84** (1951) 1232.

Salpeter 1952: E. E. Salpeter, *Phys. Rev.* **87** (1952) 328.

Sapirstein 1981: J. Sapirstein, *Phys. Rev. Lett.* **47** (1981) 1723.

Sapirstein 1983: J. Sapirstein, *Phys. Rev. Lett.* **51** (1983) 985.

Sapirstein 1984: J. Sapirstein, E. A. Terray and D. R. Yennie, *Phys. Rev.* **D29** (1984) 2290.

Schwartz 1964: C. Schwartz, *Phys. Rev.* **134** (1964) A1181.

Schweber 1961: S. S. Schweber, *An Introduction to Relativistic Quantum Field Theory* (Harper and Row, New York, 1961).

Schwinger 1948: J. Schwinger, *Phys. Rev.* **73** (1948) 416.

Schwinger 1949: J. Schwinger, *Phys. Rev.* **75** (1949) 898.

Schwinger 1951a: J. Schwinger, *Phys. Rev.* **82** (1951) 664.

Schwinger 1951b: J. Schwinger, *Proc. Nat. Acad. Sci. (U.S.A.)* **37** (1951) 452, 455.

Serber 1935: R. Serber, *Phys. Rev.* **48** (1935) 49.

Series 1986: G. W. Series, *Physica Scripta* **T12** (1986) 5.

Simon 1980: G. G. Simon, Ch. Schmitt, F Borkowski and V. W. Walther, *Nucl. Phys.* **A333** (1980) 381.

Soff 1982: G. Soff, P. Schlüter, B. Müller and W. Greiner, *Phys. Rev. Lett.* **48** (1982) 1465.

Sommerfield 1957: C. M. Sommerfield, *Phys. Rev.* **107** (1957) 328.

Soto 1966: M. F. Soto, Jr., *Phys. Rev. Lett.* **17** (1966) 1153.

Soto 1970: M. F. Soto, Jr., *Phys. Rev.* **A2** (1970) 734.

Sucher 1957: J. Sucher, *Phys. Rev.* **107** (1957) 1448.

Taylor 1985: B. N. Taylor, *J. Res. Nat. Bur. Stand.* **90** (1985) 91.

Uehling 1935: E. A. Uehling, *Phys. Rev.* **48** (1935) 55.

Van Dyck 1984: R. S. Van Dyck Jr., P. B. Schwinberg and H. G. Dehmelt, in *Atomic Physics 9*, eds. R. S. Van Dyck, Jr. and E. N. Fortson (World Scientific, Singapore, 1984.)

Weisskopf 1930: V. Weisskopf and E. Wigner, *Z. Phys.* **63** (1930) 54.

Weneser 1953: J. Weneser, R. Bersohn and N. M. Kroll, *Phys. Rev.* **91** (1953) 1257.

Wheeler 1945: J. A. Wheeler and R. P. Feynman, *Rev. Mod. Phys.* **17** (1945) 157.

Wheeler 1949: J. A. Wheeler and R. P. Feynman, *Rev. Mod. Phys.* **21** (1949) 425.

Wichmann 1954: E. H. Wichmann and N. M. Kroll, *Phys. Rev.* **96** (1954) 232; **101** (1956) 843.

Zemach 1956: A. C. Zemach, *Phys. Rev.* **104** (1956) 1771.

Zwanziger 1961: D. E. Zwanziger, *Bull. Am. Phys. Soc.* **6** (1961) 514.

Zwanziger 1964: D. E. Zwanziger, *Nuovo Cimento* **34** (1964) 77.

CHAPTER 3

THEORY OF TRANSITIONS, AND THE ELECTROWEAK INTERACTION

G. W. F. Drake

3.1 Introduction

A wide variety of experiments on atomic radiation and the interaction of atoms with external fields is made possible by the fact that the $2s_{1/2}$ state of hydrogen and hydrogenic ions is metastable. It has long been known that the $2s_{1/2}$ state is rapidly quenched to the ground state by the application of a modest electric field with the emission of $Ly - \alpha$ photons. However, as shown in Fig. 3.1, a careful study of the quenching radiation reveals a rich diversity of interference effects and quantum beat phenomena. High precision measurements of these effects provide a unique opportunity to test the theory of the radiation process in one-electron ions where accurate theoretical predictions are easily made, and to look for exotic effects such as parity non-conservation.

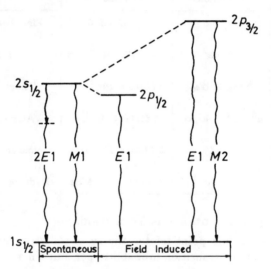

Fig. 3.1. Illustration of the spontaneous and field-induced radiative decay modes from the $2s_{1/2}$ state. The dashed lines indicate mixing with the $2p_{1/2}$ and $2p_{3/2}$ states by an external electric field, leading to field-induced $E1$ (electric dipole) and $M2$ (magnetic quadrupole) decay modes for the $2s_{1/2}$ state. Cross terms among all four single-photon decay modes produce quantum beats and inteference phenomena.

The aims of this chapter are first to review in section 3.2 the theory of spontaneous radiation from the $2s$ state of hydrogen, and then to present in section 3.3 the theory of angular and polarization-dependent asymmetries in the electric field quenching radiation. Comparison with experiment yields high precision values for the Lamb shift, electric dipole and magnetic dipole transition probabilities. Next, in section 3.4, the quantum beats in the quenching radiation and the influence of time-dependent quenching fields are discussed. Section 3.5

deals with two-photon transitions and their relationship to quenching by a static electric field. Finally, section 3.6 presents a primarily phenomenological discussion of the parity nonconserving effects to be expected in atomic physics due to the presence of weak neutral currents in the electron-nucleon interaction. This section is intended for the reader who has some familiarity with the basic ideas of quantum electrodynamics, but does not have an extensive background in elementary particle physics and the theory of beta decay.

The definitions of various physical constants used in the text and their values are summarized in Table 1.

Table 1. Definitions of Physical Constants

Symbol	Definition	Value	Name
α	$e^2/\hbar c$	$1/137.03596$	Fine structure constant
a_0	$\hbar^2/me^2$	0.529177×10^{-8}cm	Bohr radius
$\lambdabar$	$\hbar/mc \equiv \alpha a_0$	3.86159×10^{-11}cm	Compton wavelength
τ	$\hbar^3/me^4 \equiv a_0/\alpha c$	2.41888×10^{-17}s	Atomic unit of time
ε_0	e/a_0^2	5.14225×10^9V/cm	Atomic unit of field strength
Ry	$\alpha c/(4\pi a_0)$	3.289842×10^9MHz	Rydberg unit of frequency

3.2 Theory of Spontaneous Transitions

A convenient starting point for the theory of radiative transitions is Fermi's Golden Rule for the transition probability per unit time

$$w = (2\pi/\hbar)|\langle f|V_{\text{int}}|i\rangle|^2 \rho_f \tag{3.1}$$

where V_{int} is an interaction energy operator and ρ_f is the number of final states per unit energy interval. For the emission of a photon of frequency ω, polarization $\hat{e}$ and propagation vector $\mathbf{k}$ ($|\mathbf{k}| = \omega/c$), the terms in (3.1) are

$$\rho_f = \mathcal{V}k^2 d\Omega/(2\pi)^3\hbar c \tag{3.2}$$

and

$$V_{\text{int}} = e\boldsymbol{\alpha}\cdot\mathbf{A}^* \tag{3.3}$$

where ρ_f is the number of photon states of polarization $\hat{e}$ per unit energy and solid angle in the arbitrary normalization volume $\mathcal{V}$, and the photon vector potential, normalized to a field energy of $\hbar\omega$ per unit volume is given by

$$\mathbf{A} = \frac{1}{k}\left(\frac{2\pi\hbar\omega}{\mathcal{V}}\right)^{1/2}\widehat{e}e^{i\mathbf{k}\cdot\mathbf{r}} . \tag{3.4}$$

The wave functions in (3.1) are assumed to be four-component Dirac spinors and $\boldsymbol{\alpha}$ is the usual 4×4 Dirac matrix. Collecting terms, (3.1) reduces to

$$w d\Omega = \left(\frac{e^2 k}{2\pi\hbar}\right)|f|\boldsymbol{\alpha}\cdot\widehat{e}e^{-i\mathbf{k}\cdot\mathbf{r}}|i\rangle|^2 d\Omega \tag{3.5}$$

per unit time. In the non-relativistic limit, $\boldsymbol{\alpha} \rightarrow \mathbf{p}/mc$, $\exp(-i\mathbf{k}\cdot\mathbf{r}) \simeq 1$ and (3.5) becomes the familiar dipole velocity form for the transition rate (see, for example, Bethe and Salpeter, p. 249).

When (3.5) is applied to the $2s_{1/2}$ state of hydrogen, one finds that electric dipole $(E1)$ transitions to the $1s_{1/2}$ ground state are strictly forbidden by the parity selection rule, but spontaneous magnetic dipole $(M1)$ transitions are allowed when relativistic and retardation corrections are taken into account. However, the dominant decay mode in the absence of external fields is two-photon electric dipole $(2E1)$ transitions to the ground state. The $2E1$ process, which involves a second order interaction between the atom and the radiation field, is discussed separately in section 3.5.

In the presence of a small external electric field, both $E1$ and $M2$ decay processes become possible owing to the field-induced mixing of the $2s_{1/2}$ state with states of opposite parity – primarily $2p_{1/2}$ and $2p_{3/2}$. To study both spontaneous and field induced single photon decay processes, we begin by expanding the plane wave vector potential into transverse electric $(\lambda = 1)$ and magnetic $(\lambda = 0)$ multipoles (see, for example, Akhiezer and Berestetskii, 1965) according to

$$\widehat{e}e^{-i\mathbf{k}\cdot\mathbf{r}} = \sum_{LM\lambda}[\widehat{e}\cdot\mathbf{Y}_{LM}^{(\lambda)}(\widehat{k})]a_{LM}^{(\lambda)*}(\mathbf{r}) . \tag{3.6}$$

In general, there is also a longitudinal $(\lambda = -1)$ component, but this does not contribute in the transverse or Coulomb gauge in which one imposes the transversality condition $\mathbf{k}\cdot\widehat{e} = 0$. The factors $\widehat{e}\cdot\mathbf{Y}_{LM}^{(\lambda)}(\widehat{k})$ can next be written in terms of vector spherical harmonics (see, for example, Brink and Satchler, 1968), to obtain

$$\widehat{e}\cdot\mathbf{Y}_{LM}^{(1)}(\widehat{k}) = [L(L+1)]^{-1/2}k\widehat{e}\cdot\nabla_k Y_L^M(\widehat{k}) ,$$
$$\widehat{e}\cdot\mathbf{Y}_{LM}^{(0)}(\widehat{k}) = i[L(L+1)]^{-1/2}\mathbf{k}\times\widehat{e}\cdot\nabla_k Y_L^M(\widehat{k}) .$$

Substituting the above into (3.6) and retaining only the $E1, M1$ and $M2$ contributions then yields (van Wijngaarden and Drake, 1982)

$$\widehat{e}e^{-i\mathbf{k}\cdot\mathbf{r}} = (3/8\pi)^{1/2}\sum_M \{e_M \mathbf{a}_{1,M}^{(1)*} + i[\widehat{k}\times\widehat{e}]_M \mathbf{a}_{1,M}^{(0)*}$$

$$+ i(10/3)^{1/2}[\widehat{k},\widehat{k}\times\widehat{e}]_{2,M}\mathbf{a}_{2,M}^{(0)*}\} \qquad (3.7)$$

where the e_M denote the irreducible tensor components of the polarization vector

$$e_{\pm 1} = \mp\frac{1}{\sqrt{2}}(e_x \pm ie_y) , \quad e_0 = e_z \qquad (3.8)$$

and the notation $[a,b]_{2,M}$ denotes the vector-coupled product

$$[a,b]_{2,M} = \sum_{m_1,m_2} \langle 1m_1 1m_2|2M\rangle a_{m_1}b_{m_2} .$$

The $\mathbf{a}_{LM}^{(\lambda)}$ in (3.6) and (3.7) are the standard operators for electric and magnetic multipole transitions given by (in the Coulomb gauge)

$$\mathbf{a}_{LM}^{(1)} = \left(\frac{L}{2L+1}\right)^{1/2} g_{L+1}(kr)\mathbf{Y}_{LL+1M}$$

$$+ \left(\frac{L+1}{2L+1}\right)^{1/2} g_{L-1}(kr)\mathbf{Y}_{LL-1M} \qquad (3.9)$$

and

$$\mathbf{a}_{LM}^{(0)} = g_L(kr)\mathbf{Y}_{LLM} . \qquad (3.10)$$

Here, $\mathbf{Y}_{LJM}$ is a vector spherical harmonic as defined by Edmonds (1960) and the radial function $g_L(kr)$ is given by

$$g_L(kr) = 4\pi i^L j_L(kr) \qquad (3.11)$$

where $j_L(kr)$ is a spherical Bessel function with power series expansion

$$j_L(z) = \frac{z^L}{(2L+1)!!}\left\{1 - \frac{z^2/2}{1!(2L+3)} + \frac{(z^2/2)^2}{2!(2L+3)(2L+5)} + \ldots\right\} . \qquad (3.12)$$

The notation $(2L+1)!!$ means $1\times 3\times 5\times\ldots\times(2L+1)$. Since $kr = \omega r/c$, only the leading one or two terms of (3.12) normally need to be retained for low Z

atoms (the long wavelength approximation). In the non-relativistic limit, the operators $\boldsymbol{\alpha} \cdot \mathbf{a}_{LM}^{(\lambda)}$ are equivalent to

$$e\boldsymbol{\alpha} \cdot \mathbf{a}_{1M}^{(1)} \simeq \sqrt{2}e\Phi_{1M} \qquad (3.13)$$

$$e\boldsymbol{\alpha} \cdot \mathbf{a}_{1M}^{(0)} \simeq i(\nabla\Phi_{1M}) \cdot \left[\frac{e\mathbf{L}}{\sqrt{2}mc} + \sqrt{2}\boldsymbol{\mu}\right] \qquad (3.14)$$

$$e\boldsymbol{\alpha} \cdot \mathbf{a}_{2M}^{(0)} \simeq i(\nabla\Phi_{2M}) \cdot \left[\frac{e\mathbf{L}}{\sqrt{6}mc} + \sqrt{3}\boldsymbol{\mu}/\sqrt{2}\right] \qquad (3.15)$$

where $\Phi_{LM} = g_L(kr)Y_L^M(\hat{r}), \mathbf{L} = \mathbf{r} \times \mathbf{p}, \boldsymbol{\mu} = (e\lambdabar/2)\boldsymbol{\sigma}$ is the magnetic moment operator, $\lambdabar$ is the compton wavelength defined in Table 1, and the components of $\boldsymbol{\alpha}$ are the Pauli spin matrices. (See Grant, 1974, for an extensive discussion of the reduction to equivalent nonrelativistic operators.)

In the absence of external fields, only the $M1$ term $\mathbf{a}_{1M}^{(0)}$ of (3.6) contributes to the $2s_{1/2} \rightarrow 1s_{1/2}$ transition. But even this term vanishes in the nonrelativistic long wavelength approximation because, using (3.14), together with

$$i\nabla\Phi_{1M} \simeq -(4\pi/3)^{1/2}k\hat{e}_M \ , \qquad (3.16)$$

all that remains in the expression (3.5) for $wd\Omega$ is the vanishing overlap integral. The leading non-vanishing contributions come from relativistic corrections to the wave functions of order $(\alpha Z)^2$ and finite wavelength corrections (the second term of (3.12)) of order $(\omega r/c)^2$. These terms have been evaluated by Drake (1971), and Feinberg and Sucher (1971) with the result that the spin-dependent part of the equivalent non-relativistic operator becomes

$$e\boldsymbol{\alpha} \cdot \mathbf{a}_{1M}^{(0)} \simeq -(8\pi/3)^{1/2}kM_{1M} \qquad (3.17)$$

where

$$M_{1M} = \mu_M \left[1 - \frac{2p^2}{3m^2c^2} - \frac{1}{6}\left(\frac{\omega r}{c}\right)^2 + \frac{Ze^2}{3mc^2r}\right] \qquad (3.18)$$

is the effective magnetic moment transition operator acting on non-relativistic wave functions. The matrix elements for the $2s_{1/2,1/2} \rightarrow 1s_{1/2,\pm 1/2}$ transition are easily evaluated with the result

$$\langle 1s_{1/2,1/2}|M_{1,0}^*|2s_{1/2,1/2}\rangle = -(8\alpha^2 Z^2/81\sqrt{2})e\lambdabar \qquad (3.19)$$

and

$$\langle 1s_{1/2,-1/2}|M^*_{1,1}|2s_{1/2,1/2}\rangle = (8\alpha^2 Z^2/81)e\lambdabar . \tag{3.20}$$

Using the above, together with (3.17) and (3.7), the expression (3.5) for the decay rate summed over the $1s_{1/2,\pm 1/2}$ final atomic states becomes

$$wd\Omega = (k^3/2\pi\hbar)|M|^2\{|[\hat{k}\times\hat{e}]_0|^2 + 2|[\hat{k}\times\hat{e}]_1|^2\}d\Omega , \tag{3.21}$$

where $M = -(8\alpha^2 Z^2/81\sqrt{2})e\lambdabar$. This must still be summed over two linearly independent polarization vectors perpendicular to $\hat{k}$ and integrated over solid angles $d\Omega = \sin\theta d\theta d\phi$. Using

$$\sum_{\hat{e}} |[\hat{k}\times\hat{e}]_0|^2 = \sin^2\theta \tag{3.22}$$

and

$$2\sum_{\hat{e}} |[\hat{k}\times\hat{e}]_1|^2 = 1 + \cos^2\theta \tag{3.23}$$

it is apparent that the two terms in (3.21) combine to give a spherically symmetric radiation pattern, even though the initial state is assumed to be completely spin-polarized in the $2s_{1/2,1/2}$ magnetic sub-level. This rotational invariance is a general feature of all $j = 1/2 \to 1/2$ radiative transitions. The angular integration therefore just contributes an additional factor of 4π. With $k = (3/8)Z^2\alpha/a_0$, the final result is

$$w(2s_{1/2} \to 1s_{1/2}) = 4k^3|M|^2/\hbar$$
$$= (\alpha^9 Z^{10}/972)\tau^{-1} . \tag{3.24}$$

For H, equation (3.24) gives an $M1$ decay rate of only 2.496×10^{-6} s^{-1}, which is much less than the $2E1$ decay rate discussed in section 3.5. However, the process is still important for the following reasons:

(a) Since the process increases in proportion to Z^{10}, it eventually becomes the dominant radiative decay mechanism for $Z > 43$. A measurement of the total $2s_{1/2}$ decay rate in one-electron K^{35+} by Gould and Marrus (1983) is sensitive to the $M1$ contribution, and further experiments are in progress on U^{91+} (Gould *et al.* 1984). The decay rates of heavy hydrogenic ions are further discussed in section 3.5.

(b) For all helium-like ions, the $M1$ process is the dominant radiative decay mechanism for the $1s2s\,^3S_1$ metastable state. (For a review, see Sucher, 1977.)

(c) Even for light hydrogen-like ions, the $M1$ process produces interference effects which lead to observable angular asymmetries in the electric field quenching radiation as discussed in the following section.

Possible electron self-energy and vacuum polarization corrections to the basic $M1$ decay process have been studied by Lin and Feinberg (1974), Drake (1974), and Barbieri and Sucher (1978). Although individual Feynman diagrams give contributions of relative order $\alpha\ln\alpha$ and α, they exactly sum to zero for $nS \rightarrow mS$ type transitions.

3.3 Theory of Quenching Radiation Assymmetries

3.3.1 Basic formalism

The conventional method for describing the electric field quenching of the metastable $2s_{1/2}$ state of hydrogen-like systems is based on the phenomenological Bethe-Lamb quenching theory (Lamb and Retherford, 1950, 1952), which is, in turn, derived from the Wigner-Weisskopf (1930) analysis for time-dependent perturbations. In this approach, one starts from the time-dependent Schrödinger equation for an atom in an external field

$$i\hbar\frac{d\mathbf{a}}{dt} = \underline{H}(t)\mathbf{a} \tag{3.25}$$

$$\underline{H}(t) = \underline{E} + F(t)\underline{V} \tag{3.26}$$

where $\mathbf{a}$ is a column vector of state amplitudes, $\underline{E}$ is the diagonal matrix of field-free eigenvalues, $\underline{V}$ is the interaction matrix with the external field and $F(t)$ describes its time dependence. Usually, only a finite basis set of field-free states need be considered. The basic idea of phenomenological quenching theory is to replace the field-free eigenvalues E_j by $E_j - i\Gamma_j/2$ where the Γ_j are the field-free level widths. Then, in the absence of perturbations, the state amplitudes decay independently of one another according to

$$|a_j(t)|^2 = |a_j(0)|^2 e^{-\Gamma_j t/\hbar} \tag{3.27}$$

as expected. However, the solutions to (3.25) lead to much more complex decay patterns when external fields are present. Calculations have been done by many authors within this framework to describe a wide variety of different situations, as further discussed in this and subsequent sections. Recently Kelsey and Macek (1977) and Hillery and Mohr (1980) have shown from quantum electrodynamics that the Bethe-Lamb phenomenological formalism has a rigorous foundation at least to lowest relative order in α/π. The key point is that it is consistent to use relativistic wave functions for the evaluation of matrix elements, together with relativistic eigenvalues augmented by the Lamb shift and imaginary level width.

The present aim is to study asymmetries in the quenching radiation from the $2s_{1/2}$ state in the presence of a constant electric field, starting from equation (3.7) for the multipole expansion of the photon vector potential. As pointed out already, the first and last terms of (3.7) (i.e. the $E1$ and $M2$ terms) do not contribute to the $2s_{1/2} \rightarrow 1s_{1/2}$ radiative transition because of parity and (in the case of $M2$) triangular selection rules. However, in the presence of an external electric field, both terms contribute due to the field-induced mixing of s and p states. For atoms or ions such as H and He$^+$ in fields up to several kV/cm, the only significant mixing is among the manifold of states $2s_{1/2}$, $2p_{1/2}$ and $2p_{3/2}$. Since the presence of additional hyperfine structure greatly complicates the analysis, we will treat hyperfine structure separately in section 3.2 and assume for the present that it is absent. Also, we will consider time dependent switching effects in section 3.4. Here, we assume that the electric field $\mathbf{E}$ is switched on adiabatically and with the above assumptions write the perturbed $2s_{1/2}$ initial state in the form

$$\psi(2s_{1/2}, m) = a(|\mathbf{E}|)\psi_0(2s_{1/2}, m)$$

$$+ \sum_{m'}[b^{(1/2)}_{m,m'}\psi_0(2p_{1/2}, m') + b^{(3/2)}_{m,m'}\psi_0(2p_{3/2}, m')] \tag{3.28}$$

where the matrices $\underline{b}^{(j)} (j = 1/2, 3/2)$ are given by

$$\underline{b}^{(1/2)} = b_{1/2}(|\mathbf{E}|)\boldsymbol{\sigma} \cdot \mathbf{E} \tag{3.29}$$

$$\underline{b}^{(3/2)} = b_{3/2}(|\mathbf{E}|)\begin{pmatrix} -\sqrt{3}\widehat{E}_{-1} & \sqrt{2}\widehat{E}_0 & -\widehat{E}_1 & 0 \\ 0 & -\widehat{E}_{-1} & \sqrt{2}\widehat{E}_0 & -\sqrt{3}\widehat{E}_1 \end{pmatrix} . \tag{3.30}$$

The components of $\boldsymbol{\sigma}$ in (3.29) are the Pauli spin matrices and the $\widehat{E}_q(q = 0, \pm 1)$ are the irreducible tensor components of the unit vector $\widehat{E}$ in the electric field direction, as defined by (3.8). The coefficients of $\widehat{E}_q$ in (3.29) and (3.30) are obtained from the Wigner-Eckart theorem (Edmonds, 1960)

$$\langle 2p_{j,m'}|\widehat{E}_q r_q^*|2s_{1/2,m}\rangle = (-1)^{j-m} \begin{pmatrix} j & 1 & 1/2 \\ -m' & -q & m \end{pmatrix}$$

$$\times \widehat{E}_q \langle 2p_j\|\mathbf{r}\|2s_{1/2}\rangle \ . \tag{3.31}$$

The (1,1) element of (3.30) corresponds to $m = 1/2, m' = 3/2$, and the rest are numbered down the columns and across the rows to $m = -1/2, m' = -3/2$ in the (2,4) position.

Since the energies of the $2s_{1/2,\pm 1/2}$ states remain degenerate and are independent of the external field orientation, the form of equations (3.29) and (3.30) remains valid to all orders of perturbation theory. The only explicit dependence on field strength is through the overall multiplying factors $a(|\mathbf{E}|), b_{1/2}(|\mathbf{E}|)$ and $b_{3/2}(|\mathbf{E}|)$. To lowest order in the external field, they are given by

$$a = 1 + 0(|\mathbf{E}|^2) \tag{3.32}$$

$$b_{1/2} = \frac{e|\mathbf{E}|\langle 2p_{1/2}\|\mathbf{r}\|2s_{1/2}\rangle}{\sqrt{6}(\triangle_L + i\Gamma/2)} + 0(|\mathbf{E}|^3) \tag{3.33}$$

$$b_{3/2} = \frac{e|\mathbf{E}|\langle 2p_{3/2}\|\mathbf{r}\|2s_{1/2}\rangle}{\sqrt{12}(\triangle_F + i\Gamma/2)} + 0(|\mathbf{E}|^3) \ , \tag{3.34}$$

where $\triangle_L = E(2s_{1/2}) - E(2p_{1/2})$ is the Lamb shift, $\triangle_F = E(2s_{1/2}) - E(2p_{3/2})$ is the Lamb shift minus the fine structure splitting, and Γ is the level width of the $2p$ state. Higher order perturbation corrections are discussed in the Appendix. The a and b_j coefficients could also be calculated by an exact diagonalization of the Hamiltonian matrix in the $2s_{1/2}, 2p_{1/2}, 2p_{3/2}$ basis set.

The properties of the quenching radiation are determined by the matrix elements

$$A_{m,m'} = \langle 1s_{1/2}, m|\boldsymbol{\alpha} \cdot \widehat{\mathbf{e}} e^{-i\mathbf{k}\cdot\mathbf{r}}|2s_{1/2}, m'\rangle \tag{3.35}$$

between the unperturbed $1s_{1/2}$ final state and the perturbed $2s_{1/2}$ initial state as given by (3.28). Using the expansion (3.7), together with a second application of the Wigner-Eckart theorem to express matrix elements of the $\boldsymbol{\alpha} \cdot \mathbf{a}_{L,M}^{(\lambda)}$ in terms of reduced matrix elements[a], the 2×2 transition matrix $\underline{A}$ with elements $A_{m,m'}$ becomes

$$\underline{A} = V_+ \hat{e} \cdot \hat{E} \underline{1} + \boldsymbol{\sigma} \cdot [iV_-(\hat{e} \times \hat{E}) + M(\hat{k} \times \hat{e})] \qquad (3.36)$$

where

$$\begin{aligned} V_+ &= V_{1/2} + 2V_{3/2} \,, \\ V_- &= V_{1/2} - V_{3/2} + M_{3/2} \,, \\ M &= M_{1/2} + 2i(\hat{k} \cdot \hat{E})M_{3/2} \,, \end{aligned} \qquad (3.37)$$

and

$$V_{1/2} = -\frac{b_{1/2}}{4\pi^{1/2}} \langle 1s_{1/2} \| \boldsymbol{\alpha} \cdot \mathbf{a}_1^{(1)*} \| 2p_{1/2} \rangle \,,$$

$$V_{3/2} = -\frac{b_{3/2}}{4(2\pi)^{1/2}} \langle 1s_{1/2} \| \boldsymbol{\alpha} \cdot \mathbf{a}_1^{(1)*} \| 2p_{3/2} \rangle \,,$$

$$M_{1/2} = \frac{ia}{4\pi^{1/2}} \langle 1s_{1/2} \| \boldsymbol{\alpha} \cdot \mathbf{a}_1^{(0)*} \| 2s_{1/2} \rangle \,,$$

$$M_{3/2} = -\frac{b_{3/2}}{4(2\pi/3)^{1/2}} \langle 1s_{1/2} \| \boldsymbol{\alpha} \cdot \mathbf{a}_2^{(0)*} \| 2p_{3/2} \rangle \,. \qquad (3.38)$$

Numerical values for the reduced matrix elements in (3.33), (3.34) and (3.38), including the leading relativistic corrections, are summarized in Table 2. The

---

[a]The application of the Wigner-Eckart theorem is straightforward because the $\mathbf{a}_{LM}^{(\lambda)}$ are proportional to vector spherical harmonics as given by (3.9) and (3.10), and by definition

$$\boldsymbol{\alpha} \cdot \mathbf{Y}_{L\ell M} = \sum_{mq} Y_\ell^m(\theta, \phi)\alpha_q \langle \ell m 1 q | LM \rangle \,.$$

The right-hand side is a vector coupled product which transforms under rotations like Y_L^M, and hence the Wigner-Eckart theorem is

$$\langle \beta' \jmath' m' | \boldsymbol{\alpha} \cdot \mathbf{Y}_{L\ell M} | \beta \jmath m \rangle$$
$$= (-1)^{\jmath' - m'} \begin{pmatrix} \jmath' & L & \jmath \\ -m' & M & m \end{pmatrix}$$
$$\langle \beta' \jmath' \| \boldsymbol{\alpha} \cdot \mathbf{Y}_{L\ell} \| \beta \jmath \rangle \,.$$

physical significance of the above terms is as follows. $V_{1/2}$ and $V_{3/2}$ represent the amplitudes for electric field quenching of the $2s_{1/2}$ state via the admixture of $2p_{1/2}$ and $2p_{3/2}$ intermediate states respectively with the emission of an $E1$ photon, while $M_{3/2}$ is a small $M2$ correction. All three of these terms are proportional to the electric field strength through the b coefficients. The combination V_+ comes from transitions with $\Delta m = 0$ in (3.35), and the $E1$ part of V_- comes from transitions with $\Delta m = \pm 1$. $M_{1/2}$ is the amplitude for spontaneous $M1$ transitions as discussed in section 3.2. It is to a first approximation independent of field strength.

In addition to the vectors $\hat{e}, \hat{k}$ and $\hat{E}$, the quenching radiation also depends on the electron spin polarization of the $2s_{1/2}$ state. This is specified, in general, by the density matrix

$$\rho = \frac{1}{2}(1 + \boldsymbol{\sigma} \cdot \mathbf{P}) \tag{3.39}$$

where $\mathbf{P}$ is the polarization vector for the $2s_{1/2}$ state. The decay rate summed over final atomic states and averaged over initial states is then

$$w d\Omega = \frac{e^2 k}{2\pi\hbar} \quad \text{Tr}\,[\rho \underline{A}^\dagger \underline{A}] d\Omega , \tag{3.40}$$

where Tr denotes the trace. It is a straightforward, but lengthy calculation to multiply the terms in (3.40), making repeated use of the well-known identity

$$(\boldsymbol{\sigma} \cdot \mathbf{a})(\boldsymbol{\sigma} \cdot \mathbf{b}) = \mathbf{a} \cdot \mathbf{b} + i\boldsymbol{\sigma} \cdot \mathbf{a} \times \mathbf{b} \tag{3.41}$$

together with Tr $(\boldsymbol{\sigma}) = 0$.

In order to describe the polarization of the emitted radiation, it is necessary to introduce two orthogonal polarization vectors $\hat{e}_1$ and $\hat{e}_2$ both perpendicular to $\hat{k}$ such that

$$\hat{k} \times \hat{e}_1 = \hat{e}_2 , \quad \hat{k} \times \hat{e}_2 = -\hat{e}_1 . \tag{3.42}$$

An arbitrary polarization vector for the general case of elliptical polarization is then given by

$$\hat{e} = \cos\beta \hat{e}_1 + i \sin\beta \hat{e}_2 . \tag{3.43}$$

Table 2. Values of matrix elements for fine structure transitions in hydrogenic ions[1]

Matrix Element	*Value*
$\langle 2p_{1/2}\|\mathbf{r}\|2s_{1/2}\rangle$	$\dfrac{3\sqrt{2}a_0}{Z}\left[1-\dfrac{5}{12}\alpha^2 Z^2\right]$
$\langle 2p_{3/2}\|\mathbf{r}\|2s_{1/2}\rangle$	$\dfrac{-6a_0}{Z}\left[1-\dfrac{1}{6}\alpha^2 Z^2\right]$
$\langle 1s_{1/2}\|\boldsymbol{\alpha}\cdot\mathbf{a}_1^{(1)*}\|2p_{1/2}\rangle$	$\dfrac{ika_0}{Z}\left(\dfrac{2\pi}{3}\right)^{1/2}\dfrac{2^9}{3^5}\left[1-\left(\dfrac{11}{96}+\dfrac{3}{2}\ln 2-\ln 3\right)\alpha^2 Z^2\right]$
$\langle 1s_{1/2}\|\boldsymbol{\alpha}\cdot\mathbf{a}_1^{(1)*}\|2p_{3/2}\rangle$	$-\dfrac{ika_0}{Z}\left(\dfrac{4\pi}{3}\right)^{1/2}\dfrac{2^9}{3^5}\left[1-\left(\dfrac{11}{48}+\dfrac{5}{4}\ln 2-\dfrac{3}{4}\ln 3\right)\alpha^2 Z^2\right]$
$\langle 1s_{1/2}\|\boldsymbol{\alpha}\cdot\mathbf{a}_1^{(0)*}\|2s_{1/2}\rangle$	$ka_0 Z^2\alpha^3(2\pi)^{1/2}\dfrac{2^4}{3^4}[1+0.4193\alpha^2 Z^2]$
$\langle 1s_{1/2}\|\boldsymbol{\alpha}\cdot\mathbf{a}_2^{(0)*}\|2p_{3/2}\rangle$	$i\dfrac{k^2 a_0^2\alpha}{Z}\pi^{1/2}\dfrac{2^8}{3^5}[1-0.1821\alpha^2 Z^2]$
$\langle 2s_{1/2}\|z\|2p_{1/2}\rangle$	$\sqrt{3}a_0\left(1-\dfrac{5}{12}\alpha^2 Z^2\right)\Big/Z$
$\langle 2s_{1/2}\|z\|2p_{3/2}\rangle$	$-6a_0\left(1-\dfrac{1}{6}\alpha^2 Z^2\right)\Big/Z$
$\langle 1s\|z\|2p\rangle$	$2^8 a_0/(3^5\sqrt{2}Z)$
$\langle 2p\|z\|2s\rangle$	$-3a_0/Z$

In particular, $\beta = 0, \pi/2, \ldots$ corresponds to linearly polarized light and $\beta = \pi/4, 3\pi/4, \ldots$ corresponds to right or left circularly polarized light. With

---

[1] The $\mathbf{a}_{LM}^{(\lambda)}(\omega r/c)$ are evaluated at $\hbar\omega = E(2s_{1/2}) - E(1s_{1/2})$, and the reduced matrix elements are defined in terms of $3-j$ symbols by

$$\langle n'\ell' j'm'|\boldsymbol{\alpha}\cdot\mathbf{a}_{LM}^{(\lambda)}|n\ell jm\rangle = (-1)^{j'-m'}\begin{pmatrix} j' & L & j \\ -m' & M & m \end{pmatrix}$$
$$\times \langle n'\ell' j'\|\boldsymbol{\alpha}\cdot\mathbf{a}_L^{(\lambda)}\|n\ell j\rangle .$$

these conventions, equation (3.40) becomes

$$w(\hat{e}, \hat{k}, \hat{\mathbf{P}})d\Omega = \frac{e^2 k}{2\pi\hbar}[I_0 + \mathbf{P}\cdot\mathbf{J}_0 + \mathbf{P}\cdot\mathbf{J}_1 \sin 2\beta$$
$$+ \hat{E}(\hat{e}_1 : \hat{e}_1 - \hat{e}_2 : \hat{e}_2)\mathbf{J}_2 \cos 2\beta]d\Omega \qquad (3.44)$$

where

$$I_0 = 1/2|V_+|^2[1 - (\hat{k}\cdot\hat{E})^2] + 1/2|V_-|^2[1 + (\hat{k}\cdot\hat{E})^2]$$
$$+ 2\mathrm{Im}\,(M^*V_-)(\hat{k}\cdot\hat{E}) + |M|^2 \qquad (3.45)$$

$$\mathbf{J}_0 = (\hat{k}\times\hat{E})\{\mathrm{Re}\,[M^*(V_+ + V_-)] - \mathrm{Im}\,(V_-^*V_+)(\hat{k}\cdot\hat{E})\}\,, \qquad (3.46)$$

$$\mathbf{J}_1 = |V_-|^2(\hat{k}\cdot\hat{E})\hat{E} + \mathrm{Re}\,(V_-^*V_+)\hat{E}\times(\hat{k}\times\hat{E})$$
$$- \mathrm{Im}\,[M^*(V_+ + V_-)]\hat{E} + \mathrm{Im}\,[M^*(V_+ - V_-)](\hat{k}\cdot\hat{E})\hat{k}$$
$$- |M|^2\hat{k}\,, \qquad (3.47)$$

$$\mathbf{J}_2 = 1/2(|V_+|^2 - |V_-|^2)\hat{E} + \mathrm{Re}\,[M^*(V_+ - V_-)]\mathbf{P}\times\hat{k}$$
$$- \mathrm{Im}\,(V_-^*V_+)\mathbf{P}\times\hat{E}\,. \qquad (3.48)$$

The dyadic notation **b:c** in (3.44) is defined by

$$\mathbf{a}(\mathbf{b}:\mathbf{c})\mathbf{d} = (\mathbf{a}\cdot\mathbf{b})(\mathbf{c}\cdot\mathbf{d})\,.$$

The following points concerning the above equations are of particular importance.

(a) Since M is smaller than V_+ by a factor of $0(\alpha^2 Z^2)$, the dominant terms are those containing only V_+ and V_-. The angular distribution of radiation from the main polarization-independent term I_0 is illustrated by the outer elliptical curve in Fig. 3.2. The anisotropy is approximately proportional to the Lamb shift as discussed in section 3.3.3.

(b) The term $-\mathrm{Im}\,(V_-^*V_+)(\hat{k}\cdot\hat{E})\mathbf{P}\cdot\hat{\mathbf{k}}\times\hat{E}$ in $\mathbf{P}\cdot\mathbf{J}_0$ is sensitive to the imaginary level widths in the denominators of (3.33) and (3.34) as discussed in section 3.3.4. It produces the clover leaf pattern in Fig. 3.2. The remaining part $\mathrm{Re}\,[M^*(V_+ + V_-)]\mathbf{P}\cdot\hat{\mathbf{k}}\times\hat{\mathbf{E}}$ of $\mathbf{P}\cdot\mathbf{J}_0$ produces the innermost pattern in Fig. 3.2. Its measurement determines the $2s \rightarrow 1s$ magnetic dipole matrix element as discussed in section 3.3.5.

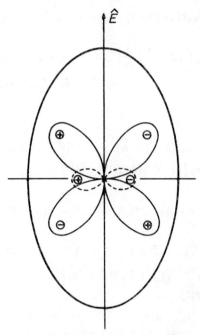

Fig. 3.2. Polar diagram (not to scale) of the contributions to quenching radiation asymmetries produced by an electric field $\mathbf{E}$ acting on the $2s_{1/2}$ state. The outer elliptical curve is the main Lamb shift anisotropy for unpolarized atoms. The clover leaf pattern is the $E1 - E1$ damping asymmetry for atoms polarized such that P points into the page, and the inner dashed curve is the corresponding $E1 - M1$ interference asymmetry. The $(+)$ and $(-)$ signs indicate positive and negative contributions to the total intensity.

(c) The $\sin 2\beta$ and $\cos 2\beta$ terms in (3.44) vanish on summing over photon polarizations, but can be observed with a polarization-sensitive detector. Although the $\sin 2\beta$ term has not been observed, the $\cos 2\beta$ linear polarization term has been measured by Ott, Kauppila and Fite (1969).

(d) The term involving $\mathrm{Im}\,(M^*V_-)(\hat{k}\cdot\hat{E})$ in (3.45) at first sight appears to violate time reversal invariance because $\hat{k}$ reverses direction for the time-reversed process of absorption. However, as discussed by Mohr (1978), $i\Gamma$ must also be replaced by $-i\Gamma$ in the denominator of V_-, making the product of both terms time-reversal invariant. The same sign reversals apply to similar terms in $\mathbf{J}_0, \mathbf{J}_1$ and $\mathbf{J}_2$, together with $\mathbf{P} \rightarrow -\mathbf{P}$.

In the electric dipole approximation where only the $V_\pm$ terms are retained, there is a general geometrical relationship between the polarization of the emitted radiation and the angular anisotropy. To see what it is, suppose that the z-axis is an axis of rotational symmetry defined, for example, by an electric field. In addition, the radiation intensities depend only on the orientation of $\hat{e}$ relative to the z-axis, and not on the direction of observation. Then the polarization in the x-direction, as illustrated in Fig. 3.3 is

$$P = \frac{I_y - I_z}{I_y + I_z} \tag{3.49}$$

and the rotational anisotropy, defined in terms of the total intensities emitted in the directions parallel $\left(I_\| = I_x + I_y \right)$ and perpendicular $\left(I_\perp = I_y + I_z \right)$ to the symmetry axis is

$$R = \frac{I_\| - I_\perp}{I_\| + I_\perp} = \frac{I_x - I_z}{I_x + 2I_y + I_z} \ . \tag{3.50}$$

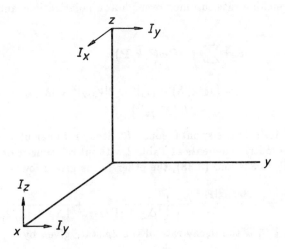

Fig. 3.3. Illustration of the relationship between the polarization of quenching radiation in the x-direction and the anisotropy in the total intensity (summed over polarizations) in the x- and z-directions. The arrows indicate photon polarization vectors.

If the z-axis is a symmetry axis, then $I_x = I_y$, and it follows immediately that

$$P = 2R/(1 - R) \ . \tag{3.51}$$

The above applies to any system for which the electric dipole approximation is valid and an axis of rotational symmetry exists. It is easy to verify that (3.51) follows from (3.44) as a special case. The fact that the quenching radiation is polarized even if $\mathbf{P} = 0$ was first found by Casalese and Gerjuoy (1969), and Fite *et al.* (1968). The corresponding anisotropy was first pointed out by Drake and Grimley (1973) and measured by van Wijngaarden *et al.* (1974).

To calculate the total quenching rate, it is instructive first to multiply out the terms in (3.45) and express them in the form

$$
\begin{aligned}
I_0 =\ & |V_{1/2}|^2 + 2|V_{3/2}|^2 + |M_{1/2}|^2 + 2/3|M_{3/2}|^2 \\
& + 2\mathrm{Im}\,[M_{1/2}^*(V_{1/2} - V_{3/2} - M_{3/2})]P_1(\cos\theta) \\
& - \{|V_{3/2}|^2 - 1/3|M_{3/2}|^2 + 2\mathrm{Re}\,[V_{1/2}^*V_{3/2} \\
& + M_{3/2}^*(V_{1/2} - V_{3/2})]\}P_2(\cos\theta)\ ,
\end{aligned}
\tag{3.52}
$$

where the $P_L(\cos\theta)$ are Legendre polynomials and $\widehat{k}\cdot\widehat{E} = \cos\theta$. Since the $P_L(\cos\theta)$ vanish on integrating over angles for $L \geq 1$, it is clear that none of the cross terms between different multipoles survive. All that remains for the total quenching rate summed over photon polarizations and integrated over angles is

$$
\begin{aligned}
\widehat{w} &= \sum_{\widehat{e}} \int d\Omega\, w(\widehat{e}, \widehat{k}, \mathbf{P}) \\
&= (4e^2k/\hbar)[|V_{1/2}|^2 + 2|V_{3/2}|^2 + |M_{1/2}|^2 \\
&\quad + 2/3|M_{3/2}|^2]\ .
\end{aligned}
\tag{3.53}
$$

The $|M_{1/2}|^2$ term is the spontaneous $M1$ decay rate calculated in section 3.2.

Using the matrix elements in Table 2 without relativistic or $M2$ corrections, together with (3.33) and (3.38), the other terms give a lowest order quenching rate of

$$
\overline{w} = \frac{3e^2 a_0^2 |\mathbf{E}|^2 \gamma_{2p}}{Z^2} \left\{ \frac{1}{|\Delta_L + i\Gamma/2|^2} + \frac{2}{|\Delta_F + i\Gamma/2|^2} \right\}
\tag{3.54}
$$

where $\gamma_{2p} = \Gamma/\hbar$ is the decay rate of the $2p$ state given by

$$
\begin{aligned}
\gamma_{2p} &= \frac{4}{3}\frac{e^2}{\hbar}k^3|\langle 1s|z|2p\rangle|^2 \\
&= (2/3)^8\alpha^3 Z^4\tau^{-1} \\
&= 6.268Z^4 \times 10^8 \mathrm{s}^{-1}\ .
\end{aligned}
\tag{3.55}
$$

The first term of (3.54) is the same as the one originally obtained by Lamb and Retherford (1950) by a different method. They solved directly the time dependent perturbation equations

$$
\begin{aligned}
i\hbar da/dt &= V^* e^{-i\omega t} b - (1/2)i\gamma_a a\ , \\
i\hbar db/dt &= V e^{i\omega t} a - (1/2)i\gamma_b b\ ,
\end{aligned}
\tag{3.56}
$$

with $\omega = [E(2p_{1/2}) - E(2s_{1/2})]/\hbar$ and $V = \langle 2p_{1/2}|e\mathbf{E}\cdot\mathbf{r}|2s_{1/2}\rangle$ for the amplitudes a and b of the $2s_{1/2}$ and $2p_{1/2}$ states respectively. The above equations follow immediately when (3.25) is re-expressed in the interaction representation, including phenomenological damping. However, their solution contains the unphysical assumption that the field is switched on suddenly at $t = 0$, leading to transient oscillations. The present calculation is based instead on the adiabatic perturbations of the $2s_{1/2}$ state as expressed by (3.28). The two methods are equivalent in the limit $t \to \infty$, provided that the $2s_{1/2}$ state is not split into sub-states which depopulate at different rates, for example by hyperfine structure or magnetic fields. In the latter case, a full time-dependent treatment which takes proper account of how the field is switched on may be necessary, as discussed in section 3.4.

The second term of (3.54) comes from mixing of the $2s_{1/2}$ state with the intermediate $2p_{3/2}$ state. It is smaller than the first term by a factor of about 100 because of the larger energy difference in the denominator. It was first obtained by Fan *et al.* (1967) by the adiabatic method, and by Kugel *et al.* (1972) using an extension of equations (3.56) to three interacting states.

3.3.2   Hyperfine structure effects for weak quenching fields

If hyperfine structure is present, the theory of Stark quenching must be formulated in terms of transitions among the various hyperfine components of the s and p states. A complete description is as complex as the general theory of angular correlations in the two-photon transitions of polarized atoms and nuclei. In addition, the angular correlations are weakly time-dependent because different hyperfine components of the $2s_{1/2}$ state are quenched at different rates(Holt and Sellin, 1972).

The field-free hyperfine structure for s-states is discussed by Series in SAH, Chapter XI. The hyperfine energy shifts for states of arbitrary angular momentum j due to the magnetic dipole moment μ_N of the nucleus can be estimated to sufficient accuracy from the formula (see, for example, Kopfermann, 1958)

$$\Delta E_{nljIF} = \frac{\alpha_{hfs}}{n^3}\left[\frac{F(F+1) - j(j+1) - I(I+1)}{j(j+1)(l+1/2)}\right] \tag{3.57}$$

where $\alpha_{hfs} = \mu_e\mu_N Z^3(1+m/M)^{-3}/(Ia_0^3)$, I is the nuclear spin and $\mathbf{F} = \mathbf{I}+\mathbf{j}$ is the total angular momentum. For example, for hydrogen $\mu_e = 1.00116 e\lambdabar/2$, $\mu_p = 2.79268(m/m_p)e\lambdabar/2$ where m_p is the proton mass and (3.57) gives a $2s_{1/2}$ state energy splitting of 177.55 MHz, which is about 1/6 of the Lamb shift.

Important simplifications occur in the quenching theory if the initial state is an incoherent mixture of all $2s_{1/2}(F, M_F)$ states with equal statistical weights, and the electric field can be taken as a weak perturbation relative to the hyperfine structure splittings. The latter condition is satisfied if the Stark shift for the $2s_{1/2}$ state, given in first order perturbation theory by

$$\Delta E_S = e^2 |\mathbf{E}|^2 |\langle 2s_{1/2}|z|2p_{1/2}\rangle|^2 / \Delta_L \qquad (3.58)$$

is much less than $\Delta E_{2,0,1/2,I,F}$; i.e.

$$|\mathbf{E}| \ll 195 Z^{4.33} (\mu_N/\mu_P)^{1/2} V/\text{cm} .$$

With the above assumptions, there is no initially preferred direction in space and we can assume that $\mathbf{E}$ points in the z-direction (the quantization axis) without loss of generality. Also, j and I remain coupled together so that the atomic states can be written in the $nljIFM_F$ coupled representation. Then, by analogy with the results of section 3.1, the differential decay rate averaged over initial states is

$$w(\hat{e}, \hat{k}) d\Omega = \frac{e^2 k}{2\pi\hbar(2I+1)(2)} \sum_{\substack{FM \\ F'M'}} |B(FM, F'M')|^2 , \qquad (3.59)$$

where

$$B(FM, F'M') = |\mathbf{E}| \sum_{j''F''} \langle 1s, F'M'|\boldsymbol{\alpha}\cdot\hat{e}e^{-i\mathbf{k}\cdot\mathbf{r}}|2p, j''F''M\rangle$$
$$\times \frac{\langle 2p, j''F''M|ez|2s, FM\rangle}{E(2s, F) - E(2p, j''F'') + i\Gamma/2} . \qquad (3.60)$$

The above describes the initial quenching radiation from the adiabatically perturbed $2s_{1/2}$ state. It is valid only for times sufficiently short that the initial statistical distribution of states is not upset by differential quenching. The full time-dependent problem is discussed in section 3.4.

The angular dependence of the terms in (3.59) can be extracted by observing that, in the electric dipole approximation, terms with $M' = M$ are proportional to

$$|e_0|^2 = |\hat{e}\cdot\hat{E}|^2$$

and terms with $M' = M \pm 1$ are proportional to

$$-e_{+1}e_{-1} = |\hat{e}\times\hat{E}|^2 .$$

A comparison with equation (3.36) shows that the decay rate is still of the same form as the polarization-independent terms in (3.44), but $|V_+|^2$ and $|V_-|^2$ must be replaced by

$$|V_+^{(HFS)}|^2 = \frac{1}{2(2I+1)} \sum_{FMF'} |B(FM, F'M)|^2 \tag{3.61}$$

$$|V_-^{(HFS)}|^2 = \frac{1}{2(2I+1)} \sum_{FMF'} [|B(FM, F'M+1)|^2 + |B(FM, F'M-1)|^2] \tag{3.62}$$

provided that magnetic dipole and quadrupole corrections are neglected. Then

$$w(\hat{e}, \hat{k})d\Omega = \frac{e^2 k}{2\pi\hbar} [I_0 + \hat{E}(\hat{e}_1 : \hat{e}_1 - \hat{e}_2 : \hat{e}_2)\mathbf{J}_2 \cos\theta]d\Omega \tag{3.63}$$

with

$$I_0 = \frac{1}{2}|V_+^{(HFS)}|^2[1 - (\hat{k}\cdot\hat{E})^2]$$
$$+ \frac{1}{2}|V_-^{(HFS)}|^2[1 + (\hat{k}\cdot\hat{E})^2] \tag{3.64}$$

and

$$\mathbf{J}_2 = \frac{1}{2}(|V_+^{(HFS)}|^2 - |V_-^{(HFS)}|^2)\hat{E} . \tag{3.65}$$

Using standard vector coupling techniques (see, for example, Edmonds, 1960), the matrix elements in (3.60) can be written in terms of $3-j$ and $6-j$ symbols as

$$\langle \gamma'j'IF'M'|T(1q)|\gamma jIFM\rangle = (-1)^{F'-M'} \begin{pmatrix} F' & 1 & F \\ -M' & q & M \end{pmatrix}$$
$$\times (-1)^{j'+I+F+1}[(2F'+1)(2F+1)]^{1/2} \begin{Bmatrix} j' & F' & I \\ F & j & 1 \end{Bmatrix}$$
$$\times \langle \gamma'j'\|T(1)\|\gamma j\rangle \tag{3.66}$$

where the reduced matrix elements are the same as those in Table 2. Sum rules for the $3-j$ and $6-j$ symbols can be used to obtain

$$|V_+^{(HFS)}|^2 \rightarrow |V_+|^2 ,$$
$$|V_-^{(HFS)}|^2 \rightarrow |V_-|^2 \tag{3.67}$$

in the limit of vanishing hyperfine structure. Some numerical values for the effects of hyperfine structure on the anisotropy of the quenching radiation are given in the following section.

3.3.3 The Lamb shift anisotropy

The Lamb shift anisotropy is the angular asymmetry predicted by equation (3.44) for an unpolarized atomic beam ($\mathbf{P} = 0$) and summed over polarization directions for the emitted photon. Under these conditions, (3.44) reduces to

$$\widehat{w}(\widehat{k})d\Omega = \frac{e^2 k}{\pi \hbar} I_0(\widehat{k}) d\Omega \ . \tag{3.68}$$

The $|M_{1/2}|^2$ term in I_0 (equation (3.45)) is completely negligible for realistic quenching fields. If further the measured intensities are averaged over the directions $+\widehat{k}$ and $-\widehat{k}$ so that the $\mathrm{Im}(M_{1/2}^* V_-)$ term drops out and the small contributions from $M_{3/2}$ are temporarily neglected (see equation (3.37)), then $I_0(\widehat{k})$ becomes (see Fig. 3.2)

$$I_0(\widehat{k}) \simeq |V_{1/2}|^2 + \mathrm{Re}\,(V_{1/2}^* V_{3/2})(1 - 3\cos^2 \theta) \\ + 1/2|V_{3/2}|^2(5 - 3\cos^2 \theta) \tag{3.69}$$

where $\widehat{k} \cdot \widehat{E} = \cos \theta$. If a dimensionless parameter ρ is defined by

$$\rho = V_{3/2}/V_{1/2} \tag{3.70}$$

then in the limit of weak electric fields, the first order perturbation expressions (3.33) and (3.34) can be used, together with the non-relativistic parts of the matrix elements in Table 2 to obtain

$$\rho \simeq \frac{\Delta_L + i\Gamma/2}{\Delta_F + i\Gamma/2} \tag{3.71}$$

and

$$I_0(\widehat{k}) = |V_{1/2}|^2[1 + \mathrm{Re}\,(\rho)(1 - 3\cos^2 \theta) \\ + 1/2|\rho|^2(5 - 3\cos^2 \theta)] \ . \tag{3.72}$$

A comparison of the intensities emitted in the directions parallel ($\theta = 0$) and perpendicular ($\theta = \pi/2$) to the electric quenching field then yields the approximate expression for the anisotropy (see Fig. 3.2)

$$R = [I_\parallel - I_\perp]/[I_\parallel + I_\perp] \simeq R^{(0)} \tag{3.73}$$

where

$$R^{(0)} = -[3\mathrm{Re}\,(\rho) + \frac{3}{2}|\rho|^2]/[2 - \mathrm{Re}\,(\rho) + \frac{7}{2}|\rho|^2] \,. \tag{3.74}$$

The above formula emphasizes that $R^{(0)}$ is determined primarily by the ratio of the Lamb shift to the $2s_{1/2} - 2p_{3/2}$ energy difference, independent of $|\mathbf{E}|$. This can be understood as follows. The electric field mixing of the $2s_{1/2}$ and $2p_{1/2}$ states alone does not produce any anisotropy of the quenching radiation since, as discussed in section 3.2, the transition to the $1s_{1/2}$ ground state is still a $j = 1/2 \rightarrow 1/2$ transition. The anisotropy comes solely from the relatively smaller mixing with the $2p_{3/2}$ state. The quantity $\mathrm{Re}(\rho) \simeq \Delta_L/\Delta_F$ describes quantitatively the ratio of the two mixing coefficients, independent of $|\mathbf{E}|$ in the limit of weak fields. Since $\mathrm{Re}(\rho) \simeq 0.08$ for low Z ions, the dominant contribution comes from the $V_{1/2}^* V_{3/2}$ cross-term in (3.69) and $R^{(0)} \simeq 0.1$ over a wide range of nuclear charge.

A measurement of R can be interpreted as a measurement of the Lamb shift because the fine structure splitting is to lowest order a known non-QED effect. However, for high precision applications, it is necessary to take into account a number of small corrections to $R^{(0)}$. These are

(a) finite electric field corrections proportional to $|\mathbf{E}|^2$,

(b) corrections due to the mixing of the $2s_{1/2}$ state with higher lying np states ($n \geq 3$), and perturbations to the $1s_{1/2}$ final state,

(c) relativistic corrections to the matrix elements,

(d) corrections due to the $2p_{3/2} - 1s_{1/2}$ magnetic quadrupole term $M_{3/2}$.

A brief discussion of these corrections follows. Further details can be found in Drake *et al.* (1979) and van Wijngaarden *et al.* (1982).

For finite field effects, it is a straightforward matter to calculate higher order perturbation corrections to the expressions (3.32) to (3.34) for the mixing coefficients $a_{1/2}, b_{1/2}$ and $b_{3/2}$. The result is a perturbation expansion for R of the form (see Appendix)

$$R_0 = R^{(0)} + |\mathbf{E}|^2 R^{(2)} + \dots \,. \tag{3.75}$$

This expansion is primarily useful for ions with zero spin nuclei. For other ions, although $R^{(0)}$ remains well defined, the Stark shifts to the energy levels may not be small compared with the hyperfine structure splittings at practical quenching field strengths. Thus the field-dependent corrections have a more complicated structure obtainable by an exact diagonalization of the Hamiltonian matrix. In addition, R becomes weakly time-dependent owing to different depopulation rates for different hyperfine states. These complications are further discussed in section 3.4.

The mixing of the $2s_{1/2}$ state with higher lying np states and perturbations to the $1s_{1/2}$ final state are described in terms of second-order electric dipole perturbation expressions analogous to $V_{1/2}$ and $V_{3/2}$ of the form (Drake and Grimley, 1973)

$$B = \sum_{n=3}^{\infty} \frac{\langle 1s|z|np\rangle\langle np|z|2s\rangle}{E(2s) - E(np)} , \tag{3.76}$$

$$C = \sum_{n=2}^{\infty} \frac{\langle 1s|z|np\rangle\langle np|z|2s\rangle}{E(1s) - E(np)} . \tag{3.77}$$

It is no longer necessary to consider the fine structure of the states because the Coulomb splittings are so much larger. The exact non-relativistic values of B and C are (see Appendix)

$$B = -25 \cdot 2^9/(3^6\sqrt{2}) \; a_0^3/e^2 ,$$
$$C = 7 \cdot 2^9/(3^6\sqrt{2}) \; a_0^3/e^2 .$$

Including the necessary multiplying factors, B and C make an additional contribution of $-ike\hat{e}\cdot\mathbf{E}(B + C)$ to the transition matrix $\underline{A}$ defined by (3.36); i.e. V_+ should be replaced by

$$V_+ - ike|\mathbf{E}|(B + C) .$$

Neglecting the level widths, the corresponding correction to R_0 is

$$\left(\frac{\delta R}{R_0}\right)_{B+C} = 2(B + C)\frac{\Delta_F}{N}\left(\frac{1 + 2\rho}{2 + \rho}\right)(1 + R_0) \tag{3.78}$$

where $\Delta_F = E(2s_{1/2}) - E(2p_{3/2})$ and

$$N = \langle 1s|z|2p\rangle\langle 2p|z|2s\rangle .$$

Using the above numerical values together with those in Table 2, the final result is

$$\left(\frac{\delta R}{R_0}\right)_{B+C} = \frac{3\Delta_F}{\hbar\omega}\left(\frac{1 + 2\rho}{2 + \rho}\right) \cdot (1 + R_0) . \tag{3.79}$$

The numerical values are approximately $-6 \times 10^{-6}Z^2$ for $Z \lesssim 20$.

Relativistic corrections to the matrix elements enter by modifying the value of ρ. Since ρ is defined by (3.70), the correction to the approximate expression (3.71) is

$$\delta\rho/\rho = \mu_{3/2} + \mu'_{3/2} - \mu_{1/2} - \mu'_{1/2} \tag{3.80}$$

where the μ_j and μ'_j are the fractional corrections of $O(\alpha^2 Z^2)$ to the matrix elements $\langle 2p_j \| r \| 2s_{1/2} \rangle$ and $\langle 1s_{1/2} \| \boldsymbol{\alpha} \cdot \mathbf{a}^{(1)*} \| 2p_j \rangle$ shown in Table 2. The values are

$$\mu_{3/2} = -\frac{1}{6}\alpha^2 Z^2, \quad \mu_{1/2} = -\frac{5}{12}\alpha^2 Z^2$$

$$\mu'_{3/2} = \left[-\frac{11}{48} - \frac{5}{4}\ln 2 + \frac{3}{4}\ln 3 \right] \alpha^2 Z^2$$

$$\mu'_{1/2} = \left[-\frac{11}{96} - \frac{3}{2}\ln 2 + \ln 3 \right] \alpha^2 Z^2 .$$

Using (3.74) and again neglecting the level widths, the corresponding change in R is

$$\left(\frac{\delta R}{R_0} \right)_{\text{rel}} = 2(\mu_{3/2} + \mu'_{3/2} - \mu_{1/2} - \mu'_{1/2}) \left(\frac{1 + 2\rho}{2 + \rho} \right) \left(\frac{1 + R_0}{1 - \rho} \right) . \tag{3.81}$$

The numerical value is approximately $1.7 \times 10^{-6} Z^2$ for $Z < 20$. The final small correction arises from the $2p_{3/2} - 1s_{1/2}$ magnetic quadrupole decay term $M_{3/2}$ first pointed out by Hillery and Mohr (1980). If the $M_{3/2}$ terms are retained in the general expression (3.52) for I_0, then they make an additional contribution of

$$\Delta I_0 = \text{Re}\,[M^*_{3/2}(V_{1/2} - V_{3/2})](1 - 3\cos^2\theta)$$

$$+ \frac{1}{2}|M_{3/2}|^2(1 + \cos^2\theta) \tag{3.82}$$

to equation (3.69). Keeping only terms linear in $M_{3/2}$ and neglecting the level widths, the corresponding correction to R is

$$\left(\frac{\delta R}{R_0}\right)_{M_2} = \frac{M_{3/2}}{V_{3/2}}\frac{(1-\rho)(1-R_0/3)}{(1+\rho/2)}$$

$$= -\frac{9\alpha^2 Z^2}{32}\frac{(1-\rho)(1-R_0/3)}{(1+\rho/2)}. \tag{3.83}$$

The numerical values are approximately $-16 \times 10^{-6}Z^2$ for $Z \lesssim 20$.

In summary, the theoretical value for the anisotropy is

$$R_T = R^{(0)}\left[1 + \left(\frac{\delta R}{R_0}\right)_{B+C} + \left(\frac{\delta R}{R_0}\right)_{\text{rel}} + \left(\frac{\delta R}{R_0}\right)_{M_2}\right]$$

$$+ R^{(2)}|\mathbf{E}|^2 + R^{(4)}|\mathbf{E}|^4 + O(|\mathbf{E}|^6) + O(\alpha^4 Z^4). \tag{3.84}$$

The fine structure splitting can similarly be calculated up to correction terms of relative $O(\alpha^4 Z^4)$ and $O(\alpha^2/\pi^2)$ from the formula

$$E(2p_{3/2}) - E(2p_{1/2}) = \frac{(Z\alpha)^4 mc^2}{32}\left\{\left[1 + \frac{5}{8}(Z\alpha)^2\right](1 - m/M)\right.$$

$$\left. + \frac{\alpha}{\pi}(1 - 2m/M) + 2\frac{\alpha}{\pi}(Z\alpha)^2 \ln Z\alpha\right\}. \tag{3.85}$$

Values of $R^{(0)}$ and the correction factors appearing in (3.84) are listed in Table 3 for a selection of zero-spin nuclei. Δ_L is the assumed value for the Lamb shift. The parameter b allows $R(0)$ to be readily calculated for a different Lamb shift $\Delta_L + \delta\Delta_L$ through the relation

$$\delta R/R^{(0)} = b\delta\Delta_L/\Delta_L \tag{3.86}$$

up to terms of first order in $\delta\Delta_L$. Values for other ions can easily be calculated from the formulas in the text and the tabulations of Drake and Lin (1976). It is clear from the smallness of the correction coefficients that equations (3.84) and (3.85) are adequate for the current level of experimental accuracy.

Table 3. Anisotropies and Small Correction (see equation (3-84))Factors for Ions with Zero Nuclear Spin

	$^4\text{He}^+$	$^{12}\text{C}^{5+}$	$^{16}\text{O}^{7+}$	$^{20}\text{Ne}^{9+}$
$\Delta_L(\text{GHz})$	14.04235	781.99	2196.21	4861.1
$R^{(0)}$	0.1179680	0.0817182	0.0726620	0.0658649
$R^{(2)}(\text{kV/cm})^{-2}$	5.822×10^{-4}	1.577×10^{-8}	1.000×10^{-9}	1.194×10^{-10}
$R^{(4)}(\text{kV/cm})^{-4}$	-0.37×10^{-5}	—	—	—
$(\delta R/R_0)_{B+C}$	-2.37×10^{-5}	-2.23×10^{-4}	-4.02×10^{-4}	-6.33×10^{-4}
$(\delta R/R_0)_{\text{rel}}$	0.64×10^{-5}	0.61×10^{-4}	1.09×10^{-4}	1.72×10^{-4}
$(\delta R/R_0)_{M2}$	-6.54×10^{-5}	-5.71×10^{-4}	-10.09×10^{-4}	-15.70×10^{-4}
b	1.0352	1.0150	1.0114	1.0091

Comparison with experiment

Measurements of R have been performed for the hydrogenic systems H (Drake, Farago and van Wijngaarden, 1975), D (van Wijngaarden and Drake, 1978) and He$^+$ (Drake, Goldman and van Wijngaarden, 1979; Patel, van Wijngaarden and Drake, 1987), using an apparatus of the type shown in Fig. 3.4. The quenching field is provided by four rods in a quadrupole arrangement, as shown in the diagram, instead of parallel capacitor plates so that the quenching radiation can at easily be observed in both the parallel and perpendicular directions. This arrangement also provides instrumental symmetry with respect to the two directions, and allows the field to be rotated in steps of 90° by simply switching the electrical potentials on the rods. The latter feature allows one to form ratios such that the relative sensitivities of the photon counters cancel out. The simultaneous measurement of $I_\parallel$ and $I_\perp$ provides a high degree of stability which allows high precision measurements to be made. The accuracy has been limited to date by photon counting statistics and non-linearities in the photon counting systems.

In addition to the above, Curnutte et. al. (1981) have measured R in O^{7+}, using an $e(\mathbf{v} \times \mathbf{B})/c$ quenching field. This is the only measurement in a heavier hydrogenic ion. The accuracy of their measurement could be improved, but has so far been limited by difficulties in correctly identifying the background noise

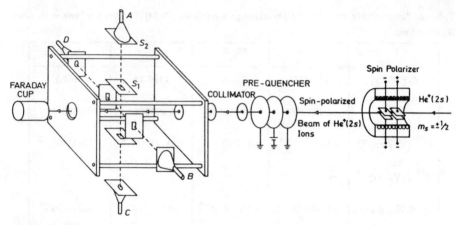

Fig. 3.4. Schematic of the apparatus used to measure quenching radiation asymmetries. For the Lamb shift anisotropy measurement, the spin polarizer part on the right is omitted.

that should be subtracted.

The results of the above measurements are compared with theory in Table 4. The theoretical values include finite quenching field effects and, in the cases of H and D, hyperfine structure corrections and time-dependent effects. It is perhaps more instructive to take the quenching theory as correct and derive from the measurements corresponding values for the Lamb shift as shown in the last column of Table 4. The values are in good agreement with the theoretical Lamb shifts, and the result for He^+ is equal in accuracy to microwave resonance measurements (Narasimham and Strombotne, 1971; Lipworth and Novick, 1957). Work is in progress to improve further the accuracy for He^+.

Table 4. Comparison of Theoretical and Experimental Anisotropies and Lamb Shifts (in GHz)

Ion	Δ_L(theory)	R^1_{theo}	R_{exp}	Δ_L(exp)
H	1.057867	0.139068	$0.13901(12)^2$	1.0574(9)
D	1.059241	0.144245	0.144259(21)	1.05936(16)
He^+	14.04235	0.118191	0.118187(12)	14.0419(15)
O^{7+}	2196.21	0.072567	0.07246(48)	2193(15)

[1] Including finite electric field and, for H and D, hyperfine structure corrections.

[2] Numbers in brackets indicate uncertainties in the final one or two figures quoted.

3.3.4 E1-E1 damping interference

If the initial $2s_{1/2}$ atomic state is spin-polarized, as described by the polarization vector $\mathbf{P}$, then the term $\mathbf{P}\cdot\mathbf{J}_0$ of (3.44) comes into play. Expanding the terms of (3.46) in terms of the V_j's and M_j's yields

$$\mathbf{J}_0 = (\hat{k}\times\hat{E})\{\mathrm{Re}\,[M_{1/2}^*(2V_{1/2}+V_{3/2}+M_{3/2})]\}$$
$$- 3(\hat{k}\cdot\hat{E})\mathrm{Im}\,[(V_{1/2}^*(V_{3/2}+M_{3/2})] . \tag{3.87}$$

The second term above is called the $E1-E1$ damping interference term because it comes primarily from the imaginary part of the cross product $V_{1/2}^*V_{3/2}$. It therefore depends on the imaginary level widths contained in the denominators of $V_{1/2}$ and $V_{3/2}$. The term $M_{3/2}$ in (3.87) can be neglected for low Z because it is smaller by a factor of $(\alpha Z)^2$.

The angular dependence of the $E1-E1$ damping term is

$$(\mathbf{P}\cdot\hat{k}\times\hat{E})(\hat{k}\cdot\hat{E}) = \frac{1}{2}|\mathbf{P}|\cos\theta\sin\theta \tag{3.88}$$

assuming that $\mathbf{P}$ is perpendicular to the $\hat{k}$, $\hat{E}$ plane, and θ is the angle between $\hat{k}$ and $\hat{E}$. The intensity distribution is illustrated by the clover leaf pattern in Fig. 3.2. Since the maximum intensity difference occurs in the directions $\theta = \pi/4$ and $\theta = 3\pi/4$, the $E1-E1$ damping asymmetry is defined by

$$A = \frac{I(\pi/4) - I(3\pi/4)}{I(\pi/4) + I(3\pi/4)}$$
$$= \frac{3\mathrm{Im}\,[V_{1/2}^*(V_{3/2}+M_{3/2})]}{2I_0(\pi/4)} . \tag{3.89}$$

Neglecting the relativistic and $M_{3/2}$ corrections, this reduces in the limit of weak fields to

$$A = \frac{3\Gamma|\mathbf{P}||E(2p_{3/2}) - E(2p_{1/2})]}{4\Delta_L^2 - 2\Delta_L\Delta_F + 7\Delta_F^2 + 11\Gamma^2/4} \tag{3.90}$$

independent of $|\mathbf{E}|$.

A high precision measurement of A for He^+ has been carried out by Drake, Patel and van Wijngaarden (1983), using essentially the same apparatus as for the Lamb shift anisotropy measurement shown in Fig. 3.4. Nearly complete spin polarization can easily be achieved by prequenching one of the $2s_{1/2,\pm 1/2}$ magnetic substates in the presence of an axial magnetic field. As shown in Fig. 3.5, a magnetic field in the z-direction causes a well-known crossing of the states $2s_{1/2,-1/2}$ and $2p_{1/2,1/2}$ near 7800 G. When the Zeeman energy shifts are taken into account, the dominant first term of equation (3.54) for the total decay rate becomes

$$\overline{w}(\pm 1/2) = \frac{3e^2 a_0^2 |\mathbf{E}|^2 \gamma_{2p}}{Z^2 [(\Delta_L \pm 4\mu_B B/3)^2 + \Gamma^2/4]} \tag{3.91}$$

where $\mu_B = e\hbar/2$ is the Bohr magneton and B is the magnetic field strength. The dashed curve in Fig. 3.5 shows the dramatic increase of the $2s_{1/2,-1/2}$ decay rate near the level crossing for a transverse electric field of 100 V/cm. Since the other magnetic sub-level $2s_{1/2,1/2}$ decays much more slowly, virtually 100% polarization can be achieved and $|\mathbf{P}| = 1$.

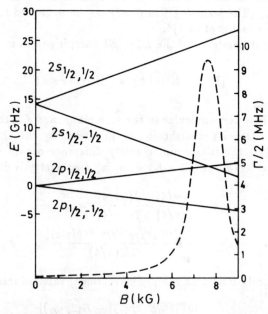

Fig. 3.5. Graph of the $He^+ 2s_{1/2}$ and $2p_{1/2}$ magnetic sublevel energies (left-hand scale) as a function of magnetic field strength. The dashed curve shows the $2s_{1/2}(m_j = -1/2)$ level width (right-hand scale) for an electric quenching field strength of 100 V/cm.

The theoretical asymmetry calculated from (3.90) with $|\mathbf{P}| = 1$ is 0.0076209 for He^+. A small finite electric field correction at the experimental quenching field strength of 246.71 V/cm reduces this to 0.0076182 — in good agreement with the observed value of 0.007603 ± 0.000020. This provides direct experimental confirmation of the Bethe-Lamb phenomenological quenching formalism since the effect depends directly upon the damping terms. Alternatively, if the experiment is interpreted as a measurement of the $2p$ lifetime $\tau = 1/(2\pi\Gamma)$, then the experimental value

$$\tau_{\exp} = (0.9992 \pm 0.0026) \times 10^{-10}\text{s}$$

falls within one standard deviation of the theoretical value

$$\tau_{\text{theo}} = 0.9972 \times 10^{-10}\text{s} \ .$$

This is the most accurate measurement of a lifetime for any hydrogen-like system.

3.3.5 $E1 - M1$ interference

As discussed in the previous section, the term $\mathbf{P}\cdot\mathbf{J}_0$ of (3.44) comes into play if the $2s_{1/2}$ state is initially spin-polarized. The first term of equation (3.87) is called the $E1 - M1$ interference term because it arises from the cross-product between the field-induced $E1$ and spontaneous $M1$ decay channels. The angular dependence of this term is

$$\mathbf{P}\cdot\widehat{k} \times \widehat{E} = |\mathbf{P}|\sin\theta \tag{3.92}$$

assuming that $\mathbf{P}$ is perpendicular to the $\widehat{k}$, $\widehat{E}$ plane and θ is the angle between $\widehat{k}$ and $\widehat{E}$. The intensity distribution gives rise to the left-right asymmetry shown by the innermost dashed curve in Fig. 3.2. Defining as before

$$A = \frac{I(\pi/2) - I(-\pi/2)}{I(\pi/2) + I(-\pi/2)}$$

then

$$A = 2\text{Re}\left[M_{1/2}^*(2V_{1/2} + V_{3/2} + M_{3/2})\right]/I_0 \ . \tag{3.93}$$

In the limit of weak quenching fields and small Z, A reduces to

$$A = 2Q\left[\frac{1 + |\rho|^2\Delta_F/(2\Delta_L)}{1 + \text{Re}\,(\rho) + 5|\rho|^2/2 + Q^2}\right] \tag{3.94}$$

where

$$Q = -\frac{243\sqrt{2}Z^2\Delta_L\overline{M}}{128e^2a_0^2|\mathbf{E}|} \tag{3.95}$$

and $\overline{M}$ is the relativistic magnetic dipole matrix element defined by equation (3.19). Q^2 is the ratio of the spontaneous $M1$ decay rate to the induced $E1$ decay rate via the $2p_{1/2}$ state. Since the $M1$ decay rate is only

$2.496 \times 10^{-6} Z^{10} \mathrm{s}^{-1}$, Q^2 is much less than unity for low Z ions at practical quenching fields and may be neglected in the denominator of (3.94). Then A increases in proportion to $1/|\mathbf{E}|$ as $|\mathbf{E}| \to 0$. It may also be necessary to include a contribution from two-photon transitions in the denominator of (3.94) if the photon detectors are sensitive to a broad band of frequencies other than $Ly - \alpha$ radiation.

The $E1 - M1$ asymmetry was first measured in He^+ by van Wijngaarden and Drake (1982), and more recently to higher precision by van Wijngaarden, Patel and Drake (1985) using an apparatus similar to the one shown in Fig. 3.4. The beam was initially spin-polarized as described in the previous section. For a quenching field of $|\mathbf{E}| = 43.63$ V/cm, equation (3.94) gives a theoretical asymmetry of 3.009×10^{-4} in good agreement with the measured value of $(2.935 \pm 0.337) \times 10^{-4}$. If the experiment is regarded as a measurement of the $M1$ matrix element, then $M = -(0.273 \pm 0.031)\alpha^2 e\lambdabar$ in agreement with the theoretical value $-0.2794\alpha^2 e\lambdabar$ obtained from (3.19).

It is difficult to attain higher precision because the effect being measured is so small. No advantage is gained by making $|\mathbf{E}|$ smaller so that A becomes larger because the experimental uncertainty comes primarily from the finite number n of photon counts recorded. Consequently, the $\pm\sqrt{n}/n$ statistical uncertainty also increases in proportion to $1/|\mathbf{E}|$ for the same total counting time as $|\mathbf{E}|$ is decreased. However, an advantage would be gained by measuring A in a heavier hydrogenic ion. If $|\mathbf{E}|$ is adjusted as a function of Z so that the total quench rate remains the same, then from equations (3.54) and (3.55)

$$|\mathbf{E}| \sim \Delta_L / Z$$

and

$$Q \sim Z^5 .$$

Using this scaling, $Q \simeq 1$ at $|\mathbf{E}| = 40$ kV/cm for $Z = 10$. However, it becomes correspondingly more difficult to achieve 100% polarization in the initial state as Z increases.

3.3.6 The total quench rate

Several measurements of the total quench rate of the $2s_{1/2}$ state in a constant electric field have been made as a method of determining the Lamb shift, beginning with the early measurement of Fan *et al.* (1967) for Li^{++}. Using (3.53) and retaining the lowest order relativistic corrections, the total decay rate is

$$\overline{w} = \overline{w}(M1) + \overline{w}(2E1) + \overline{w}(|\mathbf{E}|) \tag{3.96}$$

where

$$\overline{w}(M1) = \frac{\alpha^9 Z^{10}}{972}(1 + 1.0678\alpha^2 Z^2)\tau^{-1} , \qquad (3.97)$$

$\overline{w}(2E1)$ is the two photon decay rate discussed in section 3.5, and

$$\overline{w}(|\mathbf{E}|) = \frac{3e^2 a_0^2 |\mathbf{E}|^2 \gamma_{2p}}{Z^2} \left[\frac{1 + 2\mu_{1/2} + 2\mu'_{1/2} + \varepsilon}{|\Delta_L + i\Gamma/2|^2} + \frac{2(1 + 2\mu_{3/2} + 2\mu'_{3/2} + \varepsilon)}{|\Delta_F + i\Gamma/2|^2} \right] \qquad (3.98)$$

to lowest order in the external field. Here, γ_{2p} is the nonrelativistic $2p$ decay rate given by equation (3.55), the μ_j and μ'_j are relativistic corrections to the matrix elements as defined following equation (3.80), and $\varepsilon = 11\alpha^2 Z^2/48$ is the fractional relativistic correction to the $2s_{1/2} - 1s_{1/2}$ transition energy. The numerical values are

$$2\mu_{1/2} + 2\mu'_{1/2} + \varepsilon = (-5/6 - 3\ln 2 + 2\ln 3)\alpha^2 Z^2$$
$$\simeq -0.71555\alpha^2 Z^2$$

and

$$2\mu_{3/2} + 2\mu'_{3/2} + \varepsilon = [-9/16 - (5/2)\ln 2 + (3/2)\ln 3]\alpha^2 Z^2$$
$$\simeq -0.64745\alpha^2 Z^2 .$$

Higher order finite field corrections to (3.98) can easily be calculated as shown in the Appendix.

It is clear from equation (3.98) that if all the other parameters are assumed to be known, then a measurement of $\overline{w}(|\mathbf{E}|)$ determines the Lamb shift Δ_L. The original work of Fan *et al.* (1967) for Li^{++} was followed by several other determinations of the Lamb shift by this method for the ions C^{5+}, O^{7+}, F^{8+} and Ar^{17+} as summarized in Table 5. The experimental aspects of this work are treated in Chapter 6 of this book, together with other methods of measuring Lamb shifts in hydrogenic ions. For quench rate measurements, the basic technique is to pass a beam of metastable hydrogenic ions through a region of constant electric field and to determine $\overline{w}(|\mathbf{E}|)$ from the exponential decay of the $Ly - \alpha$ intensity as a function of position along the beam. For the heavier ions, a transverse magnetic field is used to obtain a sufficiently large $(\mathbf{v} \times \mathbf{B})/c$ motional electric field. The accuracy of this method has been limited to about $\pm 0.5\%$ because of the difficulties inherent in the measurement of lifetimes from exponential decay curves. Only the measurement of Gould and Marrus (1977, 1983) for Ar^{17+} is at a sufficiently high value of Z to provide a significant test of Lamb shift calculations. Their value of 37.89 ± 0.38 THz agrees with Mohr's

(1975) calculation of 38.250± 0.025 THz, but lies 2.7 standard deviations below Erickson's (1977) calculation of 39.01± 0.16 THz. Other more accurate methods of measuring the Lamb shift for lower Z ions are discussed elsewhere in this volume.

Table 5. Quenching Rate Measurements of the Lamb Shift[1]

Ion	Δ_L(theory)[2]	Δ_L(exp)
Li^{2+}	62.7375(66)	63.031(327)[3] GHz
C^{5+}	781.99(21)	780.1(8.0)[4]
O^{7+}	2196.21(92)	2202.7(11.0)[5]
		2215.6(7.5)[6]
F^{8+}	3343.1(1.6)	3405(75)[7]
Ar^{17+}	38.250(25)	37.89(38)[8] THz

[1] See also Chapter 6 in this volume for a more extensive tabulation of Lamb shift measurements.

[2] Mohr (1975).

[3] Fan et al. (1967).

[4] Murnick et al. (1971).

[5] Leventhal et al. (1972).

[6] Lawrence et al. (1972).

[7] Murnick et al. (1972).

[8] Gould and Marrus (1983).

3.4   Quantum Beats

3.4.1   Introduction

The phenomenon of quantum beats appears whenever the Hamiltonian describing a quantum system changes from H_0 to $H_0 + \Delta H$ over a time interval $\Delta t < \hbar/\Delta E$, where ΔE is the energy separation between eigenstates of H_0 which are coupled by ΔH. Whether the system being described is a two-level atom or a coupled pair of classical oscillators, a sudden change in the external conditions leads to a coherent superposition of the normal modes of oscillation and the appearance of beat frequencies. For atoms, the beats appear as oscillations in the normally smooth exponential decay of atomic radiation following the sudden excitation of a coherent superposition of states. Because of the uncertainty relation $\Delta E \Delta t \geq \hbar$, the line width of the emitted radiation becomes so broadened that it is no longer resolvable into separate frequency components ν_1 and ν_2, but the information that is lost from the frequency spectrum appears instead in the time spectrum in the form of quantum beats.

A coherent superposition of states can be achieved for atomic systems in many different ways such as the application of a short laser pulse whose line-width covers more than one atomic state, excitation by electron impact, the sudden entry of an atomic beam into an external field, or the passage of an atomic beam through a thin foil.

The term quantum beats was first used by Aleksandrov (1963) in connection with the coherent excitation of Zeeman sublevels by a modulated light source. As a simple example, if two states ψ_1 and ψ_2 with the same parity and decay rate γ are coherently excited at $t = 0$, then the wave function is of the form

$$\psi(t) = e^{-\gamma t/2}(C_1\psi_1 e^{-iE_1 t/\hbar} + C_2\psi_2 e^{-iE_2 t/\hbar}) . \qquad (3.99)$$

where C_1 and C_2 are constants. Since the decay rate for the emission of a photon of polarization $\hat{e}$ is proportional to an expression of the type

$$\sum_F |\langle\psi_F|\hat{e}\cdot\mathbf{r}|\psi(t)\rangle|^2$$

substitution of (3.99) yields the general form

$$I(t) \propto e^{-\gamma t}[A + B\cos(\Omega t + \phi)] \qquad (3.100)$$

for the time-dependent intensity. Here $\Omega = (E_1 - E_2)/\hbar$, and A, B and ϕ are time-independent constants. Since the oscillating part of (3.100) comes from cross-terms between ψ_1 and ψ_2, it is an interference effect which averages to zero on summing over polarizations and integrating over angles.

The excited states of hydrogen provide an opportunity to study quantum beats of a different sort. Since the spectrum contains close-lying states of opposite parity and very different lifetimes, coherent oscillations between a long-lived state (such as $2s_{1/2}$) and a short-lived state (such as $2p_{1/2}$) can be induced by the application of a weak electric field. The result is a modulated exponential decay of the total integrated intensity at the Stark-shifted $2s - 2p$ transition frequency. An early analysis of some of the effects to be expected for pulsed and modulated excitation was given by Series (1964) as a possible alternative method of measuring the Lamb shift. Since then, rich quantum beat patterns originating from the higher n states of hydrogen following beam foil excitation have been obtained by Bashkin *et al.* (1965) and Sellin *et al.* (1969). The interpretation of these experiments is complicated by the fact that the initial state populations and coherences are not accurately known as the beam emerges from the foil. However, Sellin *et al.* (1970) have been able to show that, with reasonable assumptions, one can give a good account of the general features.

The topics of quantum beats and beam-foil spectroscopy have been extensively discussed in past review articles (Andrä, 1974, 1979; Dodd and Series, 1978; Haroche, 1976). Rather than repeat this material, the intent of the present section is to study in some detail the quantum beats and time-dependent phenomena which arise when a hydrogen atom in the $2s_{1/2}$ metastable state rapidly enters an electric field. This problem has the advantage that the initial state is accurately known and experimental data are available for comparison. The techniques are illustrative of those that are useful in a wide variety of other problems.

3.4.2   Theory of field-induced quantum beats

Consider a fast beam of metastable $2s_{1/2}$ atoms as it enters a region of static electric field. In the co-moving atomic frame, the field appears to switch on as a function of time as shown in Fig. 3.6. Quantum beats will be observed in the field-induced $Ly - \alpha$ radiation as a function of position along the beam if the rise-time of the field is sufficiently fast compared with $\hbar/[E(2s_{1/2}) - E(2p_{1/2})]$.

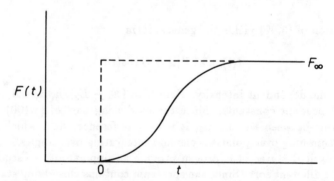

Fig. 3.6. Diagram showing the time dependence of the electric field strength in the sudden approximation (dashed curve) and for a finite rise time (solid curve).

Most discussions of quantum beats invoke the "sudden approximation" in which the perturbation is switched on as a step function at $t = 0$. One then only needs to expand the unperturbed state in the basis set of perturbed states and allow each component to propagate independently in time. In this section, we solve instead the full time-dependent problem in a finite basis set of interacting states, with the sudden approximation being a limiting case.

For fields up to about 10^4 V/cm, it is sufficient to consider only the manifold of states with principal quantum number $n = 2$. For hydrogen with nuclear spin $I = 1/2$, these are the 16 hyperfine states $2s_{1/2}(F = 0, 1)$, $2p_{1/2}(F = 0, 1)$ and $2p_{3/2}(F = 1, 2)$, each of which is $(2F + 1)$-fold degenerate in the absence of external fields. In this basis set of unperturbed states, the time-dependent Schrödinger equation to be solved is (with $\hbar = 1$)

$$i\,d\mathbf{a}/dt = \underline{U}(t)\mathbf{a} \tag{3.101}$$

$$\underline{U}(t) = \underline{E} + F(t)\underline{V} \tag{3.102}$$

where $\mathbf{a}$ is the 16-component column vector of time-dependent state amplitudes, $\underline{E}$ is the 16×16 diagonal matrix of field-free eigenvalues, $\underline{V}$ is the interaction matrix with the external electric field and $F(t)$ describes its time-dependence in the co-moving atomic frame such that $F(t) = 0$ for $t < 0$ and $F(t) \to F_\infty$ as $t \to \infty$. The diagonal matrix of field-free level widths $\underline{\Gamma}$ is included according to

$$\underline{E} = \underline{E}_0 - i\underline{\Gamma}/2 \ . \tag{3.103}$$

The components $a_i(t)$ of $\mathbf{a}$ are rapidly oscillating in time. Since (3.101) must in general be solved by numerical integration, it is computationally advantageous to transform to a new basis set with amplitudes $\mathbf{c}(t)$ defined by

$$\mathbf{c}(t) = \exp(i\underline{X}^{-1}\underline{U}_\infty\underline{X}t)\underline{X}^{-1}\mathbf{a}(t) \tag{3.104}$$

with $\underline{U}_\infty = \underline{E} + F_\infty\underline{V}$, and $\underline{X}$ is the matrix such that $\underline{X}^{-1}\underline{U}_\infty\underline{X}$ is diagonal. The columns of $\underline{X}$ are the eigenvectors of $\underline{U}_\infty$. Since $\underline{U}$ is non-Hermitian due to the inclusion of the imaginary level widths in the diagonal matrix elements, this is not a standard unitary transformation. In fact, it is readily seen that if $\underline{U}$ is symmetric, then $\underline{X}^{-1} = \underline{X}^T$ rather than $\underline{X}^\dagger$, even though $\underline{X}$ is complex.[b] The matrix $\underline{X}$ can be found, for example, by a simple generalization of Jacobi's

---

[b]In the more general case where $\underline{V}$ in (3.102) is a Hermitian matrix, write $\underline{X}$ in the form $\underline{X} = \underline{\Phi}\underline{Y}$, where $\underline{\Phi}$ is a diagonal matrix with diagonal elements

$$\Phi_{jj} = e^{i\phi_j}$$

and $\underline{Y}$ is complex. If the phases ϕ_j can be chosen to satisfy the set of equations

$$\phi_i - \phi_j = \arg(V_{ij}) = -\arg(V_{ji})$$

then the matrix $\underline{\Phi}^*\underline{V}\,\underline{\Phi}$ is brought into symmetric form and the inverse of $\underline{X}$ is

$$\underline{X}^{-1} = \underline{Y}^T\underline{\Phi}^* \ .$$

Since the atomic states have definite parity, the above choice of phases is always possible for an arbitrary configuration of electric and magnetic fields.

method for matrix diagonalization. The matrices transformed by $\underline{X}$ are denoted by

$$\underline{\tilde{E}} = \underline{X}^{-1}\underline{U}_\infty\underline{X} \tag{3.105}$$

and

$$\underline{\tilde{V}} = \underline{X}^{-1}\underline{V}\underline{X} . \tag{3.106}$$

The components of $\underline{\tilde{E}}$ are the Stark-shifted eigenvalues, including the perturbed level widths. Substituting equation (3.104) into (3.101) then gives the equation for the coefficients $\mathbf{c}(t)$

$$i d\mathbf{c}/dt = f(t)e^{i\underline{\tilde{E}}t}\underline{V}e^{-i\underline{\tilde{E}}t}\mathbf{c}(t) \tag{3.107}$$

with

$$f(t) = F(t) - F_\infty \tag{3.108}$$

and subject to the initial condition

$$\mathbf{c}(0) = \underline{X}^{-1}\mathbf{a}(0) . \tag{3.109}$$

Except that the transformation defined by $\underline{X}$ is nonunitary, the above is equivalent to the interaction representation in the basis set of field-perturbed states. Since $f(t) \to 0$ as $F(t) \to F_\infty$, equation (3.107) need only be integrated through the fringing field region to a point where $\mathbf{c}(t)$ is sufficiently close to its limiting value $\mathbf{c}(\infty)$. If the field is turned on as a step function at $t = 0$, then $d\mathbf{c}/dt = 0$ and $\mathbf{c}(t) = \mathbf{c}(0)$ independent of t. The results are then identical to the sudden approximation.

If the initial state is an incoherent mixture of atomic states described by the density matrix

$$\underline{\rho}_0 = \sum_i \alpha_i \mathbf{a}^{(i)}(0)\mathbf{a}^{(i)\dagger}(0) \tag{3.110}$$

then (3.107) must be solved separately for each of the set of initial conditions

$$\mathbf{c}^{(i)}(0) = \underline{X}^{-1}\mathbf{a}^{(i)}(0), \quad i = 1, 2, \ldots, 16 . \tag{3.111}$$

The density matrix in the transformed basis set then evolves in time according to

$$\begin{aligned}
\underline{\tilde{\rho}}(t) &= \underline{X}^{-1}\underline{\rho}(t)\underline{X}^* \\
&= \sum_i \alpha_i e^{-i\underline{\tilde{E}}t}\mathbf{c}^{(i)}(t)\mathbf{c}^{(i)\dagger}(t)e^{iE^*t} .
\end{aligned} \tag{3.112}$$

In the sudden approximation, this simplifies to

$$\tilde{\underline{\rho}}(t) = e^{-i\underline{\tilde{E}}t}\tilde{\underline{\rho}}(0)e^{i\underline{\tilde{E}}^*t} \tag{3.113}$$

or, in the basis set of field-free states

$$\underline{\rho}(t) = \underline{X}e^{-i\underline{\tilde{E}}t}\underline{X}^{-1}\underline{\rho}(0)\underline{X}^*e^{i\underline{\tilde{E}}^*t}\underline{X}^{*-1} . \tag{3.114}$$

The above provides a convenient method for writing down the exact solutions to the time-dependent Schrödinger equation for an arbitrary number of interacting states. The solution reduces to the well-known results for two interacting states (Series, 1964; Lamb, 1952) as further discussed in section 3.4.3. The quantum beat oscillations in the off-diagonal matrix elements of $\tilde{\underline{\rho}}(t)$ are explicitly displayed by writing, from (3.113)

$$\tilde{\rho}_{jk}(t) = e^{-i(\tilde{E}_j - \tilde{E}_k)t - (\tilde{\Gamma}_j + \tilde{\Gamma}_k)t}\tilde{\rho}_{jk}(0) .$$

The emission of radiation by the perturbed atom is described by the time-dependent transition probability per unit time

$$w(\hat{e}, \mathbf{k}, t)d\Omega = C\sum_i \alpha_i \sum_f |\langle\psi_f|\hat{e}\cdot\mathbf{r}|\psi_i(t)\rangle|^2 d\Omega \tag{3.115}$$

in the electric dipole approximation. If the total transition probability integrated over time is normalized to unity, then

$$C = e^2\omega^3/(2\pi\hbar c^3) .$$

The time-dependent initial states defined by the solutions to (3.99) are

$$\psi_i(t) = \begin{cases} \psi_i e^{-iE_i t} , & t \le 0 \\ \sum_{k,j}^{16} \psi_k X_{kj} c_j^{(i)}(t)e^{-i\tilde{E}_j t} , & t > 0 . \end{cases} \tag{3.116}$$

On substituting this expression into (3.115), and defining a detection matrix $\underline{D}$ by

$$D_{k',k} = \sum_f \langle\psi_{k'}|\hat{e}^*\cdot\mathbf{r}|\psi_f\rangle\langle\psi_f|\hat{e}\cdot\mathbf{r}|\psi_k\rangle \tag{3.117}$$

the result can be expressed in any one of the forms

$$\begin{aligned} w(\hat{e}, \mathbf{k}, t)d\Omega &= C\mathrm{Tr}\,(\underline{\rho}\underline{D})d\Omega \\ &= C\mathrm{Tr}\,(\underline{X}\tilde{\underline{\rho}}\underline{X}^{*-1}\underline{D})d\Omega \\ &= C\mathrm{Tr}\,(\tilde{\underline{\rho}}\underline{X}^{*-1}\underline{D}\underline{X})d\Omega \\ &= C\mathrm{Tr}\,(\tilde{\underline{\rho}}\tilde{\underline{D}})d\Omega , \end{aligned} \tag{3.118}$$

where $\tilde{D} = \underline{X}^{*-1} \underline{D} \underline{X}$. Note that this transformation is not the same as the transformation defined by (3.105) and (3.106). The matrix $\underline{D}$ is analogous to the product $\underline{A}^\dagger \underline{A}$ appearing in equation (3.114). In the above transformations, we have used the fact that the trace of a matrix product is invariant under any cyclic permutation of the factors.

In the sudden approximation, (3.114) can be used to write the transition probability in either of the forms

$$w(\hat{e}, \mathbf{k}, t)d\Omega = C\mathrm{Tr}\,[\underline{U}(t)\underline{\rho}(0)\underline{U}^*(t)\underline{D}]d\Omega$$
$$= C\mathrm{Tr}\,[\underline{\rho}(0)\underline{U}^*(t)\underline{D}\underline{U}(t)]d\Omega \qquad (3.119)$$

where

$$\underline{U}(t) = \underline{X}e^{-i\tilde{\underline{E}}t}\underline{X}^{-1} \qquad (3.120)$$

is the time-development matrix. In the first form above, the time development is attached to the density matrix, while in the second it is attached to the detection matrix.

The angular dependence of the detection matrix $\underline{D}$ can be easily extracted if there is no selection of particular final magnetic substates involved in the detection process. The final state density operator

$$\rho_F = \sum_f |\psi_f\rangle\langle\psi_f| \qquad (3.121)$$

then transforms as a scalar under rotations and the product of dipole transition operators appearing in (3.117) can be recoupled according to

$$(\hat{e}\cdot\mathbf{r})\rho_F(\hat{e}\cdot\mathbf{r}) = \sum_{K,Q}(-1)^{K+Q}T_{K,-Q}(\hat{e}^*, \hat{e})T_{K,Q}(\mathbf{r}, \rho_F\mathbf{r}) . \qquad (3.122)$$

All of the angular and polarization dependence of the emitted radiation is then contained in the irreducible tensors $T_{K,Q}(\hat{e}^*, \hat{e})$ defined by

$$T_{K,Q}(\hat{e}^*, \hat{e}) = \sum_{q_1, q_2}(\hat{e}^*)_{q_1}\hat{e}_{q_2}\langle 1q_1, 1q_2|KQ\rangle . \qquad (3.123)$$

If an arbitrary $\hat{e}$ is expressed in terms of basis vectors $\hat{e}_1$ and $\hat{e}_2$ according to

$$\hat{e} = \cos\beta\hat{e}_1 + i\sin\beta\hat{e}_2$$

as in (3.42) and (3.43), then

$$T_{K,Q}(\hat{e}^*, \hat{e}) = \cos^2\beta T_{K,Q}(\hat{e}_1, \hat{e}_1) + \sin^2\beta T_{K,Q}(\hat{e}_2, \hat{e}_2)$$
$$+ i\sin\beta\cos\beta[T_{K,Q}(\hat{e}_1, \hat{e}_2) - T_{K,Q}(\hat{e}_2, \hat{e}_1)] .$$
$$(3.124)$$

Since the only non-vanishing contributions are of the form (see, for example, Brink and Satchler, p. 54)

$$T_{0,0}(\hat{e}, \hat{e}) = -1/\sqrt{3} \tag{3.125}$$

$$T_{1,Q}(\hat{e}_1, \hat{e}_2) = (i/\sqrt{2})C_1^Q(\hat{e}_1 \times \hat{e}_2) = (i/\sqrt{2})C_1^Q(\hat{k}) \tag{3.126}$$

$$T_{2,Q}(\hat{e}, \hat{e}) = (1/\sqrt{6})C_2^Q(\hat{e}) \tag{3.127}$$

where $C_K^Q(\hat{e}) = [4\pi/(2K+1)]^{1/2}Y_K^Q(\hat{e})$, the explicit values of (3.124) are

$$T_{0,0}(\hat{e}^*, \hat{e}) = -1/\sqrt{3} \tag{3.128}$$

$$T_{1,Q}(\hat{e}^*, \hat{e}) = -(1/\sqrt{2})\sin 2\beta C_1^Q(\hat{k}) \tag{3.129}$$

$$T_{2,Q}(\hat{e}^*, \hat{e}) = (1/\sqrt{6})\{C_2^Q(\hat{e}_1) + C_2^Q(\hat{e}_2)$$
$$+ \cos 2\beta[C_2^Q(\hat{e}_1) - C_2^Q(\hat{e}_2)]\} . \tag{3.130}$$

In the above form, it is obvious that the $T_{K,Q}$ have the correct transformation properties under rotations. Recall that $\cos 2\beta$ measures the degree of linear polarization and $\sin 2\beta$ the degree of circular polarization of the emitted light. These terms vanish on summing over polarizations and only $T_{0,0}$ survives on integrating over angles. If the vectors are chosen for definiteness to be

$$\hat{k} = \sin\theta \cos\phi \hat{x} + \sin\theta \sin\phi \hat{y} + \cos\theta \hat{z}$$
$$\hat{e}_1 = -\sin\phi \hat{x} + \cos\phi \hat{y} \tag{3.131}$$
$$\hat{e}_2 = \cos\theta \cos\phi \hat{x} + \cos\theta \sin\phi \hat{y} - \sin\theta \hat{z}$$

then

$$C_2^Q(\hat{e}_1) = C_2^Q(\pi/2, \pi/2 + \phi) \tag{3.132}$$

$$C_2^Q(\hat{e}_2) = C_2^Q(\pi/2 + \theta, \phi) \tag{3.133}$$

and

$$T_{2,\pm 2}(\hat{e}^*, \hat{e}) = -(1/\sqrt{6})(1 - \cos 2\beta)C_2^{\pm 2}(\hat{k}) - (1/2)\cos 2\beta e^{\pm 2i\phi}$$
$$T_{2,\pm 1}(\hat{e}^*, \hat{e}) = -(1/\sqrt{6})(1 - \cos 2\beta)C_2^{\pm 1}(\hat{k}) \tag{3.134}$$
$$T_{2,0}(\hat{e}^*, \hat{e}) = -(1/\sqrt{6})(1 - \cos 2\beta)C_2^0(\hat{k}) - (1/\sqrt{6})\cos 2\beta .$$

The above expressions are derived in some detail because they are not equivalent to those tabulated by Andrä (1979; p 897). The coefficients of $T_{1,Q}$ and $T_{2,Q}$ which result when (3.122) is used in (3.118) are the components of the orientation vector and alignment tensor respectively as discussed by Fano and Macek (1973).

The remaining part of (3.122) is $T_{KQ}(\mathbf{r}, \rho_F \mathbf{r})$ defined by

$$T_{KQ}(\mathbf{r}, \rho_F \mathbf{r}) = \sum_{m_1, m_2} r_{m_1} \rho_F r_{m_2} \langle 1 m_1 1 m_2 | K Q \rangle . \tag{3.135}$$

The matrix elements of (3.135) are easily found by standard methods of vector coupling (see, for example, Edmonds, pp. 95 and 109) to be

$$\langle \gamma' j' m' | T_{KQ}(\mathbf{r}, \rho_F \mathbf{r}) | \gamma j m \rangle = (-1)^{j'-m'} \begin{pmatrix} j' & K & j \\ -m' & Q & m \end{pmatrix}$$
$$\times \langle \gamma' j' \| T_K(\mathbf{r}, \rho_F \mathbf{r}) \| \gamma j \rangle \tag{3.136}$$

where the reduced matrix element is

$$\langle \gamma' j' \| T_K(\mathbf{r}, \rho_F \mathbf{r}) \| \gamma j \rangle = (2K+1)^{1/2} (-1)^{j'+j+K} \sum_{\gamma_f, j_f} \begin{Bmatrix} j' & j & K \\ 1 & 1 & j_f \end{Bmatrix}$$
$$\times \langle \gamma' j' \| \mathbf{r} \| \gamma_f j_f \rangle \langle \gamma_f j_f \| \mathbf{r} \| \gamma j \rangle . \tag{3.137}$$

As a final step, one may expand $\underline{\rho}(t)$ as defined by (3.112) (or (3.114) in the sudden approximation) into time-dependent state multipoles according to

$$\rho_{\gamma j m, \gamma' j' m'}(t) = \sum_{K, Q} (-1)^{K-j'-m} \rho_{KQ}(\gamma j, \gamma' j'; t) \langle KQ | j - m j' m' \rangle \tag{3.138}$$

with

$$\rho_{KQ}(\gamma j, \gamma' j'; t) = \sum_{m, m'} (-1)^{K-j'-m} \rho_{\gamma j m, \gamma' j' m'}(t)$$
$$\times \langle j - m j' m' | K Q \rangle \tag{3.139}$$

(Edmonds, 1960; Blum, 1979). Then, using the summation properties of the vector coupling coefficients, the trace appearing in the first of equations (3.118) becomes

$$\mathrm{Tr}\,[\underline{\rho}(t) \underline{D}] = \sum (2K+1)^{1/2} (-1)^{K+Q} \rho_{KQ}(\gamma j, \gamma' j'; t)$$
$$\times T_{K,-Q}(\hat{e}^*, \hat{e}) \langle \gamma' j' \| T_K(\mathbf{r}, \rho_F \mathbf{r}) \| \gamma j \rangle \tag{3.140}$$

where the sum is over $K, Q, \gamma, \gamma', j, j'$. This provides a convenient factorization of the transition probability into the time dependence of the initial state, the angular and polarization dependence of the emitted photons, and the reduced dipole transition integrals. The form of (3.140) is similar to the theory of perturbed angular correlations in nuclear physics (Frauenfelder and Steffen, 1968; Blum, 1979). In the present case, the transformation matrix $\underline{X}$ is not invariant under rotations due to the presence of external fields, and the system can evolve into multipoles which were not initially present.

3.4.3 The two-state approximation

Considerable insight into the nature of the solutions to equation (3.101) of the previous section is gained by considering the case of just two interacting states $|a\rangle$ and $|b\rangle$ such as $2s_{1/2}$ and $2p_{1/2}$ of hydrogen in the presence of an external electric field. The matrix $\underline{U}_\infty$ in the basis set of unperturbed states then has the form

$$\underline{U}_\infty = \begin{pmatrix} E_a - i\Gamma_a/2 & V \\ V & E_b - i\Gamma_b/2 \end{pmatrix} \tag{3.141}$$

and the transformation which diagonalizes U_∞ is

$$\underline{X} = \begin{pmatrix} \cos\theta & -\sin\theta \\ \sin\theta & \cos\theta \end{pmatrix} \tag{3.142}$$

where θ is a complex angle of rotation defined by

$$\tan 2\theta = \frac{-2V}{\omega - i\Gamma/2} \tag{3.143}$$

with $\omega = E_b - E_a$ and $\Gamma = \Gamma_b - \Gamma_a$. Then

$$\sin 2\theta = -2V/r, \quad \cos 2\theta = (\omega - i\Gamma/2)/r \tag{3.144}$$

where

$$r = [(\omega - i\Gamma/2)^2 + 4V^2]^{1/2} . \tag{3.145}$$

In the notation of Series (1964), r can be separated into its real and imaginary parts according to

$$r = q - ip \tag{3.146}$$

with

$$q = [|r^2| + \mathrm{Re}\,(r^2)]^{1/2}/\sqrt{2} \tag{3.147}$$

$$p = [|r^2| - \mathrm{Re}\,(r^2)]^{1/2}/\sqrt{2} \tag{3.148}$$

where

$$\mathrm{Re}\,(r^2) = \omega^2 - \Gamma^2/4 + 4V^2 \tag{3.149}$$

$$|r^2| = [(\omega^2 - \Gamma^2/4 + 4V^2)^2 + \omega^2 \Gamma^2]^{1/2} \tag{3.150}$$

and $pq = \omega\Gamma/2$. The elements of $\underline{X}$ are then

$$\cos\theta = [(1 + \cos 2\theta)/2]^{1/2} = \left[\frac{r + \omega - i\Gamma/2}{2r}\right]^{1/2} \tag{3.151}$$

$$\sin\theta = \frac{\sin 2\theta}{2\cos\theta}$$
$$= -\frac{V}{r}\left(\frac{2r}{r + \omega - i\Gamma/2}\right)^{1/2} \equiv -\mathrm{sgn}\,(V)\left(\frac{r - \omega + i\Gamma/2}{2r}\right)^{1/2} \tag{3.152}$$

The eigenvectors are

$$|\psi^{(1)}\rangle = \cos\theta|a\rangle + \sin\theta|b\rangle$$
$$|\psi^{(2)}\rangle = -\sin\theta|a\rangle + \cos\theta|b\rangle \tag{3.153}$$

with corresponding eigenvalues

$$\tilde{E}_1 = (1/2)(E_a + E_b - i\Gamma_a/2 - i\Gamma_b/2) - r/2$$
$$\tilde{E}_2 = (1/2)(E_a + E_b - i\Gamma_a/2 - i\Gamma_b/2) + r/2\,. \tag{3.154}$$

Because of the non-unitary nature of the transformation, the eigenvectors are normalized in the sense that $\sin^2\theta + \cos^2\theta = 1$ while

$$|\sin\theta|^2 + |\cos\theta|^2 = \frac{1}{\sqrt{2}}\left[1 + \frac{1}{(1 - \delta^2)^{1/2}}\right]^{1/2} \tag{3.155}$$

with

$$\delta = 2|V\Gamma|/(\omega^2 + 4V^2 + \Gamma^2/4)\,. \tag{3.156}$$

Lamb (1952) and Series (1964) point out that an interesting feature emerges if E_a and E_b can be made to cross one another by, for example, varying an external magnetic field as shown in Fig. 3.7. For $2|V| > \Gamma/2$, r is pure real at $\omega = 0$, and $\tilde{E}_1$ and $\tilde{E}_2$ repel one another, giving a standard avoided crossing as shown in the figure. However, if $2|V| < \Gamma/2$, then r is pure imaginary at $\omega = 0$,

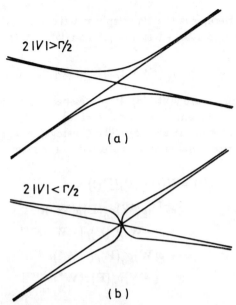

Fig. 3.7. Diagram showing the behaviour of the perturbed energy levels in the two-state approximation. The horizontal scale represents an external variable such as a magnetic field, and V is an additional electric field perturbation which mixes the states. For case (a), $2|V| > \Gamma/2$ and the real parts of the energies display the usual avoided crossing. For case (b), $2|V| < \Gamma/2$ and the real parts of the energies cross at the same point as for the unperturbed energies.

and the real parts of $\tilde{E}_1$ and $\tilde{E}_2$ continue to cross at the same point as for the unperturbed eigenvalues.

    The time-development matrix $\underline{U}(t)$ and the density matrix $\underline{\rho}(t)$ can now be immediately constructed as follows. First

$$\begin{aligned} \underline{U}(t) &= \underline{X} e^{-i\tilde{\underline{E}}t} \underline{X}^{-1} \\ &= e^{(-i\overline{E}t - \overline{\Gamma}t/2)}[\underline{1}\cos(rt/2) + i\boldsymbol{\sigma}\cdot\mathbf{W}\sin(rt/2)] \end{aligned} \tag{3.157}$$

where

$$\overline{E} = (E_a + E_b)/2$$
$$\overline{\Gamma} = (\Gamma_a + \Gamma_b)/2 ,$$

$\mathbf{W}$ is a complex vector defined by

$$\mathbf{W} = \sin 2\theta\, \hat{x} + \cos 2\theta\, \hat{z} \tag{3.158}$$

with $\mathbf{W} \cdot \mathbf{W} = 1$. Here, $\boldsymbol{\sigma}$ is the Pauli spin matrix in the abstract state vector space spanned by unperturbed states $|a\rangle$ and $|b\rangle$. In this space, the initial density matrix is

$$\underline{\rho}_0(\mathbf{P}) = 1/2[\underline{1} + \boldsymbol{\sigma} \cdot \mathbf{P}] \tag{3.159}$$

where $\mathbf{P} = \hat{z}$ specifies pure state $|a\rangle$, $\mathbf{P} = -\hat{z}$ specifies pure state $|b\rangle$, $\mathbf{P}$ in the xy-plane specifies a coherent superposition of $|a\rangle$ and $|b\rangle$, and $\mathbf{P} = 0$ describes an incoherent mixture of $|a\rangle$ and $|b\rangle$. Then, using the algebraic properties of the Pauli spin matrices, $\underline{\rho}(t)$ can be expressed in the form

$$\begin{aligned}
\underline{\rho}(t) &= \underline{U}(t)\underline{\rho}_0(\mathbf{P})\underline{U}^*(t) \\
&= e^{-\bar{\Gamma}t}[\underline{\rho}(\mathbf{W})\underline{\rho}_0(\mathbf{P})\underline{\rho}(\mathbf{W}^*)e^{pt} \\
&\quad + \underline{\rho}(-\mathbf{W})\underline{\rho}_0(\mathbf{P})\underline{\rho}(-\mathbf{W}^*)e^{-pt} \\
&\quad + \underline{\rho}(\mathbf{W})\underline{\rho}_0(\mathbf{P})\underline{\rho}(-\mathbf{W}^*)e^{iqt} \\
&\quad + \underline{\rho}(-\mathbf{W})\underline{\rho}_0(\mathbf{P})\underline{\rho}(\mathbf{W}^*)e^{-iqt}]
\end{aligned} \tag{3.160}$$

where the coefficients of $e^{\pm pt}$ and $e^{\pm iqt}$ are

$$\begin{aligned}
\underline{\rho}(\pm\mathbf{W})\underline{\rho}_0(\mathbf{P})\underline{\rho}(\pm\mathbf{W}^*) &= [\underline{A}_+ \pm \mathrm{Re}\,(\underline{B})]/4 \\
\underline{\rho}(\pm\mathbf{W})\underline{\rho}_0(\mathbf{P})\underline{\rho}(\mp\mathbf{W}^*) &= [\underline{A}_- \pm i\mathrm{Im}(\underline{B})]/4
\end{aligned} \tag{3.161}$$

with

$$\begin{aligned}
\underline{A}_\pm = \underline{\rho}_0(\mathbf{P}) \pm \{&\underline{\rho}_0(-\mathbf{P})\mathbf{W} \cdot \mathbf{W}^* \\
&+ \mathrm{Re}\,[(\boldsymbol{\sigma} \cdot \mathbf{W})(\mathbf{P} \cdot \mathbf{W}^*)] + (P_y - \sigma_y)\mathrm{Im}(\cos 2\theta \sin 2\theta^*)\}
\end{aligned} \tag{3.162}$$

$$\underline{B} = \mathbf{W} \cdot [\mathbf{P} + \boldsymbol{\sigma} + i\mathbf{P} \times \boldsymbol{\sigma}] \,. \tag{3.163}$$

This clearly reduces to $\rho(t) = \rho_0(\mathbf{P})$ at $t = 0$, but a component $\rho_0(-\mathbf{P})$ with opposite "polarization" develops with time, together with other off-diagonal coherence terms. The coefficient $\mathbf{W} \cdot \mathbf{W}^*$ of $\rho_0(-\mathbf{P})$ is

$$\mathbf{W} \cdot \mathbf{W}^* = |\cos 2\theta|^2 + |\sin 2\theta|^2 = 1/(1 - \delta^2)^{1/2} \tag{3.164}$$

with δ defined by (3.156).

To calculate the time-dependence of the emitted radiation, assume that only state $|b\rangle$ radiates so that $\Gamma_a = 0$, and the detection matrix has the form

$$\underline{D} = D\underline{\rho}(-\hat{z}) = \begin{pmatrix} 0 & 0 \\ 0 & D \end{pmatrix} . \tag{3.165}$$

One then easily obtains

$$\begin{aligned} I(t) &= \mathrm{Tr}\,[\underline{\rho}(t)\underline{D}] \\ &= \frac{De^{-\overline{\Gamma}t}}{8}\{[\alpha_+ + \mathrm{Re}\,(\beta)]e^{pt} + [\alpha_+ - \mathrm{Re}\,(\beta)]e^{-pt} \\ &\quad + [\alpha_- + i\mathrm{Im}\,(\beta)]e^{iqt} + [\alpha_- - i\mathrm{Im}\,(\beta)]e^{-iqt}\} \end{aligned} \tag{3.166}$$

with

$$\begin{aligned} \alpha_\pm &= (1 - P_z)(1 \pm |\cos 2\theta|^2) \pm (1 + P_z)|\sin 2\theta|^2 \\ &\quad \mp 2\mathrm{Re}\,[(P_x - iP_y)\sin 2\theta \cos 2\theta^*] \end{aligned} \tag{3.167}$$

$$\beta = 2(P_x - iP_y)\sin 2\theta - 2(1 - P_z)\cos 2\theta . \tag{3.168}$$

For the case $P_z = 1$ (pure state $|a\rangle$), this reduces to

$$I(t) = \frac{1}{2}De^{-\overline{\Gamma}t}|\sin 2\theta|^2[\cosh(pt) - \cos(qt)] . \tag{3.169}$$

For the case $P_z = -1$ (pure state $|b\rangle$), one obtains instead

$$\begin{aligned} I(t) =& \frac{1}{4}De^{-\overline{\Gamma}t}[|1 - \cos 2\theta|^2 e^{pt} + |1 + \cos 2\theta|^2 e^{-pt} \\ &+ 2(1 - |\cos 2\theta|^2)\cos(qt) + 4\mathrm{Im}\,(\cos 2\theta)\sin(qt)] \end{aligned} \tag{3.170}$$

in agreement with the result of Series (1964).

The above two cases are easily distinguished from one another in an observed quantum beat pattern because (3.169) vanishes at $t = 0$ while (3.170) is a maximum. However, the phase and amplitude of the oscillations alone are not sufficient to determine the **P** vector characterizing the initial state if P_x and P_y also contribute. In general, it is necessary to measure ratios involving all four coefficients in (3.166) to determine **P** uniquely.

In order to gain a better understanding of the results to be expected, assume that $\omega \gg \Gamma$ and the external field is sufficiently weak that $\omega \gg 2|V|$. Then from (3.144)

$$\sin 2\theta \simeq -2V(1 + i\Gamma/2\omega)/\omega$$
$$\cos 2\theta \simeq 1 - 2V^2(1 + i\Gamma/\omega)/\omega^2$$

and the coefficients of the oscillating terms in (3.166) are

$$\alpha_- \simeq -\frac{4V^2}{\omega^2}\left[2P_z + (1 - P_z)\frac{\Gamma}{\omega}\right] - \frac{2V}{\omega}\left(2P_x + P_y\frac{\Gamma}{\omega}\right) \qquad (3.171)$$

and

$$\mathrm{Im}\,(\beta) \simeq \frac{4V^2\Gamma}{\omega^3}(1 - P_z) - \frac{2V}{\omega}\left(P_x\frac{\Gamma}{\omega} - 2P_y\right) . \qquad (3.172)$$

The largest amplitude oscillations come from the P_x and P_y components of **P**, corresponding to the coherent excitation of states $|a\rangle$ and $|b\rangle$. These oscillations vary linearly in amplitude with field strength. If $P_x = 1$, then $\mathrm{Im}\,(\beta) \simeq \Gamma\alpha_-/\omega$ and the oscillations are predominantly $\cos(\omega t)$. Conversely, if $P_y = 1$, then $\alpha_- \simeq -\Gamma\,\mathrm{Im}\,(\beta)/\omega$ and the oscillations are predominantly $\sin(\omega t)$. The P_z terms are smaller by a factor of $2V/\omega$ in the limit of weak fields and hence the oscillations vary quadratically in amplitude with field strength. Even for an "unpolarized" beam, $\alpha_- \simeq -\mathrm{Im}\,(\beta) \simeq -4V^2\Gamma/\omega^3$, and $I(t)$ contains oscillating terms proportional to $\cos(\omega t)+\sin(\omega t)$. The linear versus quadratic dependence of the oscillation amplitude on field strength could be exploited to separate the coherent from the incoherent parts of the density matrix.

3.4.3.1 *Continuous and modulated excitation*

We consider next what happens when $I(t)$ is integrated over the process of excitation. Suppose that, instead of a sudden impulse at $t = 0$, the initial state specified by **P** is populated with probability $\sigma(t_0)dt_0$ in the time interval dt_0. Then the density matrix is

$$\bar{\rho}(t) = \int_0^t \rho(t - t_0)\sigma(t_0)dt_0 \qquad (3.173)$$

and the corresponding radiation intensity is

$$\bar{I}(t) = \int_0^t I(t - t_0)\sigma(t_0)dt_0 . \qquad (3.174)$$

Following Series (1964), two cases of special interest are excitation at a uniform rate described by $\sigma(t_0) = \sigma_0$, a constant, and modulated excitation described by

$$\sigma(t_0) = \sigma_0(1 + \cos f t_0) . \tag{3.175}$$

Under these conditions, it is appropriate to examine the steady-state situation as $t \to \infty$ and initial transient effects have died away.

For excitation at a continuous rate, equation (3.171) becomes

$$\bar{\underline{\rho}}(t) = \sigma_0 \left\{ \frac{1}{\Gamma}(1 + \sigma_x P_x) + \frac{(\omega^2 + \Gamma^2/4)}{4V^2\Gamma}(1 + P_z)(1 + \sigma_z) \right.$$
$$\left. - \frac{\omega}{2V\Gamma}[\sigma_x(1 + P_z) + P_x(1 + \sigma_z)] + \frac{1}{4V}[P_y(1 + \sigma_z) - \sigma_y(1 + P_z)] \right\} . \tag{3.176}$$

The above is obtained with the help of equations (3.144)-(3.150) by substituting (3.160) into (3.173) and integrating. The following identities are also useful

$$q^2 - p^2 = \omega^2 - \Gamma^2/4 + 4V^2$$
$$(\Gamma^2/4 - p^2)(\Gamma^2/4 + q^2) = \Gamma^2 V^2 .$$

Although (3.174) still has a complicated structure, it is immediately obvious that with $\underline{D} = D\underline{\rho}(-\hat{z})$, the steady-state radiation intensity is simply

$$\bar{I}(t) = \mathrm{Tr}\,(\bar{\underline{\rho}}\underline{D}) = \sigma_0 D/\Gamma \tag{3.177}$$

independent of ω, V and **P**. This result is to be expected because, for steady-state conditions, the rate of formation equals the total rate of decay, and the sum of the perturbed level widths equals the unperturbed level width of state $|b\rangle$.

For the case of modulated excitation, using (3.173) in (3.166) yields

$$\bar{I}(t) = \sigma_0 D/\Gamma + (\sigma_0 D/8)\{[\alpha_+ + \mathrm{Re}\,(\beta)]F(f, p) + [\alpha_+ - \mathrm{Re}\,(\beta)]F(f, -p)$$
$$+ \alpha_- \mathrm{Re}\,[G(f, q)] - \mathrm{Im}\,(\beta)\mathrm{Im}\,[G(f, q)]\} \tag{3.178}$$

where

$$F(f, p) = \frac{\sin(ft + \delta_p)}{[(\Gamma/2 - p)^2 + f^2]^{1/2}} \tag{3.179}$$

$$G(f, q) = \frac{\exp(ift + i\delta_{q-f})}{[\Gamma^2/4 + (q - f)^2]^{1/2}} + \frac{\exp(-ift + i\delta_{q+f})}{[\Gamma^2/4 + (q + f)^2]^{1/2}} \tag{3.180}$$

and the phase shifts are given by

$$\delta_p = \tan^{-1}\left(\frac{\Gamma/2 - p}{f}\right)$$

(3.181)

$$\delta_{q\pm f} = \tan^{-1}\left(\frac{q \pm f}{\Gamma/2}\right) .$$

(3.182)

The part of (3.178)in curly brackets reduces to $8/\Gamma$ in the limit $f \to 0$, and to the corresponding terms in (3.166) with the replacements

$$F(f,p) \to e^{pt}$$
$$G(f,q) \to 2e^{iqt} .$$

The analogous replacements in (3.169) and (3.170) yield the corresponding equations for the special cases $P_z = 1$ and $P_z = -1$.

As discussed by Series (1964), the $G(f,q)$ components contain resonances at $f = \pm q$. In general, the shapes are linear combinations of Lorentzian and dispersion contributions obtained by expanding

$$\text{Re}\,(G) = \cos(ft)\left[\frac{\Gamma/2}{\Gamma^2/4 + (q - f)^2} + \frac{\Gamma/2}{\Gamma^2/4 + (q + f)^2}\right]$$
$$- \sin(ft)\left[\frac{q - f}{\Gamma^2/4 + (q - f)^2} - \frac{q + f}{\Gamma^2/4 + (q + f)^2}\right]$$

(3.183)

$$\text{Im}\,(G) = \cos(ft)\left[\frac{q - f}{\Gamma^2/4 + (q - f)^2} + \frac{q + f}{\Gamma^2/4 + (q + f)^2}\right]$$
$$+ \sin(ft)\left[\frac{\Gamma/2}{\Gamma^2/4 + (q - f)^2} - \frac{\Gamma/2}{\Gamma^2/4 + (q + f)^2}\right] .$$

(3.184)

The discussion of the relative importance of $\text{Re}\,(G)$ and $\text{Im}\,(G)$ in (3.178) parallels that for the oscillating terms in (3.166). Using again the approximate expressions (3.171) and (3.172) for α_- and $\text{Im}\,(\beta)$ in the limit of weak fields, the largest amplitude signal comes from the P_x and P_y components of **P**. This part varies linearly with field strength. If $P_x = 1$, the $\text{Re}\,(G)$ term dominates, and if $P_y = 1$, the $\text{Im}\,(G)$ term dominates. The P_z terms are smaller by a factor of $2V/\omega$, and hence the resonance signal strength varies quadratically with field strength. For $P_z = 1$, only the $\text{Re}\,(G)$ term contributes, and for

$P_z = -1$ the $\text{Re}\,(G)$ term still dominates. A resonance signal persists even for an "unpolarized" beam, which is proportional to $\text{Re}\,(G) + \text{Im}\,(G)$ but smaller in amplitude by a further factor of Γ/ω. Possible experiments to detect these effects in hydrogen for the special case $P_z = -1$ are discussed by Series (1964). Closely related experiments which use the technique of modulated excitation for non-hydrogenic atoms have been done by Corney and Series (1964), Corney (1968) and Skaliński et al. (1965).

3.4.4   Comparison with experiment

Section 3.4.2 describes in detail how one can calculate the quantum beat pattern for the general case of many interacting states. For metastable hydrogen, these are the 16 hyperfine states $2s_{1/2}(F = 0, 1), 2p_{1/2}(F = 0, 1)$ and $2p_{3/2}(F = 1, 2)$. Although many quantum beat patterns have been observed by means of beam-foil spectroscopy, only a few experiments have been reported in which the initial state is accurately known. They involve passing a beam of hydrogen atoms prepared in the $2s_{1/2}$ state into an electric field region which rises from zero to its full value over a very short distance. This corresponds to a very short rise time in the co-moving atomic frame.

The first experiments of the above type were performed by Sokolov (1970) and Andrä (1974). They achieved a fast rise time by passing the atomic beam through a fine mesh which formed one plate of a parallel-plate capacitor. As the atoms pass through the mesh, they experience a sharply rising electric field parallel to the beam. The $Ly - \alpha$ intensity as a function of position along the beam contains quantum beats in good qualitative agreement with the predictions of the sudden approximation.

A more quantitative and higher resolution study of the quantum beats was performed by van Wijngaarden et al. (1976) using a different experimental arrangement. They achieved a rapidly rising transverse electric field by passing the metastable hydrogen beam through a small hole in an insulating end plate coated with a conductive carbon film as shown in Fig. 3.8. In the absence of the entrance hole, the conductive end plate completely eliminates fringing field effects because the electric field in the carbon film exactly matches that in the main body of the capacitor. The calculated field components parallel and perpendicular to the beam in the region of the entrance hole are shown in Fig. 3.9. $E_\perp$ rises rapidly to a peak value larger than the asymptotic field E_∞ over a distance of about 0.5 mm and then decays more slowly toward E_∞. There is also an important contribution from $E_\parallel$ near the entrance hole. The rise time of about 10^{-10} s is sufficiently short to excite both $2s_{1/2} - 2p_{1/2}$ and $2s_{1/2} - 2p_{3/2}$ quantum beats.

The experimental quantum beat pattern is compared with theory in Fig. 3.10 (Drake, 1977). The main oscillations correspond to the Stark-shifted $2s_{1/2} -$

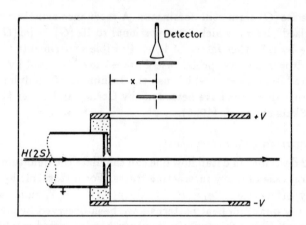

Fig. 3.8. Schematic diagram of the apparatus used to produce quantum beats by the sudden entry of a metastable hydrogen beam into an electric field. The end plate is covered with a conductive coating that generates an electrostatic charge distribution which would eliminate the fringing field in the absence of the entrance hole. (From van Wijngaarden *et. al.* 1976.)

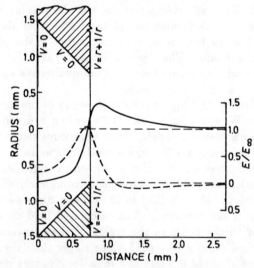

Fig. 3.9. Enlarged cross section though the quenching cell of Fig. 3.8 showing the electric field boundary conditions near the entrance hole (r is measured in units of the hole radius) and the calculated field components $E_\parallel$ and $E_\perp$ parallel and perpendicular to the horizontal beam axis. The full curve is $E_\perp/E_\infty$ and the broken curve is $E_\parallel/E_\infty$, where E_∞ is the asymptotic value of $E_\perp$.

$2p_{1/2}$ transition frequency, while the rapid oscillations correspond to the $2s_{1/2}-2p_{3/2}$ transition frequency. Not evident from the figure is a slow modulation of the beat pattern due to hyperfine structure.

Although the observed quantum beats are in rough qualitative agreement with the predictions of the sudden approximation, the finer details such as the relative phases of the oscillations and the depths of the minima depend strongly on the fringing fields shown in Fig 3.9. Since both $E_\perp$ and $E_\parallel$ are important in the fringing field region and have different time dependences, it is necessary to replace $f(t)\underline{\tilde{V}}$ in (3.107) by

$$f_\perp(t)\underline{\tilde{V}}_\perp + f_\parallel(t)\underline{\tilde{V}}_\parallel$$

where $\underline{\tilde{V}}_\perp$ contains the matrix elements of $-ez$ and $\underline{\tilde{V}}_\parallel$ contains the matrix elements of $-ex$ in the atomic frame of reference. A numerical integration through the fringing field region then yields the quantum beat pattern shown by the solid curve in Fig. 3.10. The agreement is about as good as can be expected, and provides strong support for the theoretical description of section 3.4.2.

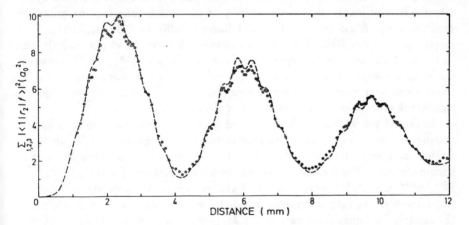

Fig. 3.10. Comparison of the theoretical quantum beat pattern for $H(2s_{1/2})$ (including instrumental broadening) with the experimental points obtained using the apparatus shown in Figs. 3.8 and 3.9. (From Drake, 1977.)

3.5 Two-Photon Transitions

3.5.1 Introduction

We have discussed in previous sections the single-photon $M1$ and induced $E1$ decay modes of the $2s_{1/2}$ state. However, in the absence of external fields, the dominant decay mechanism to the ground state for $Z < 40$ is the simultaneous emission of two $E1$ photons. The $2E1$ process arises from a second order interaction between the atom and the radiation field, as first derived by Goeppert-Meyer (1930). Non-relativistic calculations of the decay rate have been performed by several authors, beginning with the early estimates of Breit and Teller (1940) and culminating with the highly accurate values of Klarsfeld (1969) and Drake (1986). The application of more elegant mathematical techniques for performing implicit summations over intermediate states is discussed in recent papers by Tung *et al.* (1984) and Costescu *et al.* (1985).

The two-photon decay rate of neutral hydrogen is difficult to measure because the rate is only 8.229 s^{-1}. However, the emission has been observed in experiments by O'Connel *et al.* (1975), and Kruger and Oed (1975). In closely related work, Perrie *et al.* (1985) have measured the polarization correlation of the two photons emitted by metastable deuterium. This experiment is particularly significant because the results are in agreement with the predictions of quantum mechanics, but violate Bell's inequality by nearly two standard deviations. The experiment therefore helps to rule out the possibility of constructing a theory in which the indeterminacy of quantum mechanics is removed by the introduction of local "hidden variables", as inspired by the famous *gedankenexperiments* of Einstein, Podolsky and Rosen (1935), and Bohm (1951). Even larger violations of Bell's inequality have been observed by others, culminating in the meticulous work of Aspect, Dalibard and Roger (1981, 1982). However, their experiments are based on photons produced in an atomic cascade of single photon emission, rather than a true two-photon process, and may be subject to significant absorption and re-emission in the source.

In recent years, interest has centered on two-photon transitions in heavier hydrogenic ions. Since the decay rate increases in proportion to Z^6 along the isoelectronic sequence, accurate atomic-beam measurements of the decay rates become feasible. Such measurements have been performed for He$^+$, Li^{2+}, O^{7+}, F^{8+}, S^{13+} and Ar^{17+} (see Table 8 of this section). An accurate value of the two-photon decay rate is required in experiments to derive the Lamb shift from the electric field quenching rate of the $2s_{1/2}$ state in ions such as Ar^{17+} (Gould and Marrus, 1978, 1983). For these high-Z ions, relativistic effects become important. Accurate calculations, including relativistic effects to all orders, have been done by Goldman and Drake (1981), and Parpia and Johnson (1982).

In this section, the theory of two-photon transitions is first reviewed. Then methods of calculating relativistic two-photon decay rates are discussed and compared with experiment. Finally, closely related phenomena such as stimulated two-photon emission and Raman scattering are briefly mentioned, and it is shown that static electric field quenching can be viewed as the zero-frequency limit of a two-photon process.

3.5.2   Theory of two-photon transitions

The theory of two-photon transitions is best discussed in terms of the scattering matrix formalism of quantum electrodynamics. It is instructive first to re-consider the results of section 3.2 for single-photon transitions within the S-matrix formalism, and then to generalize to the two-photon case. The notations and conventions are the same as used by Akhiezer and Berestetskii (1965).

The spontaneous emission of a single photon of frequency ω and polarization $\hat{e}$ is described by the first order S-matrix element

$$S_{if}^{(1)} = -(e/\hbar) \int \psi_f(x) \widehat{A}^* \psi_i(x) d^4x \qquad (3.185)$$

where, in 4-component notation, $\widehat{A} = \gamma_\mu A_\mu$, $\overline{\psi} = \psi^* \gamma_4$ and $d^4x = d^3r\, dt$. In this and the following, a summation over repeated indices is implied. The necessary 4-vectors are defined by

$$\gamma = (-i\beta\boldsymbol{\alpha}, \beta)$$
$$x = (\mathbf{r}, ict)$$
$$k = (\mathbf{k}, i\omega/c)$$
$$A = (\mathbf{A}, iA_0)$$

with $\boldsymbol{\alpha} = \begin{pmatrix} 0 & \sigma \\ \sigma & 0 \end{pmatrix}$, $\beta = \begin{pmatrix} 0 & 1 \\ 1 & 0 \end{pmatrix}$ and $\psi_n(x) = \psi_n(\mathbf{r})e^{-iE_n t/\hbar}$ describes a stationary state of the electron with energy E_n. In the Coulomb gauge, the scalar potential A_0 is zero and the vector potential is

$$A(x) = \frac{1}{k}\left(\frac{2\pi\hbar\omega}{\mathcal{V}}\right)^{1/2} \hat{e} e^{ikx} = \mathbf{A}(\mathbf{r})e^{-i\omega t} \; . \qquad (3.186)$$

The part $\mathbf{A}(\mathbf{r})$ is the same as defined by (3.4). On substituting the above into (3.185), the integral over dt just gives a factor of $2\pi\delta(\omega - \omega_i + \omega_f)$, where $\omega_i = E_i/\hbar$. The remaining part of $S_{if}^{(1)}$ can be written in the form

$$S_{if}^{(1)} = (-2\pi i/\hbar)U_{if}^{(1)}\delta(\omega - \omega_i + \omega_f) \qquad (3.187)$$

where

$$U_{if}^{(1)} = -e \int \psi_f^*(\mathbf{r})\boldsymbol{\alpha}\cdot\mathbf{A}(\mathbf{r})\psi_i(\mathbf{r})d^3r \; . \qquad (3.188)$$

$U_{if}^{(1)}$ is the matrix element of the effective interaction energy of the electron with the electromagnetic field. It is related to the spontaneous decay rate by

$$w = (2\pi/\hbar)|U_{if}^{(1)}|^2 \rho_f(\omega)|_{\hbar\omega = E_i - E_f} \qquad (3.189)$$

which is the same as (3.1). The vertical bar notation means "evaluated at".

The simultaneous emission of two photons with vector potentials $A_1(x)$ and $A_2(x)$ corresponds to the two second order Feynman diagrams shown in Fig. 3.11. The corresponding second order S-matrix element is

$$
\begin{aligned}
S_{if}^{(2)} = {}&(e/\hbar)^2 \int\int \overline{\psi}_f(x_1)\widehat{A}_1^*(x_1)S_c^{(e)}(x_1,x_2)\\
&\widehat{A}_2^*(x_2)\psi_i(x_2)d^4x_1 d^4x_2\\
&+ (e/\hbar)^2 \int\int \overline{\psi}_f(x_1)\widehat{A}_2^*(x_1)S_c^{(e)}(x_1,x_2)\\
&\widehat{A}_1^*(x_2)\psi_i(x_2)d^4x_1 d^4x_2
\end{aligned}
\tag{3.190}
$$

where $S_c^{(e)}(x_1,x_2)$ is the electron propagator in the external field of the nucleus. It has the convenient spectral representation

$$
S_c^{(e)}(x_1,x_2) = \frac{1}{2\pi i}\int_{-\infty}^{\infty} d\omega\, e^{i\omega(t_1-t_2)}\sum_n \frac{\overline{\psi}_n(\mathbf{r}_1)\psi_n(\mathbf{r}_2)}{\omega_n(1-i\varepsilon)+\omega}.
\tag{3.191}
$$

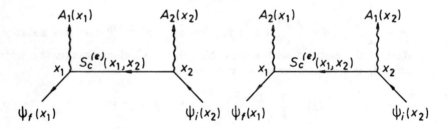

Fig. 3.11. Feynman diagrams for two-photon decay.

The summation over n includes both positive frequency (electron) and negative frequency (positron) solutions to the Dirac equation, with the poles in the denominator of (3.191) lying in the upper half plane in the former case and in the lower half plane in the latter case. This ensures the correct time ordering for both positive and negative frequency states since the integral over $d\omega$ vanishes if $t_1 < t_2$ for $\omega_n > 0$, and $t_1 > t_2$ for $\omega_n < 0$. When (3.191) and (3.186) are substituted into (3.190), the integrals over dt_1 and dt_2 can be performed with the result

$$
S_{if}^{(2)} = (-2\pi i/\hbar)U_{if}^{(2)}\delta(\omega_1+\omega_2-\omega_i+\omega_f)
\tag{3.192}
$$

where

$$U_{if}^{(2)} = -\frac{e^2}{\hbar} \sum_n \left[\frac{\langle f|\boldsymbol{\alpha}\cdot\mathbf{A}_1^*(\omega_1)|n\rangle\langle n|\boldsymbol{\alpha}\cdot\mathbf{A}_2^*(\omega_2)|i\rangle}{\omega_n + \omega_2 - \omega_i} \right.$$
$$\left. + \frac{\langle f|\boldsymbol{\alpha}\cdot\mathbf{A}_2^*(\omega_2)|n\rangle\langle n|\boldsymbol{\alpha}\cdot\mathbf{A}_1^*(\omega_1)|i\rangle}{\omega_n + \omega_1 - \omega_i} \right]$$

(3.193)

is the second order interaction energy. This is a relativistically exact expression involving a summation over both positive and negative frequency states.

The spectrum of the emitted radiation forms a continuum because the energy-conserving δ-function in (3.192) only requires that $\omega_1 + \omega_2 = \omega_i - \omega_f$. Thus only one of the two photon frequencies is independent. The triply differential emission rate in the energy interval $dE_1 = \hbar d\omega_1$ is

$$w(\omega_1,\omega_2)d\Omega_1 d\Omega_2 dE_1 = \frac{2\pi}{\hbar}|U_{if}^{(2)}|^2 \rho_f(\omega_1)\rho_f(\omega_2)dE_1$$
$$= \frac{\alpha^2 \omega_1 \omega_2}{(2\pi)^3 \hbar}|Q(\omega_1,\omega_2)|^2 d\Omega_1 d\Omega_2 dE_1$$

(3.194)

where

$$Q(\omega_1,\omega_2) = -\sum_n \left[\frac{\langle f|\boldsymbol{\alpha}\cdot\hat{e}_1 e^{-i\mathbf{k}_1\cdot\mathbf{r}}|n\rangle\langle n|\boldsymbol{\alpha}\cdot\hat{e}_2 e^{-i\mathbf{k}_2\cdot\mathbf{r}}|i\rangle}{\omega_n + \omega_2 - \omega_i} \right.$$
$$\left. + \frac{\langle f|\boldsymbol{\alpha}\cdot\hat{e}_2 e^{-i\mathbf{k}_2\cdot\mathbf{r}}|n\rangle\langle n|\boldsymbol{\alpha}\cdot\hat{e}_1 e^{-i\mathbf{k}_1\cdot\mathbf{r}}|i\rangle}{\omega_n + \omega_1 - \omega_i} \right].$$

(3.195)

The non-relativistic electric dipole approximation is obtained by making the replacement

$$\boldsymbol{\alpha}\cdot\hat{e}e^{-i\mathbf{k}\cdot\mathbf{r}} \to \mathbf{p}\cdot\hat{e}/mc$$

and restricting the sum in (3.195) to positive frequency intermediate states. The contribution from negative frequency states can be evaluated in lowest order since the denominators in (3.195) are approximately $-2mc^2/\hbar$. As discussed by Akhiezer and Berestetskii (1965, p. 489) one can then introduce projection operators onto negative frequency states and perform the sum over n by closure with the result that the negative frequency part of Q is

$$Q^-(\omega_1,\omega_2) \simeq (\hbar/mc^2)\hat{e}_1\cdot\hat{e}_2\langle f|e^{-i(\mathbf{k}_1+\mathbf{k}_2)\cdot\mathbf{r}}|i\rangle .$$

(3.196)

Since $|\mathbf{k}| = \omega/c$, the matrix element vanishes in the long wavelength approximation if the initial and final states are orthogonal. However, Q^- contributes to the relativistic corrections of relative order $\alpha^2 Z^2$ and must be included in an exact calculation.

Returning to the exact equation (3.195), it is advantageous to expand the transition operators into multipoles according to

$$\boldsymbol{\alpha}\cdot\widehat{e}e^{-i\mathbf{k}\cdot\mathbf{r}} = \sum_{LM\lambda}[\widehat{e}\cdot\mathbf{Y}_{LM}^{(\lambda)}(\widehat{k})]\boldsymbol{\alpha}\cdot\mathbf{a}_{LM}^{(\lambda)*}(\mathbf{r}) \tag{3.197}$$

where the $\mathbf{Y}_{LM}^{(\lambda)}(\widehat{k})$ are related to the vector spherical harmonics by

$$\mathbf{Y}_{LM}^{(0)}(\widehat{k}) = \mathbf{Y}_{L,L,M}(\widehat{k}) \tag{3.198}$$

$$\mathbf{Y}_{LM}^{(1)}(\widehat{k}) = -i\widehat{k}\times\mathbf{Y}_{LM}^{(0)}(\widehat{k}) \tag{3.199}$$

$$\mathbf{Y}_{LM}^{(-1)}(\widehat{k}) = \widehat{k}\mathbf{Y}_L^M(\widehat{k}) \tag{3.200}$$

and the $\mathbf{a}_{LM}^{(\lambda)}$ are the electric $(\lambda = 1)$ and magnetic $(\lambda = 0)$ multipole operators given by (3.9) and (3.10). The longitudinal $\lambda = -1$ part does not contribute in the Coulomb gauge because $\widehat{e}\cdot\widehat{k} = 0$.

Substituting the expansion (3.197) into (3.195) yields an expression for the complete angular correlation and polarization dependence of the emitted radiation for each combination of multipoles. The analysis is identical to that for the theory of angular correlations in nuclear physics. However, great simplifications occur on integrating over angles, summing over final states (including $\widehat{e}_1$ and $\widehat{e}_2$), and averaging over initial states with the result (Goldman and Drake, 1981)

$$\overline{w}(\omega_1,\omega_2)dE_1 = \frac{\alpha^2\omega_1\omega_2}{(2\pi)^3\hbar(2j_i+1)}\sum_{\substack{L_1\lambda_1 \\ L_2\lambda_2,j}}[|S^j(1,2)|^2$$

$$+|S^j(2,1)|^2 + 2\sum_{j'}(-1)^{2j'+L_1+L_2}[(2j+1)(2j'+1)]^{1/2}\begin{Bmatrix} L_1 & j' & j_f \\ L_2 & j & j_i \end{Bmatrix}$$

$$\times S^j(2,1)S^{j'}(1,2)^*]dE_1 . \tag{3.201}$$

The quantities $S^j(1,2)$ contain summations over all intermediate states with total angular momentum j for a particular combination of multipoles (L_1,λ_1) and (L_2,λ_2). They are given in terms of reduced matrix elements by

$$S^j(1,2) = \frac{(-1)^{j_f+j}}{(2j+1)^{1/2}}$$

$$\times\sum_n\frac{\langle n_f j_f\|\boldsymbol{\alpha}\cdot\mathbf{a}_{L_1}^{(\lambda_1)*}(\omega_1)\|nj\rangle\langle nj\|\boldsymbol{\alpha}\cdot\mathbf{a}_{L_2}^{(\lambda_2)*}(\omega_2)\|n_i j_i\rangle}{\omega_{n,j}+\omega_2-\omega_i} . \tag{3.202}$$

The above formulation is general and applies to any two-photon transition involving states with total angular momenta j_i, j and j_f. For the particular case of hydrogenic ions, the reduced matrix elements are all of the form (Grant, 1974)

$$\langle\alpha\|\boldsymbol{\alpha}\cdot\mathbf{a}_{LM}^{(\lambda)*}(\omega)\|\beta\rangle = (-i)^{L+\lambda+1}(-1)^{j}\alpha^{-1/2}(4\pi/(2L+1))^{1/2}$$

$$\times[(2j_\alpha+1)(2j_\beta+1)]^{1/2}\begin{pmatrix} j_\alpha & L & j_\beta \\ 1/2 & 0 & -1/2 \end{pmatrix} M_{\alpha\beta}^{(\lambda,L)} \tag{3.203}$$

where

$$M_{\alpha\beta}^{(1,L)} = \left[\left(\frac{L}{L+1}\right)^{1/2}[(\kappa_\alpha-\kappa_\beta)I_{L+1}^+ + (L+1)I_{L+1}^-]\right.$$

$$\left.-\left(\frac{L+1}{L}\right)^{1/2}[(\kappa_\alpha-\kappa_\beta)I_{L-1}^+ - LI_{L-1}^-]\right] \tag{3.204}$$

$$M_{\alpha\beta}^{(0,L)} = \frac{(2L+1)}{[L(L+1)]^{1/2}}(\kappa_\alpha+\kappa_\beta)I_L^+ \tag{3.205}$$

$$M_{\alpha\beta}^{(-1,L)} = -G[(2L+1)J_L + (\kappa_\alpha-\kappa_\beta)(I_{L+1}^+ + I_{L-1}^+) - LI_{L-1}^-$$

$$+ (L+1)I_{L+1}^-] \tag{3.206}$$

$$I_L^\pm = \int_0^\infty (g_\alpha f_\beta \pm f_\alpha g_\beta)j_L(\omega r/c)dr \tag{3.207}$$

$$J_L = \int_0^\infty (g_\alpha g_\beta + f_\alpha f_\beta)j_L(\omega r/c)dr . \tag{3.208}$$

In the above, g_α and f_α are the large and small components of the radial Dirac wave function as discussed in the following section, and κ is the Dirac quantum number. The summations in (3.201) are of course limited by the usual triangular and parity selection rules. Because of the factor (see equation (3.12))

$$j_L(\omega r/c) \simeq (\omega r/c)^L/(2L+1)!!$$

in (3.207) and (3.208), the contributions from higher multipoles decrease rapidly with increasing L for low Z, but they become increasingly important in the high Z region $(\alpha Z)^2 \to 1$.

The contribution from $M_{\alpha,\beta}^{(-1,L)}$ to $Q(\omega_1,\omega_2)$ vanishes identically if exact eigenfunctions of the Dirac equation are used and the set of intermediate states is complete (Goldman and Drake, 1981). The exact results are thus independent of the arbitrary gauge parameter G in (3.206). The replacements

$$M_{\alpha,\beta}^{(1,L)} \rightarrow M_{\alpha,\beta}^{(1,L)} - M_{\alpha,\beta}^{(-1,L)}$$
$$M_{\alpha\beta}^{(-1,L)} \rightarrow 0 \tag{3.209}$$

are equivalent to making a gauge transformation separately for each multipole. The choices $G = 0$ and $G = [(L+1)/L]^{1/2}$ in (3.209) (see equation (3.206)) correspond respectively to the velocity and length forms of the transition matrix elements in the non-relativistic limit. The degree to which the results depend on G provides an indication of the accuracy of approximate calculations.

3.5.3 Computational methods and results

The calculation of two-photon transition rates requires devising a method to perform the summations over the complete sets of intermediate states, including integrations over the positive and negative frequency continua, as shown in (3.195) or (3.202). The four-component Dirac equation to be solved is

$$H_D\psi = E\psi \tag{3.210}$$

with

$$H_D = c\boldsymbol{\alpha}\cdot\mathbf{p} + \beta mc^2 - Ze^2/r \tag{3.211}$$

and the Dirac matrices $\boldsymbol{\alpha}$ and β have their usual meanings. For any central potential, the solutions to (3.210) can be written in the form

$$\psi = \begin{pmatrix} ig(r)r^{-1}\ \Omega_{jlM} \\ -f(r)r^{-1}\ \Omega_{j\bar{l}M} \end{pmatrix} ,\bar{l} = 2j - l \tag{3.212}$$

where $g(r)$ and $f(r)$ are the large and small radial functions and Ω_{jlM} is a two-component spherical spinor defined as the vector coupled product

$$\Omega_{jlM} = \sum_{m\mu}\langle l,m,1/2,\mu|jM\rangle Y_l^m(\theta,\phi)\chi_\mu \tag{3.213}$$

with

$$\chi_{1/2} = \begin{pmatrix} 1 \\ 0 \end{pmatrix}, \quad \chi_{-1/2} = \begin{pmatrix} 0 \\ 1 \end{pmatrix}.$$

For convenience, one can define a real two component radial spinor by

$$\Phi(r) = \begin{pmatrix} g(r) \\ f(r) \end{pmatrix} . \tag{3.214}$$

Then $\Phi(r)$ satisfies the radial Dirac equation

$$H_r \Phi = E \Phi \tag{3.215}$$

with

$$H_r = \hbar c \left(-i\sigma_y \frac{d}{dr} + \sigma_x \frac{\kappa}{r} \right) + mc^2 \sigma_z - \frac{Ze^2}{r} \tag{3.216}$$

and κ is the Dirac quantum number $\kappa = \pm(j + 1/2)$ for $j = l \mp 1/2$. Exact analytic solutions to (3.216) are given, for example, by Bethe and Salpeter (1957).

The central problem is now to evaluate expressions of the form

$$T(\omega_1, \omega_2) = \frac{1}{\hbar} \sum_n \frac{\langle f|\tilde{a}_1|n\rangle \langle n|\tilde{a}_2|i\rangle}{\omega_n + \omega_2 - \omega_i} \tag{3.217}$$

where, for brevity, $\tilde{a}_1 = \boldsymbol{\alpha} \cdot \mathbf{a}_{L_1 M_1}^{(\lambda_1)}(\omega_1)$ and similarly for $\tilde{a}_2$. Two methods have been developed for this which are in principle equivalent, but are computationally very different. The first, used by Johnson (1972), and Parpia and Johnson (1982), involves re-writing (3.217) in the form

$$T(\omega_1, \omega_2) = \langle f|\tilde{a}_1|\tilde{i}\rangle \tag{3.218}$$

where $|\tilde{i}\rangle$ satisfies the inhomogeneous Dirac equation

$$(H_D + \hbar\omega_2 - E_i)|\tilde{i}\rangle = \tilde{a}_2|i\rangle . \tag{3.219}$$

Johnson solves by direct numerical integration the radial part of (3.219) analogous to (3.216). This approach is the relativistic generalization of the method of Dalgarno and Lewis (1955) and Dalgarno (1963) for calculating frequency-dependent polarizabilities. Effects of finite nuclear size are easily included, but the method has the disadvantage that equation (3.219) must be re-solved for each frequency ω_2.

The second approach, developed by Drake and Goldman (1981), and Goldman and Drake (1981), is the relativistic generalization of the method of Sturmian basis sets. In this method, the summation over intermediate states in

(3.217) is replaced by a summation over a variationally determined set of $2N$ discrete eigenvectors of the form

$$\Phi_i(r) = r^{\gamma-1} e^{-\lambda r} \sum_{j=1}^{N} r^j \left[a_{i,j} \begin{pmatrix} 1 \\ 0 \end{pmatrix} + b_{i,j} \begin{pmatrix} 0 \\ 1 \end{pmatrix} \right] \qquad (3.220)$$

with $\gamma = (\kappa^2 - \alpha^2 Z^2)^{1/2}$ and λ is an arbitrary adjustable parameter. For a given i, the $a_{i,j}$'s and $b_{i,j}$'s are $2N$ linear variational parameters determined by the conditions

$$\int_0^\infty \Phi_i^\dagger \Phi_j \, dr = \delta_{i,j} \qquad (3.221)$$

$$\int_0^\infty \Phi_i^\dagger H_r \Phi_j \, dr = \varepsilon_i \delta_{i,j} \; . \qquad (3.222)$$

The $\varepsilon_i, i = 1, \ldots, 2N$ are the discrete variational eigenvalues. The key point is that, for a Coulomb potential, the ε_i provide upper and lower bounds to the positive and negative frequency parts of the Dirac spectrum respectively, even though the spectrum is not bounded from below. In other words, a generalized version of the Hylleraas-Undheim (1930) theorem applies for basis sets constructed as shown in (3.220). The distribution of the eigenvalues as a function of the size of the basis set is illustrated in Fig. 3.12. The variational representation of the Dirac spectrum is further discussed by Grant (1982) and by Goldman (1985).

When the above variational basis set is used in place of the actual Dirac spectrum in (3.217), the result systematically improves as the basis set is enlarged. High accuracy can be obtained with $N \simeq 14$ if the arbitrary parameter λ in (3.220) is also optimized to a variational extremum. Only a single diagonalization step is required for each Z and j for all photon frequencies and combinations of multipoles, and all necessary integrals can be evaluated analytically. The above method of calculation yields results in good agreement with the revised values of Parpia and Johnson (1982) obtained by the numerical integration of (3.219).

The shape of the two-photon continuum is shown in Fig. 3.13. The quantity $\psi(y, Z)$ plotted in the figure is related to the differential decay rate by

$$dw/dy = Z^6 (9\alpha^6 / 2^{10}) \psi(y, Z) Ry \qquad (3.223)$$

where y is the fraction of the total $2s_{1/2} - 1s_{1/2}$ transition energy carried away by one of the two photons. The shape is a slowly varying function of Z, becoming somewhat more sharply peaked as Z increases owing to relativistic effects.

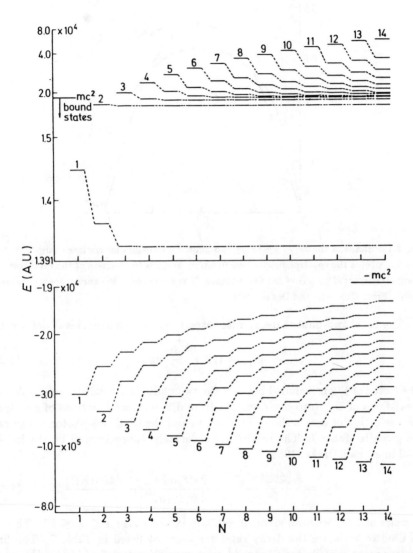

Fig. 3.12. Illustration of how the discrete variational eigenvalues of the Dirac equation change as the size of the basis set (horizontal scale) is enlarged. The upper part shows the positive energy spectrum and the lower part the negative energy spectrum. The vertical scale is logarithmic.

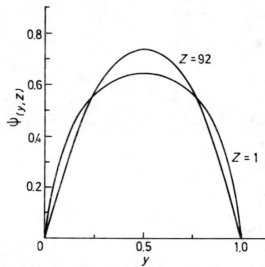

Fig. 3.13. Shape of the $2s_{1/2} \rightarrow 1s_{1/2}$ two-photon continuum for nuclear charge $Z = 1$ and $Z = 92$. $\psi(y)$ is the spectral distribution function, and y is the fraction of the total transition energy transported by one of the two photons. The areas under the curves are normalized to unity. (From Goldman and Drake, 1981.)

The total two-photon decay rate integrated over frequencies is defined by

$$\overline{w}_{2\gamma} = 1/2 \int_0^1 \frac{dw}{dy} \, dy \; . \tag{3.224}$$

The factor of $1/2$ is included because the two photons are indistinguishable. The breakdown of $\overline{w}_{2\gamma}$ into contributions from different combinations of multipoles for a selection of ions is shown in Table 6, and the total two-photon decay rates are given in Table 7. The results are accurately approximated by the formula (Goldman and Drake, 1981)

$$\overline{w}_{2\gamma} = \frac{8.22938 Z^6 [1 + 3.9448(\alpha Z)^2 - 2.040(\alpha Z)^4]}{[1 + 4.6019(\alpha Z)^2]} \tag{3.225}$$

in units of s^{-1}, with an error of $\pm 0.05\%$ in the range $1 \leq Z \leq 92$. This can be used to estimate the decay rates for ions not listed in Table 7. The finite nuclear mass contributes a reduced mass correction factor of $(1 - m/M)$ to the tabulated results and to equation (3.225). Nuclear motion in the center of mass frame introduces a further correction factor of Z_r^4, where

$$Z_r = \frac{1 + (Z - 1)m}{(m + M)}$$

Table 6. Contributions from Different Combinations of Multipoles to the $2s_{1/2}$ Two-Photon Decay Rate (in s^{-1}). (From Goldman and Drake, 1981)

	2E1	E1-M2	2M1	2E2	2M2	E2-M1
Z	$Z^{-6}\overline{w}$	$Z^{-10}\,10^{10}\overline{w}$	$Z^{-10}\,10^{11}\overline{w}$	$Z^{-10}\,10^{12}\overline{w}$	$Z^{-14}\,10^{22}\overline{w}$	$Z^{-14}\,10^{23}\overline{w}$
1	8.2291	2.5371	1.3804	4.9072	4.089	1.638
20	8.1181	2.5028	1.4046	4.8739	4.098	1.732
40	7.8096	2.4156	1.4778	4.7724	4.133	2.046
60	7.3446	2.3024	1.6005	4.5974	4.216	2.721
80	6.7440	2.1822	1.7737	4.3370	4.392	4.135
92	6.3097	2.1118	1.9030	4.1290	4.578	5.752

Table 7. Total Two-Photon Decay Rates in s^{-1} for Infinite Nuclear Mass (From Drake and Goldman, 1981)

Z	$\overline{w}^1$	Z	$\overline{w}$	Z	$\overline{w}$	Z	$\overline{w}$
1	8.2291	13	3.9491(7)	25	1.9672(9)	50	1.1869(11)
2	5.2661(2)	14	6.1547(7)	26	2.4850(9)	56	2.2980(11)
3	5.9973(3)	15	9.3017(7)	27	3.1111(9)	60	3.4282(11)
4	3.3639(4)	16	1.3686(8)	28	3.8628(9)	65	5.4387(11)
5	1.2847(5)	17	1.9668(8)	29	4.7592(9)	70	8.3139(11)
6	3.8347(5)	18	2.7682(8)	30	5.8217(9)	74	1.1404(12)
7	9.6654(5)	19	3.8242(8)	32	8.5405(9)	80	1.7701(12)
8	2.1525(6)	20	5.1956(8)	34	1.2236(10)	85	2.4824(12)
9	4.3612(6)	21	6.9530(8)	36	1.7164(10)	90	3.4021(12)
10	8.2010(6)	22	9.1785(8)	40	3.1990(10)	92	3.8361(12)
11	1.4518(7)	23	1.1966(9)	42	4.2651(10)	100	6.0045(12)
12	2.4451(7)	24	1.5423(9)	45	6.4003(10)	110	9.8152(12)

[1]Numbers in parentheses are the powers of 10 by which the numbers are to be multiplied. Rates for ions not listed can be estimated from equation (3.225). For the finite nuclear-mass correction, multiply the results by $1-m/M$ and $[1 + (Z - 1)m/(m+M)]^4$.

and $q_{\text{eff}} = -Z_r e$ is the "effective radiative charge" (Fried and Martin, 1963; Bacher, 1984). The corresponding correction factors for single photon transitions are $(1 - m/M)$ and Z_r^2.

The total theoretical decay rate of the $2s_{1/2}$ state is given by

$$\overline{w}_{\text{tot}} = \overline{w}_{2\gamma} + \overline{w}_{M1} \tag{3.226}$$

where $\overline{w}_{M1}$ is the spontaneous $M1$ decay rate discussed in section 3.2. Using (3.24) together with the relativistic correction factor listed in Table 3, $\overline{w}_{M1}$ is given to sufficient accuracy by

$$\overline{w}_{M1} = 2.4958 \times 10^{-6} Z^{10}[1 + 1.0964\alpha^2 Z^2]\text{s}^{-1} \qquad (3.227)$$

for $Z < 30$, assuming infinite nuclear mass. The values of $\overline{w}_{\text{tot}}$ are compared with the experimentally measured decay rates for several ions in Table 8. The experimental value for Ar^{17+} is sufficiently accurate to be sensitive to the small $M1$ contribution of 0.0908×10^8 s^{-1}. All of the measurements are in good agreement with theory. The experimental work is further discussed in the article by Träbert in Chapter 6.

Table 8. Comparison of Theoretical and
Experimental Total Decay Rates of the $2s_{1/2}$ State (in s^{-1}).

Ion	Theory	Experiment
He$^+$	526.61	491^{+96}_{-140}[2]
		520 ± 21[3]
		525 ± 5[4]
O^{7+}	2.1552×10^6	$(2.21\pm 0.22)\times 10^6$ [5]
F^{8+}	4.3699×10^6	$(4.22 \pm 0.28)\times 10^6$ [5]
S^{15+}	1.3964×10^8	$(1.37 \pm 0.13)\times 10^8$ [6]
Ar^{17+}	2.8590×10^8	$(2.868 \pm 0.029)\times 10^8$ [6]

[1] See also the article by Träbert in chapter 6.
[2] Kocher, Clendenin and Novick (1972).
[3] Prior (1972).
[4] Hinds, Clendenin and Novick (1978).
[5] Cocke *et al.* (1974).
[6] Gould and Marrus (1983).

3.5.4 Other two-photon processes

The closely related processes of stimulated two-photon emission and Raman scattering can be obtained by minor modifications of the results in section 3.5.2 for spontaneous two-photon emission. In stimulated two-photon emission, a photon of frequency ω_1 is absorbed and re-emitted coherently in the original direction, while a second photon of frequency $\omega_2 = \omega_i - \omega_f - \omega_1$ is emitted randomly into all available photon modes. If there are N_1 incident photons per unit volume with polarization $\hat{e}_1$, then the quantity $\rho_f(\omega_1)dE$ in (3.194) is replaced by $N_1 \mathcal{V}$, and the transition rate for the emission of photon 2 into solid angle $d\Omega_2$ with polarization vector $\hat{e}_2$ is

$$w(\omega_1, \omega_2)d\Omega_2 = (\alpha^2 c^3 \omega_2 N_1/\omega_1)|Q(\omega_1, \omega_2)|^2 d\Omega_2 . \qquad (3.228)$$

Since $N_1 c$ is the number of photons crossing unit area per unit time, the differential scattering cross section is simply

$$d\sigma/d\Omega_2 = (\alpha^2 c^2 \omega_2/\omega_1)|Q(\omega_1, \omega_2)|^2 . \qquad (3.229)$$

Raman scattering differs from the above only in that a photon of frequency ω_1 is absorbed and a single photon of frequency $\omega_2 = \omega_i - \omega_f + \omega_1$ is emitted randomly into all available photon modes. The decay rate is the same as for stimulated two-photon emission except that ω_1 is replaced by $-\omega_1$; i.e.

$$w(\omega_1, \omega_2)d\Omega_2 = (\alpha^2 c^3 \omega_2 N_1/\omega_1)|Q(-\omega_1, \omega_2)|^2 d\Omega_2 \qquad (3.230)$$

and similarly for the cross section.

Both stimulated two-photon emission and anti-Stokes Raman scattering from metastable deuterium atoms have been detected by Braunlich, Hall and Lambropoulos (1972). Although the experimental uncertainties are large, the measured cross sections are in rough agreement with calculations by Zernick (1963, 1964) and Klarsfeld (1972).

3.5.5   *Static field quenching as a two-photon process*

It is instructive to consider the quenching of hydrogenic $2s_{1/2}$ ions by a static electric field $\mathbf{E}$ as the low frequency limit of a two-photon process. This point of view establishes a close connection between the formalism of the present section and the quantum beats discussed in section 3.4.

If there is no resonance near zero frequency, then, as pointed out by Zernick (1964), one may imagine that the electric field is produced by a beam of very low frequency photons polarized so that $\hat{e}_1$ points in the field direction. The connection between the field strength and the photon density follows from the energy density equation

$$|\mathbf{E}|^2/8\pi = N_1 \hbar \omega_1 . \qquad (3.231)$$

Then the total decay rate obtained by summing the contributions from (3.228) and (3.230) is

$$w(0, \omega_2)d\Omega_2 = \frac{\alpha^2 c^3 |\mathbf{E}|^2}{8\pi\hbar} \lim_{\omega_1 \to 0} \frac{\omega_2}{\omega_1^2} |Q(\omega_1, \omega_2) + Q(-\omega_1, \omega_2)|^2 d\Omega_2 . \qquad (3.232)$$

The two amplitudes are added before squaring because in the limit $\omega_1 \to 0$ the final states are identical and the two processes of stimulated two-photon emission and Raman scattering contribute coherently.

The limit in (3.232) can be evaluated by making the following transformation. For small ω_1, $Q(\omega_1, \omega_2)$ is given by

$$Q(\omega_1, \omega_2) \simeq \sum_n \left[\frac{\langle f|\boldsymbol{\alpha}\cdot\hat{e}_1|n\rangle\langle n|\tilde{a}_2|i\rangle}{\omega_f - \omega_n + \omega_1} \right.$$
$$\left. + \frac{\langle f|\tilde{a}_2|n\rangle\langle n|\boldsymbol{\alpha}\cdot\hat{e}_1|i\rangle}{\omega_i - \omega_n - \omega_1} \right] \qquad (3.233)$$

where $\tilde{a}_2$ is the same abbreviated notation as used in (3.217). Using the Dirac equation (3.211), the matrix elements of $\boldsymbol{\alpha}\cdot\hat{e}_1$ can be re-written in the form

$$\langle f|\boldsymbol{\alpha}\cdot\hat{e}_1|n\rangle = \frac{i}{\hbar c}\langle f|[H_D, \hat{e}_1\cdot\mathbf{r}]|n\rangle$$
$$= \frac{i}{c}(\omega_f - \omega_n)\langle f|\hat{e}_1\cdot\mathbf{r}|n\rangle$$
$$= \frac{i}{c}[(\omega_f - \omega_n + \omega_1) - \omega_1]\langle f|\hat{e}_1\cdot\mathbf{r}|n\rangle \qquad (3.234)$$

and similarly

$$\langle n|\boldsymbol{\alpha}\cdot\hat{e}_1|i\rangle = \frac{-i}{c}[(\omega_i - \omega_n - \omega_1) + \omega_1]\langle n|\hat{e}_1\cdot\mathbf{r}|i\rangle . \qquad (3.235)$$

Substituting the above into (3.233) and performing sums over intermediate states by closure yields

$$Q(\omega_1, \omega_2) = \frac{i\omega_1}{c}\sum_n \left[\frac{\langle f|\hat{e}_1\cdot\mathbf{r}|n\rangle\langle n|\tilde{a}_2|i\rangle}{\omega_f - \omega_n + \omega_1} + \frac{\langle f|\tilde{a}_2|n\rangle\langle n|\hat{e}_1\cdot\mathbf{r}|i\rangle}{\omega_i - \omega_n - \omega_1} \right]$$
$$+ (i/c)\langle f|[\hat{e}_1\cdot\mathbf{r}, \tilde{a}_2]|i\rangle . \qquad (3.236)$$

Since the commutator in the last term of (3.236) vanishes, the factors ω_1 in (3.232) cancel, with the result

$$w(0, \omega_2)d\Omega_2 = \frac{\alpha^2 c\omega_2}{2\pi\hbar}|\sum_n \frac{\langle f|\mathbf{E}\cdot\mathbf{r}|n\rangle\langle n|\tilde{a}_2|i\rangle}{\omega_f - \omega_n}$$
$$+ \frac{\langle f|\tilde{a}_2|n\rangle\langle n|\mathbf{E}\cdot\mathbf{r}|i\rangle}{\omega_i - \omega_n}|^2 d\Omega_2 , \qquad (3.237)$$

where $\mathbf{E} = |\mathbf{E}|\hat{e}_1$ and $\omega_2 = \omega_i - \omega_f$. This agrees with (3.5) and (3.40) for a transition between states adiabatically perturbed by a weak external electric field. No non-relativistic approximations have been made in obtaining this result.

If the field is switched on sufficiently fast, then the above must be modified because the $2p_{1/2}$ and $2p_{3/2}$ states introduce resonances near zero frequency. The resonances have a negligible effect on the total spontaneous two-photon decay rate, but they become important if one of the photons is of low frequency. In order to demonstrate the connection with the quantum beat results of section 3.4, assume that the field is switched on as a step function at $t = 0$ so that

$$\mathbf{E}(t) = \begin{cases} 0, & t < 0 \\ \mathbf{E} & t \geq 0 \,. \end{cases} \tag{3.238}$$

Then $|\mathbf{E}(t)|$ may be written as the Fourier integral

$$|\mathbf{E}(t)| = (1/2\pi) \int_{-\infty}^{\infty} F(\omega) e^{i\omega t} d\omega \tag{3.239}$$

with

$$F(\omega) = |\mathbf{E}| \lim_{\epsilon \to 0} [i/(-\omega + i\epsilon)] \,. \tag{3.240}$$

In effect, the $2s_{1/2} - 1s_{1/2}$ transition is induced by an incident wave packet of photons with frequency distribution $F(\omega)$ centred about $\omega = 0$.

Near the $2s_{1/2} - 2p_{1/2}$ resonance frequency, the denominator of the second term in (3.233) vanishes. It is therefore necessary to include the energy shifts and widths due to the interaction with the electromagnetic field, and to treat the interaction with the external electric field exactly. The latter requirement means that the electron propagator $S_c^{(e)}(x_1, x_2)$ defined by (3.191) must be replaced by the full time-dependent retarded Green's function defined by

$$[i\hbar \frac{\partial}{\partial t_1} - H_0(\mathbf{r}_1) - V(\mathbf{r}_1, t_1)] G^{(e)}(1;2)$$
$$= i\hbar \delta(\mathbf{r}_1 - \mathbf{r}_2) \delta(t_1 - t_2) \tag{3.241}$$

where $V(\mathbf{r}_1, t_1)$ is the interaction potential with the external field which is switched on at $t_1 = 0$ as described by (3.238), and the eigenvalues of H_0 are assumed to include the radiative shifts and widths. The wave function at a time $t_1 > 0$ is then related to the wave function at an earlier time t_2 by

$$\theta(t_1 - t_2)\psi(\mathbf{r}_1, t_1) = \int G^{(e)}(\mathbf{r}_1, t_1; \mathbf{r}_2, t_2)\psi(\mathbf{r}_2, t_2) d\mathbf{r}_2 \tag{3.242}$$

where $\theta(t_1 - t_2)$ is the unit step function. For $t_2 < 0$, $\psi(\mathbf{r}_2, t_2)$ is one of the field-free incident states $\psi_n(\mathbf{r}_2, t_2)$. $G^+(1; 2)$ satisfies the well-known integral equation

$$G^{(e)}(1; 2) = S_c^{(e)}(1; 2) - i \int S_c^{(e)}(1; 3) V(\mathbf{r}_3, t_3)$$
$$G^{(e)}(3; 2) d\mathbf{r}_3 dt_3 . \tag{3.243}$$

Near a resonance, the sum over intermediate states in (3.233) is dominated by at most a few resonant terms. In this basis set of strongly interacting states, $G^{(e)}(1; 2)$ can be written as the double Fourier transform (Drake and Grimley, 1975)

$$G^{(e)}(1; 2) = \frac{1}{2\pi i} \int_\infty^\infty d\omega_1 d\omega_2 e^{i\omega_1 t_1 - i\omega_2 t_2}$$
$$\underline{\psi}^\dagger(\mathbf{r}_2) \underline{f}(\omega_1, \omega_2) \underline{\psi}(\mathbf{r}_1) \tag{3.244}$$

where $\underline{\psi}(\mathbf{r}_1)$ is a column vector of field-free eigenstates and $\underline{f}(\omega_1, \omega_2)$ is a square matrix of Fourier coefficients. Substituting into (3.243) and equating Fourier coefficients yields the integral equation

$$\underline{f}(\omega_1, \omega_2) = [\delta(\omega_1 - \omega_2) - \frac{1}{2\pi} \int_\infty^\infty d\omega_3 F(\omega_1 - \omega_3)$$
$$\times \underline{f}(\omega_3, \omega_2) \underline{V}^\dagger] \underline{\Delta}^{-1}(\omega_1) \tag{3.245}$$

where $\underline{\Delta}(\omega)$ is a square diagonal matrix with elements

$$\Delta_{n,n}(\omega) = (E_n - i\Gamma_n/2)/\hbar + \omega$$

and $\underline{V}$ is the interaction matrix with elements

$$V_{m,n} = \langle \psi_m | V(\mathbf{r}) | \psi_n \rangle .$$

If $\underline{V} = 0$, then $\underline{f}(w_1, w_2) = \delta(w_1 - w_2) \underline{\Delta}^{-1}(w_1)$ and $G^{(e)}(1; 2)$ reduces to $S_c^{(e)}(1; 2)$.

Equation (3.245) can be solved iteratively to obtain a power series expansion in the external field strength for $\underline{f}(\omega_1, \omega_2)$. However, if the Fourier coefficients $F(\omega)$ for the external field are given by (3.240), then the series can be summed to infinity with the result

$$\underline{f}(\omega_1, \omega_2) = \underline{\Delta}^{-1}(\omega_2) \left\{ \delta(\omega_1 - \omega_2) + \frac{\underline{V}^\dagger [\underline{\Delta}(\omega_1) + \underline{V}^\dagger]^{-1}}{2\pi i(\omega_2 - \omega_1 + i\varepsilon)} \right\} . \tag{3.246}$$

The integral over ω_2 in (3.244) can now be performed. For $t_2 < 0$, the contour integral must be closed in the upper half-plane where it encircles the poles in $\underline{\Delta}^{-1}(\omega_2)$ at $\hbar\omega_2 = -E_n + i\Gamma_n/2$. The result is

$$G^{(e)}(1;2) = \frac{1}{2\pi i} \int_{-\infty}^{\infty} d\omega_1 e^{i\omega_1 t_1} \underline{\psi}^{\dagger}(\mathbf{r}_2, t_2) \underline{g}(\omega_1) \underline{\psi}(\mathbf{r}_1) \qquad (3.247)$$

with

$$\underline{g}(\omega_1) = [\underline{\Delta}(\omega_1) + \underline{V}^{\dagger}]^{-1} \qquad (3.248)$$

(Drake and Grimley, 1975). For the two state problem discussed in section 4.3, the matrix $\underline{g}(\omega)$ can be written in the form

$$\underline{g}(\omega) = \frac{\hbar}{(\hbar\omega + \tilde{E}_1)(\hbar\omega + \tilde{E}_2)} \begin{pmatrix} \hbar\omega + E_b - i\Gamma_b/2 & -V_{ab} \\ -V_{ab} & \hbar\omega + E_a - i\Gamma_a/2 \end{pmatrix} \qquad (3.249)$$

where $\tilde{E}_1$ and $\tilde{E}_2$ are the perturbed eigenvalues given by (3.154). In general, for N interacting states, $\underline{g}(\omega)$ has poles at $\hbar\omega = -\tilde{E}_n, n = 1, \ldots, N$, where the $\tilde{E}_n$ are the eigenvalues of the matrix $\underline{\Delta}(0) + \underline{V}$.

We now introduce the radiation field as a first order perturbation inducing transitions from the time dependent initial state, given by (3.242), to the final state. Following the method of Hicks, Hess and Cooper (1972), if $c_f(\omega, \tau)$ is the first order probability amplitude for the atom initially in state $\psi_i(\mathbf{r}, 0)$ at $t = 0$ to be in the final state together with a photon of frequency ω, then (omitting obvious factors of $\hbar$)

$$c_f(\omega, \tau) = ie \int_0^{\tau} dt_1 \int \int d\mathbf{r}_1 d\mathbf{r}_2 \psi_f^*(\mathbf{r}_1) e^{iE_f^* t_1} \boldsymbol{\alpha} \cdot \mathbf{A}^*(\mathbf{r}_1) e^{i\omega t_1}$$
$$G^{(e)}(\mathbf{r}_1, t_1; \mathbf{r}_2, 0) \psi_i(\mathbf{r}_2, 0) . \qquad (3.250)$$

Using (3.247) for $G^{(e)}(1; 2)$, the integral over t_1 can be performed with the result

$$c_f(\omega, \tau) = \frac{e}{2\pi i} \int_{\infty}^{\infty} d\omega_1 \frac{e^{i(E_f^* + \omega + \omega_1)\tau} - 1}{E_f^* + \omega + \omega_1}$$
$$\times \sum_j (\boldsymbol{\alpha} \cdot \mathbf{A})_{fj} g_{ij}(\omega_1) . \qquad (3.251)$$

The final integral over ω_1 can be performed by contour integration. As can be seen from (3.249), the poles of $g(\omega)$ lie entirely in the upper complex plane, while $\omega_1 = -E_f^* - \omega$ lies in the lower complex plane. Consequently, $c_f(\omega, \tau) = 0$ for $\tau < 0$ as expected. For $\tau > 0$, the contour must be closed in the upper half plane, with the result

$$c_f(\omega, \tau) = e \sum_n \left(\frac{e^{i(E_f^* + \omega - \tilde{E}_n)\tau} - 1}{E_f^* + \omega - \tilde{E}_n} \times \sum_j (\boldsymbol{\alpha} \cdot \mathbf{A})_{fj} \text{Res}[g_{ij}(\omega_1)] \big|_{\omega_1 = -\tilde{E}_n} \right) \tag{3.252}$$

where Res denotes the residue of g_{ij}. This expression simplifies in the limit $\tau \to \infty$ since then the exponential term does not contribute and one may choose to close the contour in the lower complex plane where there is only a single pole at $\omega_1 = -E_f - \omega$. In this case

$$c_f(\omega, \infty) = e \sum_j (\boldsymbol{\alpha} \cdot \mathbf{A}^*)_{fj} g_{ij}(-E_f^* - \omega) \tag{3.253}$$

which no longer contains the additional summation over poles at $\omega_1 = -\tilde{E}_n$.

The quantity $2\pi |c_f(\omega, \infty)|^2 \rho(\omega)$ is the line shape $I(\omega)$ of the emitted radiation integrated over all time. In the two-state approximation, (3.249) and (3.253) yield

$$I(\omega) = \frac{2\pi e^2 |(\boldsymbol{\alpha} \cdot \mathbf{A}^*)_{fb} V_{ab}|^2}{|(\tilde{E}_1 - E_f - \hbar\omega)(\tilde{E}_2 - E_f - \hbar\omega)|^2} \rho(\omega) \tag{3.254}$$

assuming that the atom is initially in the metastable state a, while state b decays radiatively to the ground state. Using equations (3.154), the denominator of (3.254) can be re-expressed in the form

$$[(E_a - E_f - \hbar\omega)(E_b - E_f - \hbar\omega) - |V_{ab}|^2]^2$$
$$+ (E_a - E_f - \hbar\omega)^2 \Gamma^2/4$$

in agreement with the spectral distribution calculated by Fontana and Lynch (1970), starting from the Bethe-Lamb time-dependent formalism. It is clear from (3.254) that the spectrum consists of two peaks — one near $\hbar\omega = \text{Re}\,(\tilde{E}_1 - E_f)$ and one near $\hbar\omega = \text{Re}\,(\tilde{E}_2 - E_f)$.

For the extended problem in which there are two intermediate states such as $2p_{1/2}$ and $2p_{3/2}$, there are three $\tilde{E}_n$'s equal to the eigenvalues of the matrix

$$\begin{pmatrix} E_1 & V_{12} & V_{13} \\ V_{21} & E_2 & 0 \\ 0 & 0 & E_3 \end{pmatrix}$$

where 1,2,3 denote, for example, the states $2s_{1/2}, 2p_{1/2}$ and $2p_{3/2}$ respectively. The cubic secular equation for the eigenvalues is exactly the same as the cubic equation derived in the non-perturbative calculation of Holt and Sellin (1972).

A direct comparison of the present results with those of section 3.4 can be made by calculating $I(t)$ instead of $I(\omega)$, where $I(t)$ is the time dependence of the radiation integrated over all frequencies. Defining $d_f(t)$ as the Fourier transform

$$d_f(t) = (1/2\pi) \int_{\infty}^{\infty} c_f(\omega, \infty) e^{-i\omega t} \, d\omega \qquad (3.255)$$

then

$$I(t) = 2\pi\hbar |d_f(t)|^2 \rho(\bar{\omega}) \qquad (3.256)$$

where $\bar{\omega}$ is an average transition frequency. Since the contour in (3.255) must be closed in the lower half-plane for $t > 0$, $d_f(t)$ is given by

$$d_f(t) = -ie \sum_j (\boldsymbol{\alpha} \cdot \mathbf{A}^*)_{fj} e^{iE_f t}$$
$$\times \sum_n e^{i\tilde{E}_n t} \mathrm{Res}\left[g_{ij}(-E_f - \omega) \right]\Big|_{\omega = \tilde{E}_n - \tilde{E}_f} . \qquad (3.257)$$

Substituting this into (3.256) yields an expression equivalent to (3.119) [or (3.166) in the two-state approximation] for the quantum beat pattern arising from a pure initial state at $t = 0$. The above can easily be re-written in terms of a density matrix for a mixed initial state. It can be shown in general that

$$|d_f(t)|^2 = \int_{-\infty}^{\infty} \frac{d}{dt} |c_f(\omega, t)|^2 \, d\omega$$

where the right-hand side is the differential transition probability integrated over all frequencies.

In summary, the propagator formalism of the present section gives directly the frequency distribution of the emitted radiation integrated over time. The central problem is to solve equation (3.245) for the Fourier coefficients $f(\omega_1, \omega_2)$ for a given function $F(\omega)$ describing the time dependence of the external field. The method is therefore complementary to that of section 3.4 where the central problem is to solve directly the time-dependent Schrödinger equation (3.101) to find the time distribution of the emitted radiation integrated over all frequencies. The final results for the probability amplitudes are just Fourier transforms of each other.

3.6   The Electroweak Interaction
Parity Non-Conservation in Atomic Physics

3.6.1   Introduction

The purpose of this section is to introduce the reader to the basic ideas of weak interactions, and to provide a phenomenological derivation of the interaction terms responsible for parity non-conservation (PNC) in atomic physics. The guiding principle in constructing such a theory is to follow as closely as possible the example of quantum electrodynamics.

The parity operation P is defined physically by an inversion of the co-ordinate system such that $\mathbf{r} \to -\mathbf{r}, t \to t$ for every part of the system. This is equivalent to reflection through a plane, followed by a rotation through π about the normal to the plane. The interactions responsible for a physical process are said to be parity invariant if the mirror image process is equally possible. In other words, the interactions do not have an *intrinsic* handedness. Of course, experiments involving, say, circularly polarized photons may still have a preferred handedness arising from the way in which the system was initially prepared. The crucial point is whether or not the mirror image experiment is completely equivalent.

Parity has played an important role in physics ever since Laporte (1924) discovered experimentally that there is an additional selection rule governing the intensities of atomic transitions beyond the usual angular momentum selection rules. Subsequently, Wigner (1927) showed that Laporte's selection rule follows from the parity invariance of electromagnetic interactions. In the simplest non-relativistic case, if P commutes with the Hamiltonian, then the eigenvectors must be (or can be chosen to be, in the case of degeneracy) simultaneous eigenvectors of P. Since $P^2 = 1$, the simultaneous eigenvectors have either even or odd parity according to whether

$$P\psi(\mathbf{r}, t) = \psi(\mathbf{r}, t) \quad \text{or} \quad P\psi(\mathbf{r}, t) = -\psi(\mathbf{r}, t) \ .$$

The transition matrix elements $\langle f|V|i \rangle$ between states of definite parity then vanish unless the product of parities for all three factors is even. This leads to the selection rule $\Delta P = \pm 1$ for electric dipole transitions ($V_{E1} \propto \hat{e} \cdot \mathbf{r}$ is odd) and $\Delta P = 0$ for magnetic dipole transitions.

Since gravitational and nuclear strong interactions are also parity invariant, this was long thought to be an exact symmetry of nature. Only biological systems have an arbitrarily chosen preferred handedness for the three dimensional structure of complex molecules. However, in a famous experiment suggested by

Lee and Yang (1956), Wu *et al.* (1957) conclusively demonstrated that parity is not conserved in the beta decay of oriented Co^{60} nuclei induced by weak interactions. In fact, as shown in Fig. 3.14 parity conservation is violated to the maximum possible extent in that the emitted neutrinos (spin-1/2 particles of zero mass) are completely polarized in the direction opposite to their direction of travel. Since the mirror image of a left-handed screw is a right-handed screw, the mirror image experiment is not physically possible.

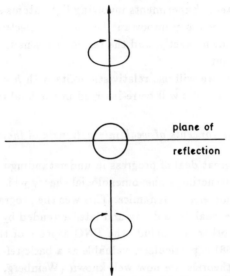

Fig. 3.14. Illustration of a process which is not equivalent to its mirror image. For the upper process, the spin is antiparallel to the direction of travel, while for the mirror image process it is parallel to the direction of travel.

The neutral current weak interaction between atomic electrons and nucleons leads to small parity non-conserving terms in the atomic Hamiltonian, and hence to violations of Laporte's rule for atomic transitions. Each atomic state acquires a small contribution from states of opposite parity. Although the effects are exceedingly small, the high precision that is attainable in atomic physics experiments with modern laser technology has made it possible to detect them in high Z atoms. The suggestion for such experiments was first made by Bouchiat and Bouchiat (1974, 1975), Moskalev (1974), Khriplovich (1974), and Sandars (1975). PNC effects have now been measured in bismuth and lead by observation of the optical rotation of light (for a review, see Fortson and Lewis, 1984), and in thallium and cesium by the technique of Stark interference

(Bucksbaum *et al.* 1981; Drell and Commins, 1984; Bouchiat *et al.* 1982 and 1984; Gilbert *et al.* 1985). Experimental work on hydrogen (Levy and Williams, 1984) is reviewed in Chapter 4 of this volume by Hinds. The precision achieved in the above measurement of Gilbert *et al.* (1985) approaches that of the best high-energy test of the electroweak theory. In addition, heavy atom atomic physics measurements complement the high-energy experiments because they probe a different energy regime and are sensitive to a nearly orthogonal set of coupling parameters. Experiments involving light atoms such as hydrogen and deuterium are particularly important because the electronic structure part of the problem is known exactly, and the results are sensitive to the complete set of coupling constants.

In this section, we will use relativistic units with $\hbar = c = 1$. However, for clarity, factors of $\hbar$ and c will be re-inserted in the final results.

3.6.2 Some basic concepts of weak interactions and their currents

One can make a great deal of progress in understanding the nature of weak interactions by constructing a phenomenologial theory which resembles as closely as possible quantum electrodynamics. This was the program initiated by Fermi (1934) to describe weak beta decay and later extended by Lee and Yang (1956, 1957) and many others to include the PNC aspects of the theory. The book by Bernstein (1968) is particularly valuable as a basic reference. Although unified electroweak theories are now well known (Weinberg, 1974, and references therein), the predictions are model dependent and the usual approach in the literature (see, for example, Feinberg and Chen, 1974; Dunford *et al.* 1978) has been to enumerate all the terms which are not forbidden by general symmetry arguments, and let the coupling constants be determined by experiment. As one example of electroweak theories, the predictions of the standard Glashow-Salam-Weinberg model are discussed in section 3.6.4. However, a complete discussion of gauge theories of weak interactions is beyond the scope of this chapter. The reader is referred to the many excellent review articles and books on the subject listed at the end. Our aim is to obtain as simply as possible the terms responsible for PNC effects in atomic physics, while side-stepping the more technical details required for an understanding of beta decay and high energy phenomena.

A phenomenological theory of weak interactions in atomic physics can be simply constructed as follows. The general way of representing interactions between particles in quantum field theories is to write down an expression of

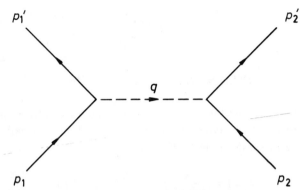

Fig. 3.15. Feynman diagram for the interaction of particles (solid lines) by means of an exchanged quantum (dashed line) carrying momentum $q = p_1 - p_1' = p_2' - p_2$.

the form

$$-1/2 \int d^4x \int d^4x' \, J_\mu^{(1)}(x) D_{\mu\nu}^c(x - x') J_\nu^{(2)}(x') \qquad (3.258)$$

for the second order S-matrix element corresponding to the Feynman diagram shown in Fig. 3.15. Here, the J_μ's are the current density operators for the particles participating in the process, and $D_{\mu\nu}^c(x)$ is a function describing the propagation of the exchanged quantum. Leaving aside the J's for the moment, $D_{\mu\nu}^c$ is most simply expressed in the momentum representation

$$D_{\mu\nu}^c(q) = \int D_{\mu\nu}^c(x) e^{-iqx} d^4x \; . \qquad (3.259)$$

By conservation of momentum, $q = p_1 - p_1' = p_2' - p_2$ if the initial and final momenta of the interacting particles are well defined (see Fig. 3.15). For a photon-like vector particle of mass M, the propagator is of the form

$$D_{\mu\nu}^c(q) = \delta_{\mu\nu} D^c(q) + q_\mu q_\nu \quad \text{terms} \qquad (3.260)$$

with $D^c(q) = -i/(q^2 + M^2 - i\varepsilon), \varepsilon > 0$, and $q^2 = \mathbf{q}^2 - \omega^2$. The gauge-dependent $q_\mu q_\nu$ terms in (3.260) do not contribute to interaction expressions of the form (3.258) because, for conserved currents,

$$q_\mu J_\mu(q) = 0 \; .$$

(This is just the equation of continuity expressed in momentum space.) $D^c(q)$ in (3.260) is the Fourier transform of the Green's function which satisfies the well-known inhomogeneous wave equation

$$(-\Box^2 + M^2) D^c(x) = -i\delta(x) \qquad (3.261)$$

with

$$\Box^2 = \nabla^2 - \frac{1}{c^2}\frac{\partial^2}{\partial t^2}$$

and

$$D^c(x) = 1/(2\pi)^4 \int D^c(q)e^{iqx}d^4q . \tag{3.262}$$

Here, $d^4q = dqd\omega$ and the term $i\varepsilon$ in the denominator of (3.260) specifies the location of poles for the ω integration (Akhiezer and Berestetskii, 1965). For arbitrary M, the integrations in (3.262) yield

$$D^c(x) = \frac{-i}{8\pi^2|\mathbf{r}|} \int_{-\infty}^{\infty} e^{-\alpha(\omega)|\mathbf{r}|-i\omega t}d\omega \tag{3.263}$$

with

$$\alpha(\omega) = \begin{cases} (M^2 - \omega^2)^{1/2} & \omega < M \\ -i(\omega^2 - M^2)^{1/2} & \omega \geq M . \end{cases}$$

For photons, $M \to 0$ and $\alpha(\omega) \to -i|\omega|$. In this case, (3.263) just describes the Coulomb interaction between charged particles (including the Breit interaction which comes from retardation corrections). However for the weak interaction, which is propagated by the massive vector bosons $W^\pm$ and Z^0, $\alpha(\omega) \to M$ at low frequencies and, from (3.263)

$$D^c(x) \simeq \frac{-ie^{-M|\mathbf{r}|}}{8\pi^2|\mathbf{r}|} \int_{-\infty}^{\infty} e^{-i\omega t}d\omega \tag{3.264}$$

which has the form of a Yukawa potential of range $|\mathbf{r}| \sim \hbar/Mc$. Since M is known to be many times larger than the proton mass, the range of the weak interaction is very short ($\sim 10^{-15}$ cm). Under these conditions, it follows from (3.264), or directly from (3.261), that $D^c_{\mu\nu}(x-x')$ in (3.258) is well approximated by

$$D^c_{\mu\nu}(x - x') = -i\delta_{\mu\nu}\delta(x - x')/M^2 . \tag{3.265}$$

The low frequency approximation used here corresponds to assuming that $q^2 << M^2c^2$ in the denominator of (3.260). This is certainly well satisfied because $q^2 \sim \alpha^2 m_e^2 c^2$ for atomic processes. Thus (3.258) becomes

$$(i/2M^2) \int d^4x J_\mu^{(1)}(x)J_\mu^{(2)}(x) \tag{3.266}$$

describing the direct interaction of two currents at a single point of space and time.

The currents in (3.266) are formed from bilinear combinations of the Dirac spinors $\overline{\psi}$ and ψ. Since each spinor has four components, exactly sixteen independent combinations can be constructed. The choice with well-defined Lorentz transformation properties is (Akhiezer and Berestetskii, 1965)

$$S = \overline{\psi}\psi$$
$$V_\mu = \overline{\psi}\gamma_\mu\psi$$
$$T_{\mu\nu} = 1/2\overline{\psi}(\gamma_\mu\gamma_\nu - \gamma_\nu\gamma_\mu)\psi$$
$$A_\mu = \overline{\psi}\gamma_\mu\gamma_5\psi$$
$$P = \overline{\psi}\gamma_5\psi \tag{3.267}$$

where

$$\gamma_5 = \gamma_1\gamma_2\gamma_3\gamma_4 = -\begin{pmatrix} 0 & 1 \\ 1 & 0 \end{pmatrix} \tag{3.268}$$

and

$$\gamma_5\gamma_\mu = -\gamma_\mu\gamma_5 . \tag{3.269}$$

Any product of γ-matrices can be written as a linear combination of the above terms. V_μ and A_μ transform as vectors under proper Lorentz transformations, while $T_{\mu\nu}$ is an antisymmetric second rank tensor with six independent components. S and P are scalar invariants.

The crucial point now is to investigate how each of the above terms transforms when the inversion operator I is applied to the spinors. For Dirac spinors, I is defined by

$$\psi'(\mathbf{r}, t) = I\psi(\mathbf{r}, t) = \eta\gamma_4\psi(-\mathbf{r}, t) \tag{3.270}$$

where η has one of the four values $\pm i, \pm 1$. The general rule for transforming the conjugate spinor $\overline{\psi}$ is as follows. If O is an arbitrary operator and

$$\psi' = O\psi$$

then

$$\overline{\psi'} \equiv \psi'^\dagger\gamma_4$$
$$= (O\psi)^\dagger\gamma_4 = \psi^\dagger O^\dagger\gamma_4 = \psi\gamma_4 O^\dagger\gamma_4 . \tag{3.271}$$

Taking $O = \eta\gamma_4$ yields

$$S' = \overline{\psi'}\psi' = |\eta|^2\overline{\psi}(\gamma_4)^4\psi = \overline{\psi}\psi . \tag{3.272}$$

Thus S is a proper scalar under inversions. On the other hand

$$P' = \overline{\psi}' \gamma_5 \psi' = |\eta|^2 \overline{\psi} \gamma_4 \gamma_5 \gamma_4 \psi \ . \tag{3.273}$$

Using (3.269), this reduces to $-P$ so that P is a pseudoscalar. Similarly, the anticommutation rule

$$\gamma_\mu \gamma_\nu + \gamma_\nu \gamma_\mu = 2\delta_{\mu\nu} \quad (\mu, \nu = 1, \dots, 4) \tag{3.374}$$

can be used to show that

$$
\begin{aligned}
V_{j'} &= -V_j, \quad V_{4'} = V_4 \\
A_{j'} &= A_j, \quad A_{4'} = -A_4
\end{aligned}
\tag{3.275}
$$

$(j = 1, 2, 3)$. Thus V_μ is a proper vector, while A_μ is a pseudovector (or axial vector because of its resemblance to angular momentum). It follows that the Dirac current defined by

$$j_\mu = ie\overline{\psi}\gamma_\mu \psi \tag{3.276}$$

has the same transformation properties as V_μ, and is therefore a proper Lorentz vector.

The experimental data for weak interactions are well described by assuming that the corresponding weak current is a mixture of vector and axial vector contributions according to

$$J_\mu = V_\mu + A_\mu \ . \tag{3.277}$$

However, (3.276) is not the most general phenomenological form that a current could have. In the momentum representation, one could also include components of the momenta along with the γ-matrices in combinations such as $p_\mu, p'_\mu, (\gamma_\nu p'_\nu)p_\mu$ etc. As discussed by Bernstein (1968), there are exactly twelve independent Lorentz vectors of this type (see Table 9), along with twelve more pseudovectors obtained by including an extra factor of γ_5. Thus the most general forms for the matrix elements of V and A taken at $t = 0$ are

$$\langle \mathbf{p}'|V_\mu|\mathbf{p}\rangle = \overline{u}(\mathbf{p}') \sum_{i=1}^{12} F_i(q^2) v_\mu^{(i)}(\mathbf{p}', \mathbf{p}) u(\mathbf{p}) \tag{3.278}$$

$$\langle \mathbf{p}'|A_\mu|\mathbf{p}\rangle = \overline{u}(\mathbf{p}') \sum_{i=1}^{12} G_i(q^2) v_\mu^{(i)}(\mathbf{p}', \mathbf{p})\gamma_5 u(\mathbf{p}) \tag{3.279}$$

where the $u(\mathbf{p})$ are free particle Dirac spinors satisfying the Dirac equation

$$(i\gamma_\mu p_\mu + m)u(\mathbf{p}) = 0 , \tag{3.280}$$

the $F_i(q^2)$ and $G_i(q^2)$ are form factors which depend only on the invariant $q^2 = (p' - p)^2$, and the $v_\mu^{(i)}$ are the twelve vectors listed in Table 9.

Table 9. A list of the twelve linearly independent lorentz vectors which can be formed from products of $p_\mu^{(\pm)} = (p' \pm p)_\mu$ and γ-matrices.

γ_μ	$\gamma_\nu p_\nu^{(\pm)} p_\mu^{(\pm)}$ [1]
$p_\mu^{(\pm)}$	$\gamma_\nu p_\nu^{(+)} \gamma_\nu p_\nu^{(-)} p_\mu^{(\pm)}$
$\sigma_{\mu\nu} p_\nu^{(\pm)}$	$\gamma_5 \epsilon_{\mu\nu\lambda\sigma} \gamma_\nu p_\nu^{(+)} p_\sigma^{(-)} p_\sigma^{(-)}$ [2]

[1] Four combinations.

[2] $\epsilon_{\mu\nu\lambda\sigma}$ is the totally antisymmetric pseudotensor with $\epsilon_{1234} = 1$.

The number of independent vectors can be reduced from twelve to three by imposing the condition that $u(\mathbf{p})$ satisfies (3.280); i.e. the particles are "on their mass shells". This neglects binding energy and radiative corrections, but gives the correct form. With an arbitrary choice of functions, the result can be written

$$\langle \mathbf{p}'|V_\mu|\mathbf{p}\rangle = \bar{u}(\mathbf{p}')[F_1(q^2)p_\mu^{(+)} + F_2(q^2)q_\mu \\ + F_3(q^2)\gamma_\mu]u(\mathbf{p}) \tag{3.281}$$

$$\langle \mathbf{p}'|A_\mu|\mathbf{p}\rangle = \bar{u}(\mathbf{p}')[G_1(q^2)p_\mu^{(+)} + G_2(q^2)q_\mu \\ + G_3(q^2)\gamma_\mu]\gamma_5 u(p) \tag{3.282}$$

with $p_\mu^{(+)} = (p' + p)_\mu$.

The above equations can be further reduced by requiring V_μ and A_μ to be Hermitian and have time reversal symmetry. (The experimental evidence strongly indicates that the weak interaction is time reversal invariant.) The Hermitian requirement for the three-vector part of (3.281) or (3.282) is

$$\langle \mathbf{p}'|V_j|\mathbf{p}\rangle = \langle \mathbf{p}|V_j|\mathbf{p}'\rangle^* , \tag{3.283}$$

Using $\bar{u}^\dagger = \gamma_4 u$, this becomes

$$\bar{u}(\mathbf{p}')v_j(\mathbf{p}',\mathbf{p})u(\mathbf{p}) = \bar{u}(\mathbf{p}')\gamma_4 v_j(\mathbf{p},\mathbf{p}')\gamma_4 u(\mathbf{p}) \qquad (3.284)$$

where $v_j(\mathbf{p}',\mathbf{p})$ refers to any of the six terms in (3.281) and (3.282). It follows that

$$
\begin{array}{ll}
F_1{}^* = F_1 & G_1{}^* = -G_1 \\
F_2{}^* = -F_2 & G_2{}^* = G_2 \\
F_3{}^* = -F_3 & G_3{}^* = -G_3 \ .
\end{array}
\qquad (3.285)
$$

Similarly, time reversal symmetry requires

$$v_j(\mathbf{p}',\mathbf{p}) = -[Tv_j(-\mathbf{p}',-\mathbf{p})\,T^{-1}]^* \qquad (3.286)$$

since currents reverse sign under time reversal. The representation $T = \gamma_1\gamma_3$ for the time reversal operator results in

$$
\begin{array}{ll}
F_1{}^* = F_1 & G_1{}^* = G_1 \\
F_2{}^* = F_2 & G_2{}^* = G_2 \\
F_3{}^* = -F_3 & G_3{}^* = -G_3 \ .
\end{array}
\qquad (3.287)
$$

(Note that γ_2, γ_4 and γ_5 are real, while γ_1 and γ_3 are pure imaginary.) Comparison of (3.285) with (3.287) shows that $F_2 = 0$, $G_1 = 0$, F_1 and G_2 are real, and F_3 and G_3 are pure imaginary. Thus the combined requirements of Hermiticity and time reversal invariance reduce the number of independent parameters from six complex functions of q^2 to two real and two imaginary ones.

The final results for V_μ can be expressed in a more conventional form by use of the identity

$$p_\mu^{(+)} = (\gamma_\nu p'_\nu - im)\gamma_\mu + \gamma_\mu(\gamma_\nu p_\nu - im) + 2im\gamma_\mu + i\sigma_{\mu\nu}q_\nu \qquad (3.288)$$

where

$$\sigma_{\mu\nu} = (1/2i)(\gamma_\mu\gamma_\nu - \gamma_\nu\gamma_\mu) \ .$$

Then, using again the Dirac equation and re-defining the form factors, the matrix elements of V_μ and A_μ are

$$\langle \mathbf{p}'|V_\mu|\mathbf{p}\rangle = i\bar{u}(\mathbf{p}')[\gamma_\mu f_v(q^2) - \sigma_{\mu\nu}q_\nu f_M(q^2)]u(\mathbf{p}) \qquad (3.289)$$

$$\langle \mathbf{p}'|A_\mu|\mathbf{p}\rangle = i\bar{u}(\mathbf{p}')[\gamma_\mu g_A(q^2) + iq_\mu g_P(q^2)]\gamma_5 u(\mathbf{p}) \qquad (3.290)$$

with

$$f_V = 2mF_1 - iF_3, \quad q_A = -iG_3$$
$$f_M = -F_1, \quad g_P = -G_2 .$$

The matrix elements of the vector V_μ can be thought of as describing either an electromagnetic or a weak current, while the axial vector A_μ enters only for weak currents. The physical significance of the terms in (3.289) and (3.290) is as follows. By comparison with the ordinary electromagnetic Dirac current (3.276), it is clear that in the limit $q^2 \to 0$,

$$f_V (q^2) \to e$$

the electric charge. This term already contains the Dirac magnetic moment, as can be seen by writing

$$u(\mathbf{p}) = \begin{pmatrix} w \\ w' \end{pmatrix} \tag{3.291}$$

where w and w' are two component Pauli spinors related by

$$w' = (\boldsymbol{\sigma} \cdot \mathbf{p}/2mc)w \tag{3.292}$$

for low momenta ($v/c \ll 1$). Assume that the current interacts with an external electromagnetic vector potential $\mathbf{A}$ chosen to satisfy the gauge condition $A \cdot (\mathbf{p}' - \mathbf{p}) = 0$. Then

$$ie\overline{u}(\mathbf{p}')\boldsymbol{\gamma} \cdot \mathbf{A}u(\mathbf{p}) \simeq ew^\dagger(\mathbf{p}') \left[\frac{\mathbf{A} \cdot \mathbf{p}}{mc} + \frac{i\boldsymbol{\sigma} \cdot \mathbf{q} \times \mathbf{A}}{2mc} \right] w(\mathbf{p}) . \tag{3.293}$$

The second term becomes $2\mu_0(\boldsymbol{\sigma} \cdot \mathbf{H}/2)$ in the co-ordinate representation, corresponding to the usual Dirac magnetic moment with a g-factor $g = 2$ ($\mu_0 = e\hbar/2mc$ is the Bohr magneton). The $f_M(q^2)$ term of (3.289) also has the form of a magnetic moment interaction, as can be seen by writing, in the same approximation

$$ie\overline{u}(\mathbf{p}')A_i\sigma_{i\nu}q_\nu u(p) \simeq -iew^\dagger(\mathbf{p}')\boldsymbol{\sigma} \cdot \mathbf{q} \times \mathbf{A}w(\mathbf{p}) . \tag{3.294}$$

In the limit $q \to 0$, the $f_M(0)$ term provides a phenomenological description of the additional tensor interaction due to the anomalous magnetic moment of the particle. If the total magnetic moment is μ, then

$$f_M(0) = \mu/2 - \mu_0 = (g - 2)\mu_0/2 . \tag{3.295}$$

For electrons, $(g - 2)/2 = \alpha/2\pi + O(\alpha^2)$.

By analogy, for weak currents, $f_V(0)$ is called the weak charge, $f_M(0)$ the weak anomalous magnetic moment and $g_A(0)$ the weak axial charge. The remaining term $g_P(q^2)$ is called the induced pseudoscalar form factor. To see the origin of this name, consider what happens when the g_P term is coupled to a second vector current as in (3.266). The result is a term of the form

$$-i[\overline{u}(p_1')q_\mu\gamma_5 g_P u(p_1)][\overline{v}_b(p_2')\gamma_\mu v_a(p_2)]$$
$$= -i[\overline{u}(p_1')\gamma_5 g_P u(p_1)][\overline{v}_b(p_2')\gamma_\mu (p_2' - p_2)_\mu v_a(p_2)] \qquad (3.296)$$

where, by conservation of momentum,

$$q = p_1 - p_1' = p_2' - p_2 .$$

Since v_a and v_b are also Dirac spinors which satisfy the Dirac equation, $\gamma_\mu (p_2' - p_2)_\mu$ can be replaced by $i(m_b - m_a)$. Thus, if v_b and v_a refer to different kinds of particles so that $m_b \neq m_a$, the coupling of currents "induces" an interaction which has the form of a pseudoscalar. This term contributes to beta decay where particles do indeed change their identities. However, it does not contribute to weak interactions in atomic physics if processes involving particle transformations are neglected. The g_P term will therefore not be further considered.

In summary, if we assume Lorentz and time-reversal invariance, together with overall current conservation, then the general forms of the matrix elements of $V_\mu^{(a)}$ and $A_\mu^{(a)}$ for fermions of type a are

$$\langle \mathbf{p}'|V_\mu^{(a)}|\mathbf{p}\rangle = i\overline{u}_a(\mathbf{p}')[\gamma_\mu f_v^{(a)} - \sigma_{\mu\nu}q_\nu f_M^{(a)}]u_a(\mathbf{p}) \qquad (3.297)$$

$$\langle \mathbf{p}'|A_\mu^{(a)}|\mathbf{p}\rangle = i\overline{u}_a(\mathbf{p}')\gamma_\mu\gamma_5 g_A^{(a)} u_a(\mathbf{p}) \qquad (3.298)$$

where a = electron, proton, neutron, etc. In the quark model of the nucleons, a can refer instead to the "up" and "down" quarks, with $p = udu$ and $n = udd$. This will be further discussed below.

3.6.3   The electron – nucleon weak interaction

We are now ready to form an expression for the effective weak interaction energy between electrons and nucleons (or quarks). On transforming back to co-ordinate space and using (3.266) and (3.260), the parity non-conserving part is

$$U_W^{(a)} = - G/\sqrt{2} \int d\mathbf{r} (A_\mu^{(e)} V_\mu^{(a)} + V_\mu^{(e)} A_\mu^{(a)})$$

$$= -G/\sqrt{2} \int d\mathbf{r} \{ i g_A^{(e)} (\bar{e} \gamma_\mu \gamma_5 e) [i f_V^{(a)} (\bar{a} \gamma_\mu a)$$

$$+ f_M^{(a)} \partial_\nu (\bar{a} \sigma_{\mu\nu} a)] + [i f_V^{(e)} (\bar{e} \gamma_\mu e)$$

$$+ f_M^{(e)} \partial_\nu (\bar{e} \sigma_{\mu\nu} e)] i g_A^{(a)} (\bar{a} \gamma_\mu \gamma_5 a) \} \tag{3.299}$$

where $\partial_\nu = \partial/\partial x_\nu$ acts on the terms $(\dots)$ to its right, e denotes the Dirac spinor $\psi_e(x)$ for an electron (and similarly for $a = n$ or p) and Fermi's constant G controls the overall strength of the interaction. If the g_A's and f_V's are taken to be dimensionless coupling constants of order of magnitude unity, then the integrand of (3.299) has dimensions of (volume)$^{-2}$. Thus G must have dimensions of (energy) $\times$ (volume). The numerical value from muon decay is

$$G = (1.02681 \pm 0.00002) \times 10^{-5} \hbar^3 / c M_p^2$$

where $M_p = 0.9382796$ GeV is the proton mass. Thus

$$G = (1.16634 \pm 0.00002) \times 10^{-5} \text{ GeV}^{-2}$$

in relativistic units and

$$G = 2.22248 \times 10^{-14} e^2 a_0^2$$

in atomic units.

Equation (3.299) can be further simplified by neglecting the weak anomalous magnetic moment term $f_M^{(a)}$ for nucleons. This is justified because

$$q_\nu f_M^{(a)} / f_V^{(a)} \simeq \alpha m / M_p$$

and so the $f_M^{(a)}$ term is presumably negligible. What remains are the three terms

$$U_w^{(a)} = + G/\sqrt{2} \int d\mathbf{r} [C_1^{(a)} (\bar{e} \gamma_\mu \gamma_5 e)(\bar{a} \gamma_\mu a)$$

$$+ C_2^{(a)} (\bar{e} \gamma_\mu e)(\bar{a} \gamma_\mu \gamma_5 a)$$

$$+ i C_3^{(a)} \partial_\nu (\bar{e} \sigma_{\nu\mu} e)(\bar{a} \gamma_\mu \gamma_5 a)] \tag{3.300}$$

where

$$C_1^{(a)} = g_A^{(e)} f_V^{(a)}$$
$$C_2^{(a)} = f_V^{(e)} g_A^{(a)}$$
$$C_3^{(a)} = f_M^{(e)} g_A^{(a)} \tag{3.301}$$

$C_1^{(a)}$ and $C_2^{(a)}$ are dimensionless, while $C_3^{(a)}$ has dimensions of $\mu_0/e = \hbar/2mc$.

In lowest order, the total interaction is obtained by summing (3.300) over all the nucleons ($a = n$ or p) or the constituent quarks ($a = u$ or d) in the nucleus. This procedure neglects radiative and QCD corrections of about 10% (Lynn, 1984). However, if it is applied in a straightforward way, then $C_1^{(a)}$ is simply additive over the nucleons with the result

$$C_1^{\text{tot}} = Z C_1^{(p)} + N C_1^{(n)}$$
$$= (2Z + N) C_1^{(u)} + (Z + 2N) C_1^{(d)} \tag{3.302}$$

where N is the number of neutrons. Results for heavy atoms are often expressed in terms of the nuclear weak charge $Q_w = -2 C_1^{(\text{tot})}$. $C_2^{(\text{tot})}$ and $C_3^{(\text{tot})}$ are much smaller because $C_2^{(\text{tot})}$ is nuclear spin dependent (and therefore not additive), and the $C_3^{(\text{tot})}$ term is reduced by a factor of α/π relative to the others. Since $N/Z \simeq 1.4$ to 1.5 in all heavy atoms, all PNC measurements in heavy atoms determine approximately the same combination of coupling constants $0.9\, C_1^{(u)} + C_1^{(d)}$ (Bouchiat and Pottier, 1984). This combination is nearly orthogonal to the quantity $2 C_1^{(u)} - C_1^{(d)}$ determined in the SLAC high energy particle experiment (Prescott *et al.* 1978), and thus provides an important supplement to the high energy work. Atomic physics measurements in light isotopes such as hydrogen and deuterium would allow separate determinations of $C_1^{(u)}$ and $C_1^{(d)}$, and would be sensitive to the C_2 and C_3 contributions as well (Lewis and Williams, 1975), as discussed in section 3.6.5.

In summary, equation (3.300) is the basic phenomenological equation for analyzing the results of PNC experiments in atomic physics and comparing the results with theoretical predictions. The predictions of one such theory — the standard unified electroweak theory — are described in the following section.

3.6.4 Coupling constants in the standard electroweak model

The Glashow-Salam-Weinberg (GSW) SU(2)×U(1) gauge theory of electromagnetic and weak interactions is now commonly called the standard theory of the electroweak interaction. Since the subject is thoroughly covered elsewhere (see Bibliography), only the relevant results will be quoted here. A discussion of other models and the parameters they predict is contained in the review of Kim *et al.* (1981).

One of the most important results of the GSW theory is the prediction of massive neutral vector bosons Z° in addition to the charged bosons $W^\pm$ presumed to be responsible for beta decay. It is the Z° neutral propagator that is responsible for current conserving PNC effects in atomic physics. The observation of the $W^\pm$ and Z° bosons at CERN (UA1 Coll. 1983; UA2 Coll. 1983, Kenyon 1985) lends considerable support to at least the basic ideas of the theory. It is characterized by the Weinberg mixing angle θ_W, which is related to the particle masses by

$$\sin^2 \theta_W = 1 - M_W^2 / M_Z^2 \equiv s^2 \ . \tag{3.303}$$

The combination $s^2 - 1/4$, which occurs frequently below, is small because $s^2 \simeq 0.215 \pm 0.014$. Following Cheng and Li (1984), the neutral currents are[c]

$$J_\mu^{(e)} = \frac{i}{2} \bar{u}_e \left[\left(-\frac{1}{2} + 2\sin^2 \theta_W \right) \gamma_\mu + \frac{1}{2} \gamma_\mu \gamma_5 \right] u_e \tag{3.304}$$

$$J_\mu^{(u)} = \frac{i}{2} \bar{u}_u \left[\left(\frac{1}{2} - \frac{4}{3}\sin^2 \theta_W \right) \gamma_\mu - \frac{1}{2} \gamma_\mu \gamma_5 \right] u_u \tag{3.305}$$

$$J_\mu^{(d)} = \frac{i}{2} \bar{u}_d \left[\left(-\frac{1}{2} + \frac{2}{3}\sin^2 \theta_W \right) \gamma_\mu + \frac{1}{2} \gamma_\mu \gamma_5 \right] u_d \ . \tag{3.306}$$

Each of the above currents can be decomposed into vector and pseudovector parts according to

$$J_\mu^{(a)} = \frac{1}{2}(V_\mu^{(a)} + \frac{1}{2} A_\mu^{(a)}) \ . \tag{3.307}$$

Then the $V - A$ cross terms in the effective low energy electron-quark interaction

$$\frac{-g^2}{\cos^2 \theta_W M_Z^2} \ J_\mu^{(e)} J_\mu^{(a)}$$

---

[c] Our γ_5 defined by (3.268) is such that $(1-\gamma_5)/2$ is the projection operator for left-handed (negative helicity) fermion states. Some authors use the reverse sign. The other γ-matrices are defined at the beginning of section 3.5.

are

$$U_W^{(a)} = \frac{-g^2}{8\cos^2\theta_W M_Z^2} \int d\mathbf{r}(A_\mu^{(e)} V_\mu^{(a)} + V_\mu^{(e)} A_\mu^{(a)}) . \tag{3.308}$$

Using (3.303), $\cos^2\theta_W M_Z^2 = M_W^2$ so that the factor in front of (3.308) corresponds to the identification

$$\frac{G}{\sqrt{2}} = \frac{g^2}{8M_W^2} \tag{3.309}$$

from the theory of beta decay. Comparing the integrand of (3.308) with (3.300) yields

$$C_1^{(u)} = 1/2 - 4s^2/3; \quad C_1^{(d)} = -1/2 + 2s^2/3$$
$$C_2^{(u)} = 1/2 - 2s^2; \quad C_2^{(d)} = -C_2^{(u)}$$
$$C_3^{(u)} = 0; \quad C_3^{(d)} = 0 . \tag{3.310}$$

If the above is extended to include the weak anomalous magnetic moment of the electron, then from (3.295) and (3.301)

$$C_3^{(a)} = C_2^{(a)} f_M^{(e)} / f_V^{(e)}$$
$$= C_2^{(a)} \alpha\hbar/4\pi mc, \quad a = u, d . \tag{3.311}$$

The above formulas provide the basis for constructing the coupling coefficients for PNC electron-nucleon interactions in terms of the quark content of the nucleons. To a good approximation ($\sim 1\%$), the C_1 coefficients are simply additive over the quarks. This follows from the conservation of the weak vector current in the good isospin limit — that is, the approximation in which the neutron and proton are degenerate in mass. In this limit, the weak charge (like the electromagnetic charge) is a constant of the motion. Then

$$C_1^{(p)} = 2C_1^{(u)} + C_1^{(d)}, \quad C_1^{(n)} = C_1^{(u)} + 2C_1^{(d)}$$
$$= 1/2 - 2s^2 \qquad\qquad = -1/2 \tag{3.312}$$

and for the deuteron

$$C_1^{(D)} = 3(C_1^{(u)} + C_1^{(d)}) = -2s^2 . \tag{3.313}$$

Thus a measurement of $C_1^{(D)}$ would provide a direct determination of the Weinberg angle in s^2.

The situation is more complicated for the C_2 and C_3 terms because, by its nature, an axial vector current is not in general conserved. (For an extensive discussion, see Bernstein, 1968.) From (3.290), the matrix element of $iq_\mu A_\mu$ between free quark states of mass M is

$$i\bar{u}(\mathbf{p}')[i\gamma_\mu(p_\mu - p'_\mu)g_A(q^2) - q^2 g_P(q^2)]\gamma_5 u(\mathbf{p})$$
$$= i\bar{u}(\mathbf{p}')[2Mg_A(q^2) - q^2 g_P(q^2)]\gamma_5 u(\mathbf{p}) . \qquad (3.314)$$

This does not in general vanish unless

$$g_P(q^2) = 2Mg_A(q^2)/q^2 \qquad (3.315)$$

which contradicts experiment. The g_A term by itself vanishes only in the limit $M \to 0$. Thus the weak axial charge $g_A(0)$ is not a constant of the motion and cannot be thought of as an adiabatic invariant which remains unchanged as the strong interaction between quarks is switched on. The renormalization of $g_A(0)$ by strong interactions can be taken into account by leaving the ratio g_A/g_V a free parameter to be determined by experiment (Lynn, 1984). From neutron beta decay $g_A/g_V = 1.255 \pm 0.006$ (Pagels, 1975). Then, proceeding as before and assuming proton-neutron symmetry with respect to strong interactions

$$C_2^{(p)} = (g_A/g_V)(1/2 - 2s^2) ,$$
$$C_2^{(n)} = -C_2^{(p)}, \quad C_2^{(D)} = 0 . \qquad (3.316)$$

The C_3's are still related to the C_2's by (3.311). Since $C_2^{(D)}$ vanishes in lowest order, a measurement of this parameter would be directly sensitive to the higher order corrections discussed below.

Radiative and other higher order corrections to the coupling constants have been calculated by Cahn and Kane (1977), Marciano and Sirlin (1983, 1984a) and Lynn (1984). In the renormalization scheme of Sirlin (1980), s^2 is *defined* by the relation

$$s^2 = 1 - M_W^2/M_Z^2 \qquad (3.317)$$

and G_μ is calculated from the theoretical muon decay rate

$$\frac{1}{\tau_\mu} = \frac{G_\mu{}^2 m_\mu{}^5}{192\pi^3}\left(1 - \frac{8m_e{}^2}{m_\mu{}^2}\right)\left[1 + \frac{3}{5}\frac{m_\mu{}^2}{M_W{}^2} + \frac{\alpha}{2\pi}\left(\frac{25}{4} - \pi^2\right) + O(\alpha^2)\right] \quad (3.318)$$

with the result (Sirlin, 1984)

$$G_\mu = (1.16634 \pm 0.00002) \times 10^{-5} \text{ GeV}^{-2} .$$

With these conventions, the predictions of the GSW standard model for the W^+ and Z masses are

$$M_W^2 = \frac{\pi\alpha}{\sqrt{2}G_\mu s^2(1-\Delta r)}$$

$$M_Z^2 = M_W^2/(1-s^2) \tag{3.319}$$

where Δr denotes radiative corrections of $O(\alpha)$ to (3.319). The numerical value is

$$\Delta r = 0.0696 \pm 0.0020$$

(Marciano and Sirlin, 1984b). Thus the G which appears in (3.300) is not G_μ but $G_\mu(1-\Delta r)$. In the calculations of Marciano and Sirlin (1983, 1984a) G is set equal to G_μ and all the $C_1^{(a)}$ coefficients multiplied by $(1-\Delta r)$.

In addition to the above overall renormalization by $(1-\Delta r)$, there are many other radiative and QCD corrections which enter in going from free electron-quark to electron-nucleon interactions. The final results obtained by Marciano and Sirlin (1984b) with $G = G_\mu$ and $s^2 = 0.216 \pm 0.014$ are

$$C_1^{(p)} = 0.006 \pm 0.028, \qquad C_2^{(p)} = 0.082 \pm 0.030$$
$$C_1^{(n)} = -0.4883 \pm 0.0030, \qquad C_2^{(n)} = -0.068 \pm 0.030$$
$$C_1^{(D)} = -0.422 \pm 0.030, \qquad C_2^{(D)} = 0.014 \pm 0.002$$

in reasonable agreement with the tabulation of Lynn (1984). In each case, $C_i^{(D)} = C_i^{(p)} + C_i^{(n)}$. The uncertainties in $C_i^{(p)}$ and $C_i^{(n)}$ are mainly due to the allowed spread in s^2 and their sensitive dependence on that parameter. The s^2 uncertainty cancels out of $C_2^{(D)}$, leaving only the uncertainties from higher order electroweak and QCD corrections. It is clear that measurements of the C_i's in hydrogen and deuterium currently in progress will be of considerable fundamental importance.

3.6.5 Calculation of matrix elements

The previous sections have discussed the general phenomenological form to be expected for PNC weak interactions in atomic physics, and the values of the coupling constants in the standard GSW model of electroweak interactions. The final step is to calculate the electronic and nuclear matrix elements appearing in (3.300). For heavy atoms, this poses a difficult problem in atomic structure theory which continues to limit the accuracy of theoretical predictions. On the other hand, hydrogenic systems have the great advantage that the atomic wave functions are the well-known analytic solutions to the Dirac (or Schrödinger) Coulomb wave equations. Finite nuclear size effects are easily taken into account by numerical integration of the wave equation.

We begin by summing equation (3.300) over nucleons and re-writing it in the form

$$U_W = U_W^{(1)} + U_W^{(2)} + U_W^{(3)} \tag{3.320}$$

where

$$U_W^{(1)} = \frac{G}{\sqrt{2}} \int d\mathbf{r} (\bar{e}\gamma_\mu\gamma_5 e)$$
$$\times \left[\sum_p C_1^{(p)} (\bar{p}\gamma_\mu p) + \sum_n C_1^{(n)} (\bar{n}\gamma_\mu n) \right] \tag{3.321}$$

$$U_W^{(2)} = \frac{G}{\sqrt{2}} \int d\mathbf{r} (\bar{e}\gamma_\mu e)$$
$$\times \left[\sum_p C_2^{(p)} (\bar{p}\gamma_\mu\gamma_5 p) + \sum_n C_2^{(n)} (\bar{n}\gamma_\mu\gamma_5 n) \right] \tag{3.322}$$

$$U_W^{(3)} = \frac{iG}{\sqrt{2}} \int d\mathbf{r} \partial_\nu (\bar{e}\sigma_{\nu\mu} e)$$
$$\times \left[\sum_p C_3^{(p)} (\bar{p}\gamma_\mu\gamma_5 p) + \sum_n C_3^{(n)} (\bar{n}\gamma_\mu\gamma_5 n) \right] \tag{3.323}$$

and the sums run over all the protons (p) and neutrons (n) in the nucleus.

The nucleons can be treated non-relativistically if terms of relative order $\alpha m_e/M$ are neglected. In this approximation

$$(\bar{p}\boldsymbol{\gamma}\gamma_5 p) = i\psi_p^\dagger \boldsymbol{\sigma}\psi_p$$
$$(\bar{p}\gamma_4\gamma_5 p) = 0$$
$$(\bar{p}\gamma_\mu p) = \psi_p^\dagger \psi_p \delta_{\mu,4} \tag{3.324}$$

where ψ_p is a two-component Pauli spinor for the proton, and similarly for the neutron terms. For the electrons, the Dirac spinors have the form

$$\begin{pmatrix} \psi_e \\ \chi_e \end{pmatrix}$$

with $\chi_e \simeq \boldsymbol{\sigma}\cdot\mathbf{p}\psi_e/2mc$ in the non-relativistic limit. With these approximations, and assuming the nuclear matter to be entirely concentrated at the origin, then the terms in (3.320) can be written in the form

$$U_W^{(1)} = \frac{-G}{\sqrt{2}}\langle\alpha'|\left[\frac{\boldsymbol{\sigma}\cdot\mathbf{p}}{2mc}, \delta(\mathbf{r})\right]_+ |\alpha\rangle C_1\delta_{\beta,\beta'} \tag{3.325}$$

$$U_W^{(2)} = \frac{G}{\sqrt{2}}\{\langle\alpha'|[\mathbf{p},\delta(\mathbf{r})]_+|\alpha\rangle\cdot\langle\beta'|\mathbf{C}_2|\beta\rangle$$
$$- i\langle\alpha'|[\mathbf{p}\times\boldsymbol{\sigma},\delta(\mathbf{r})]_-|\alpha\rangle\cdot\langle\beta'|\mathbf{C}_2|\beta\rangle\} \tag{3.326}$$

$$U_W^{(3)} = \frac{-iG}{\sqrt{2}}\langle\alpha'|[\mathbf{p}\times\boldsymbol{\sigma},\delta(\mathbf{r})]_-|\alpha\rangle\cdot\langle\beta'|\mathbf{C}_3|\beta\rangle \tag{3.327}$$

where α, α' denote the labels for the electronic wave functions, and β, β' denote nuclear spin labels. The coefficient C_1 is simply additive over the nucleons so that $C_1 = C_1^{(p)}$ for hydrogen and $C_1 = C_1^{(p)} + C_1^{(n)}$ for deuterium. $\mathbf{C}_2$ is the operator

$$\mathbf{C}_2 = \sum_j (C_2^{(p)}\Lambda^{(p)} + C_2^{(n)}\Lambda^{(n)})\boldsymbol{\sigma}_j \tag{3.328}$$

and similarly for $\mathbf{C}_3$, where $\Lambda^{(p)}$ and $\Lambda^{(n)}$ are proton and neutron projection operators, respectively, and the sum runs over all the nucleons. Since $\mathbf{I} = 1/2\sum_j\boldsymbol{\sigma}_j$ is the total nuclear spin, it follows from the Wigner-Eckart Theorem that the matrix elements of $\mathbf{C}_2$ and $\mathbf{C}_3$ can be written in the form

$$\langle\beta'|\mathbf{C}_2|\beta\rangle = 2C_2\langle\beta'|\mathbf{I}|\beta\rangle . \tag{3.329}$$

Then for hydrogen

$$C_2 = C_2^{(p)}, \quad C_3 = C_3^{(p)}$$

and for deuterium

$$C_2 = (C_2^{(P)} + C_2^{(n)})/2, \quad C_3 = (C_3^{(P)} + C_3^{(n)})/2 .$$

This neglects a small admixture of the ^3D state into the ^3S ground state of the deuteron. Note that the values of C_2 and C_3 for deuterium as defined above are smaller by a factor of 2 than $C_2^{(D)}$ and $C_3^{(D)}$ defined in section 3.6.4.

The final results are most compactly expressed in the $jIFM_F$ coupled representation ($\mathbf{F} = \mathbf{I} + \mathbf{J}$) because the matrix elements of the operators $U_W^{(1)}, U_W^{(3)}$,

$U_W^{(3)}$ are diagonal in F, M_F and independent of M_F. Using standard methods of angular momentum coupling to factor out the nuclear and electronic spin dependent parts (Edmonds, 1960), one easily obtains for the $ns_{1/2} - np_{1/2}$ and $np_{3/2}$ matrix elements

$$\langle ns, 1/2, I, F|U_W^{(1)}|np, j, I, F\rangle = -i\delta_{j,1/2}C_1\overline{V} \tag{3.330}$$

$$\begin{aligned}
&\langle ns, 1/2, I, F|U_W^{(2)} + U_W^{(3)}|np, j, I, F\rangle \\
&= 2i\overline{V}(-1)^{j+I+F}[I(I+1)(2I+1)(2j+1)]^{1/2} \left\{ \begin{matrix} F & I & 1/2 \\ 1 & j & I \end{matrix} \right\} \\
&\quad \times \left[C_2 - \frac{4(-1)^{j+1/2}}{2j+1}(C_2 + C_3/a_0) \right]
\end{aligned} \tag{3.331}$$

where

$$\begin{aligned}
\overline{V} &= \frac{\sqrt{3}\hbar G}{\sqrt{8}mc}\langle ns|\delta(\mathbf{r})\frac{d}{dz}|np\rangle \\
&= \frac{\alpha G}{\sqrt{8}\pi a_0^3}(Z/n)^4(n^2-1)^{1/2} \\
&\simeq 0.120092(Z/n)^4(n^2-1)^{1/2} \text{ Hz}
\end{aligned} \tag{3.332}$$

with $G = G_\mu$. Inserting explicit expressions for the $6-j$ symbols in (3.331) yields

$$\begin{aligned}
&\langle ns, 1/2, I, F|U_W^{(2)} + U_W^{(3)}|np, 1/2, I, F\rangle \\
&= 2i\overline{V}(C_2 + 2C_3/3a_0)[F(F+1) - I(I+1) - 3/4]
\end{aligned} \tag{3.333}$$

$$\begin{aligned}
&\langle ns, 1/2, I, F|U_W^{(2)} + U_W^{(3)}|np, 3/2, I, F\rangle \\
&= -i\overline{V}\sqrt{2}(C_3/3a_0)[(F+I+5/2)(F+I-1/2)]^{1/2}
\end{aligned} \tag{3.334}$$

independent of M_F. The signs of the above terms correspond to phases chosen as defined by Edmonds (1960) and non-relativistic radial wave functions which are positive near the origin.

Experiments to measure PNC effects in hydrogen are discussed in this volume by Hinds (Chapter 4). The experiments involve looking for PNC mixings between ns and np states near the crossings of Zeeman substates in a magnetic field. It is a straightforward matter to perform a unitary transformation which diagonalizes the complete Zeeman Hamiltonian in the basis set of coupled states, and thereby obtain from (3.330), (3.333) and (3.334) the interaction terms between Zeeman substates. The results for hydrogen are listed in Table 10, calculated at the magnetic field where each pair of states crosses. The states are labeled in the notation of Lamb and Retherford (1950) with

$$\alpha = 2s_{1/2,1/2} \qquad \beta = 2s_{1/2,-1/2}$$
$$a = 2p_{3/2,3/2} \qquad b = 2p_{3/2,1/2}$$
$$c = 2p_{3/2,-1/2} \qquad d = 2p_{3/2,-3/2}$$
$$e = 2p_{1/2,1/2} \qquad f = 2p_{1/2,-1/2}$$

together with a subscript giving the value of M_I in the uncoupled representation. As pointed out by Dunford, Lewis and Williams (1978), experiments at four crossings can determine as many as five separate data for hydrogen (or nine for deuterium) as follows:

$$575\text{G} \rightarrow C_2 + 2/3C_3$$
$$1190\text{G} \rightarrow C_1 \pm (C_2 + 2/3C_3)$$
$$4665\text{G} \rightarrow C_2$$
$$7081\text{G} \rightarrow C_3 \ .$$

The cancellation of C_3 at 4665G is an accidental consequence of magnetic field mixing, while the cancellation of C_2 at 7081G is a general consequence of (3.332) for $2s_{1/2}-2p_{3/2}$ mixing. A measurement at this crossing $(d_{-1}-\beta_{-1})$ would provide a direct determination of the anomalous part of the electron weak magnetic moment. (A similar cancellation occurs between the orbital and spin precession frequencies of an electron in a uniform magnetic field, making possible a direct measurement of $g-2$.) Combined measurements in hydrogen and deuterium would provide an overconstrained fit to the six parameters $C_i^{(p)}, C_i^{(n)}$ $(i=1,2,3)$ available in the various gauge theories of electroweak interactions.

Table 10. Matrix elements of $U_W = U_W^{(1)} + U_W^{(2)} + U_W^{(3)}$ between Zeeman substates of hydrogen for each of the level crossings. The magnetic field strength at each crossing is also listed. (From Dunford *et al.* 1978).

Matrix Element	Value $(i\overline{V})$	Magnetic Field (G)
$\langle e_0 \| U_W \| \beta_0 \rangle$	$0.00004 C_1 - 1.98591\ C_2 - 1.31020 C_3$	553.09
$\langle f_0 \| U_W \| \beta_0 \rangle$	$0.99737 C_1 + 1.10622 C_2 + 0.70358 C_3$	1156.40
$\langle f_{-1} \| U_W \| \beta_{-1} \rangle$	$0.99696 C_1 - 0.99696 C_2 - 0.62736 C_3$	1230.43
$\langle c_0 \| U_W \| \alpha_0 \rangle$	$-0.00229 C_1 + 0.60121 C_2 - 0.05244 C_3$	4677.83
$\langle d_{-1} \| U_W \| \beta_{-1} \rangle$	$-0.00212 C_1 + 0.00212 C_2 - 0.81581 C_3$	7036.43

3.6.6 Bibliography on weak interactions

The following is a list of books and review articles on the general topic of weak interactions. There is now a vast literature in this area, and the list is provided merely as a guide to the reader who wishes to pursue the subject to greater depth than the phenomenological presentation given here.

1. C. Strachan, *The Theory of Beta Decay* (Pergamon Press, Oxford, 1969).

2. J. C. Taylor, *Gauge Theories of Weak Interactions* (Cambridge University Press, Cambridge, 1976).

3. J. L. Lopes, *Gauge Field Theories: An Introduction* (Pergamon Press, New York, 1981).

4. E. Commins and P. H. Bucksbaum, *The New Weak Interactions* (Cambridge University Press, Cambridge, 1983).

5. H. Pietschmann, *Weak Interactions* (Springer Verlag, New York, 1983).

6. T. -P. Cheng and L. -F. Li, *Gauge Theory of Elementary Particle Physics* (Clarendon Press, Oxford, 1984).

7. E. Abers and B. W. Lee, "Gauge theories," *Phys. Reports* **90** (1973) 1.

8. M. A. B. Beg and A. Sirlin, "Gauge theories of weak interactions", *Ann. Rev. Nucl. Sci.* **24** (1974) 379.

9. M. A. B. Beg and A. Sirlin, "Gauge theories of weak interactions II", *Phys. Reports* **880** (1982) 1.

10. H. Harari, "Quarks and leptons", *Phys. Reports* **42C** (1978) 235.

11. H. Fritzsch and P. Minkowski, "Flavor dynamics of quarks and leptons", *Phys. Reports* **730** (1981) 67.

APPENDIX

Hydrogenic Perturbation Theory and the Stark Effect

The purpose of this Appendix is to describe some specialized techniques for calculating the properties of hydrogenic atoms in the presence of an external electric field. The general topics of the quadratic and linear Stark effects in hydrogen are thoroughly discussed by Bethe and Salpeter (1957). The chapter by Gay in this book discusses in a fundamental way the response of hydrogenic atoms to electric and magnetic fields. Here we are concerned with the following two specific problems:

(a) Calculating higher order perturbation corrections in a finite basis set of strongly interacting states.

(b) Evaluating summations over complete sets of intermediate states.

A.1 Higher order perturbation corrections

The Hamiltonian for a one-electron atom in an external electric field $\mathbf{E}$ pointing in the z-direction is

$$H = H_0 - eFz \tag{A1}$$

where $F = |\mathbf{E}|$ and H_0 is the field-free Hamiltonian. Quenching experiments are normally done in the weak field region where the quadratic Stark shift between the $2s_{1/2}$ and $2p_{1/2}$ states is less than the Lamb shift. In this region one can, to an excellent approximation, find the eigenvalues of (A1) in the finite basis set of just three strongly interacting states: $2s_{1/2}(m_j = 1/2), 2p_{1/2}(m_j = 1/2)$ and $2p_{3/2}(m_j = 1/2)$. The contribution from states of different n is smaller by factors of $O(\alpha^2 Z^2)$. Since states with the same $|m_j|$ remain degenerate to all orders of perturbation theory, the above three states are sufficient and, by rotational invariance, there is no loss of generality in assuming that an electric field applied to a $2s_{1/2}(m_j = \pm 1/2)$ state points in the z-direction. Recall that $E(2p_{3/2}) - E(2s_{1/2}) \simeq 10[E(2s_{1/2}) - E(2p_{1/2})]$ so that the influence of the $2p_{3/2}$ state is relatively weak in the low field region.

Consider first just two interacting states $2s_{1/2}$ and $2p_{1/2}$. The Hamiltonian (A1) is then a 2×2 matrix which can be diagonalized exactly. The eigenvalues are

$$E_{\pm} = E(2p_{1/2}) + \Delta_L[1 \pm (1 + x^2)^{1/2}]/2 \tag{A2}$$

where $x = 2\sqrt{3}eFa_0/\Delta_L$ and Δ_L is the Lamb shift for an ion of nuclear charge Z. A perturbation (i.e. power series) expansion of (A2) would be useful if $x^2 \ll 1$. For He^+, $\Delta_L = 14042$ MHz and the condition $x^2 \ll 1$ corresponds to

$$F^2 \ll (6.336 \,\text{kV/cm})^2 \,.$$

One would therefore expect a perturbation series expansion of (A1) for the full three state problem to be rapidly convergent for $F \simeq 500$ V/cm or less. Since $\Delta_L \sim Z^4$, the corresponding critical field strength for heavier ions increases roughly in proportion to $(Z/2)^5$.

For the full three state problem, we regard $\lambda = -eF$ as a perturbation parameter and expand the $2s_{1/2}$ eigenvalue E and wave function ψ in the form

$$E = E_0 + \lambda^2 E_2 + \lambda^4 E_4 + \dots \tag{A3}$$

$$\psi = \psi_0 + \lambda \psi_1 + \lambda^3 \psi_3 + \dots \tag{A4}$$

where

$$(H_0 - E_0)\psi_0 = 0 \tag{A5}$$

is the zero-order eigenvalue problem and the perturbation equations are

$$(H_0 - E_0)\psi_m + z\psi_{m-1} = \sum_{n=0}^{m-1} E_{m-n}\psi_n \tag{A6}$$

with $\psi_m = 0$ for m even $(m > 0)$ and $E_m = 0$ for m odd. Using the definitions

$$|S_p\rangle = \sum_{j=1/2}^{3/2} z_j|2p_j\rangle/(\Delta_j)^p$$

$$T_p = \sum_{j=1/2}^{3/2} |z_j|^2/(\Delta_j)^p$$

with

$$z_j = \langle 2s_{1/2}|z|2p_j\rangle$$
$$\Delta_j = E_0(2s_{1/2}) - E_0(2p_j) + i\Gamma/2$$

then the solutions to the perturbation equations in the $2s_{1/2}, 2p_{1/2}, 2p_{3/2}$ basis set can be written in the form

$$|\psi_{2m+1}\rangle = \sum_{p=0}^{m} c_{m,p}|S_{p+1}\rangle \qquad (A7)$$

$$E_{2m+2} = \langle\psi_0|z|\psi_{2m+1}\rangle$$
$$= \sum_{p=0}^{m} c_{m,p}T_{p+1} \qquad (A8)$$

(van Wijngaarden and Drake, 1982). The coefficients $c_{m,p}$ are determined recursively from the equation

$$c_{m,p} = -\sum_{n=p}^{m} c_{n-1,p-1}E_{2(m-n+1)}$$

starting with $c_{0,0} = 1$ and $c_{m,p} = 0$ for p negative. The above follows by substituting (A7) into (A6), using the identity

$$(H_0 - E_0)^{-1}|S_p\rangle = -|S_{p+1}\rangle$$

and equating coefficients of $|S_p\rangle$. Explicitly, the terms through fifth order are

$$|\psi_1\rangle = |S_1\rangle$$
$$|\psi_3\rangle = -E_2|S_2\rangle$$
$$|\psi_5\rangle = -E_4|S_2\rangle + E_2^2|S_3\rangle$$
$$E_2 = T_1, \quad E_4 = -T_1T_2 \ .$$

Using the above, the perturbed wave function (A4) renormalized so that $\langle\psi|\psi\rangle = 1$ up to fifth order is then

$$|\psi\rangle = (1 + \lambda^2 U_2 + \lambda^4 U_4)|2s_{1/2}\rangle$$
$$+ \lambda\sum_{j}(1 + \lambda^2 W_2^{(j)} + \lambda^4 W_4^{(j)})(z_j/\Delta_j)|2p_j\rangle \ .$$

With the definitions

$$A_2 = \sum_{j}|z_j/\Delta_j|^2$$

$$A_4 = \sum_{j}\text{Re}\,(E_2/\Delta_j)|z_j/\Delta_j|^2$$

the U's and W's are given by

$$U_2 = -A_2/2$$
$$U_4 = A_4 + 3A_2^2/8$$
$$W_2^{(j)} = -E_2/\Delta_j - A_2/2$$
$$W_4^{(j)} = E_2^2/\Delta_j^2 - E_4/\Delta_j + A_4 + 3A_2^2/8 + A_2 E_2/2\Delta_j \ .$$

The A_2 and A_4 terms come from the renormalization of ψ. It follows immediately that the perturbation expansions of the coefficients $a, b_{1/2}$ and $b_{3/2}$ in equation (3.28) are

$$a(|\mathbf{E}|) = 1 + \lambda^2 U_2 + \lambda^4 U_4 + \ldots$$
$$b_j(|\mathbf{E}|) = b_j^{(1)} \left(1 + \lambda^2 W_2^{(j)} + \lambda^4 W_4^{(j)} + \ldots \right)$$

where the $b_j^{(1)}$ are the first order values given by (3.33) and (3.34). Substituting these expansions into (3.38) gives the corresponding finite field corrections to the V_j's and M_j's, and from these, the corrections for all the radiation asymmetries discussed in section 3.3.

A.2   *Intermediate state summations*

The three state calculation of the previous section neglects the perturbation mixing of the $2s_{1/2}$ and $1s_{1/2}$ states with all other intermediate states. The corrections to the anisotropy were discussed in section 3.3.3 in terms of the quantities B and C defined in (3.76) and (3.77) to be

$$B = \sum_{n=3}^{\infty} \frac{\langle 1s|z|np\rangle\langle np|z|2s\rangle}{E(2s) - E(np)} \tag{A9}$$

$$C = \sum_{n=2}^{\infty} \frac{\langle 1s|z|np\rangle\langle np|z|2s\rangle}{E(1s) - E(np)} \tag{A10}$$

where the wave functions and energies are the simple hydrogen atom ones without fine structure. Following the methods of Dalgarno and Lewis (1955) and Schwartz and Tiemann (1959), the infinite summations in (A9) and (A10) can be performed implicitly by re-writing B and C in the form

$$B = \langle \psi_0(1s)|z|\psi_1(2s)\rangle \tag{A11}$$

$$C = \langle \psi_1(1s)|z|\psi_0(2s)\rangle \tag{A12}$$

where the ψ_0's satisfy

$$[H_0 - E_0(ns)]\psi_0(ns) = 0 \tag{A13}$$

and the ψ_1's satisfy

$$[H_0 - E_0(1s)]\psi_1(1s) + z\psi_0(1s) = 0 \tag{A14}$$

$$[H_0 - E_0(2s)]\psi_1(2s) + z\psi_0(2s)$$
$$= \langle 2p|z|2s \rangle \psi_0(2p) . \tag{A15}$$

The extra term is required on the right-hand side of (A15) because of the degeneracy of the $2s$ and $2p$ states. Using

$$\psi_0(1s) = 2e^{-r}Y_0^0(\theta, \phi)$$

$$\psi_0(2s) = \frac{1}{2\sqrt{2}}(2 - r)e^{-r/2}Y_0^0(\theta, \phi)$$

$$\psi_0(2p) = \frac{1}{2\sqrt{6}}re^{-r/2}Y_1^0(\theta, \phi)$$

$$\langle 2p|z|2s \rangle = -3$$

and $E_0(1s) = -1/2, E_0(2s) = -1/8$, the exact solutions to (A14) and (A15) are (see, e.g., Cohen and Dalgarno, 1964, 1966)

$$\psi_1(1s) = -\frac{1}{\sqrt{3}}(2r + r^2)e^{-r}Y_1^0(\theta, \phi) \tag{A16}$$

$$\psi_1(2s) = -\frac{1}{2\sqrt{6}}(30r - r^3)e^{-r/2}Y_1^0(\theta, \phi) . \tag{A17}$$

Substituting into (A11) and (A12), one immediately obtains

$$B = \frac{-25 \cdot 2^9}{3^6\sqrt{2}}a_0^3/e^2$$

$$C = \frac{7 \cdot 2^9}{3^6\sqrt{2}}a_0^3/e^2$$

as quoted in the text.

Acknowledgements

I am grateful to my colleagues for many stimulating discussions concerning this work, especially Professor Arie van Wijngaarden with whom I have collaborated for many years. His experimental skills and physical insight inspired much of the theoretical development described in sections 3.3.3 and 3.3.4. I am also grateful to Peter Mohr, Richard Marrus and Harvey Gould for generously sharing their results with me in advance of publication.

REFERENCES

Akhiezer, A. I. and Berestetskii, V. B. *Quantum Electrodynamics* (Wiley, New York, 1965).

Aleksandrov, E. B. *Opt. i Spektroskopiya* **14** (1973) 436. [*Opt. Spectry.* (USSR) **14**, 232].

Andrä, H. J. *Phy. Scr.* **9** (1974) 257.

Andrä, H. J. in *Progress in Atomic Spectroscopy,* eds. W. Hanle and H. Kleinpoppen (Plenum, New York, 1979) Part B, pp. 829-953.

Aspect, A., Grangier, P. and Roger, G. *Phys. Rev. Lett.* **47** (1981) 460.

Aspect, A., Grangier, P. and Roger, G. *Phys. Rev. Lett.* **49** (1982) 91 and 1804

Bacher, R. Z. *Phys.* **A315** (1984) 135.

Barbieri, R. and Sucher, J. *Nucl. Phys.* **B134** (1978) 155.

Bashkin, S., Bickel, W. S., Fink, D. and Wangsness, R. K. *Phys. Lett. Rev.* **15** (1965) 284.

Bernstein, J. *Elementary Particles and Their Currents* (Freeman, San Francisco, 1968).

Bethe, H. A. and Salpeter, E. E. *Quantum Mechanics of One- and Two-Electron Atoms* (Springer-Verlag, Berlin, 1957).

Blum, K. in *Progress in Atomic Spectroscopy,* eds. W. Hanle and H. Kleinpoppen (Plenum, New York, 1978) Part A pp. 71-110.

Bohm, D. *Quantum Theory* (Prentice-Hall, Englewood Cliffs, N. J., 1951).

Bouchiat, M. A. and Bouchiat, C. *J. Phys.* (Paris) **35** (1974) 899.

Bouchiat, M. A. and Bouchiat, C. *J. Phys.* (Paris) **36** (1975) 493.

Bouchiat, M. A., Guena, J., Hunter, L. and Pottier, L. *Phys. Lett.* **117B** (1982) 358.

Bouchiat, M. A., Guena, J., Hunter, L. and Pottier, L. *Phys. Lett.* **134B** (1984) 463.

Bouchiat, M. A. and Pottier, L. *in Atomic Physics 9,* eds. R. S. van Dyck, Jr. and E. N. Fortson (World Scientific, Singapore, 1984).

Braunlich, P., Hall, R. and Lambropoulos, P. *Phys. Rev.* **A5** (1972) 1013.

Breit, G. and Teller, E. *Astrophys. J.* **91** (1940) 215.

Brink, D. M. and Satchler, G. R. *Angular Momentum* (Clarendon Press, Oxford, 1968).

Bucksbaum, P. H., Commins, E. D. and Hunter, L. R. *Phys. Rev.* **D24** (1981) 1134.

Cahn, R. and Kane, G. *Phys. Lett.* **71B** (1977) 348.

Caselese, J. S. and Gerjuoy, E. *Phys. Rev.* **180** (1969) 327.

Cheng, T. -P. and Li, L. -F. *Gauge Theory of Elementary Particle Physics* (Clarendon Press, Oxford, 1984).

Cocke, C. L., Curnutte, B., McDonald, J. R., Bednar, J. A. and Marrus, R. *Phys. Rev.* **A9** (1974) 2242.

Cohen, M. and Dalgarno, A. *Proc. Roy. Soc. (London)* **A280** (1964) 258.

Cohen, M. and Dalgarno, A. *Proc. Roy. Soc. (London)* **A293** (1966) 359.

Corney, A. *J. Phys.* **B1** (1968) 458.

Corney, A. and Series, G. W. *Proc. Phys. Soc. London* **83** (1964) 213.

Costescu, A., Brândus, I. and Mezincescu, N. *J. Phys.* **B18** (1985) L11.

Curnutte, B., Cocke, C. L. and DuBois, R. D. *Nucl. Instrum. Methods* **202** (1981) 119.

Dalgarno, A. *Rev. Mod. Phys.* **35** (1963) 522.

Dalgarno, A. and Lewis, J. T. *Proc. Roy. Soc. London* **A233** (1955) 70.

Dodd, J. N. and Series, G. W. in *Progress in Atomic Spectroscopy*, eds. W. Hanle and H. Kleinpoppen (Plenum, New York, 1978) Part A, pp. 639-677.

Drake, G. W. F. *Phys. Rev.* **A3** (1971) 908.

Drake, G. W. F. *Phys. Rev.* **A9** (1974) 2799.

Drake, G. W. F. *Phys. Rev.* **A34** (1986) 2871.

Drake, G. W. F. and Goldman, S. P. *Phys. Rev.* **A23** (1981) 2093.

Drake, G. W. F., Goldman, S. P. and van Wijngaarden, A. *Phys. Rev.* **A20** (1979) 1299.

Drake, G. W. F. and Grimley R. B. *Phys. Rev.* **A8** (1973) 157.

Drake, G. W. F. and Grimley, R. B. *Phys. Rev.* **A11** (1975) 1614.

Drake, G. W. F. and Lin, C. P. *Phys. Rev.* **A14** (1976) 1296.

Drake, G. W. F., Patel, J. and van Wijngaarden, A. *Phys. Rev.* **A28** (1983) 3340.

Drell, P. S. and Commins, E. D. *Phys. Rev. Lett.* **53** (1984) 968.

Dunford, R. W., Lewis, R. R. and Williams, W. L. *Phys. Rev.* **A18** (1978) 2421.

Edmonds, A. R. *Angular Momentum in Quantum Mechanics* (Princeton University Press, Princeton, 1960).

Einstein, A., Podolsky, B. and Rosen, N. *Phys. Rev.* **47** (1935) 777.

Erickson, G. W. *J. Phys. Chem. Ref. Data* **6** (1977) 831.

Fan, C. Y., Garcia-Munoz, M. and Sellin, I. A. *Phys. Rev.* **161** (1967) 6.

Fano, U. and Macek, J. H. *Rev. Mod. Phys.* **45** (1973) 553.

Feinberg, G. and Chen, M. Y. *Phys. Rev.* **D10** (1974) 190 and 3789.

Feinberg, G. and Sucher, J. *Phys. Rev. Lett.* **26** (1971) 681.

Fermi, E. *Z. Phys.* **88** (1934) 161.

Fite, W. L., Kauppila, W. E. and Ott, W. R. *Phys. Rev. Lett.* **20** (1968) 409.

Fontana, P. R. and Lynch, D. J. *Phys. Rev.* **A2** (1970) 347.

Fortson, E. N. and Lewis L. L. *Phys. Reports* **113** (1984) 289.

Frauenfelder, H. and Steffen, R. M. in *Alpha-, Beta- and Gamma-Ray Spectroscopy*, ed. K. Siegbahn (North Holland, Amsterdam, 1968).

Fried, Z. and Martin, A.D. *Nuovo Cimento*. **29** (1963) 574.

Gilbert, S. L., Noecker, M. C., Watts, R. N. and Wieman, C. E. *Phys. Rev. Lett.* **55** (1985) 2680.

Goeppert-Mayer, M. *Ann. Phys. Lpz.* **9** (1931) 273.

Goldman, S. P. *Phys. Rev.* **A31** (1985) 3541.

Goldman, S. P. and Drake, G. W. F. *Phys. Rev.* **A24** (1981) 183.

Gould, H., Greiner, D., Lindstrom, P., Symons T. J. M. and Crawford, H. *Phys. Rev. Lett.* **52** (1984) 180.

Gould, H. and Marrus, R. *Phys. Rev. Lett.* **41** (1978) 1457.

Gould, H. and Marrus, R. *Phys. Rev.* **A28** (1983) 2001.

Grant, I. P. *J. Phys.* **B7** (1974) 1458.

Grant, I. P. *Phys. Rev.* **A25** (1982) 1230.

Haroche, S. in *High Resolution Laser Spectroscopy*, ed. K. Shimoda (Springer, Berlin, 1976), pp. 253-313.

Hicks, W. W., Hess, R. A. and Cooper, W. S. *Phys. Rev.* **A5** (1972) 490.

Hillery, M. and Mohr, P. J. *Phys. Rev.* **A21** (1980) 24.

Hinds, E. A., Clendenin, J. E. and Novick, R. *Phys. Rev.* **A17** (1978) 670.

Holt, H. K. and Sellin, I. A. *Phys. Rev.* **A6** (1972) 508.

Hylleraas, E. A. and Undheim, B. *Z. Phys.* **65** (1930) 759.

Johnson, W. R. *Phys. Rev. Lett.* **29** (1972) 1123.

Kelsey, E. J. and Macek, J. *Phys. Rev.* **A16** (1977) 1322.

Kenyon, I. *Europ. J. Phys.* **6** (1985) 41.

Khriplovich, I. B. *Zh. Eksp. Teor. Fiz. Pis. Red.* **20** (1974) 689 [*Sov. Phys. JETP Lett.* **20** (1974) 315].

Kim, J. E., Langacker, P., Levine, M. and Williams, H. H. *Rev. Mod. Phys.* **53** (1981) 211.

Klarsfeld, S. *Phys. Lett.* **30A** (1969) 382. See also B. A. Zon and L. P. Rapaport, *Zh. Eksp. Teor. Fiz.* **7** (1968) 70. [*Sov. Phys. JETP* **7** (1968) 52].

Klarsfeld, S. *Phys. Rev.* **A6** (1972) 508.

Kocher, C. A., Clendenin, J. E. and Novick, R. *Phys. Rev. Lett.* **29** (1972) 615.

Kopfermann, H. *Nuclear Moments*, translated from the Second German Edition by E. E. Schneider (Academic Press, New York, 1958).

Kruger, H. and Oed, A. *Phys. Lett.* **54A** (1975) 251.

Kugel, H. W. Leventhal, M. and Murnick, D. E. *Phys. Rev.* **A6** (1972) 1306.

Kugel, H. W. and Murnick, D. E. *Rep. Prog. Phys.* **40** (1977) 297.

Lamb, W. E. *Phys. Rev.* **85** (1952) 259.

Lamb, W. E. and Retherford, R. C. *Phys. Rev.* **79** (1950) 549.

Laporte, O. *Z. Phys.* **23** (1924) 135.

Lawrence, G. P., Fan, C. Y. and Bashkin, S. *Phys. Rev.* **28** (1972) 1612.

Lee, T. D. and Yang, C. N. *Phys. Rev.* **104** (1956) 254.

Lee, T. D. and Yang, C. N. *Phys. Rev.* **105** (1957) 1671.

Leventhal, M., Murnick, D. E. and Kugel, H. W. *Phys. Rev. Lett.* **28** (1972) 1609.

Levy, L. P. and Williams, W. L. *Phys. Rev.* **A30** (1984) 220.

Lewis, R. R. and Williams, W. L. *Phys. Lett.* **59B** (1975) 70.

Lin, D. L. and Feinberg, G. *Phys. Rev.* **A10** (1974) 1425.

Lipworth, E. and Novick, R. *Phys. Rev.* **108** (1957) 1434.

Lynn, B. W. in *Atomic Physics 9* eds. R. S. van Dyck, Jr. and E. N. Fortson (World Scientific, Singapore, 1984), pp. 212-224.

Marciano, W. J. and Sirlin, A. *Phys. Rev.* **D27** (1983) 552.

Marciano, W. J. and Sirlin, A. *Phys. Rev.* **D29** (1984a) 75.

Marciano, W. J. and Sirlin, A. *Phys. Rev.* **D29** (1984b) 945.

Mohr, P. J. in *Beam Foil Spectroscopy*, eds. I. A. Sellin and D. J. Pegg (Plenum Press, New York, 1975), Vol. 1, p. 89.

Mohr, P. J. *Phys. Rev. Lett.* **40** (1978) 854.

Moskalev, A. N. *Zh. Eksp. Teor. Giz. Pis. Red.* **19** (1974) 229. [*Sov. Phys. JETP Lett.* **19** (1974) 141, 216].

Murnick, D. E., Leventhal, M. and Kugel, H. W. in *Phys. Rev. Lett.* **27** (1971) 1625.

Murnick, D. E., Leventhal, M. and Kugel, H. W. in *Proceedings of the Third International Conference on Beam-Foil Spectroscopy*, Boulder, Colorado (1972).

Narasimham, M. A. and Strombotne, R. L. *Phys. Rev.* **A4** (1971) 14.

O'Connell, D., Kollath, K. J., Duncan, A. J. and Kleinpoppen, H. *J. Phys.* **B8** (1975) L214.

Ott, W. R., Kauppila, W. E. and Fite, W. L. *Phys. Rev.* **A1** (1970) 1089.

Parpia, F. A. and Johnson, W. R. *Phys. Rev.* **A26** (1982) 1142.

Patel, J., van Wijngaarden, A. and Drake, G. W. F. *Phys. Rev.* **A 36** (1987) in press.

Perrie, W., Duncan, A. J., Beyer, H. J. and Kleinpoppen, H. *Phys. Rev. Lett.* **54** (1985) 1790.

Prescott, C. Y. *et al.* *Phys. Lett.* **77B** (1978) 347.

Prior, M. H. *Phys. Rev. Lett.* **29** (1972) 611.

Sandars, P. G. H. *Atomic Physics 4* eds. G. zu Putlitz, E. W. Weber and A. Winnacker (Plenum, New York, 1975).

Schwartz, C. and Tiemann, J. J. *Ann. Phys.* (N.Y) **2** (1959) 178.

Sellin, I. A., Griffen, P. M. and Biggerstaff, J. A. *Phys. Rev.* **A1** (1970) 1553.

Sellin, I. A., Moak, C. D., Griffin, P. M. and Biggerstaff, J. A. *Phys. Rev.* **184** (1969) 56.

Series, G. W. *Phys. Rev.* **136** (1964) A684.

Sirlin, A. *Phys. Rev.* **D22** (1980) 971.

Sirlin, A. *Phys. Rev.* **D29** (1984) 89.

Skalinski, T., Kopystynska, A. and Ernst, K. *Bull. Acad. Pol. Sci.* **13** (1965) 851.

Sucher, J. in *Atomic Physics 5*, eds. R. Marrus, M. Prior and H. Shugart (Plenum, New York, 1977).

Tung, J. H., Ye, X. M., Salamo, G. J. and Chan, F.T. *Phys. Rev.* **A30** (1984) 1175.

UA1 Collaboration, G. Arnison *et al. Phys. Lett.* **122B** (1983) 103 and **126B** (1983) 398.

UA2 Collaboration, M. Banner *et al. Phys. Lett.* **122B** (1983) 496 and **129B** (1983) 130.

van Wijngaarden, A. and Drake, G. W. F. *Phys. Rev.* **A25** (1982) 400.

van Wijngaarden, A., Drake, G. W. F. and Farago, P. S. *Phys. Rev. Lett.* **33** (1974) 4.

van Wijngaarden, A., Patel, J. and Drake, G. W. F. *Phys. Rev.* **A33** (1986) 312.

Weinberg, S. *Rev. Mod. Phys.* **46** (1974) 255.

Weisskopf, V. and Wigner, E. *Z. Phys.* **63** (1930) 54.

Wigner, E. P. *Z. Phys.* **43** (1927) 624.

Wu, C. S., Ambler, E., Hayward, R. W., Hoppes, D. D. and Hudson, R. P. *Phys. Rev.* **105** (1957) 1413.

Zernick, K. W. *Phys. Rev.* **132** (1963) 320.

Zernick, K. W. *Phys. Rev.* **133** (1964) A117.

CHAPTER 4

RADIOFREQUENCY SPECTROSCOPY

E. A. Hinds

Introduction

The purpose of this chapter is to discuss the microwave and radiofrequency spectroscopy of hydrogen (^{1}H, ^{2}H, ^{3}H), singly ionized helium (^{3}He$^+$, ^{4}He$^+$) and muonium (M, that is μ^+e^-). The number of low frequency intervals measured in these atoms is very large; a list compiled by Beyer [BEY78] shows well over 100. Many of the intervals have been measured with exquisite precision achieved by a combination of experimental ingenuity and extreme care. The principal rôle of these precise experiments has been to test the validity of quantum electrodynamics but the strong and weak interactions have also been addressed through the effects of nuclear structure and parity violation in atoms. In order to place the experiments in this context we shall first outline the general features of the theory, but the main emphasis of this chapter is on experimental methods. Some idea of the diversity of these may be gained by glancing down the second columns of Tables 1 and 3 (pp 250 and 261). In the historical context, all these experiments pressed the contemporary theory of atomic structure to its limits. That is particularly true of those more recent experiments which the author has chosen to describe in the following pages.

4.1   Hyperfine Structure

4.1.1   *Hydrogen ground state interval*[a]

The magnetic hyperfine interaction between the electron and the proton splits the ground state of hydrogen into two levels, a singlet and a triplet. According to a simple argument due to Fermi [FER30], the splitting frequency $\Delta\nu_F(H)$ is given by

$$h\Delta\nu_F(H) = \frac{8\pi}{3}\mu_e|\psi(0)|^2\mu_p \tag{4.1}$$

in which μ_e and μ_p are the electron and proton magnetic moments and $\psi(0)$ is the wavefunction of the electron at the origin.

The value of $|\psi(0)|^2$ for hydrogen in the Schrödinger approximation with the reduced mass correction is $|\psi(0)|^2 = Z^3/\pi(a_0[1 + m/M])^3$ in which a_0 is the Bohr radius, m is the electron mass and M is the proton mass. Taking the magnetic moment of the electron to be μ_B (μ_B = Bohr magneton), the hyperfine interval may be written

$$\Delta\nu_F(H) = \frac{16}{3}\frac{\mu_p}{\mu_B}Z^3\alpha^2 cR_\infty\left(\frac{M}{M+m}\right)^3 \tag{4.2}$$

where $hcR_\infty = \alpha^2 mc^2/2$. A more exact expression should incorporate the electron magnetic moment anomaly and other radiative effects including vacuum

---

[a]See SAH, Chapter XI, and Advances, Chapter 2.

polarization. The treatment should also be relativistic and should account for nuclear recoil, size and structure (Chapter 2). The result of all this has the form [KIN84a]

$$\Delta\nu(H) = \Delta\nu_F(H)[1 + a_e + b + \varepsilon + \delta] \qquad (4.3)$$

where

$$b = \frac{3}{2}(Z\alpha)^2$$

$$\varepsilon = \alpha(Z\alpha)\left(\ln 2 - \frac{5}{2}\right)$$

$$- \frac{8\alpha}{3\pi}(Z\alpha)^2\left(\ln Z\alpha - \ln 4 + \frac{281}{480}\right)\ln Z\alpha$$

$$+ \frac{\alpha}{\pi}(Z\alpha)^2(15.38 \pm 0.29) \ .$$

Here a_e is the anomalous magnetic moment correction for the electron [KIN84b] $(a_e \simeq \alpha/2\pi)$, b is the relativistic correction of Breit [BRE30] taken to order $(Z\alpha)^2$, and ε represents various radiative corrections not included in a_e. Although we are mainly interested in hydrogen $(Z = 1)$ the more general expression above allows one to distinguish binding corrections dependent on $Z\alpha$ from other radiative corrections independent of Z. The remaining term δ incorporates nuclear recoil and nuclear structure corrections and is sensitive to the proton form factor and polarizability. These are not well established quantities at present nor have all the theoretical terms been calculated yet. The current theoretical value is [KIN84a]

$$\Delta\nu_{\text{theory}}(H) = 1\,420\,403.444(1278)\text{kHz} \qquad (4.4)$$

in which the 1 ppm (part per million) error is dominated by uncertainties related to proton structure. Other sources of error are the values of fundamental constants of which α is the least certain (1/9 ppm) [WIL79] and numerical uncertainties in the calculation of the quantum electrodynamic terms (1/30 ppm) [KIN84a].

The same interval, the ground state interval of hydrogen, is one of the most accurately measured numbers in physics [HEL70, ESS71]

$$\nu_{\text{exp}}(H) = 1\,420\,405.751\,766\,7(9)\text{kHz} \ . \qquad (4.5)$$

It is ironic that this number does not constitute a critical test of quantum electrodynamics nor does it provide a definitive value for any of the fundamental constants. At best, with some further progress in the theory it may permit a

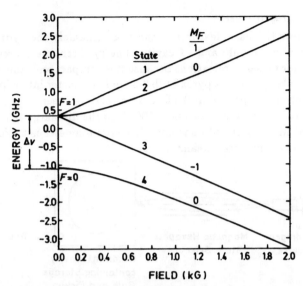

Fig. 4.1. Energy levels of the ground state of hydrogen in a static magnetic field as given by the Breit-Rabi formula (equation (4-6)).

more accurate determination of the proton polarizability. This extraordinary precision was achieved by means of a hydrogen maser, the general features of which we outline in the next section. A thorough discussion of the principles and techniques involved has been given by Kleppner *et al.* [KLE65].

Figure 4.1 shows the energy levels of the ground state of hydrogen in a static magnetic field and the transition $2 \rightarrow 4(F = 1, M = 0 \rightarrow F = 0, M = 0)$ which was measured in a very weak field $(B < 1 \text{ mG})$ using the maser. The energy levels are given by the Breit-Rabi formula [BRE31]

$$E(x) = -g_I \mu_B M B - \frac{h\Delta\nu}{4} \pm \frac{h\Delta\nu}{2}\sqrt{1 + 2Mx + x^2} \qquad (4.6)$$

where

$$x = (g_J + g_I)\mu_B B/h\Delta\nu \approx B(\text{Gauss})/507 ,$$

μ_B is the Bohr magneton and the g factors are defined with the sign convention of Bethe and Salpeter [BET57], namely $\boldsymbol{\mu}_J = -\mu_B g_J \mathbf{J}$ and $\boldsymbol{\mu}_I = +g_I \mu_B \mathbf{I}$. The negative root applies only to the lowest $(F = 0)$ level. Thus the 2-4 interval measured in the maser is

$$\frac{\Delta E}{h} = \Delta\nu\sqrt{1 \times x^2} \approx \Delta\nu + 2750B^2 \text{ Hz} \qquad (4.7)$$

where B is a weak field in Gauss.

4.1.2   The hydrogen maser

Figure 4.2 shows the basic features of the maser, namely the hydrogen source, state selector, storage bulb and rf cavity. The rf discharge source has proved most suitable for maser applications because it is compact and simple. Typically the discharge takes place in a pyrex bulb 2-3 cm in diameter into which hydrogen is admitted through a palladium leak to a pressure between 0.1 and 1 torr. An oscillator delivers rf power in excess of 100 W to the discharge at a frequency of a few hundred MHz. The bulb has an exit whose diameter is typically 500 μm through which the hydrogen beam emerges.

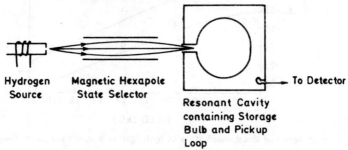

Hydrogen   Magnetic Hexapole   To Detector
Source   State Selector

Resonant Cavity
containing Storage
Bulb and Pickup
Loop

Fig. 4.2. Basic features of the hydrogen maser.

The state selector (Fig. 4.2) is a hexapole magnet in which the magnitude of the field varies as r^2, the square of the distance from the axis of the magnet. The aperture, length and maximum field of the hexapole are approximately 3 mm, 7 cm and 10 kG respectively. For fields above $x = 1$ ($B > ·507$ G) the energy levels are roughly linear in the field and hence in the hexapole the energy is proportional to r^2. It follows that atoms in levels 1 and 2 (Fig. 4.1), seeking the region of weak field, execute harmonic oscillations around the axis of the hexapole and those with the appropriate velocity are focussed into the storage bulb. Atoms in states 3 and 4, on the other hand, are defocussed, leaving a population inversion in the bulb.

When the cavity is tuned to the 2-4 transition, there will be amplification of the radiation at that frequency due to the stimulated emission of radiation from the atoms. In equilibrium the power radiated by the atoms equals the power dissipated by the cavity or coupled to external circuits.

It is straightforward to show that the radiated power $P_{\rm rad}$ is given by [BEN63]

$$P_{\rm rad} = \frac{\hbar \omega I}{2} \cdot \frac{(\mu_B B_z/\hbar)^2}{1/T_1 T_2 + (T_2/T_1)(\omega - \omega_0)^2 + (\mu_B B_z/\hbar)^2} \tag{4.8}$$

in which $\hbar \omega$ is the radiated photon energy, I is the number of atoms per second entering the bulb in state 2 minus those in state 4, B_z is the relevant component of the microwave field in the cavity and ω_0 is the resonant frequency. The relaxation times T_1 and T_2 characterize the decay of population difference ($\rho_{11} - \rho_{22}$ in the language of density matrix elements) and coherence (ρ_{12}) respectively. Near the threshold for oscillation and close to the center of the line where B_z and $(\omega - \omega_0)$ are small, we need retain only the damping term in the denominator of equation (4.8)

$$P_{\text{rad}} \approx \frac{\omega I}{2\hbar} \cdot \mu_B^2 \langle B_z \rangle_b^2 T_1 T_2 \ . \tag{4.9}$$

Here B_z is averaged over the bulb (subscript b) because of the random motion of the atoms. Note that we average before squaring because the coherence time of the decay corresponds to many bounces, i.e. an atom that begins to radiate in one part of the bulb can subsequently absorb that radiation if it bounces into a region where the sign of B_z is opposite.

The power dissipated by the cavity is

$$P_{\text{diss}} = \frac{\omega}{Q} \left[\frac{1}{2\mu_B} \langle B^2 \rangle_c V_c \right] \tag{4.10}$$

in which Q is the quality factor of the cavity and the quantity in brackets is the time average of the electromagnetic energy stored in the cavity. Now we equate P_{rad} to P_{diss} to find the threshold flux of atoms for maser oscillation

$$I_{\text{th}} = \frac{1}{Q} \frac{[\langle B^2 \rangle_c V_c / \mu_0]}{[\mu_B^2 \langle B_z \rangle_b^2]} \frac{\hbar}{T_1 T_2} \ , \tag{4.11}$$

in which the ratio in brackets is purely geometrical. When $I > I_{\text{th}}$, we cannot neglect the power broadening term $(\mu_B B_z / \hbar)^2$ in the denominator of equation (4.8). The more general result obtained from (4.8), (4.10) and (4.11) is

$$P_{\text{rad}} = \frac{\hbar \omega}{2} (I - I_{\text{th}}) \ . \tag{4.12}$$

Typical values are [KLE65] $Q = 3 \times 10^4, \langle B_z \rangle_b^2 / \langle B^2 \rangle_c = 3, V_c = 10^{-2} \text{ m}^3$, $T_1 = T_2 = 0.3$ s, $I_{\text{th}} = 10^{12}$ s$^{-1} I = 2I_{\text{th}}$ and $P = 5 \times 10^{-13}$ W. The maser power does not increase indefinitely with I, indeed, the maser stops oscillating at large values of I because of spin exchange collisions among the hydrogen atoms [KLE65].

In measurements of the hydrogen ground state splitting there are three important systematic frequency shifts to consider. First, to keep the stray magnetic field small the bulb is shielded, typically by three coaxial Moly Permalloy cylinders carefully demagnetized. Second, the 2-4 interval is perturbed when an atom bounces on the wall causing a net shift of the maser frequency which

Table 1. Summary of measurements of $\Delta\nu$ (H).

Reference	Method	Value (MHz)
NAF47	Atomic beam	1 421.3(2)
NAG47	Atomic beam	1 420.47(5)
NAF48	Atomic beam	1 420.410(6)
PRO52	Atomic beam	1 420.405 1(2)
KUS55	Atomic beam	1 420.405 73(5)
WIT56	Microwave absorption	1 420.405 72(4)
PIP62	Optical pumping	1 420.405 738 3(60)
CRA63	Hydrogen maser	1 420.405 751 800(28)
PET65	Hydrogen maser	1 420.405 751 785(16)
VES66	Hydrogen maser	1 420.405 751 786 0(20)
HEL70	Hydrogen maser	1 420.405 751 769 0(20)
ESS71	Hydrogen maser	1 420.405 751 766 7(10)

depends on the rate of bouncing (bulb size) and on the wall material. A typical shift with a Teflon coated wall is 2 parts in 10^{11}. Third, when the resonant frequency of the cavity f_c is not the same as that of the atoms f_a, the maser oscillates at a frequency f_m between the two. This frequency pulling is small because the atomic linewidth γ_a is much less than the cavity resonance linewidth γ_c, in fact

$$\left(f_m - f_a\right) = \frac{\gamma_a}{\gamma_c}(f_c - f_m) \tag{4.13}$$

in which γ_a/γ_c is typically 3×10^{-5}. Still, in order to make a measurement with 1 mHz accuracy (see equation (4.5)) the 1.4 GHz cavity resonance must be set to within 30 Hz of the atomic frequency and must remain stable at that level. Thermal and mechanical stability are achieved by making the cavity out of a quartz tube with end caps and coating the inside with silver. The mode TE_{011} is used because the current flow is azimuthal and therefore does not require good contact between the tube and the end caps.

Table 1 summarizes the measurements of $\Delta\nu(H)$.

Recently several authors have reported the operation of masers at temperatures below 1 K [HES86, HUR86, WAL86] which use a liquid helium wall coating rather than teflon. The advantages of such a maser as a frequency standard include a reproducible wall shift, smaller spin exchange contributions to the relaxation times, less thermal noise and better mechanical and magnetic stability. It is expected that eventually the cryogenic hydrogen maser may have a stability of order 10^{-18} [HUR86] averaged over several hours, a factor of 1000 better than standard technology currently provides.

4.1.3   The hyperfine anomaly δ

Interpretation of the hydrogen ground state splitting is limited by the calculation of the effects of nuclear recoil and structure. The leading nuclear correction to the Fermi splitting is the simple reduced mass factor $(1 + m/M)^{-3}$ which appears explicitly in equation (4.2). All other effects due to nuclear motion or structure are incorporated in the hyperfine anomaly δ (equation (4.3)) which amounts to some 10 ppm.

In ^{3}He and other compound nuclei, the anomaly is much larger, due mainly to the non-radial forces exerted by individual nucleons on the electron. Close to the nucleus, where the electron moves much faster than the nucleons, the electron wavefunction is best described, in a Born-Oppenheimer type of adiabatic approximation, as centered on the instantaneous positions of the protons. Sessler and Foley [SES55] have calculated this adiabatic wavefunction in ^{3}He$^+$ and have matched it to the hydrogenic solution at large distances in an attempt to determine a correction to the point nucleus approximation. Depending on their choice of the ^{3}He$^+$ nuclear wavefunction they find a contribution to δ between -183 ppm and −146 ppm. Sessler and Mills have considered the effect of the electromagnetic structure of individual nucleons [SES58] and they find an additional correction of -13 ± 2 ppm in ^{3}He$^+$ which is comparable to the proton structure correction in hydrogen. Also at this 10 ppm level are the corrections to the adiabatic approximation above. These and several smaller nuclear structure corrections [SES55] are very clearly summarized in a discussion by Rosner and Pipkin [ROS70].

Although great effort has been expended in trying to calculate the hyperfine anomalies for H and ^{3}He$^+$, our lack of knowledge about nucleon and nuclear structure has limited the theoretical accuracies to approximately 1 ppm and 10 ppm respectively with little prospect for improvement. The agreement with experiment indicates that the theory of δ is correct as far as it goes.

4.1.4   Comparison of 1S and 2S hyperfine intervals

It is possible to suppress the nuclear structure contributions to the theory to the point of irrelevance by comparing the 1S and 2S hyperfine intervals. With the uncertainty in δ practically removed, the quantum electrodynamic theory can be tested at a higher precision limited only by experiment. To the extent that the intervals $\Delta\nu_{1s}$ and $\Delta\nu_{2s}$ are proportional to $|\psi(0)|^2$ the difference $D_{21} = 8\Delta\nu_{2s} - \Delta\nu_{1s}$ is zero, but after taking small corrections into account, we may write [PRI77],

$$D_{21} = \Delta\nu_F[b_{21} + \varepsilon_{21} + \delta_{21}] \qquad (4.14)$$

where

$$b_{21} = \frac{5}{8}(Z\alpha)^2 + \frac{179}{128}(Z\alpha)^4$$

$$\varepsilon_{21} = \frac{\alpha}{\pi}(Z\alpha)^2[-3.30320\ln(Z\alpha) - 5.5515] + 0[\alpha(Z\alpha)^3]$$

and $\Delta\nu_F$ is defined by equation (4.2). Here the leading terms of equation (4.3) have been removed by the subtraction and the remaining terms are the relativistic Breit correction (b_{21}), the QED corrections (ε_{21}) and the residual hyperfine anomaly (δ_{21}). The main point of this subtraction is that the nuclear structure contributions to δ_{21} cancel to relative order $(Z\alpha)^2$. Thus in $^3\text{He}^+$ where the main nuclear structure contribution to δ is of order -160 ppm, the (uncalculated) contribution to δ_{21} is only of order 0.03 ppm and is the main uncertainty in D_{21}.

The current situation for ^1H, ^2H and $^3\text{He}^+$ is summarized in Table 2. For ^1H there is agreement between the theoretical and experimental values of D_{21} within 0.3 ppm of the ground state interval, a slightly better test of the QED

Table 2. Comparison of hyperfine intervals in ^1H, ^2H and $^3\text{He}^+$. It is worth remarking that the measurements [HEB56] and [REI56] were cited in SAH, p. 74, where a disagreement between experiment and theory was noted at the level of 1 ppm, both for ^1H and for ^2H. The agreement reported now is a result of improvements to the theory.

		$\Delta\nu$ (kHz)	D_{21} (kHz)	Theoretical D_{21}(kHz)[6]
^1H	1*s*	1 420 405.751 766 7(10)[1]	49.13(40)	48.94
	2*s*	177 556.86(5)[2]		
^2H	1*s*	327 384.352 30(25)[3]	11.16(16)	11.31
	2*s*	40 924.439(20)[4]		
$^3\text{He}^+$	1*s*	8 665 649.867(10)[5]	1 189.979(71)	1 189.80
	2*s*	1 083 354.980(88)[6]		

[1](ESS71), [2](HEB56), [3](CRA66), [4](REI56), [5](SCH69), [6](PRI77).

terms than the ground state interval itself. A similar result is found for ^2H (0.5 ppm). In $^3\text{He}^+$, however, the measurement of D_{21} is accurate to 0.008 ppm of the ground state interval and disagrees by 0.02 ppm with the theoretical value quoted. This is hardly surprising however because the residual nuclear structure contribution discussed above is not included in the theory and should be of approximately the same size as the discrepancy. Higher order QED terms, $0[\alpha(Z\alpha)^3\Delta\nu_F]$, may also be important at this level. These measurements on $^3He^+$ provide the best test of the theory of hyperfine structure and one of them is the subject of the next section.

4.1.5　*Measurement of the ground state interval in* $^3He^+$

Figure 4.3 shows the apparatus used by Schuessler, Fortson and Dehmelt [SCH69] to measure the ground state splitting in $^3He^+$. The glass vacuum envelope was pumped and a helium leak maintained a pressure of 2×10^{-8} torr of ^3He. Ions were formed inside the trap by pulsed electron bombardment of this gas. The electrodes consisted of a ring and two end caps, one of which was perforated to allow electrons from the gun to enter the trap. The inner surfaces of the electrodes were shaped to produce the quadrupole potential distribution

$$\phi(r,z) = \phi_0(r^2 - 2z^2) \tag{4.15}$$

where r and z are measured from the center of the trap and the z axis is the axis of cylindrical symmetry. The two ends caps were at the same potential and ϕ_0 is proportional to the potential difference between the end caps and the ring electrode. When the potential in such a trap is static or slowly oscillating, positive ions escape through the regions where ϕ is negative, but when the potential oscillates rapidly, the amplitude of ion motion at the driving frequency is small (micromotion). In that case the motion of the average position (macromotion) is controlled by a pseudopotential which has the form

$$\psi(r,z) = \psi_0(r^2 + 4z^2) \tag{4.16}$$

and is binding in all directions. In particular the macromotion has a harmonic oscillation along the z axis at a frequency ω_z much less than the driving frequency Ω. Schuessler *et al.* were able to excite this axial oscillation by applying a small potential difference between the end caps at frequency $\Omega + \omega_z$, a sideband of the axial oscillation generated by the superimposed micromotion. The response of the ions caused an oscillation of the image charges in the end caps and hence a current at frequency ω_z which was detected and served to indicate the number of ions in the trap.

　　Figure 4.4 shows the ground state energy levels of $^3He^+$ in a static magnetic field. Note that the hyperfine doublet is inverted compared with hydrogen (Fig. 4.1) because the nuclear magnetic moment of ^3He is negative. Also the characteristic magnetic field strength ($x = 1$) is larger (3.09 kG) because the

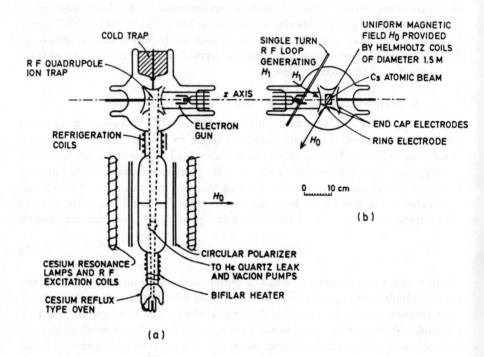

Fig. 4.3. Diagram of the apparatus used to measure the ground state hyperfine interval of $^3He^+$. (Reproduced with permission from H. A. Scheussler, E. N. Fortson and H. G. Dehmelt, *Phys. Rev.* **187** (1969) 5, (American Physical Society)).

(a) The glass vacuum chamber, ion trap and Cs atomic beam. The electron gun and trap z axis were actually rotated $45°$ out of the page around the atomic beam axis.

(b) View from top showing correct angle between trap z axis and the static magnetic field H_0. With this orientation the microwave magnetic field, which lies along the z axis, can drive both $\Delta m = 0$ and $\Delta m = \pm 1$ transitions.

hyperfine splitting is larger. The frequency measured was that of the $\Delta m = 0$ transition between states 2 and 4 labelled B in Fig. 4.4. In order to make this transition observable it was necessary to create a population difference in the ion trap between the two states and then to detect a change in that difference due to the occurrence of transitions. The experiment used an atomic beam of Cs both to create and to detect $^3He^+$ population differences in the trap.

The atomic beam was polarized by optical pumping using circularly polarized light from Cs resonance lamps. After passing through a hole in the ring electrode, the Cs atoms polarized the $^3He^+$ ions through elastic electron spin

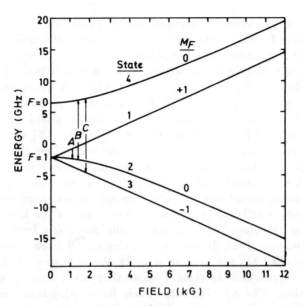

Fig. 4.4. Energy levels of the ground state of $^3He^+$ in a static magnetic field as given by the Breit-Rabi formula. Arrows A, B and C indicate the three transitions discussed in the text.

exchange collisions

$$Cs \uparrow + He^+ \downarrow \rightarrow Cs \downarrow + He^+ \uparrow$$
$$Cs \uparrow + He^+ \uparrow \rightarrow Cs \uparrow + He^+ \uparrow \ . \tag{4.17}$$

These were sufficiently frequent that the ions acquired virtually the same polarization as the atomic beam. At the same time, but at a much slower rate, the He^+ ions were neutralized by charge exchange collisions. This process is sensitive to the He^+ polarization because parallel spins required the neutral helium to be formed in a triplet state

$$Cs \uparrow + He^+ \uparrow \rightarrow He(\uparrow\uparrow) + Cs^+ + \Delta E \tag{4.18a}$$

while anti-parallel spins may form either a singlet or a triplet

$$Cs \uparrow + He^+ \downarrow \rightarrow He(\uparrow\downarrow \pm \downarrow\uparrow) + Cs^+ + \Delta E' \ . \tag{4.18b}$$

The singlet final state is favored because the energy defect $\Delta E'$ for 2^1S formation is very small (80 meV) and hence the spin antiparallel He^+ ions are more easily destroyed by charge exchange collisions. These collisions were the dominant mechanism for loss of He^+ from the trap, so the strength of the trapped ion signal was a measure of the ion polarization.

Unfortunately this detection scheme does not permit the 2-4 transition to be observed directly because in low magnetic field both states are unpolarized. Schuessler *et al.* employed the triple resonance scheme indicated in Fig. 4.4. The spin exchange elastic collisions preferentially populated state 1. A pulse of *rf* magnetic field tuned to transition $A(H_1$ in Fig. 4.3) was produced by a rigidly mounted single loop just outside the trap. This field interchanged the populations of states 1 and 2. The resulting decrease in the polarization was observed as a decrease in the trapped ion signal. Ions in state 2 were then interchanged with those in state 4 by microwave transition B, the one whose frequency was being measured. Once in state 4 the ions could not be repolarized by the elastic spin exchange collisions and a further decrease of polarization was observed. In order to enhance this B resonance signal a third transition, labelled C, was used to transfer atoms from state 4, where they were merely unpolarized, to state 3 where they could actively subtract from the polarization. The 8.6 GHz microwaves for exciting transitions B and C were introduced to the trap through a waveguide in the end cap opposite the electron gun. The trap itself was designed to resonate at this frequency in the TE_{013} cylindrical mode for which the currents are azimuthal and do not have to flow between the end caps and the ring electrode. The bandwidth of this resonance was broad enough to include both frequencies.

The drop observed in the trapped ion signal as the oscillator was swept through the 2-4 resonance was only 10 Hz wide but in order to make use of this remarkably high resolution it was necessary to measure very accurately the 7 G static magnetic field in the trap. This was done using the double quantum transition 1-3 at about 10 MHz, which was directly observable as a drop in polarization without the need for any auxiliary transitions. This frequency makes a particularly good magnetometer because it is linear in the field and independent of the hyperfine interval. The observed 2-4 transition frequency was then corrected by some 23 kHz to obtain the zero field interval.

The principal systematic source of uncertainty in this measurement was the second order Doppler shift. Since the average velocity $\langle v \rangle$ approaches zero over many cycles of a closed orbit the first order shift was very small. However, the shift $\delta \nu$ depends in second order on $\langle v^2 \rangle$ according to

$$\frac{\delta \nu}{\Delta \nu} = -\frac{\frac{1}{2}M\langle v^2 \rangle}{Mc^2} \tag{4.19}$$

and of course $\langle v^2 \rangle$ does not average to zero. The distribution of ion energies was not known but taking as a typical value 1.2 eV one finds $\delta \nu = 4$ Hz. Thus the distribution of energies and hence of second order Doppler shifts probably contributed substantially to the 10 Hz line width.

The final result was

$$\delta\nu = 8\ 655\ 649.867(10)\ \text{kHz}\ , \tag{4.20}$$

a fractional accuracy of 1 part in 10^9. This experiment was not only one of the most accurate in rf spectroscopy but surely one of the most ingenious as well.

4.1.6 Muonium ground state interval

The Fermi value for the ground state splitting of hydrogen (equation (4.2)) is readily adapted to the case of muonium (M)

$$\Delta\nu_F(M) = \frac{16}{3} \left(\frac{\mu_\mu}{\mu_p}\right) \left(\frac{\mu_p}{\mu_B}\right) \alpha^2 c R_\infty \left(\frac{m_\mu}{m_\mu + m_e}\right)^3 \tag{4.21}$$

where μ_μ is the muon magnetic moment, m_μ is the muon mass and m_e is the electron mass. The more elaborate expression for $\Delta\nu(M)$ has the same form as equation (4.3) for hydrogen,

$$\Delta\nu(M) = \Delta\nu_F(M) \left[1 + a_e + \frac{3}{2}\alpha^2 + \varepsilon + \delta\right] . \tag{4.22}$$

An analytic expression for the radiative correction term ε is given below equation (4.3) and a_e is the electron magnetic moment anomaly. The hyperfine anomaly δ, which is so troublesome in the case of normal atoms, can be calculated much more reliably in this purely leptonic system. The current expression is [KIN84a]

$$\begin{aligned}
\delta(M) = &-\frac{3\alpha}{\pi} \left(\frac{m_R}{m_\mu - m_e}\right) \left[\ln\left(\frac{m_\mu}{m_e}\right) \Big/ (1 + a_\mu)\right] \\
&+ \alpha^3 \left(\frac{m_R}{m_\mu + m_e}\right) \left[3\frac{11}{18} - 8\ln 2 - 2\ln\alpha\right] \\
&+ \left(\frac{\alpha}{\pi}\right)^2 \left(\frac{m_e}{m_\mu}\right) \left[-2\ln^2\left(\frac{m_\mu}{m_e}\right) + \frac{13}{12}\ln\left(\frac{m_\mu}{m_e}\right) + 18.18(63)\right]
\end{aligned}$$

$$\tag{4.23}$$

where $m_R = m_e m_\mu / m_e + m_\mu$ and a_μ is the anomalous magnetic correction for the muon [KIN84b] ($a_\mu \simeq \alpha/2\pi$). The first two terms involve recoil only and the third is a recoil contribution to the radiative corrections including a small hadronic vacuum polarization contribution. In this case, unlike those discussed so far, the hadronic uncertainty is negligible and the error is dominated by higher order radiative corrections.

The current theoretical value is

$$\Delta\nu_{\text{theory}}(M) = 4\,463\,304.7(1.7)(1.0) \text{ kHz} \qquad (4.24)$$

where the first error is due to uncertainties in the mass of the muon and in α while the second represents the uncalculated terms in the theory. This is to be compared with the experimental value [MAR82]

$$\Delta\nu_{\text{exp}}(M) = 4\,463\,302.88(16) \text{ kHz} \qquad (4.25)$$

which is in agreement with theory but an order of magnitude more precise. Once again the usefulness of the experiment is limited by the theory, but even so this comparison of the two constitutes one of the most stringent tests of QED today. Moreover, computation of the next set of Feynman graphs will reduce the theoretical uncertainty by an order of magnitude because of the absence of hadron structure. Given the incentive of the existing experimental result this is presumably just a matter of time, particularly computer time. This interval is likely to be of fundamental importance for the foreseeable future.

4.1.7 Measurement of muonium ground state interval

The behavior of muonium in a magnetic field is qualitatively the same as that of hydrogen (Fig. 4.1), the principal difference being one of scale. The splitting is larger by approximately the ratio of magnetic moments $(\mu_\mu/\mu_p \simeq 3.18)$ and therefore $x = 1$ corresponds here to a field of 1585 G. The energy levels obtained from equation (4.6) are

$$E_1 = -\frac{h\Delta\nu}{4} + \frac{h\Delta\nu}{2}(1+x) - g'_\mu \mu_B B$$

$$E_2 = -\frac{h\Delta\nu}{4} + \frac{h\Delta\nu}{2}(1+x^2)^{\frac{1}{2}}$$

$$E_3 = -\frac{h\Delta\nu}{4} + \frac{h\Delta\nu}{2}(1-x) + g'_\mu \mu_B B$$

$$E_4 = -\frac{h\Delta\nu}{4} - \frac{h\Delta\nu}{2}(1+x^2)^{\frac{1}{2}} \qquad (4.26)$$

where g'_μ is the g-factor for this bound state of the muon, defined like the nuclear g-factor g_I in equation (4.6) so as to be positive and $x = (g_J + g'_\mu)\mu_B B/h\Delta\nu$.

The most precise determination of $\Delta\nu(M)$ [MAR82] involved measuring the two intervals $\nu_{12} = (E_1 - E_2)/h$ and $\nu_{34} = (E_3 - E_4)/h$ in a magnetic field. The sum of these intervals is the zero field splitting, regardless of the value of x,

$$\nu_{12} + \nu_{34} = \Delta\nu \qquad (4.27)$$

while the difference

$$\nu_{34} - \nu_{12} = 2g'_\mu \mu_B B/h + \Delta\nu[\sqrt{1 + x^2} - x] , \qquad (4.28)$$

measures $g'_\mu \mu_B/h$ provided the magnetic field is known. In fact a large magnetic field was used and measured in units of the *nmr* frequency $g'_p \mu_B B/h$ for protons in water. Hence a very precise value was found for the magnetic moment ratio g'_μ/g'_p as well as for the ground state interval $\Delta\nu(M)$.

Figure 4.5 shows the apparatus used in those measurements of $\Delta\nu(M)$ [MAR82] and typical of all the experiments [HUG77]. The muons, which came in this case from the "surface muon" source (section 1.8.6) at Los Alamos National Laboratory with momentum in the range 25-30 MeV/c, were almost 100% spin-polarized along their momentum. They passed through a scintillator ($S1$) which marked their arrival and entered a pressure vessel where multiple interactions with $\frac{1}{2}$ Atm of krypton brought them to rest inside a microwave cavity. The presence of stopped muons was detected by scintillations in $S4$ caused by the decay positrons $(\mu^+ \rightarrow e^+ + \nu_e + \bar{\nu}_\mu)$ and the muon beam momentum was adjusted using a degrader so as to maximize the stopped muon rate. A solenoid created a 13.6 kG field, homogeneous throughout the cavity within a few ppm and stable to better than 1 ppm. As the muons came to rest in the microwave cavity there was very little depolarization and therefore the muonium was formed either in states 1 and 4 or in states 2 and 3 depending on the sign of the external magnetic field. In the subsequent decay, the positron was emitted preferentially along the muon spin, i.e. through scintillators $S2$ and $S3$, and the fraction of such decays was $S2 \cdot S3/S1$. A transition $1 \longleftrightarrow 2$ or $3 \longleftrightarrow 4$ induced by the microwave magnetic field was observed as a change in this fraction due to the reversal of the muon spin direction.

The dimensions of the microwave cavity were chosen so that two of the modes coincided with the two muonium frequencies $\nu_{12} = 1.918$ GHz and $\nu_{34} = 2.545$ GHz at 13.6 kG. In this way it was possible to hop rapidly from one transition to the other without disturbing the static magnetic field, and to use equation (4.27) to determine $\Delta\nu$. The profiles of the two lines were measured by varying the magnetic field keeping the microwave frequencies fixed and the signal-to-noise ratio was good enough to determine their centers within approximately 10^{-3} of their width using a detailed lineshape analysis.

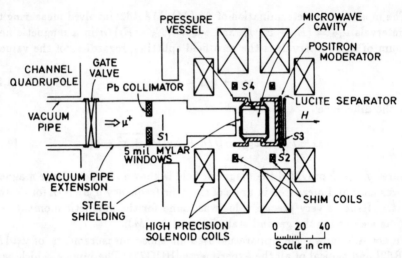

Fig. 4.5. Apparatus used in measurement of $\Delta\nu$ (M) (Reproduced with permission from F. G. Mariam *et al.*, *Phys. Rev. Lett.* **49** (1982) 993 (American Physical Society)).

The decay of the muons at a rate γ implies a typical minimum uncertainty $\hbar\gamma$ in the energy of an atom. In the simplest experiment where microwaves are on throughout the life of the atom and the decay positron is accepted at any time, the natural width of the resonance line is $\gamma/\pi = 0.145$ MHz. This linewidth can be reduced by a technique known as "old muonium", in which positron counts are vetoed for several lifetimes after the arrival of a muon, allowing only the few long lived atoms to contribute to the resonance curve [CAS75]. Of course this method reduces the height of the resonance much more than the width so it is only useful when the determination of the line center is limited not by noise but by ignorance of the line shape. Thus a modest reduction of the line width can be achieved but neither this nor any other known technique can narrow the lines dramatically; the decay rate seems to imply a fundamental limit to the accuracy of these measurements.

Table 3 summarizes the measurements of $\Delta\nu(M)$ and gives the ratio of muon and proton magnetic moments determined by the latest high field experiment [MAR82].

4.1.8 Positronium ground state interval

Positronium is mentioned here for completeness. A much fuller discussion is given by Mills in Chapter 8 of this book.

The ground state splitting due to the magnetic spin-spin interaction in positronium (Ps) can be found from equation (4.2) with the positron substituted for the proton. Thus μ_p/μ_B is replaced by 1 (ignoring the anomalous mo-

Table 3. Summary of measurements of $\Delta\nu$ in muonium. All the measurements used microwave resonance except the first. The magnetic field was either near 13.6 kG, as described in the text (high) or very low. SOF (separated oscillating fields, see section 4.2.5) and "Old M" (see this section) are line narrowing techniques.

Reference	Method	Value (MHz)
PRE61	Polarization vs. field	$2250 < \Delta\nu < 9000$
ZIO62	High field	4 461.2(2.2)
CLE64	High field	4 463.15 (40)
THO69	Low field	4 463.26 (40)
ERL69	High field	4 463.317 (21)
VOE70	High field	4 463.302 2 (89)
CRA71	Low field	4 463.311 (12)
FAV71	Low field, SOF	4 463.301 2 (23)
KOB73	Low field, SOF	4 463.304 7 (27)
CAS75	Low field, SOF and "Old M"	4 463.302 2 (14)
CAS77	High field	4 463.302 35 (52)
MAR82	High field	4 463.302 88 (16) (0.036 ppm) $\frac{\mu_\mu}{\mu_p} = 3.183\ 3461\ (11)\ (0.36\ \text{ppm})$

ment for now), $(M/M + m)$ becomes $\frac{1}{2}$ and the Fermi splitting is $\Delta\nu_F(Ps) = 2/3\alpha^2 cR_\infty$. The possibility of e^+e^- annihilation introduces another term of the same order which is, of course, absent in normal atoms and muonium. The idea is that the 1^3S_1 state spends part of the time as a photon (also spin 1). This one-photon state carries momentum $2mc$ and, since momentum is conserved, cannot be a final state for decay of the atom. Nevertheless, it exists as a virtual state accessible when the electron-positron separation is of order $2mc/\hbar$ according to the uncertainty principle and it raises the 1^3S_1 eigenvalue by $\frac{1}{2}\alpha^2 hcR_\infty$. This shift is not present in the 1^1S_0 state which, having zero spin, cannot become a single photon. Thus the lowest order estimate of the splitting is [PIR46] $\Delta\nu_0(Ps) = 7/6\alpha^2 cR_\infty$. The radiative corrections are complicated by the existence of many new diagrams associated with annihilation while the recoil corrections suffer from the absence of a small m/M expansion parameter. Only the lowest order corrections are complete although most of the next order diagrams have been evaluated [RIC81]. The result is

$$\Delta\nu(Ps) = \alpha^2 cR_\infty \left(\frac{7}{6} - \frac{\alpha}{\pi}\left[\frac{16}{9} + \ln 2 \right] + \text{higher order} \right) \qquad (4.29)$$

which, when evaluated, gives [RIC81]

$$\Delta\nu(Ps) = 203\ 381.24\ MHz \pm\ \sim 20\ MHz\ . \qquad (4.30)$$

An expression quoted in SAH , p. 78 is equivalent to equation (4.29) as far as the term in α/π. Experimental results at that time were in agreement with theory at the level of 40 MHz.

A complete review of experimental and theoretical work on Ps is given by Rich [RIC81].

4.2 Fine Structure and Lamb Shift

After the ground state, the $2S$ state of hydrogen is the most convenient starting point for low frequency spectroscopy because it is metastable. It is therefore particularly unfortunate that the $2P$ state has the shortest lifetime (1.6 ns), imparting a natural linewidth of 100 MHz to the $2S - 2P$ transitions. Nevertheless the $n = 2$ states have dominated studies of the fine structure and Lamb shift in hydrogen and are the focus of this section.

4.2.1 Energy levels

Figure 4.6 shows the $n = 2$ levels of hydrogen. In the absence of external fields and ignoring hyperfine structure for the present, the energy levels are specified by the quantum numbers $n = 2, l, j$. According to the Dirac equation the energy levels of a hydrogenic atom with a fixed, point nucleus are

$$E_D = \frac{-Z^2}{n^2} hcR_\infty \left(1 - \left[1 + \left(\frac{Z\alpha}{n-\varepsilon} \right)^2 \right]^{-\frac{1}{2}} \right) \frac{2n^2}{(Z\alpha)^2} \tag{4.31a}$$

where $\varepsilon \equiv (j + \frac{1}{2}) - [(j + \frac{1}{2})^2 - (Z\alpha)^2]^{\frac{1}{2}}$. Expressed as a power series in $Z\alpha$

$$E_D = \frac{-Z^2}{n^2} hcR_\infty \left\{ 1 + \left(\frac{Z\alpha}{n} \right)^2 \left[\frac{n}{j + \frac{1}{2}} - \frac{3}{4} \right] + \frac{1}{4} \left(\frac{Z\alpha}{n} \right)^4 \right.$$
$$\times \left. \left[\left(\frac{n}{j + \frac{1}{2}} \right)^3 + 3 \left(\frac{n}{j + \frac{1}{2}} \right)^2 - 6 \left(\frac{n}{j + \frac{1}{2}} \right) + \frac{5}{2} \right] + 0(Z\alpha)^6 \right\} . \tag{4.31b}$$

Thus the states $2S_{1/2}$ and $2P_{1/2}$ are degenerate in this approximation and the fine structure interval $2P_{3/2} - 2P_{1/2}$ is

$$\Delta E_D = \frac{Z^4 \alpha^2}{16} hcR_\infty (1 + 5/8(Z\alpha)^2 + \dots) . \tag{4.32}$$

The lowest order radiative corrections to the Dirac energies are the one-loop corrections illustrated in Fig. 4.7 and discussed theoretically in Chapter 2,

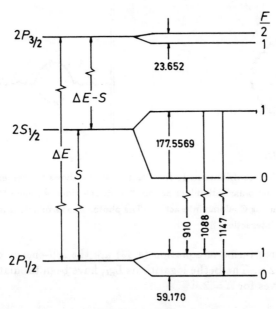

Fig. 4.6. Energy levels of the $n=2$ states of hydrogen with and without hyperfine structure. The hyperfine intervals are shown in MHz. The three $2S_{1/2} \rightarrow 2P_{1/2}$ transitions are
indicated, together with their frequencies to the nearest MHz.

sections 2.4 and 2.5. In Fig. 4.7a we show the self-energy contribution due to
the creation and absorption of a photon by the bound electron. The double lines
indicate the bound state which may have any number of Coulomb interactions
and thus the figure represents an infinite number of Feynman diagrams. The
other one-loop correction is the vacuum polarization due to the creation of a
virtual $e^+ e^-$ pair in the Coulomb field as shown in Fig. 4.7b. When these
contributions are evaluated to lowest order in α the following corrections to the
Dirac energy levels are found;
for S states:

$$\frac{8\alpha}{3\pi} \left(\frac{Z^4 \alpha^2 hc R_\infty}{n^3} \right) \left(\ln (Z\alpha)^{-2} + L_{ns} + \frac{5}{6} - \frac{1}{5} \right) \qquad (4.33)$$

for other states:

$$\frac{8\alpha}{3\pi} \left(\frac{Z^4 \alpha^2 hc R_\infty}{n^3} \right) \left(L_{nl} + \frac{3(j-l)}{4(j+\frac{1}{2})(2l+1)} \right) . \qquad (4.34)$$

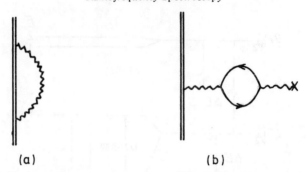

(a)   (b)

Fig. 4.7. The leading contributions to the Lamb shift. (a) shows the self energy interaction. (b) shows the vacuum polarization interaction. The double lines represent the complete propagator with continuous Coulomb interaction. The photon line terminating in a cross indicates a single Coulomb interaction.

The first three terms of equation (4.33) are the self-energy contribution to lowest order in $Z\alpha$. The Bethe logarithms L_{nl}, have been tabulated by Erickson [ERI77] who gives for $n = 2$

$$L_{2s} = -2.811\ 768\ 893\ 2(3)$$
$$L_{2p} = +0.030\ 016\ 708\ 9(3)\ .\tag{4.35}$$

The last term (-1/5) is the lowest order vacuum polarization contribution. These shifts are much smaller when $L \neq 0$ because they depend on the electron being close to the nucleus. In equation (4.34) for states having $L \neq 0$ a small Bethe logarithm is followed by a correction to the spin orbit interaction due to the anomalous magnetic moment of the electron. In this approximation then, the frequency of the Lamb shift interval, $2S_{1/2} - 2P_{1/2}$ is

$$S = \frac{\alpha}{3\pi}(Z^4\alpha^2cR_\infty)(\ln{(Z\alpha)^{-2}} - 2.208452 + 1/8)\tag{4.36}$$

and the frequency of the fine structure interval $2P_{3/2} - 2P_{1/2}$ is

$$\Delta E = \frac{1}{16}(Z^4\alpha^2cR_\infty)(1 + \alpha/\pi + 5/8(Z\alpha)^2)\ .\tag{4.37}$$

(In comparing these expressions with corresponding expressions in section 2.9 it will be helpful to remember that $R_\infty = \alpha^2mc^2/2$.)

4.2.2   The Lamb shift interval

The most accurate expressions now available also incorporate reduced mass and other recoil corrections, nuclear structure corrections, one-loop contributions of higher order in $Z\alpha$ and some diagrams with more than one loop. These matters are summarized very clearly in the review article by Erickson [ERI77]. The Lamb shift due to one-loop contributions with reduced mass corrections is [LEP84, KIN84]

$$S^{(1)} = L \left(\frac{m_R}{m}\right)^3 \left\{ \left[\ln\left(Z\alpha\right)^{-2} - 2.208452 + \ln\left(\frac{m}{m_R}\right)\right] + \frac{1}{8}\left(\frac{m}{m_R}\right) \right.$$

$$+ (Z\alpha)[2.2962\pi]$$

$$\left. + (Z\alpha)^2 \left[-\frac{3}{4}\ln^2(Z\alpha)^{-2} + 3.9184\ln\left(Z\alpha\right)^{-2} + G(Z\alpha)\right] \right\}$$

(4.38a)

where $L = (\alpha/3\pi)(Z^4\alpha^2 c R_\infty)$, m_R is the reduced mass and m is the electron mass. The last term, involving $G(Z\alpha)$, represents all the remaining Coulomb corrections (higher powers of $Z\alpha$) in the one-loop contribution to the hydrogen Lamb shift and Mohr has estimated that $G(Z\alpha) = -24.0\pm1.2$ [MOH75]. Hence

$$S^{(1)} = 1057.269(9) \text{ MHz} .$$

(4.38b)

The reduced mass corrections in equations (4.38) account for nuclear recoil only to the extent that the radial scale of the wavefunction has been adjusted but there are also dynamical recoil effects, the most important of which is a modification of the Coulomb potential around a recoiling proton due to the finite speed of light. This contribution to the Lamb shift is [LEP84]

$$\Delta S_{\text{Recoil}} = L(4.860Zm/M) = 0.359 \text{ MHz} .$$

(4.39)

There is also a purely electrostatic shift of the S states due to the finite size of the proton. This is readily derived if we let the classical electron charge density ρ_0 be constant throughout the volume of the proton (an adequate approximation) and equal to the non-relativistic, point nucleus value $-e|\psi(0)|^2$. This amounts to the usual first order approximation that the wavefunction is unchanged by a small perturbation. The potential within a spherical region of constant charge density has the form $V = V_0 - 4\pi\rho_0 r^2/6$. Thus the electrostatic energy of the electron-proton interaction is

$$\int \rho_n(V_0 - 4\pi\rho_0 r^2/6)d\tau$$

(4.40)

where ρ_n is the charge density of the proton. The result for a point nucleus is just the first term, so the level shift is

$$\frac{4\pi}{6}Ze^2|\psi(0)|^2\langle R^2\rangle$$

(4.41)

where $\langle R^2 \rangle$ is the mean square charge radius of the nucleus. In order to compare with other contributions to the Lamb shift this can be recast as a frequency proportional to L

$$\Delta S_{\text{proton}} = \frac{\pi}{2\alpha} L \frac{\langle R^2 \rangle}{(\hbar/mc)^2} = 0.195 \langle R(\text{fm})^2 \rangle \text{MHz} \ . \qquad (4.42a)$$

At the moment it is not clear what value to take for the proton size. The most recent result [SIM80], $\langle R^2 \rangle^{1/2} = 0.862(12)$ fm, is in disagreement with the previously accepted value [HAN63], $\langle R^2 \rangle^{1/2} = 0.805(11)$ fm, for no apparent reason. We therefore take the mean value and assume an error of $1/\sqrt{2}$ times the difference; $\langle R^2 \rangle^{1/2} = 0.834(40)$. Thus

$$\Delta S_{\text{proton}} = 0.136(13) \text{ MHz} \ . \qquad (4.42b)$$

Two-loop QED corrections are of comparable size and have all been calculated. Higher order Coulomb corrections, which are presumably smaller by a factor of order $Z\alpha$, have not. The two-loop contribution is [KIN84]

$$S^{(2)} = L \left(\frac{m_R}{m} \right)^3 \left\{ 0.323 \frac{\alpha}{\pi} + 0(\alpha Z\alpha) \right\}$$
$$= 0.101 \text{ MHz} + 0(10 \text{ kHz}) \ . \qquad (4.43)$$

It is possible that uncalculated radiative corrections to the recoil terms are also of order 10 kHz [KIN84].

This is as far as theory has progressed. A predicted value can be obtained simply by adding the contributions $S^{(1)}$, ΔS_{recoil}, ΔS_{proton} and $S^{(2)}$, with the result

$$S_{\text{theory}} = 1057.865(16) \text{ MHz} \qquad (4.44)$$

in which the uncertainty comes partly from the numerical computation of the one-loop radiative correction to all orders in $Z\alpha$ but principally from the measurements of the proton size. In addition there are uncalculated terms of order 10 kHz. (The result in Table 2.1 is in complete agreement with this except that the proton size is taken from SIM 80.)

4.2.3 The 2P fine structure interval

The most accurate expression available for this interval incorporates reduced mass and other recoil terms and higher order radiative corrections than those in equation (4.37). The result is

$$\Delta E = \frac{Z^4 \alpha^2 c R_\infty}{16} \left(\frac{m_R}{m} \right) \left\{ 1 + \frac{5}{8}(Z\alpha)^2 + \ldots \right.$$

$$+ 2a_e \left(\frac{m_R}{m} \right) - \left(\frac{m_R}{M} \right)^2$$

$$\left. + \frac{\alpha}{\pi}(Z\alpha)^2 \left[\ln (Z\alpha)^{-2} + \frac{11}{24} \pm \frac{3}{2} \right] \right\}$$

$$= 10\,969.044(3) \text{ MHz} \tag{4.45}$$

where m and M are electron and proton masses, and m_R is the reduced mass. The first line in parentheses comes from the power series expansion of the Dirac energies (equation (4.32)). The second line corrects for the anomalous electron moment; of which the lowest order part is the α/π in equation (4.37). There is also a new term, not suggested by equation (4.37), associated with the proton recoil [ERI77]. The last line contains the bound state QED corrections which contribute only 14 kHz to the total interval in contrast to the Lamb shift where they are the principal part. This correction can be understood by noting that the small component of the Dirac wavefunction for $2P_{1/2}$ is an S state while that for $2P_{3/2}$ is not. Hence the correction has the form of an S state shift (expression (4.33)) but is suppressed by an extra factor of $(Z\alpha)^2$. The quoted uncertainty in this part contributes an uncertainty of 2 kHz to the interval. This is combined with an independent 2 kHz error due to the uncertainty in α (0.11 ppm).

4.2.4   The slow beam Lamb shift measurements

In 1938 the most complete picture of the hydrogen atom was that provided by the enormously successful Dirac theory of the electron. However, several independent measurements of the Balmer lines in H and D had suggested that the $J = 1/2 - J = 3/2$ interval in the $n = 2$ states was smaller than the Dirac theory predicted (SAH, Chapter VII). Pasternak [PAS38] noticed that those experimental results could be reconciled if the $2S_{1/2}$ energy level were shifted upwards by about 1 GHz but this point of view remained controversial until 1947 when Lamb and Retherford demonstrated the shift of the $2S_{1/2}$ level beyond any doubt using rf spectroscopy on an atomic beam [LAM47].

There followed a classic series of experiments in which Lamb and his coworkers measured the $2S_{1/2} - 2P_{1/2}$ Lamb shift interval S in hydrogen and deuterium and the $2S_{1/2} - 2P_{3/2}$ interval $\Delta E - S$ in deuterium [LAM50, 51, 52a, 52b, TRI53, DAY53]. Their work was described in SAH, Chapter VIII. The existence of the shift was also confirmed during this period by optical spectroscopy [KOP49, MUR49, MAC50, KUH50; see also SAH, Chapter X]. Lamb produced a slow atomic hydrogen beam by thermal dissociation and excited the $2S$ state by electron impact. This state lived much longer than the time of flight through the apparatus and was therefore effectively stable. He used a microwave electric field to drive $2S - 2P$ transitions which were followed by rapid spontaneous decay to the ground state. The most natural approach would have been to measure the $2S - 2P$ intervals in zero magnetic field by sweeping the oscillator frequency and plotting the lineshape, but in 1947 this was not technically feasible. Instead Lamb applied a variable magnetic field in order to Zeeman shift the levels into resonance with an oscillator of fixed frequency. The width of the transition was 100 MHz (corresponding typically to 200 G) because of the short $2P$ lifetime (1.6 ns). Faced with this enormous linewidth Lamb determined the theoretical form of the lineshape and by careful fitting of the data was able to deduce the intervals to better than one thousandth of the linewidth.

There follows an outline of the many different assaults made over the last 35 years on the $n = 2$ intervals. A remarkable number of methods have been invented to make these measurements but in the face of the natural linewidth the accuracy has improved very slowly.

The first precise measurement of $n = 2$ structure subsequent to Lamb was in 1965 by Robiscoe [ROB65]. A slow beam of metastable atoms, produced by thermal dissociation and electron bombardment, entered a region of magnetic field parallel to their direction of travel. Figure 4.8 shows the Zeeman shifts of the $2S_{1/2}$ and $2P_{1/2}$ levels induced by the field, and in particular the level crossing between states β_- and e_-. The purpose of Robiscoe's experiment was to measure the magnetic field at which this crossing occurred and by means of theory to infer a value for the Lamb shift. A weak electric field (~ 1 V/cm) was used to admix some of the unstable $2P$ character of state e_- into the metastable $2S$ state β_- thereby inducing decay of the metastable beam and causing a reduction of the signal at the detector. As the magnetic field was varied to take the levels β_- and e_- through the crossing at 605 G, a dip 65 G wide was observed in the beam intensity due to the increased state mixing when the levels came close together.

As an aside it is interesting to ask whether the levels actually crossed once they were coupled by an electric field. If the states involved do not decay it is well known that the levels anticross, the closest approach being $2|V|$ where V is the matrix element of the interaction coupling them. Lamb showed [LAM52a] that when the levels decay they can cross provided $4|V| \leq |\Gamma_1 - \Gamma_2|$, Γ_1 and Γ_2 being the decay rates of the two levels. The weak field in Robiscoe's experiment amply satisfied this condition and there was indeed a crossing between the coupled levels. (The theory is given in section 3.4.3).

Robiscoe analyzed the level crossing dip measured in his experiment and

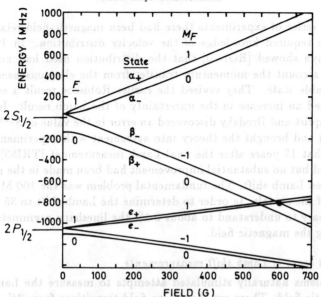

Fig. 4.8. The Zeeman shift of the $n = 2$ levels of hydrogen. The labels α, β and e follow the nomenclature of Lamb. The subscripts $\pm$ refer to $m_I = \pm\frac{1}{2}$ in the large field limit. A dot marks the level crossing studied by Robiscoe.

found a value for the Lamb shift which disagreed with Lamb's result. The difference was 300 kHz while the one standard deviation errors were 35 kHz (Lamb) and 50 kHz (Robiscoe). Subsequent experiments on the $\beta_+ - e_+$, crossing and on deuterium seemed to confirm the discrepancy [ROB66a,b]. One improvement in this later work concerned the motional electric field $\mathbf{v} \times \mathbf{B}$ which was not zero due to imperfect alignment of the magnetic field along the beam velocity and could add to or subtract from the applied electric field. As the magnetic field increased through the level crossing the variation in the net electric field would add a sloping background to the level crossing signal. In an attempt to cancel this effect the applied electric field was periodically reversed (30 Hz). It was not until two years later than an even more subtle magnetic field effect was discovered [ROB68]. The $\mathbf{v} \times \mathbf{B}$ electric field modified the velocity distribution of the atomic beam by preferentially stimulating the fast atoms to decay throughout the length of the apparatus. Hence as the field was increased, the average velocity of the metastable beam decreased and the metastable atoms spent longer in the electric field region. The resulting increase in decay had generated a line asymmetry. Once this correction was made, Robiscoe's results came close to agreement with Lamb's. Unfortunately the experimental value exceeded the theoretical one by three standard deviations [PAR67].

In both series of experiments there had been magnetic-field-related corrections which required knowledge of the velocity distribution. In 1970 Robiscoe and Shyn showed [ROB70] that the distribution used had not properly taken into account the momentum transfer from the electrons used to excite the metastable state. They revised the earlier Robiscoe result a second time and proposed an increase in the uncertainty of the Lamb result. In the same year Appelquist and Brodsky discovered an error in the value of $S^{(2)}$ (see equation (4.43)) and brought the theory into agreement with experiment [APP70]. So it was that 17 years after the final Lamb measurement [TRI53] much had been learned but no substantial improvement had been made in the precision of the measured Lamb shift. The fundamental problem was the 100 MHz natural linewidth of the signals. In order to determine the Lamb shift to 35 kHz it had been necessary to understand to about 0.1% the lineshape asymmetries caused by sweeping the magnetic field.

4.2.5   Fast beam rf Lamb shift measurements

These problems naturally stimulated attempts to measure the Lamb shift in zero magnetic field. There are three zero field transitions from $2S_{1/2}$ to $2P_{1/2}$ and these are shown in Fig. 4.6. A simple approach is to tune an oscillator through one of these lines and to determine the line center by fitting to the known profile. Of course the linewidth is still 100 MHz and it is now necessary to understand the microwave power and field distribution within 0.1% over a 100 MHz sweep. These constraints had driven Lamb to use a static magnetic field and even today demand a technical tour de force. The required techniques were developed during the 1970's and led eventually to a fourfold improvement in accuracy [LUN75, AND76, NEW79, LUN81].

The experiments used hydrogen atoms formed from a fast beam of protons (20-100 keV) by charge capture in a target of H_2 or N_2 gas. This method produced much higher metastable intensity than the slow beam, principally because the metastable fraction is of order 1% while the efficiency of slow beam electron impact excitation is of order 10^{-6}. The fast beam could not have been used in the earlier experiments because of the motional electric field $\mathbf{v} \times \mathbf{B}$. The beam then passed through a microwave field tuned midway between the overlapping $F = 1 \rightarrow 1$ and $1 \rightarrow 0$ transitions (see Fig. 4.6) which caused virtually all the $F = 1$ sublevels of $2S_{1/2}$ to decay via the $2P_{1/2}$ state. The surviving $F = 0$ metastable beam was ready to be used in a precise determination of the $F = 0 \rightarrow 1$ transition frequency at 910 MHz. A dip in metastable intensity was observed as the 910 MHz oscillator was swept through the resonance.

In the experiment of Andrews and Newton [AND76, NEW79] the $F = 0 \rightarrow 1$ transition was induced by an oscillating electric field in a coaxial transmission line. The line consisted of a cylindrical center conductor and a rectangular outer conductor as shown in Fig. 4.9. The atomic beam entered the line near a corner, where the electric field is zero, and travelled close to the long face of the outer conductor leaving at the opposite corner. The vanishing of the field in the corners allowed the entrance and exit holes to be cut in the line without affecting the impedance significantly. The microwave power was measured at the end of the line using a precise attenuator and a power meter. In the absence of reflections this would have been a measure of the power density at the position of the atomic beam, but since reflections could not be entirely eliminated there was a small standing wave whose size and spatial distribution varied as the frequency was swept. The consequent uncertainty in the strength of the microwave field at the beam proved to be the principal source of uncertainty in their determination of the Lamb shift, contributing $\pm$ 17 kHz out of a total of $\pm$ 20 kHz.

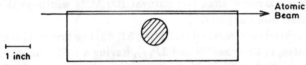

Fig. 4.9. Cross section through the transmission line used by Andrews and Newton.

A number of systematic corrections had to be made to the measured resonance frequency in order to determine the Lamb shift. (i) First, and by far the largest of these was a subtraction of the hyperfine level shifts. The $2S_{1/2}$ hyperfine interval was known from experiment to be 177.5569 MHz [HEB56] but the $2P_{1/2}$ level shifts had to be calculated. In a first approximation the hyperfine intervals are proportional to $(2L+1)^{-1}(J+1)^{-1}$ and hence the $2P_{1/2}$ splitting is just one third of the $2S_{1/2}$ splitting, but using the theory of Brodsky and Parsons [BRO67, 68], which includes radiative, relativistic and recoil corrections, the $2P_{1/2}$ hyperfine interval was found to be 59.1721 MHz, somewhat less than one third of the $2S_{1/2}$ splitting. In addition the energy of the $2P_{1/2}F = 1$ state is lowered by 2.5 kHz through the off-diagonal hyperfine coupling to the $2P_{3/2}F = 1$ state. Taking all these hyperfine shifts into account, the $F = 0 \rightarrow 1$ interval is less than the Lamb shift by 147.958 MHz with no appreciable uncertainty at this level of accuracy. (ii) The second correction was to account for the shifts of the resonant frequency due to the presence of the microwave field. A field $\varepsilon \cos \omega t$ has both a positive and a negative frequency component in its spectrum; $\frac{1}{2}\varepsilon e^{i\omega t}$ and $\frac{1}{2}\varepsilon e^{-i\omega t}$. In a two-level atom at resonance with one component of the field, the level spacing is perturbed by the other component and the resonant frequency is shifted. This is the Bloch-Siegert shift. In this real, many-level, atom there is an additional shift of the same general type due to the off- resonance coupling between the $2S$ and $2P_{3/2}$ states. This correction was of course power dependent and varied from run to run. A typical correction was 90 kHz. (iii) The Doppler shift of the resonance was cancelled to first order by reversing the direction of propagation of the microwaves and averaging the measurements. However, the second order shift, due to relativistic time dilation, was not suppressed by this method and required a correction of +20 kHz (see equation (4.19)). (iv) Small corrections were also made for the Zeeman and Stark shifts caused by stray fields, and for the line asymmetries caused by the wings of nearby transitions. The latter would have been a large effect had not the $F = 1$ atoms been removed by the first microwave field.

The final result was uncertain by 20 kHz, that is 200 ppm of the natural linewidth, a factor of 2 improvement in precision over the early measurements.

At the same time Lundeen and Pipkin [LUN75, LUN81] were pursuing a very similar experiment with one important difference; they used two separate but coherent microwave fields to drive the 910 MHz transition. The idea was to obtain a line narrower than the natural 100 MHz width so that the center could be located more accurately.

In the separated field method [RAM50, 51, 63] the first field creates a superposition of states, in this case $2S$ and $2P_{1/2}$, having a well defined relative phase. During the unperturbed flight of the beam towards the second microwave field the $2S$ and $2P$ states propagate with time dependences $e^{-i\omega_s t}$ and $e^{-i\omega_p t}$ and their relative phase therefore oscillates at a beat frequency $\omega_s - \omega_p$ determined by the energy difference. When the atoms arrive, after time of flight T, the accumulated phase of the atomic oscillation is $(\omega_s - \omega_p)T$ and that of the second field is $\omega T + \phi_0$, where ϕ_0 indicates a possible phase difference between the two fields. At resonance $\omega = \omega_s - \omega_p$ the field and the atomic coherence have acquired the same phase shift (assuming $\phi_0 = 0$), so the $2S - 2P$ transition continues where it left off. But if the oscillator is moved off resonance by an amount $\delta\omega = \pi/T$ the field presented to the atom in the second region is the reverse of that in the first and coherent $2P$ admixture will be undone. Thus the apparatus is, in effect, an interferometer of fixed path difference in which the frequencies ω and $\omega_s - \omega_p$ are compared. The interference fringes which appear in the lineshape are often called Ramsey fringes after the pioneer of the separated field technique.

The application of Ramsey's method to decaying states and in particular to a measurement of the hydrogen Lamb shift was first discussed by Hughes [HUG60]. He noted that the fringe linewidth $1/2T$ could be less than the natural linewidth $\gamma/2\pi$, where γ is the decay rate of atoms in the $2P$ state. He also noted that the size of the fringes is proportional to the size of the $2P$ state amplitude in the second field and hence to $e^{-\frac{1}{2}\gamma T}$. Thus a threefold reduction of the linewidth ($\gamma T = 3\pi$) requires a hundredfold loss of signal. The first demonstration of this method was a measurement of the $3S_{1/2} - 3P_{1/2}$ interval in hydrogen by Fabjan and Pipkin [FAB72] who observed the expected line narrowing.

That was followed by the measurement of interest here, the Lundeen-Pipkin measurement of the $n = 2$ Lamb shift [LUN75, 81], the general features of which have already been outlined at the beginning of this section. They measured the fraction of $F = 0$ metastable atoms $Q(\phi_0)$ quenched by the $F = 0 - 1$ transition in separated fields of relative phase ϕ_0. As the frequency was varied, this gave a broad line profile in which there were small Ramsey fringes. In order to extract the fringe pattern they measured the signal with the two fields in phase $Q(0)$ and out of phase $Q(\pi)$ and took the difference. The result is shown in Fig. 4.10. The width of the line was only 33 MHz, 1/3 of the natural width and the signal to noise ratio was still good enough to take advantage of the narrowing.

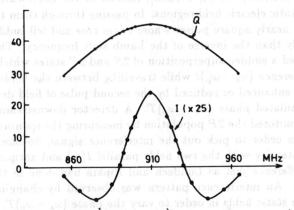

Fig. 4.10. Average resonance signal $\frac{1}{2}[Q(0) + Q(\pi)]$ and Ramsey pattern $Q(0) - Q(\pi)$. Note the change of scale; the peak of the Ramsey pattern is 1%. The radius of each plotted point is 10 standard deviations. (Reproduced with permission from S. R. Lundeen and F. M. Pipkin, *Phys. Rev. Lett.* **46** (1981) 232, (American Physical Society)).

The new parameter ϕ_0 brings with it the possibility of systematic errors. The Ramsey pattern is symmetric about the center of the line for $\phi_0 = 0$ but as ϕ_0 increases the fringes move across the line, becoming completely antisymmetric when $\phi_0 = \pi/2$. Hence the center of the fringe pattern is systematically shifted if ϕ_0 switches between δ and $\delta + \pi$ rather than 0 and π. The phase offset δ could not be made exactly zero so Lundeen and Pipkin interchanged the first and second regions in order to reverse its sign. Since the systematic error is proportional to δ, an average taken over both arrangements eliminates it. The direction of propagation of the fields in the transmission lines was also reversed in order to average out the first order Doppler shift.

The largest single uncertainty was due to an apparent variation of power at the beam as the frequency was swept. It was detected by studying the resonance lineshapes in detail. Although the mechanism was not found explicitly this would have been the effect of small standing waves such as those that limited the measurement of Andrews and Newton. The remaining important corrections were for hyperfine structure, microwave power shifts and the second order Doppler shift all of which have been discussed above.

The final result [LUN86] was uncertain by 9 kHz, that is 90 ppm of the natural linewidth, a factor of 4 improvement in precision over the early measurements.

4.2.6 Lamb shift measurements using static electric fields

An interesting variant of fast beam, separated field experiment was carried out by Sokolov [SOK73, 79, 84; PAL85]; instead of the microwave fields he used separated static electric field regions. In passing through them the atoms experienced two nearly square pulses whose edges rose and fell suddenly, i.e. much more rapidly than the inverse of the Lamb shift frequency. Thus the first region prepared a sudden superposition of $2S$ and $2P$ states which freely evolved a phase difference $(\omega_s - \omega_p)t$ while travelling between the fields. The $2P$ population was enhanced or reduced by the second pulse of field depending on the total accumulated phase $(\omega_s - \omega_p)T$. A detector downstream from the second field monitored the $2P$ population by measuring the spontaneous Lyman-α emission. In order to pick out the interference signal, Sokolov measured the Lyman-α intensity with the two fields parallel $I(+)$ and antiparallel $I(-)$ and took the difference just as Lundeen and Pipkin had done in their microwave experiment. An interference pattern was observed by changing the distance between the static fields in order to vary the phase $(\omega_s - \omega_p)T$.

Thus Sokolov's apparatus was, in effect, an interferometer of variable path difference measuring the fixed frequency $\omega_s - \omega_p$. It is a fundamental difference between this and the microwave experiment that this interference pattern was measured against a distance rather than a frequency. Consequently the Lamb shift could be determined only after an auxiliary experiment to measure the beam velocity which had to be well defined and stable.

For the auxiliary measurement the first electric field alone was used to produce $2P$ atoms which then delayed spontaneously. The decay length l_0 was measured by plotting the fluoresence intensity versus distance using a moveable detector. The velocity was then deduced using the relation $v = \gamma l_0$, where γ is the decay rate.

The statistical uncertainty in Sokolov's Lamb shift measurement was only 2 kHz [SOK84; PAL85], the smallest ever reported. Unfortunately the systematic error is not easily assessed. The beam velocity was measured before and after each run and most of the runs were rejected because it had changed. For the remaining runs it was assumed that the velocity had remained constant, and indeed the set of Lamb shift results so chosen lay in a single group of full width 10 kHz. Still there is the possibility that a systematic error might have affected the whole group. This could have occurred in the determination of the decay length from the exponential fluorescence curve, a notoriously difficult type of measurement [HIN78] generally only precise to a few percent and never before to a few ppm. Equally, there might be an error in the calculated value $\gamma = 6.264\ 881\ 2(20) \times 10^8$ s^{-1} taken for the decay rate which is also at an unprecedented level of precision [PAL85]. At present it must be said that the total uncertainty of this measurement is unknown.

In 1975 Drake, Farago and van Wijngaarden demonstrated yet another method of measuring Lamb shifts [DRA75]. They induced Lyman-α radiation from a metastable hydrogen beam using a static electric field $E\hat{z}$ and measured the intensities I_z and I_x emitted parallel and perpendicular to the field. In this way they measured the asymmetry

$$R = \frac{I_z - I_x}{I_z + I_x} \qquad (4.46)$$

of the intensity distribution. The idea [DRA73] was that the decay proceeds mainly through two coherent channels, $2S - 2P_{1/2} - 1S$ and $2S - 2P_{3/2} - 1S$ and that the value of R depends on their relative amplitudes because the former is spherically symmetric and the latter is not. Thus a measurement of R determines the relative amplitudes of the two decay channels. According to perturbation theory these amplitudes depend on the energy denominators S and $\Delta E - S$ (Fig. 4.6) so that R is related to the ratio $S/\Delta E$. Although the precision of the method has not been able to compete with rf methods in hydrogen the principle is rather general. The method and its applications are discussed in more detail by Drake in Chapter 3 of this volume.

4.2.7 Present status of the Lamb shift

Figure 4.11 shows the results of the Lamb shift measurements described in the last three sections, together with the theoretical expectation. Perhaps the most striking feature of this diagram is the modesty of the improvement since Triebwasser, Dayhoff and Lamb published their result 34 years ago. It stands as a testament to the monumental achievement of Lamb and his co-workers.

The experimental evidence is clearly consistent with the theoretical expectation. At the 10-20 kHz level quantum electrodynamics seems to be correct and the Lamb shift in hydrogen seems to understood. Further progress requires advances on several fronts. The proton radius, the one-loop binding corrections and the higher order QED diagrams would all have to be determined more accurately to improve the theory substantially. Also, new experimental ideas are probably needed if the measurement is to progress much beyond the 10 kHz level of precision.

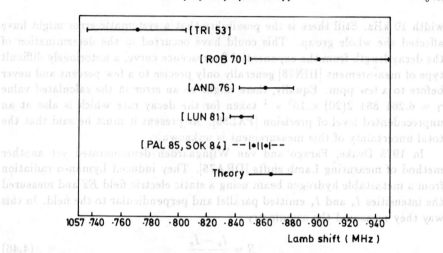

Fig. 4.11. Measured and theoretical values of the Lamb shift in hydrogen. The error bars are 1σ. The final limits of error given by TRI53 (3σ) and ROB70 (2σ) have been scaled accordingly. The dotted extension of the error bar given by SOK84 indicates unknown systematic errors. The theoretical value is taken from equation (4.44) which uses Mohr's computation of the Coulomb correction [MOH75].

Radiative level shifts have been measured in many other states of hydrogen and also in D, T, ^{3}He$^+$ and ^{4}He$^+$. This wealth of information is comprehensively reviewed in an article by Beyer [BEY78] and more recent developments are described by [BOL84]. Recently the Lamb shift has also been observed in muonium [ORA84, BAD84]. The author has focussed somewhat narrowly on the $n = 2$ Lamb shift interval in hydrogen because from the beginning this was the definitive testing ground for the theory of radiative level shifts and it has remained so to this day. Also, the remarkable range of experimental techniques employed to determine this one interval has allowed the author to touch on many of the principles of modern rf spectroscopy within a single theme.

Heavier systems in hydrogen-like and helium-like states of ionization have recently become important in the study of radiative level shifts [MOH83], particularly because the binding corrections ($G(Z\alpha)$ in equation (4.38a)) grow more rapidly with Z than the nuclear structure effects. These systems are discussed by Träbert in Chapter 6 of this book.

4.2.8 *Measurements of other $n = 2$ intervals: the fine structure constant*

The fine structure interval ΔE between $2P_{3/2}$ and $2P_{1/2}$ levels can be expressed in terms of the fine structure constant α with very little theoretical uncertainty (see section 4.2.3) and this has made it interesting for metrology.

The first precise value of α to be obtained in this way was based on the measurements of S and $\Delta E - S$ in deuterium by Triebwasser, Dayhoff and Lamb [TRI53, DAY53] (they measured only S in hydrogen). Unfortunately, the value was wrong because of an error in $\Delta E - S$ and a long period of confusion followed as a reconciliation was sought between theory and experiment. A fascinating account of this period is given by Kleppner [KLE71]. The resolution of the problem began in 1967 when Parker, Langenberg and Taylor announced a new experimental value for e/h which, together with the gyromagnetic ratio of the proton and other well-known constants, gave a new more precise value for α differing by five standard deviations from the old value [PAR67].

This situation prompted Wing and Franken [WIN68] and Metcalf, Brandenberger and Baird [MET68, BAI72] to measure ΔE directly using level crossing spectroscopy (see section 1.6.2). Metcalf *et al.* flowed atomic hydrogen gas through a cell where a magnetic field of approximately 3484 G defined a $\hat{z}$ axis. Light from a Lyman-α lamp incident along the $\hat{x}$ axis excited a number of $2P$ levels including $(J, m_J) = (3/2, -3/2)$ and $(1/2, -1/2)$ which were close to crossing. A detector monitored the fluorescence along the $\hat{y}$ axis. When the two excited levels were separated by more than their natural width, each contributed independently to the total scattering cross section which varied smoothly as the magnetic field was swept. But when the levels crossed, the scattering amplitudes interfered coherently and a resonant change occured in the angular distribution of the scattered light [FRA61]. As the magnetic field was swept a Lorentzian resonance signal was observed in the fluorescence detector, centered on the magnetic field at which the levels crossed. The theory of the Zeeman effect [BRO67, 68] was then used to deduce the fine structure interval in zero magnetic field. The result,

$$\Delta E = 10\,969.13(10) \text{ MHz}, \qquad (4.47)$$

was consistent with the new value of α and inconsistent with the interval deduced by combining S with $\Delta E - S$. This is the only direct measurement of a P state fine structure interval in hydrogen or any of its isotopes to have been published.

A re-measurement of $\Delta E - S$ was clearly required, and three independent measurements in hydrogen were almost immediately forthcoming, followed recently by a fourth using a fast beam and separated microwave fields [SAF84]. These results are shown in Fig. 4.12. The theory using the current value of α [WIL79] is confirmed by three of the four measurements. One [KAU71] disagrees with the theory and we do not know why.

Today the three most accurate values of the fine structure constant are completely unrelated to atomic fine structure (Chapter 2). They are derived from

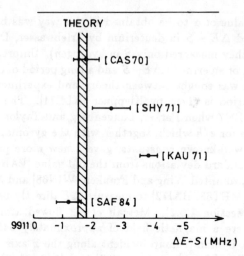

Fig. 4.12. Measured and theoretical values of the interval $\Delta E - S \left(2P_{3/2} - 2S_{1/2}\right)$ in hydrogen.

measurements of the anomalous magnetic moment of the electron [DYC84], the ac Josephson effect [WIL79, TAY85] and the ground state hyperfine interval $\Delta\nu$ in muonium [MAR82, SAP83]. In that order the values are

$$\alpha^{-1} = 137.035\ 994(9)(0.06\ \text{ppm})$$
$$137.035\ 981(12)(0.09\ \text{ppm})$$
$$137.035\ 988(20)(0.15\ \text{ppm})\ . \tag{4.48}$$

In the field of high precision measurements it is quite remarkable that three such different experiments are so concordant in their results and one is compelled by the agreement to have considerable confidence in this value of α. The fine structure interval $2^3P_0 - 2^3P_1$ in ^{4}He provides the fourth most accurate value (0.8 ppm) [FRI81] and the recently discovered quantized Hall effect also promises a high precision determination of α [BRA84].

It seems unlikely that hydrogen has any more to offer in this area because the linewidth associated with the lifetime is such a large fraction (1%) of the fine structure interval.

4.3   Weak Interactions

The following brief account of the theory is given to provide a setting for the sections on experimental methods which follow. A more complete account of the theory is to be found in Chapter 3.

4.3.1 Theory

Weak interactions are mediated by the exchange of virtual W^+, W^- and Z^0 particles, which together with the photon are the gauge bosons of the standard electroweak theory of Glashow, Weinberg and Salam [COM83]. In stable atoms the electrons have no charged interactions (exchange of W^+ or W^-) in first order (that would constitute β-decay) and the second order effects are, of course, very feeble. On the other hand Z^0 exchange does not affect the charges of the constituent particles and contributes in first order to atomic structure. The range of the virtual Z^0 is very short ($\sim 10^{-18}$ m) because it is massive (~ 100 GeV/c^2) and for our purposes it is adequate to consider the interaction as a simple current-current interaction at a point. Assume, as in the standard theory, that the currents contain only vector (V) and axial vector (A) components. Then there are four main terms to consider in the effective Hamiltonian

$$H_{\text{eff}} = \frac{G_F}{\sqrt{2}} \sum_N (C_{VV}^{eN} \overline{\psi}_e \gamma^\mu \psi_e \overline{\psi}_N \gamma_\mu \psi_N + C_{AA}^{eN} \overline{\psi}_e \gamma^\mu \gamma_5 \psi_e \overline{\psi}_N \gamma_\mu \gamma^5 \psi_N$$
$$+ C_{AV}^{eN} \overline{\psi}_e \gamma^\mu \gamma_5 \psi_e \overline{\psi}_N \gamma_\mu \psi_N + C_{VA}^{eN} \overline{\psi}_e \gamma^\mu \psi_e \overline{\psi}_N \gamma_\mu \gamma^5 \psi_N) .$$
$$(4.49)$$

ψ_N and ψ_e are field operators, e refers to the electron and N to the muon in muonium, the proton in hydrogen or the nucleons in heavier isotopes. The C's are coupling constants, γ_μ and γ_5 are the usual Dirac matrices ($\mu = 1, 2, 3, 4; \gamma_5 = \gamma_1 \gamma_2 \gamma_3 \gamma_4$) and G_F is the Fermi coupling constant. Of course the fundamental couplings are to the quarks but at low energy it is convenient to consider the nucleons as fundamental. There are also terms proportional to the momentum transfer but they are small in atoms. The values of the coupling constants according to the standard model are given in Table 4.

Table 4. Coupling constants for an electron exchanging a Z^0 with various particles at low momentum transfer, as defined by equation (4.49). The values are those given by the standard electroweak theory. W stands for $(1-4\sin^2 \theta_W)$ where θ_W is the electroweak mixing angle, the free parameter of the standard theory. η is the ratio of axial vector to vector amplitude in neutron β-decay and has the value 1.250(9) [KEL80].

N	C_{VV}^{eN}	C_{AA}^{eN}	C_{AV}^{eN}	C_{VA}^{eN}
p	$-W^2/2$	$-\eta/2$	$W/2$	$\eta\, W/2$
n	$W/2$	$\eta/2$	$-1/2$	$-\eta\, W/2$
μ^+	$-W^2/2$	$-1/2$	$W/2$	$W/2$

The first two terms in equation (4.49) are of even parity because they involve the product of two odd currents (the C_{VV} term) and of two even currents (the

C_{AA} term). These interactions cause shifts of the energy levels of an atom. The last two terms are of odd parity and according to first order perturbation theory they do not shift the levels; they are the terms responsible for parity violation in hydrogen.

Bég and Feinberg [BEG74] have calculated the parity-conserving weak interaction contributions $\delta(\Delta\nu)$ to the ground state hyperfine interval in hydrogen or muonium. To a good approximation the vector current $\overline{\psi}_N\gamma_\mu\psi_N$ is independent of nuclear spin while the axial current $\overline{\psi}_N\gamma_\mu\gamma^5\psi_N$ is proportional to it. Consequently the C_{AA} term contributes to the ground state splitting but the C_{VV} term does not. The result obtained by Bég and Feinberg is

$$\delta(\Delta\nu) = 140C_{AA}\text{Hz} . \tag{4.50}$$

In hydrogen this amounts to a reduction of the interval by approximately 90 Hz according to the standard model. Although this is large compared with the accuracy of measurements the existence of the shift has not yet been demonstrated because of the theoretical uncertainty due to proton structure. In muonium, where $C_{AA} = -1/2$, neither theory nor experiment is adequate to reveal the shift. Nevertheless, the requirement that theory and experiment should agree yields the most stringent model-independent constraints available;

$$|C_{AA}^{ep}| \leq 50$$
$$|C_{AA}^{e\mu}| \leq 50 . \tag{4.51}$$

These limits are much larger than the values of the coupling constants in the standard model and further, more accurate measurements would be very desirable.

The parity-violating terms in (4.49) may be replaced to a good approximation by a non-relativistic potential H_{PV} for an electron interacting with a point nucleus at $r = 0$. In atomic units

$$H_{PV} = \frac{\alpha G_F}{2\sqrt{2}}[-C_{AV}^{eN}(\boldsymbol{\sigma}_e\cdot\mathbf{p})\delta(\mathbf{r})+C_{VA}^{eN}(\boldsymbol{\sigma}_e\cdot\mathbf{p})(\boldsymbol{\sigma}_e\cdot\boldsymbol{\sigma}_N)\delta(\mathbf{r})+\text{Hermitian conjugate}]$$

$$\tag{4.52}$$

[DUN78]. The matrix of H_{PV} between the $nS_{1/2}$ and $nP_{1/2}$ states of hydrogen is diagonal in the hyperfine quantum numbers F and M and the elements are

$$\langle nP_{1/2}FM|H_{PV}|nS_{1/2}FM\rangle = iV\big(C_{AV}^{ep} - C_{VA}^{ep}[2F(F+1)-3]\big) \tag{4.53a}$$

where

$$V = \frac{\alpha G_F}{2\sqrt{2}}\cdot\frac{Z^4}{\pi n^4}\sqrt{n^2-1} . \tag{4.53b}$$

In the case of hydrogen with $n = 2$, V corresponds to a frequency of 0.0128 Hz.

4.3.2 General principles of parity measurements

The general principles guiding measurements of C_{AV} and C_{VA} are as follows. In an atom without parity violation the eigenstates have definite parity and electromagnetic transitions between a particular pair of states involve either even or odd atomic operators but not both. An example of particular significance in the next section is a transition between S states, which can be driven by a magnetic dipole interaction ($M1$, even operator) but not by an electric dipole ($E1$, odd operator). However, since the parity violating interaction (4.52) imparts a small amount of P admixture to the S states, odd and even transition amplitudes can occur simultaneously. To a good approximation the small $E1$ amplitude is proportional to the weak interaction strength G_F and to the electric dipole matrix element rE, while the $M1$ amplitude depends on the magnetic dipole matrix element μB. The interference between these two amplitudes can generate a contribution to the transition rate which may usefully be regarded as the product of two factors. One, involving the transition moments, is linear in the weak interaction strength G_F and, being proportional to r, has odd parity. The other, involving the external fields, is proportional to E and also has odd parity. Since the rate of a purely electromagnetic transition is invariant under parity transformations it is quite general for the interference between odd and even moments to involve an odd parity combination of external fields. This property of the interference term allows us to distinguish it experimentally from the (generally) much larger parity conserving rate. Of course one might search for a purely parity violating rate (electric dipole in the case of this example) [ROB68, HIN78] but since that depends on G_F^2 it is very much smaller and furthermore the parity violation is not manifest in the external fields.

4.3.3 Experiments to measure parity violation

Parity violation in hydrogen was discussed as early as 1959 by Zel'dovich [ZEL59]. Later many practical aspects of making a measurement were explored by Michel [MIC65], but the subject really came to life with the discovery of weak neutral currents in 1973/74 [HAS73, BEN74]. Almost immediately Feinberg published two important papers on the weak circular polarization of spontaneous decay radiation from hydrogenic atoms [FEI74a, b] and Lewis and Williams discussed the rate of excitation of hydrogen atoms by polarized light [LEW75]. It was subsequently suggested that the best hope of detecting parity violation in hydrogen lay in microwave transitions among the $2S_{1/2}$ levels [HIN77] and the three experiments on hydrogen now underway are all based on such transitions. As yet there are no results at a significant level of accuracy. Dunford [DUN81] has proposed a similar scheme with He^+.

One experiment, that of Hinds and collaborators at Yale, involves the $2S_{1/2}$ and $2P_{1/2}$ states shown in Fig. 4.6. The strength of the PNC interaction is measured by a study of hyperfine transition [HIN77] from $(2S_{1/2}F = 0)$ to $(2S_{1/2}F = 1)$ at 178 MHz using an atomic beam of hydrogen.

According to first order perturbation theory the usual $(2S_{1/2}F)$ state $|SF\rangle$ is modified by an admixture of the $(2P_{1/2}F)$ state $|PF\rangle$ to become

$$|S'F\rangle = |SF\rangle + \frac{\langle PF|H_{PV}|SF\rangle}{E_{SF} - E_{PF}}|PF\rangle \tag{4.54}$$

where E_{SF} and E_{PF} are the eigenvalues of those two states and we have neglected the smaller effect of $2P_{3/2}$ and other more distant states. Hence the electric dipole matrix element between the states normally labelled $(2S_{1/2}F = 0)$ and $(2S_{1/2}F = 1)$ is

$$\langle S'1|z|S'0\rangle = \langle S1|z|P0\rangle\frac{\langle P0|H_{PV}|S0\rangle}{E_{S0} - E_{P0}} + \frac{\langle S1|H_{PV}|P1\rangle}{E_{S1} - E_{P1}}\langle P1|z|S0\rangle . \tag{4.55}$$

The weak interaction matrix elements are given by equations (4.53). The energy denominators are both approximately equal to the Lamb shift S; 969 MHz for $F = 0$ and 1088 MHz for $F = 1$. Consequently the terms involving C_{AV}^{ep} almost cancel and the electric dipole matrix element is approximately

$$\langle S'1|z|S'0\rangle = \frac{4\sqrt{3}iV}{S}C_{VA}^{ep} . \tag{4.56}$$

Of course the main point is that this matrix element is not zero and that an electric dipole hyperfine transition can therefore be excited.

The apparatus is illustrated in Fig. 4.13. The main features are a source of metastable hydrogen atoms, a series of interaction regions and a detector that does not register ground state atoms. A high intensity $2S$ atomic beam is produced by passing a 500 eV beam of protons through a caesium target. The atoms are prepared in the $F = 0$ state using a microwave state selector [EDW81] to stimulate decay of the $F = 1$ atoms by driving resonant transitions to the short lived $2P_{1/2}$ levels. The beam now enters the main interaction region in which it passes sequentially through two separated parallel, coherent oscillating fields. The first field (β) is magnetic and drives the normal $F = 0 \rightarrow F = 1$ magnetic dipole transition. The second field is electric (ϵ) and drives the parity-forbidden electric dipole transition $F = 0 \rightarrow F = 1$ through the transition moment discussed above. The $F = 1$ amplitudes induced in the two regions may be written

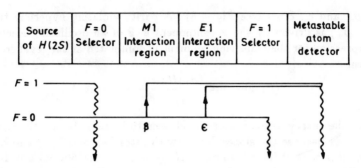

Fig. 4.13. Schematic diagram of the apparatus used by Hinds and collaborators to measure parity violation in hydrogen.

$$\mathbf{A}_{M1} = \mu\boldsymbol{\beta}$$
$$\mathbf{A}_{E1} = id\boldsymbol{\epsilon}e^{i(\phi_0+\phi)} \qquad (4.57)$$

where μ and d are real and proportional respectively to the magnetic dipole and (parity-forbidden) electric dipole transition moments. The angle ϕ_0 is a quantum mechanical phase and ϕ is the phase angle between $\boldsymbol{\epsilon}$ and $\boldsymbol{\beta}$. The state amplitudes are polarized along the field directions because the initial $F = 0$ state is spherically symmetric. Thus the probability of transitions to the $F = 1$ state is

$$P = (\mathbf{A}_{M1} + \mathbf{A}_{E1})^* \cdot (\mathbf{A}_{M1} + \mathbf{A}_{E1})$$
$$= \mu^2\beta^2 + d^2\epsilon^2 + 2\mu d\boldsymbol{\beta}\cdot\boldsymbol{\epsilon}\sin(\phi_0 + \phi) \ . \qquad (4.58)$$

Here we see first the allowed $M1$ term, then the negligible forbidden $E1$ term and last the interference term to be measured, which is proportional to the weak interaction through d and depends on the odd combination of fields $\boldsymbol{\beta}\cdot\boldsymbol{\epsilon}$.

Atoms driven into the $F = 1$ state are picked out by passing the beam through a second state selector (Fig. 4.13) which stimulates $F = 0$ atoms to decay, but transmits $F = 1$ atoms. The latter are then detected by measuring the flux of emitted Lyman-α photons when the beam enters a region of strong static electric field. Thus the number of detected atoms is proportional to P (equation (4.58)). The interference term of interest is modulated by chopping the phase of the rf electric field between ϕ and $\phi + \pi$ to change its sign and is extracted from the total detector signal by phase sensitive detection.

Another version of the experiment by Williams and collaborators [LEV82a] is performed in a static magnetic field $B\hat{z}$ of approximately 553 G where the levels β_+ and e_- cross (see Fig. 4.8). A parity violating hyperfine transition $(\alpha_- \to \beta_+)$ is driven by the z component $\boldsymbol{\varepsilon} \cdot \hat{z}$ of an oscillating electric field. The expression for the transition moment is dominated at the level crossing by one term

$$\langle \beta_+ |z| \alpha_- \rangle = \frac{\langle \beta_+ |H_{PV}| e_- \rangle}{E_{\beta_+} - E_{e_-}} \langle e_- |z| \alpha_- \rangle \qquad (4.59)$$

in which the energy denominator would seem to be zero. In fact the eigenvalues are complex because the states decay (with rates Γ_{2P} and Γ_{2S}) and hence the denominator goes at its minimum to $i(\Gamma_{2P} - \Gamma_{2S})/2$ or $50i$ MHz in frequency units. (When decay is considered in this way the Hamiltonian is not Hermitian and the derivation of the transition moments is much more delicate than the present discussion might suggest [BEL77], although the results given are correct.) Since Γ_{2S} is negligible

$$\langle \beta_+ |z| \alpha_- \rangle = \frac{4\sqrt{3V}}{\Gamma_{2P}} C_{VA}^{ep} . \qquad (4.60)$$

In Williams' experiment the parity allowed transition amplitude is induced not by a separate oscillating magnetic field but by another component $\boldsymbol{\varepsilon} \cdot \hat{x}$ of the same microwave electric field in conjunction with a static electric field $E\hat{x}$. The three fields $B\hat{z}, E\hat{x}$ and $\boldsymbol{\varepsilon}$ are coplanar and are arranged as shown in Fig. 4.14. This "Stark-induced" transition moment is dominated by one term near the $\beta - e$ level crossings

$$\langle \beta_+ |x| \alpha_- \rangle = \frac{\langle \beta_+ |xE| e_+ \rangle}{E_{\beta_+} - E_{e_+}} \langle e_+ |x| \alpha_- \rangle . \qquad (4.61)$$

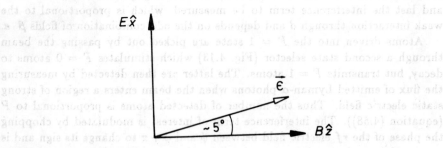

Fig. 4.14. Configuration of fields used in the experiment of Williams and colloborators to measure parity violation in hydrogen.

Interference between the Stark-induced amplitude and the parity forbidden one gives rise to a term in the transition rate proportional to the weak interaction and to $(\varepsilon \cdot \mathbf{B})(\varepsilon \cdot \mathbf{E})$, which can be identified experimentally because it changes sign when either $\mathbf{E}$ or $\mathbf{B}$ is reversed.

Under the parity operation, electric fields change sign but magnetic fields do not. Hence the two signatures $\beta \cdot \varepsilon$ (equation (4.58)) and $(\varepsilon \cdot \mathbf{B})(\varepsilon \cdot \mathbf{E})$ are suitably odd under parity. It is interesting to note however that they also appear to be odd under time reversal (T) which reverses magnetic fields but not electric fields. So it would seem that these experiments do not measure the standard weak interaction, which is even under time reversal, but a more exotic P and T violating effect. Of course that is not so and we consider the two experiments in turn. At zero magnetic field the $M1$ and $E1$ transition moments are $\pi/2$ out of phase. This is a consequence of time reversal invariance and is true to the extent that we neglect the imaginary parts of the energy denominators in equation (4.55). It follows that ε and β must also be in quadrature if the two transition amplitudes are to interfere. Thus the apparently T-odd nature of the interference signal is simply due to a hidden time ordering in the experiment $-\beta$ leads (or lags) ε by $\pi/2$. A more accurate version of its signature is Im $(\beta \cdot \varepsilon)$ which is even under $T(\beta$ changes sign and complex numbers are conjugated) and therefore characterizes a P-odd, T-even interaction. At the $\beta - e$ level crossing the two transition moments used by Williams *et al.* are given by equations (4.59) and (4.61). Again the numerators are $\pi/2$ out of phase and if the energy denominators were real, the transition amplitudes would be in quadrature and unable to interfere. But now there are substantial phase shifts due to the complex eigenvalues and a more complete form of the signature is

$$\text{Im} \left(\frac{\varepsilon \cdot \mathbf{B}}{E_{\beta_+} - E_{e_-}} \cdot \frac{\varepsilon \cdot \mathbf{E}}{E_{\beta_+} - E_{e_+}} \right)$$

which is even under T. Again the T-odd signature expresses a hidden time ordering — the direction of spontaneous decay — and not an intrinsic T violation within the Hamiltonian of the atom. Lévy and Williams have confirmed experimentally that spontaneous decay can produce formally T-odd contributions to a transition rate [LEV82b].

Two key parameters in these experiments are (i) the ratio R of the interference term in P (proportional to H_{PV}) to the total probability P, and (ii) the signal-to-noise ratio S of the interference term to the total noise in the detector. It follows from equation (4.58) for the zero field experiment that when ϕ is chosen to give $\phi_0 + \phi = \pi/2$ and when β is parallel to ε, these are given by

$$R = \frac{1}{\sqrt{P}} \cdot 2d\varepsilon \qquad (4.62)$$

and

$$S = \sqrt{N} \cdot 2d\varepsilon \qquad (4.63)$$

where NP is the number of atoms detected and we have neglected small terms in P. The noise is assumed to be $\sqrt{NP}$ due to Poisson statistics. Evidently it is advantageous to make P small (weak field β) so that the interference signal will be a larger fraction of the total (equation (4.62)) but not so small that other sources of noise dominate $\sqrt{NP}$. It is also advantageous to make $d\varepsilon$ as large as possible. Similar general conclusions apply to the experiment of Williams *et al.* at the level crossing. They suppress P by making E small and by making ε almost perpendicular to $\mathbf{E}$.

Equations (4.62) and (4.63) indicate that $d\varepsilon (d\varepsilon \cdot \hat{z}$ in the case of Williams *et al.*) is an important figure of merit for these experiments. A comparison of equations (4.56) and (4.60) shows that the parity violating transition moment is larger at the level crossing than at zero magnetic field by the ratio of energy denominators $S : \Gamma_{2p} = 10$. All other things being equal (e.g. interaction times) the value of d will be correspondingly an order of magnitude larger at the level crossing. Unfortunately, the proximity of the P states which enhances the parity violating transition moment also increases the rate of the quenching process $2S - 2P - 1S$ and ε has to be reduced in order to avoid losing the metastable beam altogether. It turns out that these effects are mutually compensating so that the experiments have approximately equal sensitivity to the parity violating coupling constant C^{ep}_{VA}.

A third experiment on hydrogen by Adelberger and coworkers is described in [FOR84]. It is also performed near the $\beta - e$ level crossing and involves the interference of Stark-induced and parity violating transitions in separated oscillating fields.

The main source of systematic error in these experiments is stray electric field. A stray, Stark-induced transition amplitude can masquerade as a parity violating one, although it does not have the same phase or polarization, and some discriminant such as reversal of the static magnetic field is then required to distinguish the spurious parity violating signal from a real one. In practice this limits the acceptable stray electric field to a few mV/cm if C^{ep}_{VA} is to be measured at the level of unity. It is not at all easy to keep stray fields that small. In the experiments that use a 553 G magnetic field there is the additional problem that the atoms experience a motional electric field $\mathbf{v} \times \mathbf{B}$ in their rest frame (beam velocity $v = 3 \times 10^5$ m/s). In order to control that field adequately it is necessary to make the magnetic field very homogeneous and to control the angle between $\mathbf{v}$ and $\mathbf{B}$ at the 10^{-5} rad. level. These problems are thought to be tractable but at present the best published result is $C^{ep}_{VA} < 620$ [LEV82a]. A new result, shortly to be published by Hinds and collaborators is

$C^{ep}_{VA} < 300$. The experiments are enormously difficult but that is nothing new for hydrogen. The work continues, as it always has, because hydrogen being the most fundamental atom is a uniquely important system in which to test fundamental theories.

At present several experiments have measured C^{eN}_{AV} in various heavy atoms at the 10% level of accuracy but C^{eN}_{VA} has not been detected in any atom. For a review of parity violation in atoms, see [FOR84].

Acknowledgement

The author is indebted to G. K. Woodgate for reading the manuscript and making many useful suggestions and to S. Clemas for typing it.

REFERENCES

AND76   D. A. Andrews and G. Newton, *Phys. Rev. Lett.* **37** (1976) 1254.

APP70   Appelquist and Brodsky, *Phys. Rev.* **A2** (1970) 2293.

BAD84   A. Badertscher, S. Dhawan, P. O. Egan, V. W. Hughes, D. C. Lu, M. W. Ritter, K. A. Woodle, M. Gladisch, H. Orth, G. zu Putlitz, M. Eckhause, J. Kane, F. G. Mariam and J. Reidy, *Phys. Rev. Lett.* **52** (1984) 914.

BAI72   J. C. Baird, J. Brandenberger, K. -I. Gondaira and H. Metcalf, *Phys. Rev.* **A5** (1972) 564.

BEG74   M. A. B. Bég and G. Feinberg, *Phys. Rev. Lett.* **33** (1974) 606; Errata, **35** (1975) 130.

BEL77   J. S. Bell and G. Karl, *Nuovo Cimento* **41A** (1977) 487.

BEN63   P. L. Bender, *Phys. Rev.* **132** (1963) 2154.

BEN74   A. Benvenutti *et al.*, *Phys. Rev. Lett.* **32** (1974) 800.

BET57   H. Bethe and E. Salpeter, *Quantum Mechanics of One- and Two-Electron Atoms* (Springer Verlag, Berlin, 1957).

BEY78   H.-J. Beyer, in *Progress in Atomic Spectroscopy*, eds. W. Hanle and H. Kleinpoppen, (Plenum Press, 1978) p. 529.

BOL84   J. J. Bollinger, S. R. Lundeen and F. M. Pipkin, *Phys. Rev.* **A30** (1984) 2170.

BRA84   E. Braun, P. Gutmann, G. Hein, F. Melchjert, P. Warnecke, S. Q. Xue and K. v. Klitzing, *Prec. Meas. and Fund. Consts. II*, ed. B. N. Taylor and W. D. Phillips, *Nat. Bur. Stand. (U. S.), Spec. Publ.* 617 (1984).

BRE30   G. Breit, *Phys. Rev.* **35** (1930) 1447.

BRE31   G. Breit and I. Rabi, *Phys. Rev.* **38** (1931) 2082.

BRO67   S. J. Brodsky and R. G. Parsons, *Phys. Rev.* **163** (1967) 134.

BRO68   S. J. Brodsky and R. G. Parsons, *Phys. Rev.* **176** (1968) 423 (Errata).

CAS75 D. E. Casperson, T. W. Crane, V. W. Hughes, P. A. Souder, R. D. Stambaugh, P. A. Thompson, H. Orth, G. zu Putlitz, H. F. Kaspar, H. W. Reist and A. B. Denison, *Phys. Lett.* **B59** (1975) 397.

CAS77 D. E. Casperson, T. W. Crane, A. B. Denison, P. O. Egan, V. W. Hughes, F. G. Mariam, H. Orth, H. W. Reist, P. A. Souder, R. D. Stambaugh, P. A. Thompson and G. zu Putlitz, *Phys. Rev. Lett.* **38** (1977) 956.

CLE64 W. E. Cleland, J. M. Bailey, M. Eckhause, V. W. Hughes, R. M. Mobley, R. Prepost and J. E. Rothberg, *Phys. Rev. Lett.* **13** (1964) 202.

COM83 E. D. Commins and P. H. Bucksbaum, *Weak Interactions of Leptons and Quarks* (Cambridge University Press, 1983).

COS70 B. L. Cosens and T. V. Vorberger, *Phys. Rev.* **A2** (1970) 16.

CRA66 S. B. Crampton, H. G. Robinson, D. Kleppner and N. F. Ramsey, *Phys. Rev.* **141** (1966) 55.

CRA71 T. Crane, D. Casperson, P. Crane, P. Egan, V. W. Hughes, R. Stambaugh, P. A. Thompson and G. zu Putlitz, *Phys. Rev. Lett.* **27** (1971) 474.

DAY53 E. S. Dayhoff, E. Treibwasser and W. E. Lamb, *Phys. Rev.* **89** (1953) 106.

DRA73 G. W. F. Drake and R. B. Grimley, *Phys. Rev.* **A8** (1973) 157.

DRA75 G. W. F. Drake, P. S. Farago and A. van Wijngaarden, *Phys. Rev.* **A11** (1975) 1621.

DUN78 R. W. Dunford, R. R. Lewis and W. L. Williams, *Phys. Rev.* **A18** (1978) 2421.

DUN81 R. W. Dunford, *Phys. Lett.* **99B** (1981) 58.

DYC84 R. S. van Dyck Jr., P. B. Schwinberg and H. G. Dehmelt, in *Atomic Physics 9*, eds. R. S. van Dyck and E. N. Fortson (World Scientific, Singapore, 1984) p. 53.

EDW81 J. Wm. Edwards, G. L. Greene and E. A. Hinds, *Nucl. Instrum. Methods* **197** (1981) 581.

ERI77 G. W. Erickson, *J. Phys. Chem. Ref. Data.* **6** (1977) 831.

ERL69 R. O. Erlich, H. Hofer, A. Magnon, D. Stowell, R. A. Swanson and V. L. Telegdi, *Phys. Rev. Lett.* **23** (1969) 513.

ESS71 E. Essen, R. W. Donaldson, M. J. Baugham and E. G. Hope, *Nature* **229** (1971) 110.

FAB72 C. W. Fabjan and F. M. Pipkin, *Phys. Rev.* **A6** (1972) 556.

FAV71 D. Favart, P. M. McIntyre, D. Y. Stowell, V. L. Telegdi, R. DeVoe and R. A. Swanson, *Phys. Rev. Lett.* **27** (1971) 1336.

FEI74a   G. Feinberg and M. Y. Chen, *Phys. Rev.* **D10** (1974) 190; Errata **10** (1974) 3145.

FEI74b   G. Feinberg and M. Y. Chen, *Phys. Rev.* **D10** (1974) 3789.

FER30   E. Fermi, *Z. Phys.* **60** (1930) 320.

FOR84   E. N. Fortson and L. L. Lewis, *Phys. Reports* **113** (1984) 290.

FRA61   P. A. Franken, *Phys. Rev.* **121** (1961) 508.

FRI81   W. Frieze, E. A. Hinds, V. W. Hughes and F. M. J. Pichanick, *Phys. Rev.* **A24** (1981) 279.

HAN63   L. N. Hand, D. J. Miller and R. Wilson, *Rev. Mod. Phys.* **35** (1963) 335.

HAS73   F. J. Hasert *et al., Phys. Lett.* **46B** (1973) 121; *Phys. Lett.* **46B** (1973) 138.

HEB56   J. W. Heberle, H. A. Reich and P. Kusch, *Phys. Rev.* **101** (1956) 612.

HEL70   E. Hellwig, R F. C. Vessot, M. W. Levine, P. W. Zitzewitz, D. W. Allen and D. J. Glaze, *IEEE Trans. Instrum. Meas.* **IM19** (1970) 200.

HES86   H. F. Hess, G. P. Kochanski, J. M. Doyle, I. J. Greytak, and D. Kleppner, *Phys. Rev.* **A34** (1986) 1602.

HIN77   E. A. Hinds and V. W. Hughes, *Phys. Lett.* **B67** (1977) 487.

HIN78   E. A. Hinds, J. E. Clendenin and R. Novick, *Phys. Rev.* **A17** (1978) 670.

HIN80   E. A. Hinds, *Phys. Rev. Lett.* **44** (1980) 374.

HUG60   V. W. Hughes, in *Quantum Electronics*, ed. C. H. Townes (Columbia University Press, New York, 1960) p. 582.

HUG77   V. W. Hughes and T. Kinoshita, in *Muon Physics*, eds. V. W. Hughes and C. S. Wu (Academic Press, New York, 1977) p. 11, Vol. 1.

HUR86   M. D. Hürlimann, W. N. Hardy, A. J. Berlinsky, and R. W. Cline, *Phys. Rev.* **A34** (1986) 1605.

KAU71   S. L. Kaufman, W. E. Lamb, K. R. Lea and M. Leventhal, *Phys. Rev.* **A4** (1971) 2128.

KEL80   R. L. Kelly, C. P. Horne, M. J. Losty, A. Rittenberg, T. Shimada, T. G. Trippe, G. G. Wohl, G. P. Yost, N. Barasch-Schmidt, C. Bricman, C. Dionisi, M. Mazzucato, L. Montanet, R. L. Crawford, M. Roos and B. Armstrong, *Rev. Mod. Phys.* **52** Part II S1 (1980).

KIN84a   T. Kinoshita and J. Sapirstein, *Atomic Physics 9*, eds. R. S. Van Dyck and E. N. Fortson (World Scientific, Singapore, 1984) p. 38.

KIN84b   T. Kinoshita, R. S. Van Dyck Jr., P. B. Schwinberg and H. G. Dehmelt, *Atomic Physics 9*, eds. R. S. Van Dyck and E. N. Fortson (World Scientific, Singapore, 1984) p. 53.

KLE62   D. Kleppner, H. M. Goldberg and N. F. Ramsey, *Phys. Rev.* **126** (1962) 603.

KLE65   D. Kleppner, H. C. Bess, S. B. Crampton and N. F. Ramsey, R. F. C. Vessot, H. E. Peters and J. Vanier, *Phys. Rev.* **A138** (1965) 972.

KLE71   D. Kleppner, in *Atomic Physics and Astrophysics*, eds. M. Chrétien and E. Lipworth (Gordon Breach, 1971) p. 1.

KOB73   H. G. Kobrak, R. A. Swanson, D. Favert, W. Kells, A. Magnon, P. M. McIntyre, J. Roehrig, D. Y. Stowell, V. L. Telegdi and M. Eckhause, *Phys. Lett.* **B43** (1973) 526.

KOP49   H. Kopferman, H. Krüger and H. Olmann, *Z. Phys.* **126** (1949) 760.

KUH50   H. Kuhn and G. W. Series, *Proc. R. Soc. London* **A202** (1950) 127.

LAM47   W. E. Lamb and R. C. Retherford, *Phys. Rev.* **72** (1947) 241.

LAM50   W. E. Lamb and R. C. Retherford, *Phys. Rev.* **79** (1950) 549.

LAM51   W. E. Lamb and R. C. Retherford, *Phys. Rev.* **81** (1951) 222.

LAM52a W. E. Lamb, *Phys. Rev.* **85** (1952) 259.

LAM52b W. E. Lamb and R. C. Retherford, *Phys. Rev.* **86** (1952) 1014.

LEP84   G. P. Lepage and D. R. Yennie, in *Precision Measurements and Fundamental Constants II*, eds. B. N. Taylor and W. D. Phillips; *Nat. Bur. Stand. (USA), Spec. Publ.* **617** (1984) 185.

LEV82a L. P. Lévy and W. L. Williams, *Phys. Rev. Lett.* **48** (1982) 607.

LEV82b L. P. Lévy and W. L. Williams, *Phys. Rev. Lett.* **48** (1982) 1011.

LEW75   R. R. Lewis and W. L. Williams, *Phys. Lett.* **59B** (1975) 70.

LUN75   S. R. Lundeen and F. M. Pipkin, *Phys. Rev. Lett.* **34** (1975) 1368.

LUN81   S. R. Lundeen and F. M. Pipkin, *Phys. Rev. Lett.* **46** (1981) 232.

LUN86   S. R. Lundeen and F. M. Pipkin, *Metrologia* **22** (1986) 9.

MAC50   J. E. Mack and N. Austern, *Phys. Rev.* **77** (1950) 745.

MAR82   F. G. Mariam, W. Beer, P. R. Bolton, P. O. Egan, C. J. Gardner, V. W. Hughes, D. C. Lu, P. A. Souder, J. Orth, J. Vetter, U. Moser and G. zu Putlitz, *Phys. Rev. Lett.* **49** (1982) 993.

MET68   H. Metcalf, J. C. Baird and J. R. Brandenberger, *Phys. Rev. Lett.* **21** (1968) 165.

MIC65   F. C. Michel, *Phys. Rev.* **B138** (1965) 408.

MOH75   P. J. Mohr, *Phys. Rev. Lett.* **34** (1975) 1050.

MOH83   P. J. Mohr, in NATO ASI *Relativistic Effects in Atoms, Molecules and Solids*, ed. G. Malli (Plenum Press, New York, 1983).

MUR49   K. Murukawa, S. Suwa and T. Karmer, *Phys. Rev.* **76** (1949) 1721.

NEW79   G. Newton, D. A. Andrews and P. J. Unsworth, *Philos. Trans. R. Soc.* **290** (1979) 373.

ORA84   C. J. Oram, J. M. Bailey, P. W. Schmor, C. A. Fry, R. F. Kiefl, J. B. Warren, G. M. Marshall and A. Olin, *Phys. Rev. Lett.* **52** (1984) 910.

PAL85   V. G. Pal'chikov, Yu. L. Sokolov and V. P. Yakovlev, *Metrologia* **21** (1985) 99. A different decay rate $\gamma = 6.264\ 938 \times 10^8 \mathrm{s}^{-1}$ given by [SOK84] yields a different value for the Lamb shift.

PAR67   W. H. Parker, D. N. Langenberg and B. N. Taylor, *Phys. Rev. Lett.* **81** (1967) 287.

PAS38   S. Pasternak, *Phys. Rev. Lett.* **54** (1938) 1113.

PIR46   J. Pirenne, *Arch. Sci. Phys. Nat.* **28** (1946) 233.

PRE61   R. Prepost, V. W. Hughes and K. Ziock, *Phys. Rev. Lett.* **6** (1961) 19.

PRI77   M. H. Prior and E. C. Wang, *Phys. Rev.* **A16** (1977) 6.

RAM50   N. F. Ramsey, *Phys. Rev.* **78** (1950) 695.

RAM51   N. F. Ramsey and H. B. Silsbee, *Phys. Rev.* **84** (1951) 506.

RAM63   N. F. Ramsey, *Molecular Beams* (Oxford University Press, 1963).

REI56   H. A. Reich, J. W. Heberle and P. Kusch, *Phys. Rev.* **104** (1956) 1585.

RIC81   A. Rich, *Rev. Mod. Phys.* **53** (1981) 127.

ROB65   R. T. Robiscoe, *Phys. Rev.* **A138** (1965) 22.

ROB66a   R. T. Robiscoe and B. L. Cosen, *Phys. Rev. Lett.* **17** (1966)69.

ROB66b   R. T. Robiscoe, *Bull. Am. Phys. Soc.* **11** (1966) 62.

ROB68   R. T. Robiscoe, *Phys. Rev.* **168** (1968) 4.

ROB70   R. T. Robiscoe and T. W. Shyn, *Phys. Rev. Lett.* **24** (1970) 559.

ROS70   S. D. Rosner and F. M. Pipkin, *Phys. Rev.* **A1** (1970) 571.

SAF84   K. A. Safinya, K. K. Chan, S. R. Lundeen and F. M. Pipkin, in *Precision Measurements and Fundamental Constants II*, eds. B. N. Taylor and W. D. Phillips, *Nat. Bur. Stand. (USA) Spec. Publ.* **617** (1984) 127.

SAH   See [SER57].

SAP83   J. Sapirstein, *Phys. Rev.* **A51** (1983) 985.

SCH69   H. A. Scheussler, E. N. Fortson and H. G. Dehmelt, *Phys. Rev.* **187** (1969) 5.

SER57   G. W. Series, *Spectrum of Atomic Hydrogen* (Oxford University Press, 1957).

SES55   A. M. Sessler and H. M. Foley, *Phys. Rev.* **98** (1955) 6.

SES58   A. M. Sessler and R. L. Mills, *Phys. Rev.* **110** (1958) 1453.

SHY71   T. W. Shyn, T. Rebane, R. T. Robiscoe and W. L. Williams, *Phys. Rev.* **A3** (1971) 116.

SIM80   G. G. Simon, C. Schmitt, F. Borkowski and V. H. Walther, *Nucl. Phys.* **A333** (1980) 381.

SOK73   Yu. L. Sokolov, *Sov. Phys. JETP* **36** (1973) 243.

SOK79   Yu. L. Soklov, in *Atomic Physics 6* ed. R. Damburg (Plenum Press, New York, 1979) p. 207.

SOK84   Yu. L. Sokolov, in *Precision Measurements and Fundamental Constants II*, eds. B. N. Taylor and W. D. Phillips, *Nat. Bur. Stand. (USA) Spec. Publ.* **617** (1984) 135.

TAY85   B. N. Taylor, *J. Res. Nat. Bur. Stand.* **90** (1985) 91.

THO69   P. A. Thompson, J. J. Amato, P. Crane, V. W. Hughes, R. M. Mobley, G. zu Putlitz and J. E. Rothberg, *Phys. Rev. Lett.* **22** (1969) 161.

TRI53   S. Triebwasser, E. S. Dayhoff and W. E. Lamb, *Phys. Rev.* **89** (1953) 98.

VOE70   R. DeVoe, P. M. McIntyre, A. Magnon, D. Y. Stowell, R. A. Swanson and V. L. Telegdi, *Phys. Rev. Lett.* **25** (1970) 1779.

WAL86   R. L. Walsworth, I. F. Silvera, H. P. Godfried, C. C. Agosta, F. C. Vessot and E. M. Mattison, *Phys. Rev.* **A34** (1986) 2550.

WIL79   E. R. Williams and P. T. Olsen, *Phys. Rev. Lett.* **42** (1979) 1575.

WIN68   W.H. Wing, Ph. D. Thesis, University of Michigan (unpublished) (1968).

WIT56   J. D. Wittke and R. H. Dicke, *Phys. Rev.* **103** (1956) 620.

ZEL59   Ya. B. Zel'dovich, *Zh. Eksp. Teor. Fiz.* **36** (1959) 964.

ZIO62   K. Ziock, V. W. Hughes, R. Prepost, J. Bailey and W. Cleland, *Phys. Rev. Lett.* **8** (1962) 103.

CHAPTER 5

OPTICAL SPECTROSCOPY

G. W. Series and T. W. Hänsch

5.1 Before The Laser Era

The enormous impact made upon precision spectroscopy in the visible and ultraviolet by the invention of tunable lasers has been described in an earlier chapter. SAH was written, of course, before the laser revolution had even been conceived: the optical spectroscopy of that earlier volume was the spectroscopy of prisms, gratings and interferometers. Spectra were registered photographically: the era of pressure-scanned interferometers with photoelectric recording was in its infancy and had not yet been applied to hydrogen. The battle against Doppler broadening was fought with liquid gases used as refrigerants for gas discharges. The practicability of beams of atomic hydrogen had been discussed, but no useful spectroscopic measurements on such beams had been made.

Nevertheless the work that was done between the publication of SAH and the laser era was not without value, particularly as much of it was directed towards determinations of the Rydberg constant, whose importance in the context of the fundamental atomic constants is emphasized in Chapter 10. Lest this work be forgotten, we devote a few paragraphs to it.

5.1.1 Helium$^+$; resolution of the fine structure

In the nineteen sixties a series of very thorough studies of the transition $n = 4-3$ at 468.6 nm (in those days, still 4686 Å) was carried out by Mack and his younger colleagues Roesler, Berry, DeNoyer and Kessler at the University of Wisconsin [ROE 64a, ROE 64b, BER 70] and at the Bureau of Standards, Washington [KES 71] and by Larson and Stanley at Purdue University (LAR 67). Other lines in He$^+$, ($n = 6-4$ and $n = 5-4$) were also studied. Liquid nitrogen cooling was generally used after it was established that a significant contribution to linewidth resulted from recoil of the atoms in the excitation process; it was argued that the extra effort required for liquid hydrogen cooling would not be worthwhile.

A systematic Doppler shift arising from motion of the radiating ions in hollow cathode discharges was investigated quantitatively [ROE 64b], and this, indeed, provides an interpretation of the anomaly reported in SAH (p. 59). As they travel towards the hollow cathode, ions are accelerated towards the observer. At the low pressure used by Series those ions in the longer-lived fine-structure states achieved higher velocities than did those which radiated more quickly, giving rise to a *differential* Doppler effect within the fine-structure complex. But at the higher pressure used by Herzberg the more frequent collisions led to thermalization of the ionic velocities and eliminated the differential effect.

In all this work the Dirac theory, supplemented by QED, was entirely vindicated, though, of course, the accuracy of measurement of the fine structure intervals fell far short of the standards of microwave spectroscopy. Indeed, the logical conclusion of this and other work in optical spectroscopy was that one should concentrate, not on the fine structure intervals, but on absolute measurements of wavelength against a standard, and, by assuming the validity of the theory, to use the result to provide an experimental value of the Rydberg constant. This argument prevails today.

5.1.2   The Rydberg constant: renewed interest

The time was indeed ripe for a new determination of the Rydberg. Cohen [COH 52] had uncovered a discrepancy between earlier determinations and had tentatively ascribed this to the use of an unreliable standard of wavelength by one group of workers. This deduction was entirely vindicated [SER 59], and it become apparent that the value currently in use for this important constant of physics was based on experimental work which, already by 1970, had been rendered obsolete by advances in the techniques of high resolution spectroscopy. The conference at Gaithersburg on Precision Measurement and Fundamental Constants [KES 71] where this point was elaborated [SER 71] included a report of Kessler and Roesler's study of the He^+ lines, and particularly of 468 nm, presented as preliminary work directed towards a determination of the Rydberg, and an account by Matsui [MAT 71] of a determination based on a study of Balmer-α, excited in a liquid nitrogen cooled source and spectroscopically analysed by a Möbius-band interferometer (a two-beam interferometer with variable path difference). At that time also a value for the Rydberg had been obtained by Csillag [CSI 68], using a Fabry-Perot interferometer. He studied the first seven members of the Balmer series in deuterium.

Following the conference the study of D_α and also of T_α was taken up at the National Physical Laboratory, UK, by Kibble and others [KIB 73], who advanced the technique by using liquid *helium* to cool the hydrogen discharge tubes. By this time also techniques in photoelectric recording and in computing had advanced so that it was practicable to analyse the spectral line profiles by computer, fitting them to theoretically-calculated curves with relatively few parameters chosen to describe the separation and intensities of overlapping fine-structure components. Numerical evaluation of the Stark effect was necessary to determine which components were most sensitive to electric fields, and whether microfields and collisions in the discharge might have influenced the results [BLA 73]. It was concluded that they had had no perceptible effect.

The work at the Bureau of Standards was pushed forward [KES 73], and the final result was based on measurements of the lines 468 nm ($n = 4 - 3$), 656 nm ($n = 6 - 4$) and 1012 nm ($n = 5 - 4$). The results obtained by all these workers have now been superseded by laser spectroscopic measurements, but they are included, for completeness, in Chapter 10.

5.1.3   Student laboratories

Before entering the (expensive) field of laser spectroscopy we remark that significant and instructive spectra may be obtained for hydrogen and deuterium using simple and inexpensive equipment appropriate to an undergraduate teaching laboratory. A series of experiments is described by Series and Stacey [SER 82] wherein the Rydberg constant is determined to an accuracy of 5 parts in 10^5, the electron-proton mass ratio to 1.5 parts in 10^4, and the fine structure constant to 4%.

5.2   Precision Laser Spectroscopy: First Round of Experiments

A series of experiments, remarkable for the dramatic improvement on earlier work, for the promise of future achievement, for the stimulus they gave to atomic spectroscopists, and for the short period of time in which they were carried out was undertaken in the years 1973-79 by T. W. Hänsch and his distinguished group of young collaborators at Stanford University. We are grateful to Professor Hänsch and to the US National Bureau of Standards for permission to reproduce, as part of the text of this chapter, an article written for the second of the conferences on Precision Measurement and Fundamental Constants held at Gaithersburg in 1981, in which this work is summarized. (Unfortunate delays prevented the publication of the Proceedings until 1984 [TAY 84].)

Work at Stanford under Hänsch's direction continued until 1986 when he departed for Munich. Many of his young colleagues have initiated work elsewhere or associated themselves with work which was already in progress.

Reprinted from *Precision Measurement and Fundamental Constants II*, eds. B. N. Taylor and W. D. Phillips, Nat. Bur. Stand (U.S.), Spec. Publ. 617 (1984) 111-115.

5.2.1   Precision laser spectroscopy*

T. W. Hänsch

Department of Physics, Stanford University, Stanford, CA 94305

Precision laser spectroscopy of atomic hydrogen and deuterium will be reviewed. The Balmer-α line has been studied by Doppler-free saturated absorption spectroscopy, polarization spectroscopy, optical- radiofrequency double quantum spectroscopy, and by laser-quenching of a beam of metastable atoms. These experiments have led to an eightyfold improvement in the accuracy of the Rydberg constant. Two-photon spectroscopy of the $1S$-$2S$ transition has made possible an accurate

*Work supported by the National Science Foundation under Grant PHY80-10689 and by the U. S. Office of Naval Research, contract ONR N00014-78-C-0403.

measurement of the ground state Lamb shift and further advances in resolution promise new stringent tests of quantum electrodynamic theory.

Key words: fundamental constants; high resolution spectroscopy; hydrogen; lasers; precision measurements; quantum electrodynamics.

I. Introduction

Precision laser spectroscopy of simple quantum mechanical systems can be a powerful tool to determine better values of fundamental constants and to probe the limits of quantum electrodynamic theory. At present, only the hydrogen atom and its isotopes, the heavier hydrogen-like ions, and certain hydrogen-like exotic atoms permit sufficiently accurate model calculations for such an approach. Future theoretical advances should make it possible to add three-body systems such as the neutral helium atom or the hydrogen molecular ion to the list of promising candidates.

Of these candidates, only hydrogen and deuterium have so far been studied extensively by precision laser spectroscopy [1], and we will limit our discussions to the work on these two simplest of the stable atoms. We will review the experiments reported to date, and we will investigate some of the possible directions for future progress.

II. Precision Laser Spectroscopy of the Hydrogen Balmer-α Line and the Rydberg Constant

The simple and regular Balmer spectrum of atomic hydrogen can be readily observed in the light emitted by a glow discharge. The exploration of this spectrum has played a crucial role in the development of atomic physics and quantum mechanics [1]. More than once, seemingly minute discrepancies between experiment and theory have led to major revolutions in our understanding of quantum physics. However, no classical spectroscopic observation has ever succeeded in fully resolving the intricate fine structure of the Balmer lines. The spectra always remained blurred by Doppler broadening due to the rapid thermal motion of the light hydrogen atoms.

II.1 Saturation spectroscopy

Dramatic progress in spectral resolution became possible only with the advent of highly monochromatic tunable dye lasers together with techniques of Doppler-free laser spectroscopy. The prominent red Balmer line was the first hydrogen line to be studied by Doppler-free saturated absorption spectroscopy [2]. In this technique, two monochromatic laser beams are sent in opposite directions through the absorbing gas sample. When the laser is tuned to the center of a Doppler broadened line both beams can interact with the same atoms, those with zero axial velocity. The signal is observed as a bleaching of the absorption of the probe beam, caused by the saturating beam.

Even though only a relativity simple pulsed dye laser was available in our initial experiments at Stanford, we were thrilled by the spectra that could be recorded in a simple Wood type gas discharge. As illustrated in Fig. 1, we were able to resolve single fine structure components of the Balmer-α line, and the $n=2$ Lamb shift could be observed directly in the optical spectrum. In 1974, Munir Nayfeh at Stanford [3] completed an absolute wavelength measurement of the strong $2P_{3/2} - 3D_{5/2}$ component of hydrogen and deuterium which yielded an eightfold improved value of the Rydberg constant, compared to the value recommended in the 1973 adjustment of the fundamental constants [4].

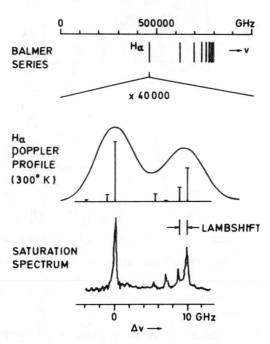

Fig. 1. *Top:* Balmer spectrum of atomic hydrogen. *Center:* Doppler profile of the red Balmer-α line at room temperature and theoretical fine structure components. *Bottom:* Doppler-free spectrum of Balmer-α, recorded by saturated absorption spectroscopy with a pulsed dye laser [2,3].

Since then, frequency stabilized cw dye lasers, whose resolution is not limited by pulse length, have become readily available. Moreover, new techniques of saturation spectroscopy have been developed, which can reach shot-noise limited sensitivity despite laser intensity fluctuations [5,6]. In polarization spectroscopy [5], the probe beam monitors the dichroism and birefringence of the sample induced by a circularly or linearly polarized saturating beam. The technique takes advantage of the fact that small changes in light polarization can be detected with higher sensitivity than changes in intensity. Because of its higher sensitivity, the method permits measurements at lower atom densities and lower laser intensities, so that pressure broadening, power broadening, and related problems can be much reduced.

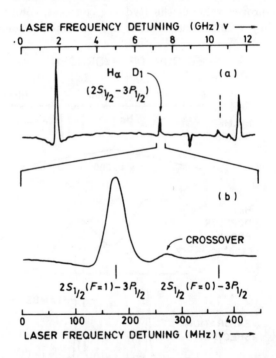

Fig. 2. Doppler-free polarization spectrum of the Balmer-α line, recorded with a cw dye laser in a mild He-H$_2$ discharge [7].

In 1978, J. E. M. Goldsmith *et al.* at Stanford [7] completed a new measurement of the Rydberg constant, observing the hydrogen Balmer-α line in a mild helium-hydrogen discharge by polarization spectroscopy with a cw dye laser. As shown in Fig. 2, the resolution of the weak but narrow $2S_{1/2} - 2P_{1/2}$ fine structure component was at least 5 times better than in the earlier pulsed experiments. The absolute wavelength of this component was measured relative to the ith hyperfine component of the $^{127}I_2$ - $XR(127)11$ - 5 transition at 632.8 nm, using a near-coincident I_2 line (the ith hyperfine component of the $^{127}I_2$ B - $XR(73)5$ - 5 transition) as an intermediate reference [8]. Systematic line shifts due to the discharge plasma were studied very carefully in a series of measurements which yielded several results interesting in their own right [9,10]. For instance, anomalous pressure shifts have been observed which could be explained in terms of collisional decoupling of the 3P hyperfine structure [9]. The final evaluation of all measurements gave a threefold improvement in the accuracy of the Rydberg value. Another twofold improvement could be obtained immediately with a more accurate absolute wavelength measurement of the intermediate iodine reference line.

The two Stanford measurements, as originally published, used values for the iodine reference wavelength (^{20}Ne: $^{127}I_2$, i) that differed slightly from the rounded value of 632.991 399 nm recommended by the Committee for the Definition of the Meter [11]. The results in Table 1 and Fig. 3 have been adjusted to be consistent with this rounded value. Specifically, the 1974 Rydberg value has been adjusted downwards by 0.0024 ppm, and the 1978 value has been adjusted upwards by 0.0013 ppm. The 1974 result is a composite of two somewhat differing Rydberg values for deuterium and hydrogen. The adjusted 1974 deuterium value, $R_\infty = 109\ 737.314\ 8 \pm 0.001\ 0$ cm^{-1} is in excellent agreement with the 1978 measurement. The adjusted 1974 hydrogen value $R_\infty = 109\ 737.312\ 8 +$ 0.001 0 cm^{-1}, is almost 0.02 ppm smaller. However, the latter includes substantial systematic corrections for the larger $2P_{3/2}$ hyperfine splitting of the light isotope, and the uncertainty of these corrections in the presence of collisions may have been underestimated in 1974.

Table 1 and Fig. 3 also give the results of an independent measurement of the Rydberg constant, reported in 1980 by B. W. Petley *et al.* [12]. The Balmer-α line of hydrogen was observed in a Wood type gas discharge by saturated absorption spectroscopy with a cw dye laser. The result is in good agreement with the earlier Stanford values.

II.2   Laser spectroscopy of an atomic beam

The accuracy of the best measurements in hydrogen discharges [7] does not appear to be limited by pressure shifts or Stark effect in the discharge plasma. Nonetheless, it has long been obvious that a collisionless beam of metastable hydrogen atoms would be a more ideal sample for precision laser spectroscopy [3].

S. R. Amin *et al.* of Yale University are the first who have succeeded with such an atomic beam experiment, and they report on a new Rydberg measurement elsewhere in these proceedings [13]. The Balmer-α line of hydrogen

Table 1. Measurements of the Rydberg Constant.

Year	Authors	$R_\infty [\text{cm}^{-1}]$
73	COHEN, TAYLOR [6]	109 737.317 70 $\pm$ 0.008 30
74	HANSCH NAYFEH, LEE, CURRY, SHAHIN[5]	109 737.314 10 $\pm$ 0.001 00
78	GOLDSMITH, WEBER, HANSCH [10]	109 737.314 90 $\pm$ 0.000 32
79	PETLEY, MORRIS, SHAWYER[15]	109 737.315 13 $\pm$ 0.000 85
81	AMIN, CALDWELL, LICHTEN[16]	109 737.315 21 $\pm$ 0.000 11

THE RYDBERG CONSTANT

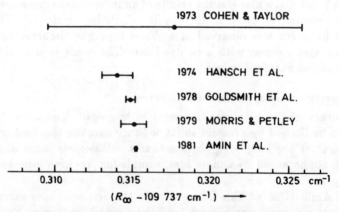

Fig. 3. Recent measurements of the Rydberg constant.

and deuterium is observed by exciting the metastable $2S(F = 1)$ atoms with cw dye laser beams which cross the atomic beam at a right angle. Most of the excited $3P$ atoms quickly decay into the $1S$ ground state, and the resulting quenching of the $2S$ state can be observed with a detector for metastable atoms, just as in the classical Lamb-Retherford experiment [14]. Such linear atomic beam spectroscopy requires fewer systematic corrections than nonlinear saturation spectroscopy, and a wavelength measurement by direct comparison with an iodine stabilized He-Ne laser has yielded a new Rydberg value accurate to one part in 10^9, as shown in Table 1 and Fig. 3.

II.3   Two-photon spectroscopy

Although the accuracy of linear laser spectroscopy of a metastable hydrogen beam appears amenable to further improvements, the resolution of the narrowest Balmer-α components will always be limited by the short lifetime of the upper $3P$ state to no better than 29 MHz.

D. E. Roberts and E. N. Fortson [15] were the first to point out that narrower lines can be obtained if an additional radiofrequency field is applied so that radiofrequency optical double quantum transitions are induced from the $2S_{1/2}$ level to the longer living $3S_{1/2}$ and $3D_{1/2}$ level. E. W. Weber and J. E. M. Goldsmith [16] have observed lines as narrow as 20 MHz by applying this technique to hydrogen atoms in a gas discharge, and they have been able to measure the small $3P_{3/2} - 3D_{3/2}$ Lamb shift directly by comparing single- and double-quantum signals. C. E. Wieman and collaborators [17] have recently begun to apply the same technique to a beam of metastable hydrogen atoms, and they expect to reach a resolution better than 1 MHz, corresponding to the natural width of the $3S$ level.

The same narrow lines could also be observed by excitation with two laser photons of equal frequency. If the two photons come from opposite directions, first order Doppler broadening is automatically eliminated without any need to select slow atoms, because from a moving atom the two photons have equal but opposite Doppler shifts, so that their sum-frequency is constant [18]. Doppler-free two-photon spectroscopy of the Balmer-α transition has so far been stifled by the lack of suitable highly monochromatic tunable lasers in the near infrared. But visible dye lasers should make it possible to study transitions from $2S$ to high Rydberg levels by this technique.

By comparing the wavelengths of such transitions with that of the Balmer-α line, one might detect, for instance, some small deviations from Coulomb's law, which may exist within atomic dimensions, but which may have escaped detection in the past.

III.   Two-Photon Spectroscopy of Hydrogen $1S - 2S$

There is another, even more intriguing transition in hydrogen which can be studied by Doppler-free two-photon spectroscopy: the transition from the $1S$ ground state to the metastable $2S$ state. The $1/7$ s lifetime of the upper level implies an ultimate natural linewidth as narrow as 1 Hz.

There is no intermediate near resonant level which would enhance the two-photon transition rate. However, even small numbers of excited $2S$ hydrogen atoms can be detected with high sensitivity, by monitoring the vacuum ultra-violet Lyman-α radiation emitted after conversion to the $2P$ state by collisions or external fields, or by photoionizing the $2S$ atoms and observing charged particles.

Unfortunately, however, two-photon excitation of $1S - 2S$ requires mono-chromatic ultraviolet radiation near 243 nm, where there are still no good tun-able laser sources available. Intense coherent radiation at this wavelength can be generated by frequency doubling of a pulsed dye laser in a non-linear optical crystal, and $1S - 2S$ two-photon spectra have been observed at Stanford with such sources [19-21]. But the resolution remained limited by the bandwidth of the pulsed lasers.

III.1   Spectroscopy with pulsed lasers

The best $1S - 2S$ spectra so far have been recorded by C. E. Wieman, [21] who reached a resolution of 120 MHz (FWHM at 243 nm) with the help of a blue single-mode cw dye laser oscillator with nitrogen-pumped pulsed dye laser am-plifier chain and lithium niobate frequency doubler. The hydrogen atoms were generated in a Wood type discharge tube and carried by gas flow and diffusion into the observation chamber, where they were excited by two counterpropa-gating beams from the laser system. The emitted vacuum ultraviolet Lyman-α photons were observed through a magnesium fluoride side window by a solar blind photomultiplier. The spectral resolution was sufficient to resolve the hy-perfine doublets (Fig. 4), although another 100 million-fold improvement should ultimately be possible. But even these crude spectra permitted a measurement of the 671 GHz H-D isotope shift to within 6.3 MHz and provided a first quali-tative confirmation of the predicted small 11.9 MHz relativistic correction due to nuclear recoil [21].

In the same series of experiments [19-21], the $1S - 2S$ energy interval was compared with the $n = 2 - 4$ interval, by simultaneously observing the Balmer-β line with the visible dye laser output. If the simple Bohr theory were correct, the $n = 1—2$ would be exactly four times the $n = 2 - 4$ interval, and both transitions would be observed at exactly the same laser frequency. In reality, this degeneracy is lifted by relativistic and quantum electrodynamic corrections,

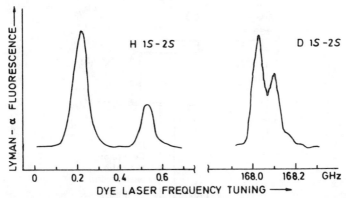

Fig. 4. Doppler-free two-photon spectrum of the $1S$ - $2S$ transition in hydrogen and deuterium with resolved hyperfine splittings [21].

and we expect line splittings and displacements as illustrated in Fig. 5. By measuring the separation of the $1S - 2S$ resonance from one of the Balmer-β components, one can determine an experimental value of the Lamb shift of the $1S$ ground state, which cannot be measured by radiofrequency techniques, because there is no $1P$ state which could serve as a reference.

The predicted 8149.43± 0.08 MHz Lamb shift of the hydrogen $1S$ state has been confirmed within 0.4%, by comparing the $1S - 2S$ spectrum with a polarization spectrum of the Balmer-β line, observed with the cw dye laser output in a Wood type gas discharge [21]. The uncertainty in this experiment was dominated by laser frequency shifts due to rapid refractive index changes in the pulsed dye amplifiers. Such chirping introduces unknown phase parameters into the calculation of the line-shape of the two-photon signals. Pressure shifts of the Balmer-β spectrum were the next largest source of error.

Both problems can be overcome with mere technical improvements, and the intrinsic narrow natural linewidths of the observed transitions make it appear likely that measurements of the $1S$ Lamb shift will eventually reach a higher accuracy than radiofrequency measurements of the $2S$ Lamb shift [22]. The latter provide one of the most stringent current tests of quantum electrodynamics, but they are plagued by a 100 MHz natural linewidth due to the short lifetime of the $2P$ state. Persisting discrepancies between experiment and the predictions of different computational approaches [23, 24] make further accurate Lamb shift measurements highly desirable.

III.2   *Current efforts towards higher resolution*

In order to avoid the limitations of a pulsed laser source, A. I. Ferguson, J. E. M. Goldsmith, B. Couillaud, A. Siegel, J. E. Lawler and other collabora-

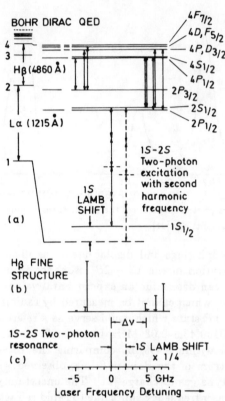

Fig. 5. *Top:* Simplified diagram of hydrogen energy levels and transitions. The Dirac fine structure and QED corrections are shown on an enlarged scale: hyperfine structure has been ignored. *Bottom:* Fine structure spectrum of the Balmer-β line and relative position of the $1S$-$2S$ two-photon resonance, as recorded with the second harmonic frequency. The dashed line gives the hypothetical position of the $1S$-$2S$ resonance if there were no $1S$ Lamb shift.

tors at Stanford have invested considerable effort into an experiment designed to observe the hydrogen $1S - 2S$ two-photon transition with low power cw ultraviolet radiation. While known non-linear optical crystals do not permit efficient 90 degrees phase matched second harmonic generation down to 243 nm, cw ultraviolet radiation at this wavelength can be produced as the sum frequency of a violet krypton ion laser and a yellow rhodamine 6G dye laser in a crystal of ammonium dihydrogen phosphate (ADP), cooled close to liquid nitrogen temperature [25]. With frequency stabilized single frequency lasers of 0.6 and 2.5 watts power, respectively, focussed to a waist diameter of 100 μm inside a 5 cm long ADP crystal, about 700 μ W of tunable cw ultraviolet power have been produced. A frequency-locked external passive enhancement cavity increases this power to several mW at the sample. Such a power should be sufficient for a resonant signal of several hundred Lyman-α photons per second under the chosen experimental conditions.

Unfortunately, however, the ADP crystal is damaged within less than a minute under these conditions. We speculate that the ultraviolet light produces color centers, perhaps associated with heavy-ion impurities, and that these centers then absorb enough of the intense primary beams to damage the crystal. We are presently investigating whether mixing of more widely separated primary wavelengths in ADP near room temperature can provide a cure for this problem, and we are actively pursuing alternative approaches. Once the crystal damage problem is solved, it should be possible to reduce the bandwidth of such a cw ultraviolet source with the help of fast servo controls to a few kHz.

However, other causes of line broadening, in particular pressure broadening, transit broadening, and transverse Doppler broadening have to be overcome before such a resolution can be approached in the $1S-2S$ two-photon spectrum [25]. A beam of ground state hydrogen atoms, cooled close to liquid helium temperature [26], and interacting with nearly collinear counterpropagating laser beams would minimize such problems, and it appears technically quite feasible to observe the $1S-2S$ two-photon transition with a line width of a few tens of kHz, or a resolution approaching one part in 10^{11}. The line center could then be determined to within 1 part in 10^{13} or better, once accurate frequency standards become available in the visible and ultraviolet region.

Such a precise measurement of the $1S-2S$ frequency could, of course, be used to determine a still better value of the Rydberg constant. However, the current uncertainty of the electron/proton mass ratio [27] (about 0.14 ppm) limits the accuracy of such a Rydberg value to about 1 part in 10^{10}. Considerable improvements of direct measurements of the electron/proton mass ratio have been predicted [27]. Alternatively, a better mass ratio could be determined from a precision measurement of the 671 GHz H-D isotope shift of the $1S-2S$ frequency. However, uncertainties of the fine structure constant and of the mean square radii of the nuclear charge distributions would still impose error limits of about 4 parts in 10^{11} for the Rydberg.

If the electron/proton mass ratio and the fine structure constant can be measured independently with improved accuracy, then a precise measurement of the $1S-2S$ H-D isotope shift could provide an accurate probe for nuclear structure and recoil shifts.

To determine a precise Rydberg value that is not limited by nuclear structure corrections, one could combine the $1S-2S$ measurement with a precise measurement of a two-photon transition from $2S$ to one of the higher nS levels. Taking advantage of the fact that the lowest order nuclear structure corrections scale with the inverse cube of the principal quantum number n, one can easily construct differences of transition frequencies which are no longer sensitive to the exact nuclear sizes.

Neither a measurement of the $1S - 2S$ frequency nor of the isotope shift by itself can provide a very stringent test of quantum electrodynamics, because we are free to adjust the values of fundamental constants until the calculations agree with the observations. However, if we form the ratio of the $1S - 2S$ frequency to the frequency of a different hydrogenic transition, such as a Balmer transition, or a transition to or between Rydberg states, we arrive at a dimensionless quantity, which, to lowest order, no longer depends on the Rydberg constant, and which can be calculated with very high precision. An accurate measurement of such a frequency ratio would permit a very interesting test of theory, and if the past is any guide, it may well lead to some surprising discovery.

REFERENCES (Section 5.2.1 only)

[1] T. W. Hänsch, G. W. Series, and A. L. Schawlow, Sci. Am. **2403**, 94 (March 1979).
[2] T. W. Hänsch, I. S. Shahin, and A. L. Schawlow, Nature **235**, 63 (1972).
[3] T. W. Hänsch, M. H. Nayfeh, S. A. Lee, S. M. Curry, and I. S. Shahin, Phys. Rev. Lett. **32** 1336 (1974).
[4] E. R. Cohen and B. N. Taylor, J. Phys. Chem. Ref. Data **2**, 663 (1973).
[5] C. E. Wieman and T. W. Hänsch, Phys. Rev. Lett. **36**, 1170 (1976).
[6] R. K. Raj, P. Bloch, J. J. Snyder, G. Camy, and M. Ducloy, Phys. Rev. Lett. **44**, 1251 (1980).
[7] J. E. M. Goldsmith, E. W. Weber, and T. W. Hänsch, Phys. Rev. Lett. **41**, 940 (1978).
[8] J. E. M. Goldsmith, E. W. Weber, F. V. Kowalski, and A. L. Schawlow, Appl. Opt. **18**, 1983 (1979).
[9] E. W. Weber and J. E. M. Goldsmith, Phys. Lett. **70A**, 95 (1979).
[10] E. W. Weber, Phys. Rev. **A20**, 2278 (1979).
[11] Comité Consultatif pour la Définition du Mètre, 5ᵉ Session–1973, Bureau International des Poids et Mesures, p. M23.
[12] B. W. Petley, K. Morris, and R. E. Shawyer, J. Phys. B: Atom. Molec. Phys. **13**, 3099 (1980).
[13] S. R. Amin, C. D. Caldwell, and W. Lichten, these proceedings; Phys. Rev. Lett. **47**, 1234 (1981).
[14] W. E. Lamb and R. C. Retherford, Phys. Rev. Lett. **79**, 549 (1950).
[15] D. E. Roberts and E. N. Fortson, Phys. Rev. Lett. **31**, 1539 (1973).
[16] E. W. Weber and J. E. M. Goldsmith, Phys. Rev. Lett. **41**, 940 (1978).

[17] C. E. Wieman, private communication; D. Shiner and C. Wieman, these proceedings.

[18] N. Bloembergen and M. D. Levenson, in *High Resolution Laser Spectroscopy* (Topics in Applied Physics, Vol. 13), Ed. by K. Shimoda (Springer-Verlag, Berlin, 1976), p. 315.

[19] T. W. Hänsch, S. A. Lee, R. Wallenstein, and C. E. Wieman, Phys. Rev. Lett. **34**, 307 (1975).

[20] S. A. Lee, R. Wallenstein, and T. W. Hänsch, Phys. Rev. Lett. **35**, 1262 (1975).

[21] C. E. Wieman and T. W. Hänsch, Phys. Rev. **A22**, 192 (1980).

[22] S. R. Lundeen and F. M. Pipkin, Phys. Rev. Lett. **46**, 232 (1981).

[23] G. W. Erickson, Phys. Rev. Lett. **27**, 780 (1971).

[24] P. J. Mohr, Phys. Rev. Lett. **34**, 1050 (1975).

[25] A. I. Ferguson, J. E. M. Goldsmith, T. W. Hänsch, and E. W. Weber, in *Laser Spectroscopy IV*, Ed. by H. Walther and K. W. Rothe (Springer Series in Optical Sciences, Vol. 21, Springer-Verlag, Berlin, 1979) p. 31.

[26] S. B. Crampton, T. J. Greytak, D. Kleppner, W. D. Phillips, D. A. Smith, and A. Weinrib, Phys. Rev. Lett. **42** (1979).

[27] R. S. Van Dyck, Jr. and P. B. Schwinberg, Phys. Rev. Lett. **47**, 395 (1981); these proceedings.

5.3   Precision Laser Spectroscopy: Later Work

5.3.1   Levels of accuracy; units and standards of measurement

It is first to be appreciated that the measurements we shall now be concerned with are of extraordinarily high accuracy, and that the period 1979-1986 embraces the year 1983 when the metre was re-defined as a unit of the SI subsidiary to the second (by fixing the value of the speed of light as 299 792 458 ms^{-1} [GIA 83, BAI 82]). In the earlier work it was customary to report the *wavenumber* of spectral lines, the number of oscillations per unit of length, in vacuo, and measurements were made with reference, strictly speaking, to the definition of the metre in terms of the krypton (gas-discharge) standard, but in practice, to an agreed value for the wavenumbers of certain secondary standards, normally lines from iodine-stabilized helium-neon lasers. These lines are still used as standards, though nowadays it is customary to characterize spectral lines by their *frequencies*, the conversion to wavenumber being made by using the fixed value for the speed of light.

A consequence of the change is that the values now agreed for the wavenumbers of the secondary standards do not coincide with the values used *before* the re-definition of the metre, the difference being of the order of 2 parts in 10^9.

Since measurements are now (1986) being quoted with accuracies of 6 or 7 parts in 10^{10}, the reader should be aware that pre-1983 values may need adjustment before they are compared with post-1983 values. The numbers we shall quote in describing particular experiments will be those quoted by the authors themselves in their historical context, except that in tabulations the laser results will be shown as post-1983 corrected values.

In this connection we may refer to a group of secondary standards which were not used before but which have been used in some recent spectroscopic studies of hydrogen because they fall across the Balmer-β structure, so leaving only a small spectroscopic interval to be measured in the determination of the frequencies of Balmer-β, and, through the application of frequency-doubling and two-photon spectroscopy, of components of Lyman-α in hydrogen and in deuterium. These standard lines occur in the spectrum of molecular tellurium, and their frequencies have been established by interferometric comparison with iodine-stabilized helium-neon lasers to an accuracy of about 4 parts in 10^{10} [BAR 85].

5.3.2 Objectives of experiments and reduction of measurements

Whereas radiofrequency techniques are well adapted to the study of fine structures, and so are especially favorable to the determination of the fine-structure constant and Lamb shifts, precision laser spectroscopy spans a spectral region where transitions occur in hydrogen between levels of different n, particularly the lower values of n, and so are favorable for the determination of the Rydberg constant, R_∞. Indeed, the precision measurements that have been made so far in this spectral region are normally published as 'determinations of the Rydberg', though the $1S - 2S$ transition is an exception, as we shall see below.

Each advance in technique offers the possibility of measuring, with higher accuracy than formerly, one or more fine structure or hyperfine components of a transition of, say, Balmer-α. To obtain from this a value of R_∞ it is necessary to convert the measurement to that appropriate to a nucleus of infinite mass and to take account of fine and hyperfine structure and radiative corrections.

The reduction from hyperfine to fine structure transition frequencies is straightforward since the hyperfine coupling between levels of different n is negligible: the application to each level of the formula $\sum_i g_i \delta_i = 0$, where g_i is the statistical weight of a hyperfine level and δ_i its displacement, yields the corrected fine structure transition frequencies. (If the accuracy of a measurement were to call for it, corrections could readily be applied by standard methods for coupling between fine structure levels of the same n, for example, between $2P_{1/2}$ and $2P_{3/2}$.)

The major part of the reduced mass correction is normally applied through the non-relativistic relation $R_M = R_\infty/(1 + m_e/M)$, where m_e is the mass of the electron and M that of the nucleus. This leaves relativistic reduced mass corrections to be incorporated into the smaller, radiative corrections. We shall have more to say, in a moment, about the non-relativistic correction.

There remain the fine structure and radiative contributions to be evaluated before the Rydberg constant can be determined. For these it is customary to rely on the theoretical calculations of Erickson [ERI 77] or Johnson and Soff [JOH 85]. The various terms scale with R_∞ and have been evaluated with use of a particular value. A value from the new measurements is obtained simply by re-scaling. Though this procedure would appear to rely heavily on QED theory, an alternative logic is to observe that the fine and hyperfine intervals have been sufficiently checked by experiment so that the formula $R_\infty(1/n^2 - 1/m^2)$ can be fitted to an experimentally adjusted determination of this interval, and the Rydberg that is deduced is essentially the combination $m_e e^4 \mu_0^2 c^3 / 8h^3$ that occurs in Bohr's theory, or in Schrödinger's (non-relativistic) theory. (The units in this expression are reciprocal length: the Rydberg constant is conventionally quoted in cm^{-1} or in m^{-1}. The conversion to units of frequency introduces no uncertainty since 1983 since the value of c is fixed).

The non-relativistic reduced mass correction has, historically, been important as a means of determining the electron/proton mass ratio, m_e/m_p. The method was to measure the displacement between corresponding lines in different isotopes — conventionally, hydrogen and deuterium, but tritium has also been used. With relatively imprecise knowledge of the nuclear mass ratios the m_e/m_p ratio can be determined from the isotopic displacement, assuming that this is entirely due to the difference in masses. In the nineteen thirties this was the best method for determining m_e/m_p, and it appeared likely to be so again when the first laser measurements were made of isotopic displacements in Lyman-α. But advances in the study of trapped ions and electrons [VAN 86] have overtaken this expectation and in recent work the reduced mass correction has been evaluated by taking a value of m_p/m_e, 1836.152701 (37), from this work. Meanwhile, the precision of measurement achieved in the most recent determinations of isotopic displacement [BIR 86] is within sight of the point where it will become necessary to take account of nuclear volume effects. We may therefore look forward to a period when measurements of these displacements will bear on the question of the nuclear charge radius of the proton, the deuteron and the triton.

We return now to the reduction of measurement of the $1S - 2S$ transition frequency. The transition is exceptional for the exceedingly high potential accuracy with which it can be measured, and also because there is no other fine structure level belonging to $n = 1$ against which the position of the $1S$ level can be measured. The Lamb shift of the $1S$ state, about 8 GHz, cannot therefore be experimentally investigated by radiofrequency measurements within the

$1S$ manifold but must be ascertained with reference to $n = 2$ or higher levels, which associate it necessarily with the much larger Bohr interval of order R_∞. Measurements of the $1S - 2S$ transition frequency may therefore be used *either* to determine R_∞ using theoretical values of the $1S$ Lamb shift, *or* to determine the $1S$ Lamb shift using values of R_∞ from other sources. In either case the fine structure displacements in $n = 2$ must be known and a value taken for m_p/m_e.

An ingenious method used by Hänsch and his colleagues (5.2.1, section III.1) avoids the dependence on the Rydberg. One and the same laser simultaneously generates the spectrum of Balmer-β at 486 nm, and , with frequency-doubled radiation, the two-photon spectrum of Lyman-α at 121 nm. The fine structure of Balmer-β being known, one can effectively use its spectrum to fix the Bohr level for $n = 1$ in relation to the Lyman-α complex. The fine structure displacement of the $1S$ term (Dirac shift plus Lamb shift) is thereby ascertained. Attractive though the scheme is, its drawback is that the accuracy of determination of the $1S$ Lamb shift is limited by the intrinsic linewidth of components of Balmer-β which is so much greater than that of the $1S - 2S$ transition. Nevertheless, the possibility exists of using a two-photon scheme (similar to that described in section 5.3.4.4) to generate $2S - 4S$ transitions as components of Balmer-β, thereby reducing the limiting linewidth by virtue of selecting the longer-lived $4S$ rather than $4P$ state as upper level in the transition. In the concluding section of this chapter we show how other combinations of transition frequencies may be used to separate the fine from the gross structure terms.

5.3.3 Design of experiment

We have emphasized the enormous advance made by freeing optical spectroscopy of the Doppler broadening of spectral lines. In considering how best to take advantage of this, the other factors which limit accuracy need to be considered.

We turn first to the finite linewidth imposed by the radiative lifetime of excited states, or the interaction time should that be less than the radiative lifetime, as it necessarily is for the ground state $1S$ and almost always is for the metastable state $2S$. This immediately points to the importance of studying the $1S - 2S$ transition, though the difficulties of doing so are formidable (section 5.3.5). The lower P states have the shortest lifetimes and therefore the greatest intrinsic width. Thus, although Balmer-α is a convenient line to study, and particularly the transitions $2S_{1/2} - 3P_{1/2,3/2}$ (section 5.3.4.2), the 29 MHz width of the $3P$ levels is a limitation on accuracy to be avoided if possible. Numerical values of some lifetimes and radiative widths are given in Table 2 below.

Table 2. Lifetimes and radiative widths of some excited levels in hydrogen.

	$2S^1$	$3S$	$4S$	$5S$	$6S$	$\ldots$	$10S$
$\tau/10^{-8}$s	1/8s	16	23	36	57	$\ldots$	198
Δ/MHz	1.3 Hz	0.99	0.69	0.44	0.28	$\ldots$	0.08
	$2P$	$3P$	$4P$	$5P$	$6P$		
$\tau/10^{-8}$s	0.16	0.54	1.24	2.40	4.1		
Δ/MHz	99	29	13	6.6	3.9		
	$3D$	$4D$	$5D$	$6D$		$8D$	
$\tau/10^{-8}$s	1.56	3.65	7.0	12.6	$\ldots$	28.9	
Δ/MHz	10	4.36	2.3	1.2	$\ldots$	0.55	

[1]The units written under $2S$ override the units labelling the respective rows.

Notice the long lifetimes of the S levels by comparison with the P levels. Notice also that the D levels have substantially longer lifetimes than the P levels. The relatively narrow $2S - nS$ and $2S - nD$ transitions have been explored by two-photon spectroscopy (section 5.3.4.3).

Other contributions to linewidth which are, to a greater or lesser extent, under the control of the experimenter are:

(*a*) perturbation of the atoms — particularly in higher states — by collisions with electrons in gas discharges and by incipient Stark broadening arising from ionic microfields;

(*b*) residual first-order Doppler effect in beam experiments, arising from incomplete collimation of atomic beams and laser beams;

(*c*) spectral width of laser radiation used to scan a resonance;

(*d*) power broadening by the laser. This is a form of 'interaction time' broadening, in that increase in the transition rate implies a shortening of time spent in a particular state.

Associated with (*a*) is the possibility of a shift of the resonance frequency. Both shift and broadening can often be investigated experimentally by varying the current and pressure in the gas discharge. In studying Balmer-α, for example, one finds that certain components are more sensitive than others to electric fields. If the investigative studies are made on these components, one may feel confident that the perturbation of other components chosen for the definitive measurement will be below a quantifiable level.

Effect (*b*) is quantifiable in terms of the geometry of the apparatus. There is additionally the possibility of a shift of resonance frequency arising from misalignment, and even with perfect alignment there is a shift and a contribution to linewidth arising from the second order Doppler effect.

The desirability of reducing laser linewidth for spectroscopical and metrological applications (effect (c)) has been responsible for a tremendous effort in many laboratories. Not only in commercial and government-sponsored laboratories, but also in some university laboratories, the linewidth of tunable lasers has been reduced to a few hundred kHz.

With power broadening by the laser, effect (d), we may associate also a shift known as the 'light shift'. It is a shift of atomic energy levels (strictly; of the system, atoms plus light field) which occurs when the laser frequency is slightly off-resonance for some particular transition, say $a - b$. If, then, one is exploring the resonance $a - c$, where b lies near to c the presence of b will yield a resonance frequency $a - c$ which does not correspond exactly to the interval $a - c$ in the free atom. But this effect is well understood and can be calculated [BAR 61, PAN 66].

Though effects giving rise to systematic shifts can often be assessed, sometimes quite accurately, there is generally an uncertainty associated with the calculation, both of the shift and of the associated broadening.

The signal-to-noise ratio is, of course, a relevant factor in aiming for the highest accuracy. For a given transition it depends on the number density of atoms that can be attained, on the intensity of radiation that can be generated at the required frequency and on the transition matrix element, on the efficiency of the detection system, and on the skill of the experimenter in eliminating noise not necessarily associated with the signal. Generally speaking, a considerable degradation of signal relative to noise is an acceptable sacrifice in exchange for a substantial reduction in linewidth. But to take full advantage of a good signal-to-noise ratio, knowledge of the spectral profile is essential, for then — even with asymmetric lines — the atomic resonance frequency can be determined to an accuracy represented by a small fraction of the resonance linewidth.

A most important feature of the spectroscopy of any era is that of calibration of the spectra. The traditional technique of high resolution spectroscopy is interferometric comparison of the radiation under investigation with a standard radiation. In its most developed form the technique will entail comparison by computer of digital records of scans of interferometer fringes representing the standard and the test radiations. Corrections can be made for non-linear scanning. The interferometric method is by no means obsolete: it is essentially a comparison of wavelengths inside a common cavity. But the cavity length is wavelength-dependent because of frequency-dependent phase changes on reflection of light from the metallic- or dielectric-coated end mirrors, so this systematic effect must be assessed and corrected for.

It is avoided in the method of comparison of frequencies by measuring the beat frequency between two radiations falling on a common detector. The spectral purity of the radiations needs to be sufficient to allow a beat note to be formed, so it is a technique that is not feasible with conventional light sources. Moreover, since optical frequencies are so high compared with the frequencies to which photodiodes can respond, the radiations being compared must lie close together in the spectrum. That is why the tellurium secondary standard (section 5.3.1) has been chosen: it lies within the Balmer-β complex, but finds its principal application in determining the frequency of Lyman-α resonances at four times the frequency of the driving radiation (section 5.3.5). In the use of tellurium as a frequency standard, the vapor in a sealed cell provides absorption lines on to which a laser can be locked.

A different method of solving the problem is being developed by Meisel and colleagues in Karlsruhe [BUR 84]. The large gap in frequency (about 17×10^3 GHz) between a commonly used standard (iodine-stabilized helium-neon laser) and the hydrogen resonance to be explored will be bridged by a technique of stepping, using auxiliary lasers. The number of steps contemplated is about 200.

For every new experiment that is contemplated these various contributions to linewidth and shift, and the uncertainties associated with them, and with the proposed method of calibration, need to be assessed and balanced, one against another. The later sections in this chapter will give an account of what has been accomplished, and of some of the proposals which have been made for making progress beyond the present limits.

5.3.4   Determinations of the Rydberg constant

The most refined experiment of this type at Stanford in the early set of experiments was that of Goldsmith *et al.* [GOL 78] which is referred to and illustrated in section 5.2.1. The spectral line in question, Balmer-α in hydrogen, was investigated by polarization spectroscopy. The advantage of this over saturation spectroscopy, it will be recalled, is that it calls for a lower intensity of radiation and has an improved signal-to-noise ratio.

While this work was in progress, Petley and colleagues at the National Physical Laboratory were applying saturation spectroscopy to the same line [PET 80]. We give an account of this work because, although the accuracy claimed for the final result is less than that of some determinations, the great care taken in assessing and allowing for systematic effects earns it a place in any record of work at the highest level of expertise.

5.3.4.1   Determination of the Rydberg at the National Physical Laboratory, UK: Balmer-α

The atoms were prepared and sampled in an electrical discharge in molecular hydrogen flowing at low pressure through the optical absorption tube. The pressure and discharge current were varied, resulting in detectable changes in linewidth and in shifts dependent on pressure. The results were extrapolated to zero pressure. There was some asymmetry in the line profiles which was allowed for in the estimation of errors.

It will be recalled that the method requires the light from a tunable laser to be divided to produce a saturating beam and a probe beam, counterpropagating through the sample. In this case two probe beams were used, one intersecting the saturating beam, the other displaced from it. The probe beams were modulated, photo-detected, the difference taken and recorded by phase-sensitive methods. The signal was displayed on a pen- recorder as well as stored in a computer for numerical processing.

The spectra were calibrated by interferometric methods against an iodine-stabilized helium-neon laser.

The components of the fine structure chosen for measurement were the three strongest, $2P_{3/2} - 3D_{5/2}$; $2S_{1/2} - 3P_{3/2}$; $2P_{1/2} - 3D_{3/2}$ (see the term diagram on page 56 of SAH).

The measurements were corrected for the following effects:

(a) difference in effective length of the etalon between the standard and sample wavelengths;

(b) Stark effect arising from microfields in the discharge;

(c) pressure shift;

(d) unresolved 'crossover' signals;

(e) partially-resolved hyperfine structure.

The size of these corrections were various, but all of order 10^{-2} m^{-1}; the estimated statistical error of the measurements at one standard deviation was 3×10^{-3} for the strongest component (on which most measurements were taken) and 6×10^{-3} m^{-1} for each of the others. The spectral linewidth, of order 180 MHz, was substantially broader than the natural width of the P-states, the excess width arising in part from the spectral width of the laser, but mostly from broadening attributable to microfields and electron collisions in the discharge.

To obtain a value for the Rydberg constant the method of scaling against Erickson's calculations was used (section 5.3.2). The final result was quoted as

$$R_\infty = 10\ 973\ 731.513(85)\text{m}^{-1} \ .$$

The value quoted by Goldsmith *et al.* [GOL 78] was

$$R_\infty = 10\ 973\ 731.476(32)\text{m}^{-1} \ .$$

The reader will find values different from these quoted in recent publications, also in section 5.4 and in Chapter 10 — these values have been corrected as explained in section 5.3.1.

5.3.4.2 Determination of the Rydberg at Yale: Balmer-α and Balmer-β

These determinations by Amin *et al.* [AMI 81, ZHA 86, ZHA 87] differ in many respects from that of the preceding section. They were carried out on beams of H atoms excited by electron impact to the $2S$ state and irradiated transversely with tunable laser light. They make no use of non-linear techniques to eliminate Doppler broadening. Resonance at the Balmer-α and Balmer-β frequencies was detected by monitoring the number of metastable atoms surviving in the beam, just as in the Lamb-Retherford experiment, the difference being that here the atoms were removed to the $3P$ and $4P$ states rather than to the $2P$ state. Doppler broadening was brought within acceptable limits by the use of directed beams of light and atoms and particular attention was paid to accurate transverse alignment of the laser against the atomic beam. Resonance linewidths were in the region 40-50 MHz, reflecting the 30 MHz natural linewidth of the P-states.

In this experiment the spectrum is simplified in that the atoms are, by use of magnetic quenching before the interaction region, prepared in one hyperfine state only of $2S$, and the only allowed transitions are to the $3P$ and $4P$ states, where the hyperfine structure is unresolved. There was some stray magnetic field in the irradiated section of the beam, leading to incipient Zeeman structure, which was allowed for.

The wavelength calibration was done interferometrically. (The phase shift correction applicable to the dielectric mirrors called for special consideration and was a relatively large effect.) The wavelength standard used was, at first, a local version of an iodine-stabilized helium-neon laser which, in turn, needed to be calibrated against a Bureau of Standards laser. Later the Bureau of Standards lasers were used directly.

Among the corrections were the not-obvious items: optical pumping (arising from partial resolution of Zeeman components) and diffraction (in making comparisons between lasers of different geometrical characteristics), and photon recoil.

After the first publication the entire apparatus was rebuilt, some corrections were eliminated and others better controlled. The final result derived from Balmer-α was

$$R_\infty = 10\ 973\ 731.569(7)\text{m}^{-1} \quad [\text{ZHA 86}]$$

and from Balmer-β:

$$R_\infty = 10\ 973\ 731.573(3)\text{m}^{-1} \quad [\text{ZHA 87}]$$

These are averages of results obtained from hydrogen and from deuterium. They belong to the post-1983 era.

5.3.4.3 Determination of the Rydberg in Paris: avoiding the P states

A group under the general direction of Professor Cagnac has avoided the problem of P-state level widths by studying two-photon transitions from $2S$ to nS and nD states [BIR 85a, BIR 85b, BIR 87]. The states $n = 8$ and 10 were first explored, though there is the possibility of going to higher states with the home-made tunable laser (730-780 nm) which was employed. The two-photon technique, it will be recalled, (with the two photons of equal frequency in the laboratory frame), eliminates the Doppler effect in first order.

The atomic hydrogen was prepared as a beam and excited to the $2S$ states by electron impact (which deviated the atoms). The metastable beam was irradiated *longitudinally* along its path, interacting with the forward and return beams from the laser along about 20 cm, the beam being within a Fabry-Perot cavity to enhance the excitation efficiency, The cavity length was variable over small displacements, being locked to the laser frequency. Transitions out of the $2S$ states were monitored by quenching the atoms, after they had passed out of the shielded interaction region, with a static electric field and by measuring the Lyman-α fluorescence.

The frequency calibration of the spectra was accomplished via a thermally-controlled Fabry-Perot étalon locked to the laser radiation, variations in whose length changed the frequency of the laser. An auxiliary He-Ne laser was also mode-matched into this cavity. The difference frequency was measured between the light from a standard laser and this auxiliary laser by measuring the beats generated when the radiations were mixed on a suitable diode.

A spectrum showing transitions to the $8D_{3/2,5/2}$ states is reproduced in Fig. 5.1. Notice the linewidth, 1.4 MHz, associated with the interval $2S - 8D$, 7.7×10^8 MHz; a fractional linewidth, therefore, of 1.8×10^{-9}. The value of the Rydberg constant determined from these measurements reflects this accuracy:

$$R_\infty = 10\ 973\ 731.569(6) \text{m}^{-1} (\text{post-1983 value}) .$$

It is to be noticed that, although there is no first order Doppler effect in this method, a correction of about 0.05 MHz needed to be applied for H (and half that amount for D) for the Doppler effect in second order. The hyperfine structure of $2S$ is well resolved, but not that of the D-states.

The linewidth recorded in the spectra is almost three times as great as the radiation width of the $8D$ levels. The contributions to line-broadening include second order Doppler effect (spread of velocities), finite interaction time (atomic trajectories oblique to the laser beams) and saturation of the transition. Light-shift associated with the latter was allowed for. The laser linewidth — less than 1 MHz — also contributes. Further contributions come from collisions and from

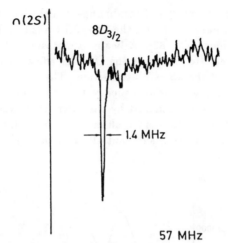

57 MHz

Fig. 5.1. 2*S*-8*D* transitions in H (Reproduced with permission from F. Biraben *et al.*, *Europhys. Lett.* **2**, N°12, (December 1986) 925 (Les Èditions de Physique)).

parasitic electric fields. There remains scope for improvement, and there is no doubt that determinations of higher accuracy will eventually be reported by this group.

But the value of the experiments goes beyond the presentation of an improved value for the Rydberg. It was possible also to obtain spectra of deuterium, and a value of the isotopic shift of the transition $2S - 8D_{5/2}$ was obtained:

$$209\ 691.29(7)\ \text{MHz}\ ,$$

which agrees with a theoretical value based on the current values of mass ratios. It is hoped that this value can also be improved. At the next order of magnitude the measurement may be expected to show evidence of differential nuclear volume effects between the proton and the deuteron.

Moreover, in addition to the resonances $2S-8D_{3/2,5/2}$, signals were recorded for the transition $2S - 8S$ and experimental values of the interval $8D_{3/2} - 8S_{1/2}(F = 1)$ used to determine the Lamb shift of $8S$ [BIR 85a]. The result, 17.7 (9) MHz is in satisfactory agreement with the theoretical value, 16.7 MHz. This measurement, taken within the fine structure of the $n = 8$ levels, is valuable in that it supports the theoretical evaluation of the relative positions of the $n = 2$ and $n = 8$ levels on which the determination of the Rydberg by scaling depends.

5.3.4.4 Further studies to avoid the P states

It is, of course, not necessary that the two beams of light in two-photon excitation should originate from one and the same laser: a number of authors have considered the advantages of combining optical with radiofrequency excitation. Some preliminary results on the $2S - 3S$ transition have been obtained by Wieman and Shiner [WIE 83]. A beam of metastable hydrogen atoms was irradiated transversely (to eliminate the Doppler effect) with laser light at (or near) the frequency of the $2S - 3P$ transition. The second step, $3P - 3S$ was brought about by tuning a radiofrequency field through the appropriate frequency range. The beam, quenched by the laser light, was partially restored when the radiofrequency field was simultaneously applied, the resonance width for the double transition being substantially less than for the laser transition alone (see Fig. 5.2).

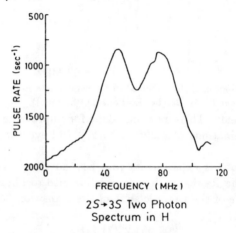

$2S\rightarrow3S$ Two Photon
Spectrum in H

Fig. 5.2. $2S$-$3S$ spectrum in H (Reproduced with permission from C. E. Wieman, in *Quantum Metrology and Fundamental Physical Constants*, eds. Paul H. Cutler and A. A. Lucas (Plenum Publishing Corporation, 1983) p. 403).

It is expected that this work will be taken further, though it has suffered a temporary disruption.

5.3.5   *The 1S–2S transition*

We have pointed out the special nature of this transition: since $1S$ is the ground state and $2S$ is metastable the potential linewidth, 1.3 Hz, is many orders of magnitude less than that of any other interval in hydrogen. The frequency itself is about 2.5×10^{15} Hz; the sharpness of resonance, therefore, that offers itself for exploitation, is potentially 2×10^{15}, which is greater than that of the present primary standard of frequency. Measurement of the interval offers a direct route to the determination of the Lamb shift in the $1S$ state, given values of the Rydberg constant, the fine structure constant and the Lamb shift of the $2S$ state.

The $1S − 2S$ transition is an obvious candidate for two-photon spectroscopy, but the difficulties are formidable. First, as concerns the matrix element of the transition. The concept of two-photon electric dipole transition requires the existence of a level of opposite parity to serve as an intermediate (non-resonant) state, and the two-photon matrix element depends inversely on the distance of the energy level of the intermediate state from a level halfway between the initial and final states. For the $1S − 2S$ transition the $2P$ levels serve as the most important intermediate states, and since these are far from the halfway position the matrix element is exceptionally small. Secondly, the transition rate depends on the square of the light intensity, and that is much less for a cw laser than for a pulsed laser of the same average power. In that it has proved exceedingly difficult to generate even moderate amounts of power in stable, cw radiation at the required frequency, 243 nm, the first successful experiments were carried out with pulsed radiation, notwithstanding the inferior spectral purity. The tremendous efforts that have been put into generating cw radiation have ultimately been rewarded, but the lower power requires that the radiation be matched into an optical cavity to enhance the intensity.

For the interaction of hydrogen atoms with this radiation, the desirable objectives are long interaction times and a clean environment, and some progress towards this end has been made. But in most of the attempts so far, experimenters have satisfied themselves with atoms in, or issuing directly from a gas discharge — albeit at a pressure as low as will allow a sufficient density of atoms. In this experiment the appearance of a signal under any conditions is indeed a landmark.

Gas discharges automatically provide a solution to the problem of detecting the excitation of atoms into the $2S$ state. The unavoidable collisions of atoms in this state result in a transfer into the $2P$ states, from which they rapidly decay with the emission of Lyman-α. Observation of this radiation monitors the resonance.

A more recently developed monitor of resonance which does not rely on collisions is the detection of ions produced from $2S$ atoms by photoionization. This is the technique that has been used successfully in monitoring the corresponding transition in positronium.

The provision of a standard against which to measure the frequency is raised in an acute form by the $1S − 2S$ transition. The development of the tellurium standard (section 5.3.1) was undertaken to meet this need. The stabilized laser frequency used in the most recent work [BEA 87], 616 515 272.91 MHz, was quoted with an uncertainty of 0.36 MHz, while the linewidth of the observed resonance was 3.7 MHz and the uncertainty claimed for the final result was 1.5 MHz. Clearly, further progress in reducing the linewidth is pointless unless improved methods can be devised for measuring the frequency or for comparing it with some other fundamental interval. A desirable objective would be to develop a frequency chain in which the $1S - 2S$ frequency is linked directly to the primary standard. This, indeed, has been achieved for radiations at the red end of the visible spectrum (a review article is [BAI 82]). We return to the second alternative (comparison with some other interval) at the end of this chapter.

Precision measurements of the $1S - 2S$ frequency were first reported from Stanford [HIL 86] and from Southampton [BAR 86]; it is known that work on this transition is in progress elsewhere (see *Note* at the end of this section).

There is a great deal of common ground and shared experience between the work at Stanford and the work at Southampton, and the publications were made simultaneously. As light source, each group used a tunable cw laser at 486 nm, amplified this light to sufficient intensity for frequency doubling by means of a pulsed laser amplifier, and filtered the resulting radiation before directing it to the frequency-doubling crystal (urea). The hydrogen atoms were prepared by the Stanford group in the form of a beam, the Southampton group in a low-pressure microwave discharge. Both groups used the method of photoionization as a detector of excitation to $2S$. Both groups calibrated their spectra by interferometric comparison of the primary, cw radiation with a tellurium reference standard. The transition frequency was determined as

$$2\ 466\ 061\ 395.6(4.8)\text{MHz at Stanford, and as}$$
$$2\ 466\ 061\ 397.(25.)\text{MHz at Southampton},$$

(both being post-1983 values). The spectra illustrating the two publications show linewidths of about 150 MHz (Stanford) and 200 MHz (Southampton), but the greater uncertainty of the Southampton result appears to arise from a problem relating to the tellurium reference. There were also difficulties with the frequency determination at Stanford and, as we shall see in a moment, the result quoted above has been superseded. The principal contribution to the linewidth in each case was the spectral width of the pulsed, frequency-doubled radiation (but it is to be noticed that resonance widths calibrated at the transition frequency are twice as broad as when calibrated against the frequency of the radiation).

Work at Southampton is continuing, but no further publication has been made at the time of writing.

The Stanford group has succeeded in carrying out the experiment with cw radiation [FOO 85, BEA 87], reducing the linewidth to about 15 MHz (at the transition frequency) and enabling them to quote the more reliable transition frequency $2\ 466\ 061\ 413.8\ (1.5)$ MHz, the substantial difference from the earlier work being ascribed to the inadequately understood distortion (chirping) of the 486 nm radiation on transmission through the pulsed amplifier. In this improved version of the experiment the cw radiation at 243 nm was generated by the sum-frequency mixing of ultraviolet light from a fixed-frequency ion laser with red light from a variable- frequency dye laser. Both lasers, of course, are highly stabilized. The desired radiation, tunable in the region around 243 nm, and of only 2 MHz spectral width, was formed by the non-linear interaction of the primary beams with a crystal of KDP, thermally controlled to secure efficient phase-matching. Output powers of a few mw were achieved 'routinely, over periods of many hours'.

The atoms in this experiment were generated in a microwave discharge in a cell placed inside a cavity formed between concave mirrors, whose purpose was to enhance the light field of the 243 nm radiation. The cavity length was servo-locked to the frequency of the radiation as it was varied through resonance. The detection of resonance was by observation of collision-induced Lyman-α radiation.

The frequency of the 243 nm radiation was calibrated directly in this experiment by the heterodyne method of beat-frequency measurement against laser radiation deriving from the tellurium standard, and then frequency- doubled in a crystal of urea. The uncertainty in the standard frequency, reported as about 5 parts in 10^{10}, constituted the greatest uncertainty in the final result. The linewidth (extrapolated to zero pressure) was below 8 MHz (at the transition frequency), and limited by the spectral widths of the fundamental lasers and by transit-time broadening. The noise on the signal imposed a limitation on the precision with which the experimental resonance curves could be fitted to an ideal (Lorentzian) shape.

Figure 5.3 shows a sketch of the apparatus used for pulsed excitation, and a resonance curve.

Note. Work at the Clarendon Laboratory, Oxford, on the $1S$–$2S$ transition, which has been in progress for a number of years, has recently come to fruition [BOS87]. The resonance was observed, cw, using 486 nm radiation frequency-doubled by a crystal of β-barium borate, which will allow the investigation to continue as a direct determination of the small intervals $4(H_\beta)$-(Ly_α). The present results 2 466 061 414.13 (79) MHz(H) and 2 466 732 408.55 (73) MHz (D) were published as the final proofs of this manuscript were being read.

5.4   Experimental Results: Comparison with Theory

5.4.1   The Rydberg constant

The method whereby measurements of transition frequencies are reduced to values of the Rydberg constant has been described in section 5.3.2. Table 3 presents the results of the determinations described in this chapter, except that

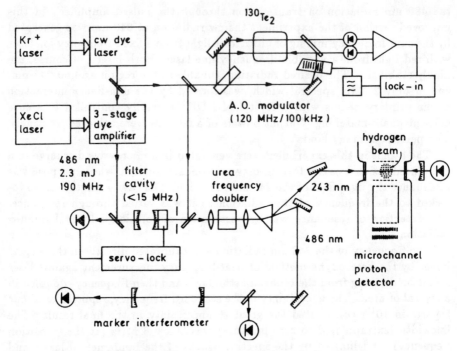

(a) Experimental arrangement used for pulsed excitation.

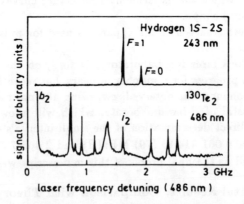

(b) Resonance curves for H, and tellurium reference lines.

Fig. 5.3. The $1S$-$2S$ transition studied at Stanford (Reproduced with permission from E. A. Hildum *et al.*, *Phys. Rev. Lett.* **56** (1986) 576 (American Physical Society)).

measurements later described as inferior by their authors have been omitted. The remarkable agreement, to a level better than 1 part in 10^9, of the latest measurements is impressive, particularly in that they were made on different transitions. The value 109 737.315 7 cm^{-1} satisfies them all and falls within Petley's measurement with its uncertainty, though not within Goldsmith's. The CODATA (1986) value, with its uncertainty, does not cover the recent measurements but reflects the position over the previous decade (see Chapter 10).

Table 3. Experimental values of $(R_\infty - 109737)$cm^{-1}. (determined with use of $m_p/m_e = 1836.152701$ and $c = 299\ 792\ 458$ ms^{-1}).

Authors	Value	Transition
Goldsmith *et al.* [GOL 78]	.31500 (32)^2	Balmer-α
Petley *et al.* [PET 80]	.31521 (64)^2	Balmer-α
Zhao *et al.* [ZHA 86]	.31569 (7)	Balmer-α (H and D)
Zhao *et al.* [ZHA 87]	.31573 (3)	Balmer-β (H and D)
Biraben *et al.* [BIR 86]	.31569 (6)	2S-8D, 10D (H and D)
CODATA 86	.31534 (13)	

[2] Quoted by Biraben *et al.* as corrected for m_p/m_e and c values given above.

5.4.2   The 1S-2S transition

The two precision measurements of this interval in H are

$$2\ 466\ 061\ 397(25)\text{MHz}, \quad \text{Southampton [BAR 86], and}$$
$$2\ 466\ 061\ 413.8(1.5)\text{MHz}, \quad \text{Stanford, revised [BEA 87],}$$

which are consistent with one another.[a] The two sets of authors compute the 1S Lamb shift L_1 from their measurements in slightly different ways, using different values of R_∞. The results are sensitive to the value chosen for R_∞: roughly speaking, a difference $3\Delta R/4$ is introduced into the value of L_1 by a difference ΔR in the value of R_∞. Results given by the authors are:

---

[a] See additionally the measurements recently reported from Oxford, (*Note* appended to section 5.3.5).

Spectroscopic determination of the $1S$ Lamb shift L_1 in H.

	R_∞/cm^{-1}	L_1/MHz	L_1/MHz (theor)
Southampton	109737.31534[3]	8182 (25)	8173.248 (81)
Stanford	109737.31573	8173.3 (1.7)	8173.06 (16)

[3] In Table 2 of BAR 86 the headings ΔE and ΔR appear to have been transposed. The difference between the two values of R_∞, 0.00039 cm^{-1}, corresponds to a difference of about 0.9 MHz in the evaluation of L_1, which is barely significant. It is gratifying that both results agree equally well with the theoretical values and with one another.

5.4.3   Spectroscopic isotope shift

The earliest reported direct measurement by precision laser spectroscopy of isotopic displacement were those of Wieman and Hänsch [WIE 80], who determined the H-D shift in the $1S - 2S$ transition as 670 992.3± 6.3 MHz, compared with the theoretical value 670 994.96 ± 0.81 MHz.[e] The measurement was notable for demonstrating the need to include in the theory the relativistic reduced mass correction, $-(m/M\alpha^4 mc^2/8n^4$, equal to -23.81 MHz for H and -11.92 MHz for D. (This term was called 'lowest order relativistic nuclear recoil' in the paper.) The figures may be compared with contributions for the nuclear size effect, the distribution of nuclear charge over a finite volume. The effect is well known in the spectroscopy of the heavier elements, but usually considered negligible for hydrogen. It is much more important for states having s-electrons than for others. Its value for H is 1.00± 0.05 MHz and for D, 6.78±0.09 MHz, less than the mass-dependent recoil term, but not very much less.

These authors also measured the H-D isotope shift in the transition $2S_{1/2} - 4P_{1/2}$ of Balmer-β as 167 783.9±1.2 MHz, compared with the theoretical value 167 783.7±0.2 MHz. Since this transition also involves an s-electron, and since the size effect scales as n^{-3}, the differential effect here between H and D would be about 0.6 MHz, that is, at this level of accuracy the experiment would not have been sensitive to the size effect.

The Paris group [BIR 86] reported a measurement of the H-D shift in the $2S_{1/2} - 8D_{5/2}$ transition as 209 691.29 (7) MHz, to be compared with 209 691.32 MHz, Erickson's value, modified to reflect the current value of m_e/m_p. This theoretical value would have included the relativistic reduced mass and nuclear recoil corrections as well as the non-relativistic reduced mass; also the nuclear size effect. The agreement between experiment and theory is gratifying. Here, as in the experiments described above, the contribution from the nuclear

---

[e]Boshier *et al.* [BOS 87] obtain 670 994.4(9) MHz.

size effect is at a level comparable with the uncertainty of measurement. But at the next order of accuracy the main uncertainty in the theory will be the uncertainty in this nuclear size effect.

For obvious reasons, tritium has been neglected in laser spectroscopic work, but the group at Oxford in the Clarendon Laboratory has recently completed a measurement, by saturation spectroscopy, of isotopic displacements between H, D and T in Balmer-α [STA 88]. Very careful analysis of the line profile and study of the discharge conditions allowed displacements to be determined to an accuracy of 2 MHz. The results are in satisfactory agreement with predictions based on the known mass ratios, but are not sensitive to the nuclear size effect.

5.4.4   Indications for future measurements

Hänsch and his colleagues have pointed out [HAN 86, FOO 85, also section 5.2.1] the advantages of taking certain combinations of measured transition frequencies. On the one hand, quantum electrodynamic predictions can be checked with greater sensitivity; on the other, some uncertainties arising from theory can be eliminated in the determination of the Rydberg constant.

Consider, for example, the experiment of Hänsch and his colleagues (5.2.1, section III.1) in which the frequency of the $1S - 2S$ transition was compared directly (by use of one and the same laser) with that of the $2S - 4P$ transition. Disregarding fine structure, we have $\nu_{12} = 3R/4, \nu_{24} = 3R/16$. By forming the quantity $\Delta = 4\nu_{24} - \nu_{12}$ the gross structure cancels out and with it uncertainties in the Rydberg constant. The experimental value of Δ is a measure of fine structure and QED effects only.

This particular combination suffered from imprecision arising from the large width of the $4P$ levels. But these could be greatly reduced by working with S-states. Consider, then, the combination $\nu(1S - 2S) - 3\nu(2S - nS) = 3R/n^2 +$ (fine structure and QED terms). By taking n sufficiently large, the uncertainty in the Rydberg is reduced to insignificance, and the combination frequency affords a precision test of QED: the uncertainties on the theoretical side are now dominated by the nuclear size effect and high order QED effects of magnitude less than 100 kHz. As Hänsch points out, the usefulness of this approach in testing QED is not limited by deficiencies in the frequency calibrations or in the imprecision in the primary standard itself.

If, on the other hand, our intention is to reduce the dependence of the Rydberg on QED uncertainties we may take advantage of the n^{-3} dependence of the dominant QED contributions. Consider the following list of contributions to energy levels, scaled in units of R_M:

Energy of levels $n =$ gross structure $\sim 1/n^2$

                         +Dirac fine structure $\sim \alpha^2/n^3$

                         +self-energy (leading term) $\sim \alpha^3/n^3$

+vacuum polarization (leading term) $\sim \alpha^3/n^3$
+higher order radiative effects $\sim \alpha^4/n^3$
+nuclear size effects $\sim \alpha^2/n^3$
+nuclear size correction
    to nuclear size effects $\sim \alpha^4/n^3$
+relativistic recoil $\sim \alpha^3/n^3$
+very small residuals   ...

These factors carry the principal $n-$ and α-dependence of the respective contributions. There is, additionally, a relativistic reduced mass contribution $-(m_e/M)m_e c^2(\alpha/8n^4)$, but this can be taken care of by correcting the transition frequency. This done, we may separate the n^{-3} dependence in the following way. Write

$$\nu_{\text{corr}}(n, m) = R_M \left[\left(\frac{1}{m^2} - \frac{1}{n^2} \right) - a \left(\frac{1}{m^3} - \frac{1}{n^3} \right) \right]$$

which, though not strictly accurate, has taken care of the principal relativistic and radiative contributions to the energy. Given two experimental values for this quantity we may solve the equations for a or, alternatively, for R_M. For example, let us combine the $1S - 2S$ and $2S - nS$ transitions to obtain

$$\delta = \nu_{\text{corr}}(2, 1) - 3\nu_{\text{corr}}(n, 2) = 3R_M \left[\frac{1}{n^2} - a \left(\frac{1}{2} + \frac{3}{n^3} \right) \right] .$$

We use this to eliminate a from $\nu_{\text{corr}}(2, 1) = R_M[3/4 - 7a/8]$ to obtain

$$\frac{3R_M}{4} = \left[\nu_{\text{corr}}(2, 1) \left(1 + \frac{6}{n^3} \right) - \frac{7\delta}{4} \right] \div \left(1 - \frac{7}{n^2} + \frac{6}{n^3} \right) ,$$

whence R_M is in terms of the frequencies $\nu_{\text{corr}}(2, 1)$ and $\nu_{\text{corr}}(n, 2)$.

And, further, composites of isotopic shifts may be used to determine the electron/proton mass ratio independently of knowledge of the nuclear size. We shall hear more of such studies as the precision of measurement of transition frequencies advances beyond the level of parts in 10^{10}.

Acknowledgments

The author expresses his deep appreciation to Dr. D. N Stacey for his well-informed comments on a first draft of this paper, also to Dr. Foot for his assistance in providing information, to Dr. Wieman, Dr. Ferguson, Professor Cagnac, Dr. Meisel, Dr. Petley and Professor Lichten for furnishing offprints and papers in advance of publication.

REFERENCES

AMI 81   S. R. Amin, C. D. Caldwell and W. Lichten, *Phys. Rev. Lett.* **47** (1981) 1234.

BAI 82   K. M. Baird, *Philos. Trans. Roy. Soc.* **A307** (1982) 673.

BAR 85   J. R. M. Barr, J. M. Girkin, A. I. Ferguson, G. P. Carwood, P. Gill, W. R. C. Rowley and R. C. Thompson, *Opt. Commun.* **54** (1985) 217.

BAR 86   J. R. M. Barr, J. M. Girkin, J. M. Tolchard and A. I. Ferguson, *Phys. Rev. Lett.* **56** (1986) 580.

BAR 61   J. P. Barrat and C. Cohen-Tannoudji, *J. Phys. Radium* **22** (1961) 329.

BEA 87   R. G. Beausoleil, D. H. McIntyre, C. J. Foot, B. Couillaud, E. A. Hildum and T. W. Hänsch, *Phys. Rev.* **A35** (1987) 4878.

BER 70   H. G. Berry and F. L. Roesler, *Phys. Rev.* **1A** (1970) 1504.

BIR 85a   F. Biraben and L. Julien, *C. R. Acad. Sci. Paris* **300** (1985) 161.

BIR 85b   F. Biraben and L. Julien, *Opt. Commun.* **53** (1985) 319.

BIR 87   F. Biraben, J. C. Garreau and L. Julien, *Europhys. Lett.* **2** (1987) 925.

BLA 73   J. A. Blackman and G. W. Series, *J. Phys. B* **6** (1973) 1090.

BOS 87   M. G. Boshier, P. E. G. Baird, C. J. Foot, E. A. Hinds, M. D. Plimmer, D. N. Stacey, J. B. Swan, D. A. Tate, D. M. Warrington and G. K. Woodgate, *Nature* **330** (1987) 463.

BUR 84   B. Burghardt, H. Hoeffgen, G. Meisel, W. Reinert and B. Vowinkel, in *Precision Measurement and Fundamental Constants,* eds. B. N. Taylor and W. D. Phillips, NBS Special Publication 617 (1984) p. 49.

COD 86   CODATA Bulletin No. 63, Nov.(1986). Reprinted in *Rev. Mod. Phys.* Oct. 1987.

COH 52   E. R. Cohen, *Phys. Rev.* **88** (1952) 353.

CSI 68   L. Csillag, *Acta Phys. Acad. Sci. Hung.* **24** (1968) 1.

ERI 77   G. W. Erickson, *J. Phys. Chem. Ref. Data* **6** (1977) 831.

FOO 85   C. J. Foot, B. Couillaud, R. G. Beausoleil and T. W. Hänsch, *Phys. Rev. Lett.* **54** (1985) 1913.

GIA 83   P. Giacomo, *Eur. J. Phys.* **4** (1983) 190.

GOL 78   J. E. M. Goldsmith, E. W. Weber and T. W. Hänsch, *Phys. Rev. Lett.* **41** (1978) 940.

HAN 86   T. W. Hänsch, R. G. Beausoleil, U. Boesl, B. Couillaud, C. J. Foot, E. A. Hildum and D. H. McIntyre, in *Methods of Laser Spectroscopy* eds. A. Ben-Reuven and M. Rosenbluh (Plenum, 1986) p. 163.

HIL 86   E. A. Hildum, U. Boesl, D. H. McIntyre, R. G. Beausoleil and
            T. W. Hänsch, *Phys. Rev. Lett.* **56** (1986) 576.

JOH 85   W. R. Johnson and G. Soff, *At. Data Nucl. Data Tables* **33** (1985)
            405.

KES 71   E. G. Kessler, Jr. and F. L. Roesler, in *Precision Measurement and
            Fundamental Constants*, eds. D. N. Langenberg and B. N. Taylor,
            NBS Special Publication 343 (1971).

KES 73   E. G. Kessler, Jr., *Phys. Rev.* **A7** (1973) 408.

KIB 73   B. P. Kibble, W. R. C. Rowley, R. E. Shawyer and G. W. Series,
            J. Phys. **B 6** (1973) 1079.

LAR 67   H. P. Larson and R. W. Stanley, *J. Opt. Soc. Am.* **57** (1967) 1439.

MAT 71   T. Matsui, in same publication as KES 71 (1971).

PAN 66   S. Pancharatnam, *J. Opt. Soc. Am.* **56** (1966) 1636.

PET 80   B. W. Petley, K. Morris and R. E. Shawyer, *J. Phys.* **B 13** (1980)
            3099.

ROE 64a   F. L. Roesler and J. E. Mack, *Phys. Rev.* **135** (1964) 58.

ROE 64b   F. L. Roesler and L. K. DeNoyer, *Phys. Rev. Lett.* **12** (1964) 396.

SAH   *Spectrum of Atomic Hydrogen*, Part I of this book.

SER 59   G. W. Series and J. C. Field, in *Symposium on Interferometry*,
            paper 2-4; also contributions by W. C. Martin and J. Terrien,
            Nat. Phys. Lab. Teddington (1959).

SER 71   G. W. Series, in same publication as KES 71 (1971).

SER 82   G. W. Series and D. N. Stacey, *Eur. J. Phys.* **3** (1982) 129.

STA 88   D. N. Stacey *et al.*, *J. Phys. B* (1988).

TAY 84   *Precision Measurement and Fundamental Constants II*, eds. B. N.
            Taylor and W. D. Phillips, NBS Special Publication 617 (1984).

VAN 86   R. S. Van Dyck Jr., F. L. Moore, D. L. Farnham and P. B. Schwin-
            berg, *Bull. Am. Phys. Soc.* **31** (1986) 244.

WIE 80   C. E. Wieman and T. W. Hänsch, *Phys. Rev.* **A22** (1980) 192.

WIE 83   C. E. Wieman, in *Quantum Metrology and Fundamental Physical
            Constants*, eds. Paul H. Cutler and A. A. Lucas (Plenum, 1983)
            p. 403.

ZHA 86   P. Zhao, W. Lichten, H. P. Layer and J. C. Bergquist, *Phys. Rev.*
            A34 (1986) 5138.

ZHA 87   P. Zhao and W. Lichten, H. P. Layer and J. C. Bergquist, *Phys.
            Rev. Lett.* **58** (1987) 1293.

CHAPTER 6

SPECTROSCOPY OF ONE-ELECTRON IONS OF INTERMEDIATE AND HIGH Z

E. Träbert

6.1 Introduction

The basic methods employed to study hydrogen-like ions with nuclear charges $Z \gg 1$ are fundamentally the same as those employed for hydrogen: there are direct spectroscopic observations, there are experiments studying the response of hydrogen-like ions exposed to static fields (quench experiments) and there are studies which make use of oscillating fields (resonance experiments). The production of the highly charged ions and the methods of inducing or detecting a signal, however, are sufficiently distinct from and, as far as the parameters are concerned, very different from the cases of H^0 and He^+ which are discussed in other chapters of this book. A separate treatment is therefore worthwhile.

What one is interested in concerning highly ionized atoms are the systematic trends of the atomic structure which depend on the nuclear charge Z, also some non-trivial decay modes of certain long-lived excited levels. There are relativistic effects and effects which are understood in the framework of quantum electrodynamics (QED). Hydrogen-like ions are of particular interest because there exists the Dirac formalism to describe a one-electron system with an infinitely heavy nucleus with high precision. Precise experiments can test the validity limits of this approximation and of its various extensions. For high Z systems the higher order terms in the QED treatment of various corrections become important or even predominant, some of which cannot be expected ever to be measurable in hydrogen or helium.

In this chapter a survey of the state of the art of various lines of experimental development is attempted, with specific examples given for ions ranging from $Z = 3$ (Li^{2+}) to $Z = 92$ (U^{91+}). This presentation has been aided by a number of earlier surveys [50, 55, 60, 61, 29]. Among these a review paper by Kugel and Murnick [50] is highly recommended for additional reading and for more detailed discussions of some of the experimental techniques.

As it happens, the different experimental methods described below relate to particular topics in the study of the atomic structure. More or less classical spectroscopy is used to observe the $n = 1 - n' = 2$ transitions and to deduce information on the ground state Lamb shift $(S(n = 1))$ and on the $2p_{1/2} - 2p_{3/2}$ fine structure interval. External fields (stationary or oscillating ones) are used to gain information on the details of the structure of $n = 2$ levels, in particular the $2p_{1/2} - 2s_{1/2}$ term difference which will, for simplicity, be dubbed the $n = 2$ Lamb shift $(S(n = 2))$ in the following discussion. Details of the make-up of the total QED-related level shifts are to be found in Chapter 2 of this book.

Table 1 Results of Precision Wavelength Measurements on Hydrogen-like Heavy Ions

Element	Nuclear Charge	Transition	Lamb shift	Method	Ref.	Result	Fractional precision of transition energy	Lamb shift test
Li	3		2s	QE	[32]	63031 ± 327 MHz		0.5%
		2s − 2p₁/₂	2s	MW	[56]	62765 ± 21 MHz		0.034%
			2s	MW	[21]	62790 ± 70 MHz		0.11%
C	6		2s	QB	[69] [48]	781.1 ± 8.0 GHz		0.5%
O	8		2s	QB	[52]	2215.6 ± 7.5 GHz		0.35%
			2s	QB	[54]	2202.7 ± 11.0 GHz		0.5%
		1s − 2p	2s	A	[13]	2192 ± 15 GHz		0.7%
F	9	2s − 2p₃/₂	2s	L	[49]	3339 ± 35 GHz	0.05%	1%
P	15	2s − 2p₃/₂	2s	L	[73]	20.12 ± 0.20 THz	0.04%	1%
S	16		2s	QB	[91]	25.14 ± 0.24 THz		1%
		2s − 2p₃/₂	2s	L	[35]	25.27 ± 0.06 THz	0.01%	0.25%
		1s − 2p	2p fs	S	[75]	2.99 ± 0.2 eV		

Table 1 (cont'd.)

Element	Nuclear Charge	Transition	Lamb shift	Method	Ref.	Result	Fractional precision of transition energy	Lamb shift test
Cl	17	$2s - 2p_{1/2}$	$2s$	L	[90]	31.19 ± 0.22 THz	0.7%	0.7%
		$1s - 2p$	$1s$	ST	[44]	0.85 ± 0.10 eV		12%
						from $1s - 2p_{3/2}$		
						0.84 ± 0.12 eV		
		$1s - 2p$	$1s$	T	[74]	from $1s - 2p_{1/2}$		12%
						0.90 ± 0.10 eV		11%
			$2p\,fs$		[74]	3.889 ± 0.030 eV		
		$1s - 2p$	$1s$	S	[18]			$\approx$ 15%
Ar	18		$2s$	QB	[38]	37.89 ± 0.38 GHz		1%
		$1s - 2p$	$\left\{ \begin{array}{l} 1s \\ 2p\,fs \end{array} \right.$	S	[9]	$\left\{ \begin{array}{l} 1.0 \pm 0.5 \text{ eV} \\ 5.1 \pm 0.3 \text{ eV} \end{array} \right.$	150 ppm	50%
		$1s - 2p$		SR	[6]	± 0.4 eV	120 ppm	
		$1s - 2p$		SR	[7]	1.145 ± 0.016 eV	5 ppm	1.5%
		$1s - 2p$	$\left\{ \begin{array}{l} 1s \\ 2p\,fs \end{array} \right.$	ST	[58]	$\left\{ \begin{array}{l} 1.154 \pm 0.030 \text{ eV} \\ 4.78 \pm 0.04 \text{ eV} \end{array} \right.$	11 ppm	3%

Table 1 (cont'd.)

Element	Nuclear Charge	Transition	Lamb shift	Method	Ref.	Result	Fractional precision of transition energy	Lamb shift test
Ti	22	$1s-2p$	$2p\,fs$	S	[22]	10.8 eV	100 ppm	
		$1s-2p$	$2p\,fs$	ST	[8]	11.05 eV		
Fe	26	$1s-2p$	$1s$	S	[10]		90 ppm	
						from $1s-2p_{3/2}$		18%
						3.4 ± 0.6 eV		
						from $1s-2p_{1/2}$		
						4.13 ± 0.7 eV		
		$1s-2p$	$2p\,fs$	S	[41]	21.5 ± 0.4 eV		
		$1s-2p$	$2p\,fs$	S	[34]	21.1 eV		
		$1s-2p$	$2p\,fs$	S	[82]	21.2 ± 0.2 eV		
		$1s-2p$	$1s$ and $2p\,fs$	S	[77a]	(see footnote, section 6.2.5)		
Zn	30	$1s-2p$	$2p\,fs$	S	[41]	38.2 ± 0.8 eV		
Kr	36	$1s-2p$	$2p\,fs$	S	[11]	80.8 ± 1.4 eV		
			$1s$	S	[81]	11.95 ± 0.5 eV	36 ppm	4%

Methods: Q Quench experiments with electric (E) or (motional electric) magnetic (B) fields; A asymmetry measurements; L laser; MW microwave experiments; S spectroscopy of fast ion beams except for experiments at tokamaks (T) or of fast-beam-excited recoil ions (R); fs finestructure measurements of Ly_α line doublet.

All presently available results of experiments on one-electron systems as discussed below are given in Table 1. Where a particular experiment has been improved over the years, only the final results have been quoted.

6.2 Light Sources

The ideal light source for high precision studies of hydrogen-like ions would provide selectively-excited one-electron ions (no satellite line problem [26]), with the ions being at rest (no Doppler broadening or shift), not interacting with each other or the walls (very high vacuum), nevertheless being plenty (for sufficient brightness). Of course, no real light source meets all requirements at once, but unfortunately none does more than approximately satisfy even only one of them. The advantages and drawbacks of a number of light sources are shortly discussed in the following section. Specific features are dealt with when describing particular experiments.

6.2.1 Gas discharges

Gas discharges as employed for the early studies of hydrogen are not suitable for the production and excitation of multiply ionized atoms. Vacuum sparks in their various modifications (condensed spark, sliding spark, low inductance spark) have been developed to yield spectra of highly ionized species (see [25, 28]), but they have hardly been used at all for hydrogen-like systems. This is not a matter of precision: early spectra of hydrogen-like and helium-like ions were precise enough to indicate the presence of deviations from the atomic structure as calculated from non-QED theory [25, 27, 30]. The problem lies rather in the removal of the second most tightly bound electrons and the excitation of the remaining one. Furthermore, lines of hydrogen-like spectra have been used as calibration lines assuming their wavelengths to be known from theory.

In the modern giant gas discharge experiments like tokamak fusion test reactors, ions in high charge states are readily excited and offer themselves as suitable subjects for precision spectroscopy [71], but only very recently has such a device been used for precision measurements on one-electron systems [44]. The presence of such systems is well known, but they are normally used as plasma diagnostic tools because of the susceptibility of their degenerate high-lying levels to perturbations by electric fields. Realizing the potential for precision spectroscopy not only of line profiles (for diagnostic purposes) but of some large term differences which are scarcely affected by external fields, one may even envisage high-resolution laser spectroscopic experiments on highly ionized species in a tokamak plasma [46].

In more conventional experiments the Alcator C tokamak (at MIT) [44] and the Princeton large torus (PLT) [8] have been used. Typical temperatures (given as thermal energy) are in the range 1 to 2 keV, with electron densities of order $1-3\times10^{14}$ cm^{-3} and magnetic fields up to $B \simeq 8 - 10$ T. The plasma rotates rapidly along the torus in these devices and in some less regular ways, but neither these motions nor the Doppler broadening limit the attainable precision for the determination of the $1s - 2p$ transition energies. Problems arise from the presence of ions in lower charge states which cause the appearance of satellite lines very close to and, for many of them, not resolvable from the Ly$_{\alpha1,\alpha2}$ lines of interest. The brightness of the light source on the other hand was sufficient to determine the required wavelengths of Cl^{16+} [44] with a precision of 34 ppm from about 100 discharges. This precision corresponds to 12% of the groundstate ($1s$) Lamb shift and was among the best achieved for ions with $Z > 2$ as of in early 1984. Although an improvement by a factor of 7 appeared feasible, other types of experiments have, within a year, surpassed the expected limit of that method (see below).

However, work on spectroscopy at Alcator C continued and meanwhile yielded part of the promised capability for improvements: In 1986 a study on Ar^{17+} was published [58] which reports data more precise than the aforementioned experiment by a factor of three, reaching an uncertainty of 11 ppm on the wavelengths of the $1s - 2p$ transitions. Although this falls short of the precision of the recoil ion experiment [7] discussed later, it provides a very useful comparison to the latter in that the tokamak spectra suffer less from a major source of error, the number of unresolved satellite lines within the Ly$_{\alpha}$ line profile.

6.2.2 Laser-produced plasmas

With the advent of high-power lasers another method to produce very hot but well-localized small plasmas became feasible: A multi-megawatt or gigawatt pulsed laser (Nd: Yag, CO$_2$, etc.) with pulse energies in the kJ range is focused onto a target. The energy of the light pulse heats the material until rapid evaporation and ionization occurs. In the plume of released material the atoms experience high electric fields (laser light and other charged particles) and collisions with fast electrons which result from ionization; subsequently very high stages of ionization may be reached in the plume which expands with velocities in the km/s range. Spectroscopy is usually done with a line of sight along the surface of the irradiated material and thus perpendicular to the mass motion of the ionized atoms. Because of the dynamics of the process and the spatial/temporal separation of the reaction products this light source offers information on the charge states of the radiating ions [33, 47]. However, the fairly high particle density, the high field strengths, the only partially collimated ion plume with subsequent Doppler shifts of the spectral lines and the ubiquitous electrons appear to prevent this otherwise most useful light source from being of interest for the study of one-electron ions.

6.2.3 Ion sources

There are a number of devices under development which involve electron cyclotron resonance (ECR) or strong electron beams (EBIS, CRYEBIS) to ionize atomic species to very high charge states. These machines are planned to serve as ion sources for accelerators where they will be very useful. Their yield of hydrogen-like ions, however, is not yet sufficient for spectroscopic purposes, and the energy spread of their output is fairly large; $q \times (20 - 50)$eV, with q the charge of the ion, which seems to be typical. For a recent review on the technical achievements see [1].

6.2.4 Recoil ions

If highly charged fast heavy ions from an accelerator pass through a gas cell, they can effect multiple ionization of the target gas atoms even under single collision conditions and at fairly large collision parameters. This implies that the projectile trajectory is changed very little whilst the target atom is kicked to move almost at right angles to the projectile and with a very low recoil velocity. For the production of few-electron Ar ions by the impact of very highly charged U ions (U^{44+} and higher) typical recoil energies of as little as about 20 eV have been found [5]. The spectroscopic potential of such highly charged ions almost at rest was first realized by Sellin *et al.* [76]. A recent application to the measurement of the $1s - 2p$ transitions of Ar^{17+} resulted in the then most precise determination of the $1s$ Lamb shift ($\pm$ 1.5% [7], see below). A major problem lies in the density of the gas target: Ions in all charge states are produced; a majority of the target atoms remains neutral, but provides electrons for secondary charge exchange with the highly charged ions of interest, with very high cross sections and plenty of neutral collision partners. The result is again the appearance of satellite lines in the spectra. Although systematical variations of gas pressure and excitation conditions help to understand the satellite line spectrum [17], the unresolvable blends with satellite lines provide the main obstacle to further reduction of the error limits.

If, however, one extracts the highly charged recoil ions from the region with high gas density, one can study them in a clean environment; trap them for better localization and storage or do laser spectroscopic studies on them [62]. Experiments towards these goals are being pursued in various laboratories.

The target substance is most easily ionized and excited if it consists of single atoms, i.e. a rare gas. Hydrogen-like Ar^{17+} or even bare Ar nuclei have been produced in quantity by the impact of U ions in very high charge states (up to 66+) [85]. The prospects for producing hydrogen-like Kr in this way are not very bright. The only source of extremely highly ionized projectiles in sufficient quantities will be storage rings for heavy ions as planned as an extension at GSI Darmstadt. Even with beams of U^{92+} available, the recoil ion source will not be able to produce hydrogen-like ions beyond Kr.

6.2.5 Fast ion beams

In 1963 two nuclear physicists Kay and Bashkin [45, 3] independently and in parallel discovered that a beam of fast ions from an accelerator radiates visible light after passing through a thin target foil of arbitrary material. This observation has led to a wide range of experiments on gross and fine structure and lifetimes of singly and multiply-excited states of ions of many charge states [2] (Fig. 6.1).

After traversal of a zone of gaseous or solid matter, fast ions show a distribution of charge states due to a multitude of ionization, excitation and electron capture processes occurring in the interaction with the target material. The mean charge state of this distribution increases with the ion beam energy. This implies that for the production of hydrogen-like ions of high Z very high beam energies, and consequently very large accelerators are a prerequisite. Whereas hydrogen-like $Li^{2+}(Z = 3)$ is easily produced at energies of about 0.1 MeV/nucleon [84], this is not the optimum for the one-electron system, and satellite lines arising from lower charge states are present e.g. near Ly_α (Fig. 6.2). At energies of about 1 MeV/nucleon the hydrogen-like spectrum is much cleaner, although by no means free of satellites (Fig. 6.3). In order to produce hydrogen-like $Kr^{35+}(Z = 36)$ energies of order 20 MeV/nucleon are necessary, and for the production of hydrogen-like $U^{91+}(Z = 92)$ which has most recently been demonstrated [39] energies of order 500-1000 MeV/nucleon are required.

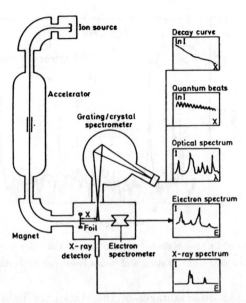

Fig. 6.1. General arrangement of fast beam experiments. A beam of ions is generated in an ion source, accelerated, magnetically selected for charge state and energy (by momentum after acceleration) and directed to an experimental chamber. Here the beam may be excited by interaction with a foil or gas target or a laser beam and spectroscopy of photons or electrons may be carried out.

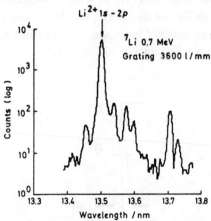

Fig. 6.2. Spectrum of Li near $Li^{2+}Ly_\alpha$. The ion beam energy was 0.1 MeV/nucleon and favored the Li^+ charge state. The spectrum was recorded with a 2.2 m grazing incidence spectrometer equipped with a holographically produced grating of 3600 l/mm. See [84].

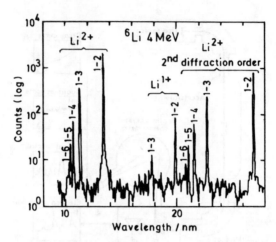

Fig. 6.3. Spectrum of Li at a higher ion beam energy than the one in Fig. 6.2. The same apparatus but with a 600 l/mm grating and hence lower spectral resolving power was used.

Considering the disadvantages of the stationary light sources mentioned above, it may seem surprising that this light source should prove to be useful for precision spectroscopy, although one of the aforementioned requirements, that is that the ions be at rest, is clearly not satisfied. The clue lies in the fact that fast ion beams from accelerators are extremely well (angle-)collimated and (energy-)stabilized so that observations with well-collimated detection systems may suffer Doppler shifts but can keep Doppler broadening a minor nuisance.

The general arrangement of fast-beam spectroscopy is shown in Fig. 6.1, the geometry of observation is depicted in Fig. 6.4.

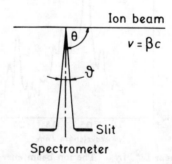

Fig. 6.4. Geometry of observation. ϑ is the acceptance angle of the detection system.

The equal velocity of all ions of such a beam enables the experimenter to select a wavelength range of his detector and follow the excited and then decaying ions along the ion beam to obtain lifetime data. Ions of 0.5 MeV/nucleon travel about 1 cm in one nanosecond ($\beta = v/c \simeq 0.01$), ions of 50 MeV/nucleon travel 10 cm/ns ($\beta \simeq 0.10$). This compensates a little for the shorter lifetimes of high Z systems, but the effects of special relativity begin to play a rôle, too.

The light of wavelength λ_0 emitted by a fast ion of velocity $v = \beta c$ is observed in the laboratory rest frame at

$$\lambda = \lambda_0 \gamma (1 - \beta \cos \theta) \tag{6.1}$$

with γ the relativistic time dilation (or second order Doppler shift which does not depend on the observation angle θ as does the first order Doppler shift, the second term in the bracket).

For any observation a detector of finite size spans a solid angle. This implies that light rays under slightly different observation angles and hence with slightly different Doppler shifts contribute to the signal; the spectra show Doppler broadened lines. This broadening is at minimum for an observation along the beam (as in collinear laser spectroscopy of atomic or ionic beams), but the Doppler shift is at a maximum. There the precision of the wavelength measurement critically depends on the determination of the beam energy which is rarely achieved to less than $\Delta E/E \simeq 10^{-4} (\Delta v/v \simeq 0.5 \times 10^{-4})$. If the observation angle is 90° ($\pi/2$), the first order Doppler shift vanishes, but a precise alignment procedure of ion beam and detection system to intersect at $\theta = 90°$ is required. At this angle the γ factor of course is still present, and one still needs to know the ion velocity, but it is less critical for the precision of the experiment. One can easily see from equation (6.1) that at an angle $\theta = \frac{\pi}{2} - \frac{\beta}{2}$ the Doppler shifts of first and second order cancel, but to determine that particular angle, the same procedures as for a setting at 90° have to be performed. A variation of the beam energy and comparison of the observed wavelengths is usually done in order to check the angle setting of the detection system with respect to the ion beam.

High precision work often requires high spectral resolution, too. This entails a cutting down of the detector solid angle to reduce the Doppler broadening. Although this entails a significant loss of signal, the improvement of the precision is often obvious. Fig. 6.5 shows an example [67]: The H_α line of $O^{7+}(Z = 8)$ was observed with a grazing incidence monochromator equipped with a high resolution grating; the fast ion beam was viewed at (almost) right angles. The

H_α profile is obtained with better resolution and thus appears to arise from a cooler light source than the corresponding line of hydrogen from a discharge at room temperature [89] or a cooled one [77] before the advent of the laser. Of course this demonstration is aided by the fact that the finestructure intervals on a wavelength scale are almost independent of Z, that means that the H_α wavelength for O^{7+} is about 1/64 of the value for H^0, but the wavelength intervals are about the same.

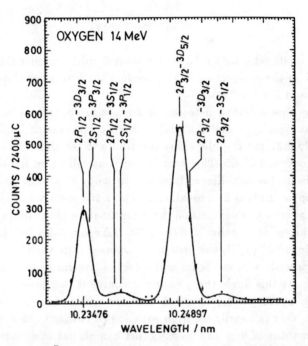

Fig. 6.5. "H_α" line of O^{7+} observed with the same apparatus as used for Fig. 6.2. The beam energy was 14 MeV. Further reduction of the Doppler broadening was achieved by a reduction of the angle of acceptance of the spectrometer from the normal $0.8°$ (f/80) to about $0.4°$ (f/160) by introduction of an auxiliary slit near the exciter foil. (Reproduced with permission from H. R. Muller, P. H. Heckmann and E. Trabert, *Z. Phys.* **A308** (1982) 283 (Springer-Verlag)).

Similar considerations make the study of high Z hydrogenic systems so attractive and promising. For example, Wieman and Hänsch [86] carried out a laser spectrometric study of the H $1s$ Lamb shift by intercomparing the $1s - 2s$ two-photon transition with H_β as the latter line has almost exactly four times the wavelength of the former. The ground state Lamb shift scales with Z

roughly like the finestructure, that is $\sim Z^4$. It causes a wavelength shift of the Ly_α lines in the fourth diffraction order by roughly 1.7 pm (0.0017 nm). This is $\lambda/\Delta\lambda \simeq 10^5$ for hydrogen but only $\lambda/\Delta\lambda \simeq 200$ for Kr $(Z = 36)$. For high Z systems, which cannot be excited or significantly influenced by lasers yet, this latter value is a spectroscopic resolving power which is in reach; an experiment along these lines is in preparation [12, 63]. Even without an absolute wavelength calibration, the intercomparison of $1s-2s$ in fourth diffraction order with Balmer-β in first order permits a Lamb shift difference measurement which ought to yield 7/8 of the ground state Lamb shift (n^{-3} dependence, difference of $S(n = 1)$ and $S(n = 2)$; the $1s - 2s$ single-photon M1 transition dominates for high Z).[a]

One of the setbacks of the fast-beam light source is the low luminosity. At low to medium energy accelerators beam currents of the order of 0.3 μA (particles) can be achieved. This corresponds to 2×10^{12} ions per second traversing a cross section of the beam. The average distance of the ions from each other then is about 50 nm along the beam, not counting the lateral displacement in typical cross sections of a few mm^2. The beams are delivered in beam tubes with high vacuum ($p < 10^{-8}$ bar) in order to avoid collision losses along the beam line from the accelerator. Obviously they do not disturb each other, are hardly affected by the environment but they are also excited by the passage through the target with low yield only; estimates give a fraction of one per thousand ions leaving the target in an excited state. Typical solid angles of detection systems, reflectivities of diffraction gratings or crystals (for EUV and soft X-rays) and conversion efficiencies of detectors combine to bring detection efficiencies down to 10^{-8}. Thus the signal rates are low. For high energy accelerators the particle currents are lower by several orders of magnitude, but in the X-ray regime the detection systems may be of higher efficiency.

Considering advantages and disadvantages, the fast ion beam is a logical counterpart for highly ionized systems of the atomic beams used for hydrogen experiments, and indeed most precision experiments on high Z systems have been done by employing these beams, as is detailed in the following sections.

6.3 More or Less Classical Spectroscopy: $n=1$ Lamb Shift

In order to learn about the ground state Lamb shift one has to observe ground state transitions. The strongest of these and the most easily studied of the

---

[a] A recent publication [77a] reports a measurement by this method applied to Fe^{25+}. The values 1.77815(19)Å, 1.78364(19)Å were obtained for $1s-2p P_{3/2}$, $P_{1/2}$ respectively.

$1s - np$ transitions is $1s - 2p$ (Ly$_\alpha$). For hydrogen the transition energy is about 10.2 eV, and it increases with Z^2. The ground state Lamb shift is of order 3.5×10^{-5} eV and increases roughly with Z^4. The ratio of transition energy to ground state Lamb shift becomes more favorable with increasing nuclear charge, and for a given experimental level of precision the measurement for the highest accessible Z gives the most information on $S(n = 1)$.

For $Z \lesssim 5$, Ly$_\alpha$ is in the EUV where it is most difficult to perform high precision experiments because of the low efficiency of grating spectrometers. In the soft X-ray regime (for $Z \simeq 6 - 10$) the situation is not much better because of the available Bragg crystals with large lattice spacings. More promising is the range beyond about $Z = 14$ (Si), and it is this range where all significant experiments of this type are done.

The basic principles of such measurements are simple: produce the desired ionic charge state by foil or gas excitation of a suitably fast ion beam, determine the trajectory and the velocity of the ions, observe the ions at a well-determined angle and compare the measured wavelength to a precise standard. Schleinkofer *et al.* [75] first demonstrated the feasibility of such procedures for He-like $S^{14+}(Z = 16)$ with a precision of 15% of the QED contribution to the $1s$ Lamb shift. For Ly$_\alpha$ of S^{15+} a nearby calibration standard was not available, and therefore only the finestructure interval of Ly$_\alpha$ (that is the $2p_{1/2} - 2p_{3/2}$ term difference) was measured, but not the absolute wavelength of the $1s - 2p$ transitions. This leads to the question of the calibration method and available standards. Formerly the X-ray wavelength standards were tied to the meter via a lengthy chain of intercomparisons involving a.o. the measurement of the same soft X-ray lines with Bragg reflectors of large lattice spacings and with ruled gratings. Deslattes, Kessler and Henins at NBS in the late 1960s developed an interferometric technique involving a direct link of X-ray and visible wavelength ranges by visible-laser interferometric control of the displacements of an X-ray interferometer [15, 16]. Precise lattice spacings of crystalline materials and hence precise X-ray wavelengths then could be determined. Many inconsistencies were found in the older X-ray wavelength data, and present precision experiments of the type to be discussed here will be able to give accurate results only after the particular calibration standards will have been covered and updated by the NBS wavelength standard redetermination program which aims at an overall accuracy of about 0.5 ppm.

These wavelength standards are K_α-lines of various elements, excited under standard conditions and detected by almost standard equipment also developed at NBS. The spectrometers with plane or curved crystals are aligned using elaborate optical polygon beam deflectors to determine the $90°$ observation angle, extremely precise goniometers (better than $0.0003°$) and position sensitive proportional counters as detectors. The positional sensitivity of such detectors sometimes is not directly used to measure the spectrum — which would involve a precise determination of the detector linearity etc. — but to center the observed and the reference lines onto the same position on the detector, so that the crystal rotation angle may be precisely determined.

Several experiments on $1s - 2p$ transitions have — within about two years — proceeded from demonstrating a precision which left an error amounting to about a half of the QED contributions to the ground state term value [10] to the latest experiments in which a precision corresponding to less than one percent of the $1s$ Lamb shift can be achieved [18, 19, 79]. Similarly there are determinations of the $2p$ finestructure interval — which is more easily done that the measurement of absolute transition energies — which began with observations [41, 34, 22] in agreement with theoretically calculated values [29] and arrived at precisions in the 1% range [74, 82]. All available results are contained in Table 1, but a few of the experiments will be explained in some detail.

Three experiments concerned the spectroscopy of impurities in tokamaks. The study by Bitter *et al.* [8] on Ti $(Z = 22)$ is mainly concerned with the satellite lines appearing in the spectra which give evidence of plasma conditions. A number for the $2p$ finestructure interval is given but not elaborated. Another experiment, by the Källnes and their coworkers [44, 78] is on Cl $(Z = 17)$, and is the first of its kind directed to precision spectroscopy of the $1s - 2p$ transitions with the aim of determination of the $1s$ Lamb shift. A Bragg crystal spectrometer was directed at one of the viewing ports of the Boston Alcator C tokamak, but a gap of 25 cm was left to accommodate a removable X-ray source (Ar K_α, Ag L_α) for *in-situ* wavelength calibrations. The Ar K_α reference light source gave a signal of about 50 cps whereas the count rate of the Cl emission during the tokamak pulses went up to a thousand times this value. With careful attention to systematic effects caused by the plasma and by the detection system, a wavelength precision of 34 ppm was achieved. The next experiment in this series, on Ar $(Z = 18)$, reached a wavelength precision of 11 ppm [58]. After further improvements the method is expected to permit precisions as good as 5 ppm. This would, if achieved, correspond to less than 2% of $S(n = 1)$.

Practically the same precision as in the tokamak experiment has been achieved in a fast-beam experiment by Richard *et al.* [74] at Brookhaven. In this experiment scientists from NBS took part (Deslattes *et al.*) who had contributed to the development of X-ray standards [16], and the publication includes the report on a remeasurement of the Ar $K_{\alpha1,2,4}$ reference lines to 3-20 ppm accuracy which was necessary for this experiment. The Ar K_α radiation was excited in a gas cell by an electron beam. The gas cell could be positioned at the same location as otherwise the fast Cl ion beam which would be excited by passage through a thin carbon foil. The radiation from the gas cell or from the fast ion beam could be dispersed alternately by a $R = 1$ m curved crystal (Si(111)) onto a position sensitive detector. The actual observation angle was set to 90° but checked by varying the ion beam energy in the range 70-175 MeV ($\beta \simeq 0.065 - 0.11$). The remaining uncertainties (extrapolation to $\beta = 0$) of the Doppler shift (alignment problem) and of the calibration as well as, to a minor extent, the presence of satellite lines in the Ly_α line profiles (which could be reduced but not eliminated) contribute most to the accuracy limits achieved, some 10% of $S(n = 1)$ and 1% of the finestructure interval $2p_{1/2} - 2p_{3/2}$ (see Figs. 6.6 and 6.7).

Similar but somewhat less precise measurements have been done for Ar ($Z = 18$) [10] at Orsay, for Ti ($Z = 22$) [22], Fe ($Z = 26$) [9] and for Kr ($Z = 36$) [11] at GSI Darmstadt and [81] at GANIL Caen. Measurements for U ($Z = 92$) can be envisaged since not only the production of U^{91+} has been reported [39] but the first X-ray spectra (of not yet sufficient resolution) have been observed at Berkeley, with plenty of signal contributing to a clear survey spectrum in only 15 minutes of beam time [40]. For U^{91+} the $1s - 2p_{3/2}$ transition energy reaches 102.2 keV, it includes a ground state Lamb shift of about 475 eV [40]. Fig. 6.8 indicates the features of hydrogen-like U^{91+}. Among them is the attraction of direct spectroscopic access to the $2s_{1/2} - 2p_{3/2}$ transition [40]. First spectroscopic experiments on helium-like U^{90+} have been reported from the Berkeley Super HILAC accelerator [68] so that hydrogen-like U^{91+} seems within reach.

In the above beam-foil experiments the alignment error of the detection system causes an error via the Doppler shift of the observed light. A light source which is basically free of this error is the recoil ion light source made up of gas atoms which are multiply ionized and excited by distant collisions with highly ionized heavy projectiles (see above). The potential for precision spectroscopy of hydrogen-like Ar^{17+} has been tested in an experiment at GSI Darmstadt [6, 17, 7]: U^{66+} ions of 5.9 MeV/nucleon were directed at a gas cell filled with Ar. A Ti X-ray tube excited a K X-ray fluorescence source, the light of which could enter the spectrometer — except for being blocked by a chopper wheel (Fig. 6.9)— by the same light path as the light from the

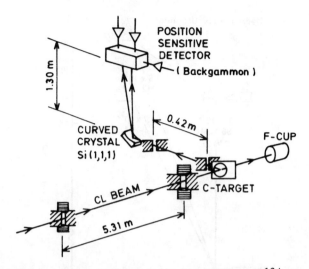

Fig. 6.6. Schematic of experimental set-up for the measurement of Cl^{16+} Ly_α. (Reproduced with permission from P. Richard *et al.*, *Phys. Rev.* **A29** (1984) 2939, (American Physical Society)).

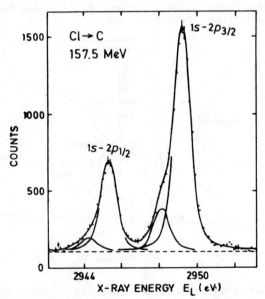

Fig. 6.7. Profile of the Ly_α lines of Cl^{16+} as obtained with the set-up shown in Fig. 6.6. The distortions of the line profile are due to the presence of satellite lines. These satellites are represented here by just one component for each of the $Ly_{\alpha1,\alpha2}$ lines. (Reproduced with permission from P. Richard *et al.*, *Phys. Rev.* **A29** (1984) 2939, (American Physical Society)).

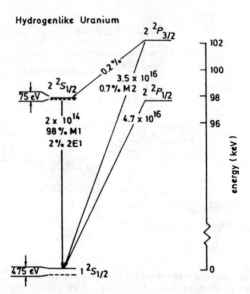

Fig. 6.8. Level scheme of hydrogen-like U^{91+} with transition rates, branching ratios and QED level shifts indicated. (Reproduced with permission from H. Gould, *Nucl. Instrum. Methods* **B9** (1985) 658 (North-Holland Physics Publishing)).

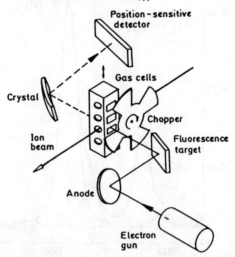

Fig. 6.9. Schematic view of the recoil ion experiment on Ar^{17+}. The calibration X-rays from the fluorescence target pass through the gas target during the dark periods of the pulsed heavy-ion beam and a Johann-type crystal spectrometer is used to measure the heavy-ion-induced X-rays relative to the calibration standard. (Reproduced with permission from H. F. Beyer *et al.*, *J. Phys.* **B18** (1985) 207, (IOP Publishing Ltd.)).

gas cell. The ubiquitous satellite line distribution was studied by variation of the gas pressure (Fig. 6.10) and excitation conditions, but in the end caused a significant uncertainty. Again, the X-ray wavelength standard initially was not precise enough, but finally a precision of the wavelength determination of the Ly_α line components of 5 ppm was achieved, testing the $1s$ Lamb shift to 1.5%. Agreement with theory [64-66] was claimed.

As the reduction (to zero) of the Doppler shifts was achieved at the cost of contamination by spectator lines which are not understood with the necessary precision, another experiment returned to fast ion beams, but with a grip on the satellite lines [18]. At Heidelberg fast Cl ions were produced with energies of the order of 100 MeV. These ions were then stripped of most of their electrons, and the charge state fraction with bare nuclei (Cl^{17+}) was selected by a magnet. This selected beam was then decelerated in a linear accelerator tuned to slow down the ions of the beam to energies of about 20-80 MeV. The slow naked ions were passed through a windowless cell containing He gas; some of the ions would then capture an electron, some (fewer) even two of them. The ions then are observed by the same techniques with precision spectrometers as mentioned above. Now there are two major advantages with this method. First, the beam dynamic range, the range of velocities of the ions, can be varied between wide limits, and velocities much lower than optimum for the production of hydrogen-like ions can be reached. Second, the electron capture process is both strongly energy-dependent (and can hence be checked for its influence on the spectator line spectrum) and fairly selective so as to populate preferentially low angular momentum states (and hence fewer of the long-lived levels than are populated in the beam-foil interaction). The combination of these advantages permits an accuracy of the $1s - 2p$ transition wavelength determination which is better than with recoil ion experiments. The presently expected precision of the $1s$ Lamb shift determination is substantially less than one percent.

This method is amenable to ions with higher nuclear charge provided the accelerator-decelerator arrangement is available. With increasing precision and increasing nuclear charge the problem of the radiative width of the $2p_{1/2,3/2}$ levels becomes significant even for the $1s - 2p$ transitions. However, there is the $2s_{1/2}$ level which for low Z primarily decays via emission of two E1 quanta to the ground level. The other major decay channel, by magnetic monopole (M1) radiation, is insignificant at low Z, but the transition probability increases at a higher power of Z (Z^{10} vs. Z^6 of 2E1) [60]. At $Z = 18$ (Ar) it contributes 3.2% to the decay of the $2s_{1/2}$ level as confirmed by a lifetime measurement at Berkeley [59, 36-38]. (With a 1% precision of the lifetime determination,

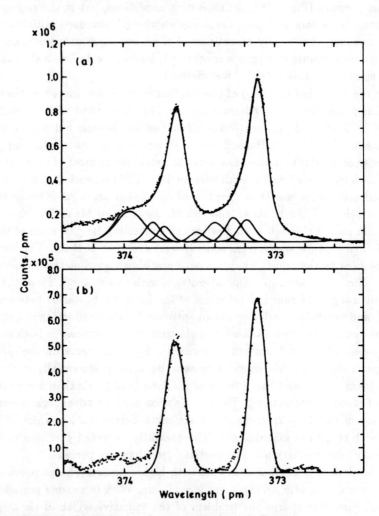

Fig. 6.10. Spectrum obtained using the apparatus shown in Fig. 6.9.
(a) Experimental Lyman-α profile (dots) and empirical fit (full curve) using seven satellite
   components.
(b) Experimental Lyman-α profile with theoretical satellite background subtracted (dots)
   and single-component fit to the lines (full curve).
The remaining discrepancy between data and Ly$_\alpha$ profile is ascribed to satellite lines from
states with spectator electrons in high-lying levels. (Reproduced with permission from H. F.
Beyer *et al.*, *J. Phys.* **B18** (1985) 207 (IOP Publishing Ltd.)).

this was the first positive test of the M1 contribution, and the decay curve was constructed from the signals of the 2E1 decay continuum.) At $Z \simeq 45$ the M1 decay takes over to become the major decay branch. In any case, up to rather high values of Z the $2s_{1/2}$ lifetime remains longer than the $2p_{1/2,3/2}$ lifetimes (which vary in proportion to Z^{-4}), and hence the radiative width of the $2s_{1/2}$ level and the subsequent broadening of the $1s_{1/2} - 2s_{1/2}$ transition remain smaller. The next series of accel-decel experiments therefore will aim at this transition to be observed in Kr $(Z = 36)$. (A theoretical treatment of these competing transition processes is given in Chapter 3.)

6.4   Application of Constant External Fields: Quench Experiments

The symmetries of the hydrogen-like ions and hence the properties of the excitation states are altered when the ions are exposed to external electric or magnetic fields, that is with fields not matching the spherical symmetry of the Coulomb field of the unperturbed ion. Laboratory magnetic fields hardly affect the structure of low-lying levels in multiply ionized atoms. Electric fields on the other hand, can be made strong enough so that the breakdown of spherical symmetry influences terms which before showed metastability, and the loss of longevity then can be used to study properties of the atom, in particular the $2s$ Lamb shift.

The basic mechanism is the Stark effect mixing of states with different parity: The $2s$ and $2p$ states are no longer states of different kinds, but with the electric fields new states are established which are linear combinations of the former ones, each bearing part of the properties of the unperturbed levels. Besides the level shift there is a marked effect on the lifetime: In the limit of high field strengths the former $2s_{1/2}$ and $2p_{1/2}$ levels become completely mixed, that is, they are shifted and become almost equal in decay probability. This high field decay probability is close to one half of the sum of the $2s_{1/2}$ and $2p_{1/2}$ zero field decay probabilities. This implies a change by a factor of two from zero to high field for the $2p_{1/2}$ level and a variation by many orders of magnitude for the $2s_{1/2}$ level. The lifetime variation is related to the term separation, and a series of lifetime measurements of "$2s_{1/2}$" at various field strengths then allows one to determine the zero field term separation which is the Lamb shift. The theoretical treatment has to include the $2p_{3/2}$ term as well, although its influence is rather small due to the larger term separation. The basic Bethe-Lamb and Lüders [57] theory for this Stark quenching was improved and detailed by Holt and Sellin [42]. A full treatment is given in Chapter 3 of this book.

The pioneering experiment for heavy ions was carried out by Fan *et al.* [31, 32] on Li^{2+}, followed by a series of studies of C^{5+} [48, 69, 53], O^{7+} [52, 54], S^{15+} [91] and Ar^{17+} [36-38]. The general lay-out of the experiment of Fan *et al.* is shown in Fig. 6.11. A fast ion beam containing hydrogen-like ions is prepared and then exposed to a (transverse) electric field up to about 10 kV/cm. A fixed detector sensitive to Ly_α radiation monitors the signal at the beginning of the field region (spontaneous decays), a second one can be moved along the ion beam to record decay curves for a number of field strength values. The first set of experiments was sufficient to obtain an experimental $n = 2$ Lamb shift value with 0.5% precision [32] for Li^{2+}.

To effect the same mixing in higher Z ions as achieved for Li^{2+}, much higher electric field strengths are necessary, but they are most difficult to achieve and maintain technically. An elegant way out of this dilemma is provided by the high velocity of the fast ions: a magnetic field is experienced by the rapid ions as a simultaneous electric field, a so-called motional electric field. In the combination of fast heavy ions and quite usual magnetic fields up to about 2T motional electric field strengths can be produced to permit measurements up to $Z = 18$ [38] and probably for another few elements beyond.

The experiment reported in most detail on the techniques employed and on the checks for systematic errors is at the same time the one which is most significant for the test of higher order contributions to the $2s$ Lamb shift: Gould and Marrus in a series of experiments [36-38] achieved a precision of 1% in the determination of $S(n = 2)$ in Ar^{17+}. Major topics in their discussion and treatment of systematic errors are the determination of the ion beam energy (8.5 MeV/nucleon) by the energy loss in a surface-barrier detector, cascades from high-lying excited states (which might affect the measured lifetimes), interference from the spectra of helium-like ions, collisions of the fast ions with atoms of the residual gas, spectator electron effects and field quenching effects. As concerns the lifetime of $2s_{1/2}$, the signal of the single photon M1 line or the two-photon 2E1 continuum was recorded. The $1s - 2s$ M1 signal could not be resolved from $1s - 2p$ Ly_α, and the resulting decay curves were spoilt by the slow cascade repopulation of the intrinsically rapidly decaying $2p$ levels. Resorting to the 2E1 continuum the decay curves showed clean single exponential behavior, although the 2E1 continuum arising from the decay of the $2\ {}^1S_0$ level of helium-like Ar^{16+} is to be found in the same spectral region. Therefore care had to be taken to keep the Ar^{16+} charge state fraction small, a goal that was achieved by letting a beam of bare Ar^{18+} ions from the Berkeley Super-HILAC pick up electrons in a very thin foil only (Fig. 6.12). If the target is sufficiently

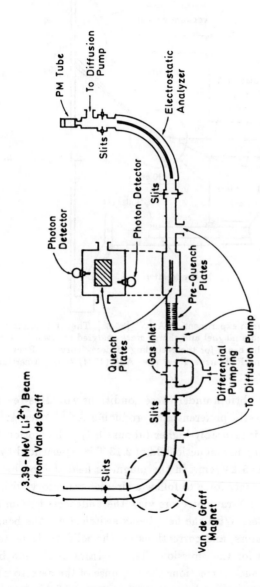

Fig. 6.11. Principal arrangement of quench experiments. (Reproduced with permission from C. Y. Fan, et al., *Phys. Rev.* **161** (1967) 6, (American Physical Society)). The ion beam may be quenched in the experimental chamber to record decay curves at various strengths or in the pre-quench section to determine the background signal. After passage through the field region the velocity of the ion beam is measured with an electrostatic analyzer.

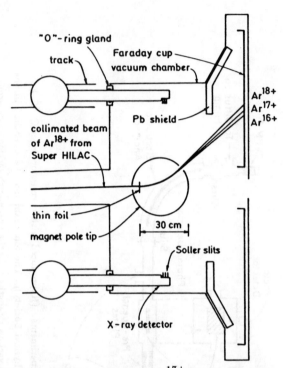

Fig. 6.12. Lay-out of the quench experiment on Ar^{17+} [36-38]. The ion beam is excited by a foil, quenched and deflected in a magnet and charge state analyzed afterwards. Two movable X-ray detectors can track the beam for the recording of decay curves. (Reproduced with permission from H. Gould and R. Marrus, *Phys. Rev.* **A28** (1984) 2001, (American Physical Society)).

thin (here $8\mu g/cm^2$, equilibrium under these conditions would be reached with about 400 $\mu g/cm^2$ foils, still preferentially producing Ar^{17+} over Ar^{16+}), the ratio of Ar^{17+} to Ar^{16+} is favorably higher (about 16:1). The motional electric field produced by a maximum magnetic field of 2.13 T is experienced by the ions ($\beta = 0.134$) as high as 865 kV/cm. In the magnetic field the ion beam is deflected because of the Lorentz force to follow a circular trajectory with a radius of only 0.5 m. Two Si(Li) X-ray detectors view the undeflected beam symmetrically from opposite sides. With the field being switched on, the beam moves towards one of the detectors, and corrections for the solid angle subtended by the detectors as well as for the emission characteristics of the ion beam are necessary. After various such corrections the response of the detectors could be linearized and decay curves could be measured at various field strengths. For magnetic fields in the range 1.47 T- 2.13 T, the lifetime of the quenched $2s_{1/2}$

state varies from 1.87 ns to 1.29 ns (Fig. 6.13); the effective magnet pole diameter of about 22 cm allowed the authors to measure 2.5 decay lengths on average. The unperturbed $2s_{1/2}$ level lifetime is (3.487 ± 0.036) ns. A measurement with the same fractional precision on the $2s$ Lamb shift but at lower nuclear charge has most recently been done by Zacek *et al.* [91] on $S(Z = 16)$. The perturbation of the symmetry of the Coulomb field of the atom by an external electric field leads not only to the aforementioned level shifts and the shortening of the lifetime of long-lived states but also to asymmetries of the angular distribution of the radiative emission (section 3.3.3). Whereas a fast ion beam excited by passage through a perpendicularly arranged foil normally displays cylindrical symmetry of the radiation pattern, sometimes notable variations occur which depend on θ (see Fig. 6.4) and are caused by alignment, that is the different population of magnetic sublevels depending on the absolute value of the magnetic quantum number. With an electric field across the beam direction, however, the electron orbits can be deformed so that an asymmetry of the emission along and perpendicular to the field direction can be observed [23]. The value of this asymmetry $A = (I_{\parallel} - I_{\perp})/(I_{\parallel} + I_{\perp})$ (with $I_{\parallel}, I_{\perp}$ the observed signal parallel and perpendicular to the field) is a measure of the Stark mixing and thus, as above, of the term separation and the Lamb shift. Whereas for He^+ such measurements have been extended to great detail [24, 87, 88] there are very few experiments for heavier ions [4, 13, 14]. The excitation of the ion beams in these experiments has been effected by thin foils. Foil excitation, however, is non-selective and populates almost any excitation state. The higher-lying levels are then subject to the same external field and become aligned. This alignment can, at least partially, be transferred by cascades to the levels of interest. The result would be an alignment of these levels and subsequent radiation asymmetry which is not entirely due to the $n = 2$ level structure. The systematic errors which are unsatisfyingly large in the present experiments may perhaps be reduced when the excitation takes place in a gas, because the population of levels in ion-atom collisions is much better understood than the beam-foil interaction, levels of low angular momentum are populated almost exclusively and cascade modeling can be done with some confidence.

In order to illustrate the experimental problems, some information on the Manhattan experiment [13] may be in order: The required electrical field strength in preliminary experiments was shown to bring about field emission and residual gas ionization at untolerable levels. Therefore the coils of a magnetic quadrupole lens used for the ion beam transport system were rewired to give a dipole field, and the necessary electric field was produced as a motional

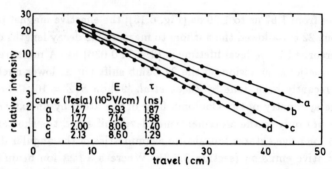

Fig. 6.13. Decay curves obtained in the quench experiments depicted in Fig. 6.12. (Reproduced with permission from H. Gould and R. Marrus, *Phys. Rev.* **A28** (1984) 2001, (American Physical Society)).

field. The asymmetries to be measured are very small ($R = 0.07246 \pm 0.00048$ in this case) which implies that the signals $I_\perp$ and $I_\parallel$ have to be measured with high statistical precision and with great care for the efficiency of the detectors and for systematic errors. The statistical uncertainty of the data was much smaller than the final result reflects, but background problems occurring when the quench field was turned on limited the precision of the experiment. Later improvements [14] appear to have reduced this source of errors only in part.

6.5 Application of Oscillating External Fields: Resonance Experiments

Lamb and Retherford [51] detected the $n = 2$ Lamb shift by inducing a transition from the metastable $2s_{1/2}$ level to the $2p_{3/2}$ levels of H atoms by irradiation with radiofrequency (rf) electromagnetic radiation. A magnetic field was used to tune Zeeman level term differences to the fixed-frequency radiation, until a change in the number of metastable atoms could be detected after the interaction zone.

Because of the Z^4 increase of the finestructure interval $2p_{1/2} - 2p_{3/2}$ and the almost similar increase of the $2s$ Lamb shift it is obvious that the matching frequency of the radiation field increases as rapidly with Z. For He$^+$ the microwave method worked to high precision, but for Li^{2+} the problems were already formidable [20, 21]: A fast ion beam (^{6}Li^{2+}, 3-7 MeV) was used. The necessary microwave radiation frequency reached 55 GHz and the tuning field necessary had a strength of 0.2-0.8 T. Because of the high demand on the performance of many components of the apparatus (see discussion in [50]) it was a major achievement to obtain a $2s$ Lamb shift value with a precision of about 0.1% although the original goals may have been even higher.

Leventhal [56], however, has managed to obtain considerably lower error limits by ionizing lithium vapour inside the field of a large magnet. The ionized atoms were then partially confined (trapped) by electric fields and subjected to microwave radiation of fixed frequency. The resonance condition was matched by varying the external magnetic field and tested by observation of the Ly_α radiation emitted in case of successful pumping from the $2s$ to the $2p_{3/2}$ level (Fig. 6.14). The precision reached 335 ppm of the $n = 2$ Lamb shift, a precision that has not been matched by any Lamb shift determination in any other heavy ion. Higher order terms in the Lamb shift calculation become more important at higher nuclear charge, hence even the less precise data for higher Z ions are of appreciable value.

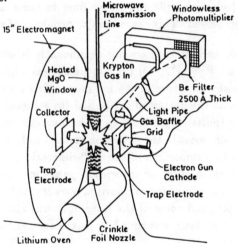

Fig. 6.14. Arrangement of the experiment by Leventhal on Li^{2+}. A combination of microwave-optical-pumping, ion trapping and Zeeman tuning is employed. (Reproduced with permission from M. Leventhal, *Phys. Rev.* **A11** (1975) 427, (American Physical Society)).

Beyond Li^{2+} the microwave frequencies required can be produced but not the proper radiation power. But for F^{8+} the range of powerful (infrared) lasers is entered, and a pioneering study employing an HBr laser has been done by Kugel *et al.* [49]. As in the case of hydrogen, the radiative width of the $2p$ levels amounts to about $1/10$ of the $2p_{1/2} - 2s_{1/2}$ interval. Scanning the profile of the resonance was achieved by changing the angle of intersection of laser and ion beams. As in all these experiments, the repopulation of the short-lived $2p$ levels by the induced transition from the long-lived $2s$ level is monitored via detection of the subsequent Ly_α emission in the X-ray range, usually carried out with large proportional counters.

A laser resonance experiment on ions almost at rest is being prepared at Oxford [62]. There an F-center laser is to be directed at hydrogen-like Ne^{9+} ions extracted from a recoil ion source, a gas cell bombarded with highly-charged heavy ions. Tuning will have to be done by tuning the laser frequency.

More laser resonance experiments have been done on P^{14+} [72, 73] and on S^{15+} [80] employing a tunable dye laser and on Cl^{16+} [70, 90] with a CO_2 laser. The lower frequency (longer wavelength) of the latter device matches the $2s_{1/2} - 2p_{1/2}$ interval, whereas the other experiments measured the $2s_{1/2} - 2p_{3/2}$ interval. The CO_2 laser is not continuously tunable, so a combination of Doppler tuning (by variation of the intersection angle of laser and fast ion beam) and line tuning (selecting different lasing lines in the many-line spectrum of CO_2) had to be employed. This caused major problems of signal normalization.

The other laser experiments are not significantly less difficult. In the Louvain experiment on P^{14+} [72, 73] the ion beam was supplied by a cyclotron. This implies a pulsed structure of the ion beam. The 5 ns beam pulses had to be synchronized with the output of the pulsed (to achieved high peak powers of 100 kW) dye laser (pulse length 7 ns) to make ion beam pulses and light pulses intersect and actually meet in the detection zone. The laser wavelength calibration was achieved by comparison of the 45th diffraction order of the laser light with the 9th diffraction order of absorption peaks in water vapor. The spectrometer and the light source for comparison were located about 300 m from the cyclotron laboratory, and some of the laser light was sent to the spectrometer via optical fibres. The laser was tuned to match the resonance condition.

The S^{15+} experiment [80, 35] used a 130 MeV dc ion beam from the Strasbourg tandem accelerator and a tunable cw dye laser. The ion and laser beams intersected at a small angle to keep the intersection zone long and to increase the efficiency of the pumping process from $2s$ to $2p_{3/2}$ (Fig. 6.15). One of the resulting resonance curves is shown in Fig. 6.16.

All these laser experiments have achieved precisions of the order of 1% of the $2s$ Lamb shift; the experiment on S^{15+} [80, 35] was finally quoted with an uncertainty as small as 0.25%. This then is the most precise experiment on the $n = 2$ Lamb shift yet, for moderately heavy ions. The many obstacles towards further improvements are outlined in that work [35] by sorting the about 30 recognized sources of error into various groups which need their own specific treatment.

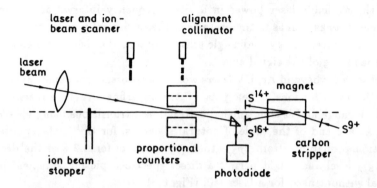

Fig. 6.15. Geometry of the laser resonance experiment on S^{15+} [80]. Ion and laser beam are counterpropagating and intersect at a small angle. (Reproduced with permission from H. D. Sträter *et al.*, *Phys. Rev.* **A29** (1984) 1596, (American Physical Society)).

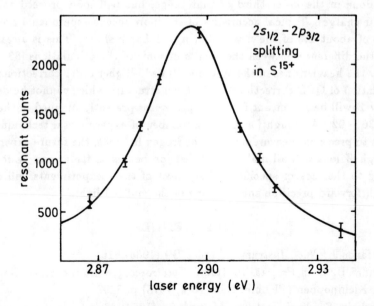

Fig. 6.16. Resonance curve obtained in the experiment depicted in Fig. 6.15. (Reproduced with permission from H. D. Sträter *et al.*, *Phys. Rev.* **A29** (1984) 1596, (American Physical Society)).

In so far as the $2s_{1/2} - 2p_{3/2}$ interval has been determined, the $2p_{1/2} - 2p_{3/2}$ fs interval is taken for granted from extended Dirac theories with allowances for the different $2p_{1/2}, 2p_{3/2}$ Lamb shifts. Limitations of the method lie in the available laser power in a given frequency interval and in low ion beam currents as well as in the obstacle of the intrinsically wide resonance. Ion beam divergence, energy and angle scattering and the like would cause much less broadening of the signal curve.

Not very far beyond Ar, UV lasers would be necessary for experiments of this type. However, for ions as heavy as Tl, the hyperfine structure reaches values in the eV range and is thus amenable to induced transitions using visible laser light. At the end of the table of natural elements, for U^{91+}, the spontaneous transition probability from $2p_{3/2}$ to $2s_{1/2}$ accounts for 0.2% of the decays of the $2p_{3/2}$ level and might permit a direct spectroscopic measurement of that interval without need for a laser [40] (Fig. 6.8).

6.6   Conclusion

Experiments on one-electron ions by now span the entire periodic table of elements occurring in nature. Experiments with a bearing on QED effects have been done in the lower third of that range, but will soon proceed to higher nuclear charges. Typical accuracies of the more recent experiments are in the range of about 1% of the $n = 2$ and $n = 1$ Lamb shifts. This is already less than the difference between the results of some of the calculations [83, 29, 64-66] for the heavier ones of the systems studied. Higher order corrections in the calculation of QED corrections to the term structure which cannot be detected at low Z will be significant for the experiments presently planned in the range $Z = 36 - 92$. Although for low Z a number of experimental techniques give access to precision measurements on hydrogen-like ions, the future experiments for high Z ions will all make use of fast ion beams as far as can be foreseen. Owing to the lack of specific probes, most of these experiments will employ straightforward precision spectroscopy of the emitted light.

REFERENCES

[1] Arianer J., *Nucl. Instrum. Methods* **B9** (1985) 516.

[2] Andrä H. J., in *Progress in Atomic Spectroscopy*, Part B eds. W. Hanle and H. Kleinpoppen (Plenum, New York, 1978) p. 829.

[3] Bashkin, S., *Nucl. Instrum. Methods* **28** (1964) 88.

[4] Betz H.-D., Rothermel J., and Bell F., *Arbeitsbericht EAS* **1** (1980) 77.

[5] Beyer H. F., Mann R., Folkmann F. and Mokler P.H., *J. Phys.* **B15** (1982) 3583.

[6] Beyer H. F., Mokler P.H., Deslattes R. D., Folkmann F. and Schartner K. -H., *Z. Phys.* **A318** (1984) 249.

[7] Beyer H. F., Deslattes R. D., Folkmann F. and LaVilla R. E., *J. Phys.* **B18** (1985) 207.

[8] Bitter M., von Goeler S., Cohen S., Hill K. W., Sesnic S., Tenney F., Timberlake J., Safronova U. I., Vainshtein L. A., Dubau J., Loulergue M., Bely-Dubau F. and Steenman-Clark L., *Phys. Rev.* **A29** (1984) 661.

[9] Briand J. P., Tavernier M., Indelicato P., Marrus R. and Gould H., *Phys. Rev. Lett.* **50** (1983) 832.

[10] Briand J.P., Mossé J. P., Indelicato P., Chevallier P., Girard-Vernhet D., Chetioui A., Ramos M. T. and Desclaux J. P., *Phys. Rev.* **A28** (1983) 1413.

[11] Briand J. P., Indelicato P., Tavernier M., Gorceix O., Liesen D., Beyer H. F., Liu B., Warczak A. and Desclaux J. P., *Z. Phys.* **A318** (1984) 1.

[12] Brown J. S., Duval B. P., Klein H. A., Laursen J., McClelland A. F., Mokler P. H. and Silver J. D., *GSI Sci. Rep.* **84-1** (1983) 161.

[13] Curnutte B., Cocke C. L. and Dubois R. D., *Nucl. Instrum. Methods* **202** (1982) 119.

[14] Curnutte B. and Cocke C. L., *Bull. Am. Phys. Soc.* **28** (1983) 789.

[15] Deslattes R. D. and Henins A., *Phys. Rev. Lett.* **31** (1973) 972.

[16] Deslattes R. D., Kessler E. G., Sauder W. C. and Henins A., *Ann. Phys.* (N.Y.) **129** (1974) 378.

[17] Deslattes R. D., Beyer H. F. and Folkmann F., *J. Phys.* **B17** (1984) L689.

[18] Deslattes R. D., Schuch R. and Justiniano E., *Phys. Rev* **A32** (1985) 1911.

[19] Deslattes R. D., *Nucl. Instrum. Methods* **B9** (1985) 668.

[20] Dietrich D. D., Ph.D. thesis, State University of New York, Stony Brook (1974).

[21] Dietrich D. D., Lebow P., de Zafra R. and Metcalf Ḣ., *Bull. Am. Phys. Soc.* **21** (1976) 625.

[22] Dohmann H. D., Liesen D. and Pfeng E., Abstracts XIIIth ICPEAC, Berlin, p. 467 (1983).

[23] Drake G. W. F. and Grimley R. B., *Phys. Rev.* **A8** (1973) 157.

[24] Drake G. W. F., Goldman S. P. and van Wijngaarden A., *Phys. Rev.* **A20** (1979) 1299.

[25] Edlén B., *Nova Acta Req. Soc. Sci. Upsaliensis*, Ser. IV, Vol. 9 (1934).

[26] Edlén B., and Tyrén F. S. *Nature* **143** (1939) 940.

[27] Edlén B., *Ark. Fys.* **4** (1952) 441.

[28] Edlén B., *Phys. Scripta* **T3** (1983) 5.

[29] Erickson G. W., *J. Phys. Chem. Ref. Data.* **6** (1977) 831.

[30] Eriksson H. A. S., *Nature* **161** (1948) 393.

[31] Fan C. Y., Garcia-Munoz M. and Sellin I. A., *Phys. Rev. Lett.* **15** (1965) 15.

[32] *−Phys. Rev.* **161** (1967) 6.

[33] Fawcett B. C., in *Advances in Atomic and Molecular Physics*, eds. D. R. Bates and B. Bederson (Academic Press, New York, 1974) Vol. 10 p. 223.

[34] Fortner R. J., Howell R. H. and Matthews D. L., *J. Phys.* **B13** (1980) 35.

[35] Georgiadis A. P., Müller D., Sträter H.-D., Gassen J., von Brentano P., Sens J. C. and Pape A., *Phys. Lett.* **A115** (1986) 108.

[36] Gould H. and Marrus R., *Phys. Rev. Lett.* **41** (1978) 1457.

[37] *−J. Phys. (Paris) Coll.* **40** (1979) C1-30.

[38] *−Phys. Rev.* **A28** (1984) 2001.

[39] Gould H., Greiner D., Lindstrom P., Symons T. J. M. and Crawford H., *Phys. Rev. Lett.* **52** (1984) 180.

[40] Gould H., *Nucl. Instrum. Methods* **B9** (1985) 658.

[41] Hailey C. J., Stewart R. E., Chandler G. A., Dietrich D. D. and Fortner R. J., *J. Phys.* **B18** (1985) 144.

[42] Holt H. K. and Sellin I. A., *Phys. Rev.* **A6** (1972) 508.

[43] Johnson W. R. and Soff G., *At. Data Nucl. Data Tables* **33** (1985) 405.

[44] Källne E., Källne J., Richard P. and Stöckli M., *J. Phys.* **B17** (1984) L115.

[45] Kay L., *Phys. Lett.* **5** (1963) 36.

[46] Knize R. J., *Phys. Rev.* **A27** (1983) 2258.

[47] Kononov E. Ya, *Phys. Scripta* **27** (1983) 117.

[48] Kugel H. W., Leventhal M. and Murnick D. E., *Phys. Rev.* **A6** (1972) 1306.

[49] Kugel H. W., Leventhal M., Murnick D. E., Patel C. K. N. and Wood II O. R., *Phys. Rev. Lett.* **35** (1975) 647.

[50] Kugel H. W. and Murnick D. E., *Rep. Prog. Phys.* **40** (1977) 299.

[51] Lamb W. E. and Retherford R. C., *Phys. Rev.* **72** (1947) 241.

[52] Lawrence G. P., Fan C. Y. and Bashkin S., *Phys. Rev. Lett.* **28** (1972) 1612.

[53] Leventhal M. and Murnick D. E., *Phys. Rev. Lett.* **25** (1970) 1237.

[54] Leventhal M., Murnick D. E. and Kugel H. W., *Phys. Rev. Lett.* **28** (1972) 1609.

[55] Leventhal M., *Nucl. Instrum. Methods* **110** (1973) 343.

[56] Leventhal M., *Phys. Rev.* **A11** (1975) 427.

[57] Lüders G., *Z. Naturforsch.* **5a** (1950) 608.

[58] Marmar E. S., Rice J. E., Källne K., Källne E. J. and La Villa R. E., *Phys. Rev.* **A33** (1986) 774.

[59] Marrus R. and Schmieder R. W., *Phys. Rev.* **A5** (1972) 1160.

[60] Marrus R., *Nucl. Instrum. Methods* **110** (1973) 333.

[61] Marrus R., in *Topics in Current Physics: Beam-foil Spectroscopy*, ed. S. Bashkin (Springer, Heidelberg, 1976) p. 209.

[62] McClelland A. F., Palmer C. W. P., Finch E. C. and Silver J. D., *Nucl. Instrum. Methods* **B9** (1985) 710.

[63] McClelland A. F., Brown J. S., Duval B. P., Finch E. C., Klein H. A., Laursen J., Lecler D., Mokler P. H. and Silver J. D., *Nucl. Instrum. Methods* **B9** (1985) 706.

[64] Mohr P. J., *Phys. Rev. Lett.* **34** (1975) 1050.

[65] -in *Beam-Foil Spectroscopy*, eds. Sellin I. A. and Pegg D. J. (Plenum, New York, 1976) Vol. I, p. 89.

[66] -*At. Data Nucl. Data Tables* **29** (1983) 453.

[67] Müller H. R., Heckmann P. H. and Träbert E., *Z. Phys.* **A308** (1982) 283.

[68] Munger C. T. and Gould H., *Phys. Rev. Lett.* **57** (1986) 2927.

[69] Murnick D. E., Leventhal M. and Kugel H.W., *Phys. Rev. Lett.* **27** (1971) 1625.

[70] Murnick D. E., Patel C. K. N., Leventhal M., Wood II O. R., and Kugel H. W., *J. Phys.* (Paris) *Coll.* **40** (1979) C1-34.

[71] Peacock N. J., Stamp M. F. and Silver J. D., *Phys. Scripta* **T8** (1984) 10.

[72] Pellegrin P., El Masri Y., Palffy L. and Prieels R., *Phys. Rev. Lett.* **49** (1982) 1762.

[73] Pellegrin P., El Masri Y and Palffy L., *Phys. Rev.* **A31** (1985) 5.

[74] Richard P., Stöckli M., Deslattes R. D., Cowan P., La Villa R. E., Johnson B., Jones K., Meron M., Mann R. and Schartner K., *Phys. Rev.* **A29** (1984) 2939.

[75] Schleinkofer L., Bell F., Betz H. -D., Trollmann G. and Rothermel J., *Phys. Scripta* **25** (1982) 917.

[76] Sellin I. A., Elston S. B., Forester J. P., Griffin P. M., Pegg D. J., Peterson R. S., Thoe R. S., Vane C. R., Wright J. J., Groeneveld K. -O., Laubert R. and Chen F., *Phys. Lett.* **61A** (1977) 107.

[77] Series G. W., *The Spectrum of Atomic Hydrogen* (Oxford University Press, Oxford, 1957) (reprinted in this book).

[77a] Silver J. D., McClelland A. F., Laming J. M., Rosner S. D., Chandler G. C., Dietrich D. D. and Egan P. O., *Phy. Rev. A* (1987) August.

[78] Stöckli M. P., Richard P., Deslattes R. D., Cowan P., La Villa R. E., Johnson B., Jones K., Meron M., Mann R., Schartner K., Källne K. and Källne J., *Bull. Am. Phys. Soc.* **29** (1984) 781.

[79] Stöckli M. P. and Richard P., *Nucl. Instrum. Methods* **B9** (1985) 727.

[80] Sträter H. D., von Gerdtell L., Georgiadis A. P., Müller D., von Brentano P., Sens J. C. and Pape A., *Phys. Rev.* **A29** (1984) 1596.

[81] Tavernier M., Briand J. P., Indelicato P., Liesen D. and Richard P., *J. Phys.* **B18** (1985) L327.

[82] Tavernier M., Briand J. P., Indelicato P., Desclaux J. P., Marrus R. and Prior M., (1985) (to be published).

[83] Taylor B. N., Parker W. H. and Langenberg D. N., *Rev. Mod. Phys.* **41** (1969) 375.

[84] Träbert E., Blanke J. H., Hucke R and Heckmann P. H., *Phys. Scripta* **31** (1985) 130.

[85] Ullrich J., Cocke C. L., Kelbch S., Mann R., Richard P. and Schmidt-Böcking H., *J. Phys.* **B17** (1984) L785.

[86] Wieman C. and Hänsch T. W., *Phys. Rev.* **A22** (1980) 192.

[87] van Wijngaarden A. and Drake G. W. F., *Phys. Rev.* **A25** (1982) 400.

[88] van Wijngaarden A., Helbing R., Patel J. and Drake G.W. F., *Phys. Rev.* **A25** (1982) 862.

[89] Williams R. C., *Phys.Rev.* **54** (1938) 558.

[90] Wood II O. R., Patel C. K. N., Murnick D. E., Nelson E. T., Leventhal M., Kugel H. W. and Niv Y., *Phys. Rev. Lett.* **48** (1982) 398.

[91] Zacek V., Bohn H., Brum H., Faestermann T., von Feilitzsch F., Giorginis G., Kienle P. and Schuhbeck S., *Z. Phys.* **A318** (1984) 7.

CHAPTER 7

HYDROGENIC SYSTEMS IN
ELECTRIC AND MAGNETIC FIELDS

J. C. Gay

7.1 Introduction

Whereas most of this book is concerned with the fine detail of the hydrogen spectrum and refined theoretical analysis, the theme of this chapter is completely different. It highlights the role of the main binding interaction in hydrogen, the Coulomb field. As shown in previous chapters, any system — especially hydrogen — is best studied by noting its response to perturbations. But when electric and magnetic perturbations become so strong that they are comparable to the binding field itself, the structure of the atom drastically changes. How to understand and describe such change is our main concern in this chapter on hydrogenic systems in electric and magnetic fields.

By hydrogenic systems, we mean those in which the Coulomb interaction represents the dominant part of the binding potential. Obviously this concerns a wide variety of elementary objects in physics, excitons and impurities in solid state physics, ions in plasma physics, charged clusters, and of course, Rydberg atoms.[1,2,3] As will be shown in the following, Rydberg atoms are a nearly perfect realization of Coulomb-driven systems in three dimensions. Rydberg atoms are those that follow a $R/(n-\delta)^2$ law for term values. For hydrogen, of course, $\delta = 0$ and the law expresses the quantization of binding in a Coulomb field, in the non-relativistic approximation. For the alkali series of low orbital quantum number (S, P and D terms), the non-zero δ expresses the deviation of the binding from coulombic. For the series of higher l, the Coulomb law is again, to an approximation sufficient for this chapter, applicable.

What is there new to say about the action of electric and magnetic fields on the atom, about the Zeeman and Stark effects? Every elementary textbook deals with these topics. They provide beautiful examples for the application of perturbation theory, and the Zeeman effect is a splendid proving ground for the quantum theory of angular momentum.[4] Yes, indeed, but what is new is the realization that the elementary treatments are powerless in the face of perturbations of significant magnitude, comparable to the magnitude of the Coulomb field. Underlying this is the fact that the Hamiltonian of the perturbed system is usually non-separable in conventional coordinate systems. This is the key reason for which the structure and dynamics of atoms in external fields is only now beginning to be understood, although it has been standard practice for years to apply fields to atoms.

The dynamics of systems with a non-separable Hamiltonian has attracted considerable interest for a long time. These questions were addressed by Poincaré before 1900[5] in a classical context, and by Einstein[6] around 1917 for the purpose of quantizing such systems. They are still at the heart of modern physics although numerous developments have occurred in the meantime. Obviously the matter is difficult. But it is also of importance. Most systems in physics are of non-integrable type, except for the harmonic oscillator and the Kepler problem. Of special interest, among various questions, is that of the quantum analog of classical chaos, or what are the characteristics of the spectrum of a system for which the classical trajectories are chaotic? It turns out that recent advances on Rydberg atoms in external fields allow us to address this point and may provide, unexpectedly, experimental examples.[7] An atom in a magnetic field, a problem extensively discussed in the following sections, and for which the non-separable character of the Hamiltonian froze the theory for more than 60 years, is such an example.

Advances in understanding the physics of atoms in external fields were stimulated by the results of numerous experiments which became possible owing to technical advances, among which the invention of tunable lasers is the most important. It has become common-place to excite atoms — especially alkali atoms — into high lying Rydberg states, though it is still difficult to do so for hydrogen itself. As the Coulomb binding energy scales as $-R/n^2$, electric or magnetic energies comparable to the Coulomb energy can be achieved with electric or magnetic fields currently available in the laboratory. A convenient choice of the n value allows one to reach the non-perturbative regime of the spectra, in which the character of the Hamiltonian becomes essentially non-separable.

Experiments, then, revealed regularities which did not depend on the atomic species, pointing therefore to the existence of general laws which previous studies of Zeeman and Stark effects had given no hint of. Clearly, something essential was missing from our rather crude understanding of the behavior of atoms in external fields. The discovery of the quasi-Landau resonances in 1969 is such an example[8] which triggered a re-examination of the theory of the magnetic problem. Further experimental investigations showed that the phenomena which occur close to the zero-field threshold are continuous, in a diabatic sense, with those taking place in the regime where the external field perturbation is small compared to the Coulomb one.

The way of conceiving this low field regime turns out to be of extreme importance. The conventional thinking has been that it is the regime in which the n^2 degeneracy of the n Coulomb shell is removed by the action of external fields. The angular momentum quantum number l is no longer defined, as L^2 does not commute with the external field perturbation. But angular momentum plays an important role governing selection rules in optical excitation processes.

In this regime l "forbidden" lines appear, from which follows the label of "inter-l mixing regime". But such a label is merely descriptive and does not address the fundamental question of how such mixing occurs, which is the only one of importance.

Considered from a more positive view point, this regime is the one in which the Coulomb symmetry is broken by the action of the external field. To say this is to do more than to re-label a regime. It is to offer a new concept which leads to a simple understanding of the physics of atoms in external fields. But what is the Coulomb symmetry and what are its specific features?

Elementary manifestations have been well-known for years. They are the closed or symmetrical character of the classical trajectory, as well as the n^2 degeneracy of the energy shell with principal quantum number n. The origin of these features was recognized early; they express the rotational invariance of the orbital Coulomb interaction in a four-dimensional space. This is the first level of symmetry established by Fock around 1935. Further studies up to 1970 gave rise to a full branch of theoretical physics in the context of high energies. The amazing thing is that these fundamental insights into atomic structure are mostly ignored in the field to which they really belong.

Our purpose here will be to re-discover these concepts and their usefulness for understanding the physics of atoms in external fields. A study of selected experimental results will allow us to stress the limits of the standard analysis. At this stage, we recall some basic results on the orbital structure of the Coulomb problem. The most prominent features of the experimental spectra then become quite simple to understand. This amounts to re-discovering that the symmetries of the Coulomb interaction govern the dynamical behavior in the atom.

But, we will begin with simple phenomenological considerations which make clear the suitability of Rydberg atoms for experimentally studying these questions.

7.2 Elementary Considerations

The most striking features of the physical systems under consideration can be derived from simple phenomenological models. This is highly relevant to the fact that these models pose general questions which it should be possible to answer from simple and general concepts. These questions are, what are the characteristics of the electron's motion when submitted to the joint action of several forces with different symmetries which may have comparable orders of magnitude? In the present atomic physics context, these are, for example, the

Coulomb, Lorentz and electric field forces:

$$\mathbf{F}_c = -e^2\mathbf{r}/r^3$$
$$\mathbf{F}_1 = q\mathbf{v}\Lambda\mathbf{B}$$
$$\mathbf{F}_E = q\mathbf{E} \tag{7.1}$$

which in the Hamiltonian formulation, in the symmetrical gauge ($\mathcal{A} = \frac{1}{2}\mathbf{B}\Lambda\mathbf{r}$) and assuming the proton infinitely massive, leads to ($e^2 = q^2/4\pi\varepsilon_0$):

$$H = \frac{p^2}{2m} - \frac{e^2}{r} - \frac{q}{2}\mathbf{B}\cdot\mathbf{L} + \frac{q^2B^2}{8m}(x^2 + y^2) - q\mathbf{E}\cdot\mathbf{r} . \tag{7.2}$$

The magnetic contributions are respectively the paramagnetic interaction (responsible for orbital Zeeman effect) and the diamagnetic interaction proportional to B^2. The latter also depends on the degree of excitation of the electron's motion.

7.2.1 Coupling constants — orders of magnitude

An estimate of the coupling constants can be drawn from the ratio of the magnetic or electric forces to the Coulomb force. The evaluation of the atomic dimensions (or the degree of excitation of the electron's motion) which are needed for the estimate is carried out in the Coulomb limit with $r \simeq n^2a_0$ (a_0 the Bohr radius). It follows immediately:

$$\eta = F_1/F_c = Bn^3/B_c = \gamma \cdot n^3, \quad \xi = F_E/F_c = E \cdot n^4/E_c = \beta \cdot n^4 , \tag{7.3}$$

where $\gamma = B/B_c$ and $\beta = E/E_c$. $B_c = m \cdot e^2/\hbar qa_0$; $E_c = e^2/q \cdot a_0^2$ are atomic units of magnetic and electric fields. They correspond to the huge values $B_c = 2.35 \times 10^9$ Gauss and $E_c = 5.14 \times 10^9$ V/cm. Under laboratory conditions, values of γ greater than 10^{-4} can be reached only with difficulty. But, owing to the fact that the coupling parameters η or ξ depend on the degree of excitation n, conditions such that η or $\xi >> 1$ can easily be achieved. This makes it clear why Rydberg atoms excited to high principal quantum numbers are so important since the critical fields (for which η or $\xi \cong 1$) scale as B_c/n^3 or E_c/n^4.

These considerations have an obvious counterpart in energy. For the magnetic problem, the high field limit of (7.2) is Landau motion: the motion of a charged particle in a magnetic field. The unit of energy is $\hbar\omega_c$ where ω_c is the cyclotron frequency:

$$\omega_c = qB/m \tag{7.4}$$

which is twice the Larmor frequency. The motion is in some sense equivalent to a 2-dimensional harmonic motion. The magnetic contribution to the energy in (7.2) is thus of the order of $n\hbar\omega_c$ while the Coulomb binding energy is $-R/n^2$ (R the Rydberg constant). With $\gamma = \hbar\omega_c/2R = B/B_c$, one deduces that, when $\eta \cong 1(F_1 \cong F_c)$ the total (magnetic plus coulombic) energy of the electron is nearly zero. The equality of the Coulomb and magnetic forces also implies the equality of the classical Bohr frequency, $2R/n^3$ to $\hbar\omega_c$. Hence, using the correspondence principle, one deduces that the spacing of resonances close to the zero-field threshold should be proportional to $\hbar\omega_c$, as it is.

The same kind of arguments can be used successfully in the electric field problem, replacing ω_c with the linear Stark frequency (in atomic units):

$$\omega_E = \frac{3}{2}n \cdot E/E_c \tag{7.5}$$

upon which, close to the zero field threshold, the spacing of the resonances is found to scale as $E^{3/4}$. This is the characteristic of the Stark resonances discovered in 1977.[10,11]

Obviously these are crude considerations. They completely disregard the non-compatible geometrical symmetries of the forces which underly the Hamiltonian (7.2), which is usually non-separable. The unexpected matching of physical reality with this simple model is one of the questions we will further address. But for a long time, in the magnetic problem, it was as far as theoretical understanding had gone.

7.2.2 The atomic spectrum in external fields—the three gross regimes

From the consideration of the relative strengths of the forces,[12] there are three regimes of atomic motion. They are with increasing energy:

− *The low field Coulomb regime* $(\gamma n^3 << 1$ or $\beta n^4 << 1)$

The external field force is small compared to the Coulomb one. This takes place for negative energies and the structure is Coulomb-like. The role of the external force is to break the Coulomb symmetry and to remove the n^2 degeneracy of the n Coulomb shell. How this occurs is indeed a key point for understanding the physics of the other regimes. Physical manifestations are Zeeman effect, linear Stark effect or the appearance of a diamagnetic band.

− *The strong mixing regime* $(\gamma n^3 \cong 1$ or $\beta n^4 \cong 1)$

The external and Coulomb forces are of comparable strengths. This is essentially a non-perturbative regime where the effects of the non- separability of the

Hamiltonian (7.2) should be at a climax. One may infer that the electron's motion becomes featureless and disorganized. But experiments have proved just the opposite. This is the regime which takes place close to zero energy. Physical manifestations are quasi-Landau resonances and Stark resonances. The behavior is no longer of the usual atomic type.

– *The high field regime* $\left(\gamma n^3 >> 1 \text{ or } \beta n^4 >> 1\right)$

This takes place above the zero field threshold $(E > 0)$ in the continuum region. The Coulomb force is a small perturbation to the external field ones. The regime is basically of Landau type in the magnetic problem with a quantization law $W_n \sim \left(n + \frac{1}{2}\right) \hbar \omega_c$, far different from the one obeyed in the strong mixing regime. The quantization in the electric field problem is $W_n \sim E^{2/3}(n+3/4)^{2/3}$ (triangular potential well). In both cases, the atomic continua no longer exist and are replaced with resonance spectra.

7.2.3 Orders of magnitude and further remarks

The previous characteristics are hardly those to be expected from an atom. In the magnetic problem, the spectrum evolves between a Coulomb and Landau signature and this is not a *matter of absolute strength of the field*. As a consequence of the γn^3 scaling law, the three gross regimes do exist and the phenomena will be qualitatively the same for fields in the microgauss or in the Tesla range. What matters indeed is the quality of experimental detection. Actually, as discussed in the following sections, the basic features of these atoms dressed with static fields have been seen under wide experimental conditions including accelerators and white dwarfs (!) but also in excitonic systems in solid state physics. The latter systems can be thought of as an electron-hole pair in which the effective Rydberg constant is far smaller than in the atomic physics case, due to screening effects. The effective B_c^* critical field can be in the range of a few kilogauss which allows the high field regime to be easily reached. Most of the experiments on semiconductors and magnetism have been performed under such conditions for more than 30 years.[2,13]

Some orders of magnitude are given in Table 7.1 in the atomic physics situation. This illustrates the suitability of Rydberg atoms for studying these problems. Actually, the η parameter can be varied by more than 10^3 in practical experimental conditions.

But in addition to these scaling laws, Rydberg atoms have another quality. They are a nearly perfect realization of Coulomb-driven systems in three dimensions. From the point of view of symmetries, this turns out to be fundamental for the class of questions we deal with.

Table 7.1. Some orders of magnitude of the binding energies, Bohr frequencies and critical electric and magnetic fields according to the principal quantum number in Rydberg atoms.

n	1	4	30	100	300
$E_n = -R/n^2$	10^5 cm^{-1}	6.25×10^3 cm^{-1}	100 cm^{-1}	10 cm^{-1}	30 GHz
$\omega_n = 2R/n^3$			220 GHz	6 GHz	220 MHz
$E_n = E_c/n^4$	5.14 GV/cm	20 Mv/cm	630 kV/cm	51 V/cm	0.63 V/cm
$B_n = B_c/n^3$	2.35×10^5 T	3.67×10^3 T	87 kG	2.35 kG	87 G

7.2.4 The failure of traditional methods

We have remarked that the Hamiltonian (7.2) is usually non-separable and lacking any constant of the motion. The magnetic field problem, in fact, received a quantum solution only around 1985. Actually, conventional theoretical methods are most inefficient for tackling such a situation. The diamagnetic Hamiltonian $H_D = \frac{\gamma^2}{8}(x^2 + y^2)$ couples almost every state in the Coulomb ladder, including continua (which play an essential role). It is thus hopeless, even with the use of huge computers, to generate a solution in the strong mixing regime.

The requirement of interpreting simple experimental results in simple terms thus leads to a change in the angle of view, to the introduction of more powerful concepts. They are those pertaining to the symmetries of the Coulomb field. Not only does this give the solution to basic physical problems but also it casts a new light on the dynamical properties of the atom.

7.3 Basic Aspects of Experiments

Before discussing some sample experimental results which have played a role in the evolution of these topics, we perform a critical review of experimental conditions. Perhaps one might begin by saying that the ideal experiment is still to be performed. It would be on a hydrogen atomic beam, using single mode c. w. dye laser excitation. For dealing with the magnetic field problem, the use of trapped (neutral) hydrogen atoms would be in fact the perfect solution. Even the motion of the center-of-mass does not separate in the usual sense.[3,14]

Experiments to date have been performed on alkali Rydberg atoms. In turn they have stimulated the development of theoretical models in the simplest non-relativistic hydrogen-like situation. Preliminary results of two experiments on hydrogen itself have just been obtained[15,16,17] and look promising.

There are basic choices to be made in these experiments, what kind of optical excitation, whether to use beam or vapour conditions; and these are inter-related.

7.3.1   The choice of the atomic species and optical state selection

Since we are concerned with unsolved problems, experiments should be designed for the simplest situation. The situation in alkali Rydberg atoms differs from that in hydrogen owing to quantum defect corrections. These are important for penetrating orbits $l \leq 2 (S, P$ and D series). They are also smaller (at fixed l) when the atom is lighter.[18,19]

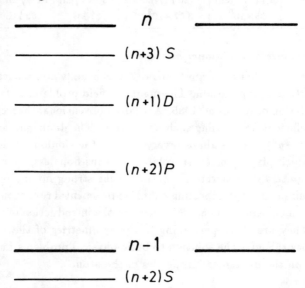

Fig. 7.1. Energy level structure of alkali Rydberg atoms (Rubidium case). Owing to their quantum defects, the S, P and D series are energetically separated from the others. States with $l \geq 3$ are quasi-degenerate. They build an "incomplete" manifold compared to the pure Coulomb situation as the degeneracy is $(n^2 - 9)$ instead of n^2. The low-field behavior for each class of state is very different.

As shown schematically in Fig. 7.1, the situation for the Rydberg states of the alkalis differs from the pure Coulomb case, in that the S, P, D series are out of the manifold of quasi-degenerate states (with $n^2 - 9$ states only). Hence a very simple way of achieving hydrogen-like situations is first to excite those states having a nearly zero quantum defect. But, as only the S, P and D series can be conveniently optically excited (through 1 or 2 photons) from the ground state, this requires exotic excitation schemes. But this alone is not sufficient. Additionally the field mixing should exclude the components of the (nS, nP, nD) series. This can be realized from consideration of the symmetries of the perturbing potential. An example is furnished by the magnetic field problem. L_z is a constant of the motion (see section 7.5). The excitation of states with $l = 3$ and $m = 3$ in alkali atoms results in the S, P and D series having no role in the diamagnetic mixing of the spectrum. Such a situation is nearly hydrogenic. It has been realized on Caesium atoms using the so-called hybrid resonance optical process [19,20] which populates the F series with $m = 3$. The maximum value of the quantum defect is that of the F state $\delta(F) = 0.033$ which only generates very small departures from the hydrogen behavior.[21,22]

Obvious generalizations would be to use 3-photon excitations of $(l = 3, m = \pm 3)$ states or, better, excitation of $m = 4, G$ type series... . The use of electric quadrupolar transitions is also possible.[23] For some light atoms (Li, Na), the use of stepwise excitation with two photons turns out to be sufficient.

It is clear, then that the combined choice of atomic species and optical state selection is most important. It makes alkali Rydberg atoms as perfect a coulombic system as desired. This indeed explains why diamagnetic patterns of a hydrogenic type were obtained experimentally, long before any experiment had been done on hydrogen itself!

The proper interest of experiments on hydrogen atoms lies really in other directions, tests of relativity and Q.E.D., or studies designed to test the breaking of symmetries.

It is also worth mentioning that non-hydrogenic effects also have great interest, leading to new classes of mechanism. And they are to be found in most atoms...

7.3.2 Atomic beam experiments

The choice of atomic beam or vapor phase generates in turn severe constraints on the atomic species, the laser source and the method of detection. But it is no bad thing to have had a variety of experimental results obtained under far different experimental conditions; it is a good indicator of the reliability of the work, done on difficult topics, before any guiding theory had been established.

Most experiments have been done on alkali atoms, using pulsed laser excitation and detection of the ions through field ionization. A typical set-up is shown in Fig. 7.2. Stepwise excitation using one, two or three photons is used to excite atoms into Rydberg states which in some cases are of quasi-hydrogenic type. This is the case for the sodium D series or the lithium P series. Playing with the polarization and frequencies of the lasers, a wide selection of Rydberg series is indeed possible,[24,25,26] tuning the situation from hydrogenic to non-hydrogenic.

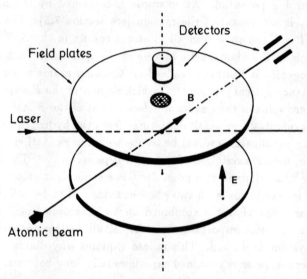

Fig 7.2. Sketch of the experimental set-up in beam experiments. Pulsed dye laser excitation is combined with electric field ionization techniques.

The pulsed dye lasers are usually pumped by the second or third harmonic of a YAG laser, or a nitrogen laser which allows one to have huge optical powers at one's disposal. This may be required due to the rather small densities in atomic beams, in the range 10^8 to 10^{10} at/cm^3. From which may follow one of the drawbacks in this scheme: the fluctuating character of the intensities and laser lineshape. The typical linewidth of a grazing incidence type laser [27] is about 0.1 cm^{-1} (3 GHz). This is 1000 times the standard linewidth for a single mode c. w. dye laser. This is also 1000 times the natural linewidth of Rydberg atoms with $n \cong 30$ in a beam. Hence the light source may spoil the potentially high resolution character of this class of experiments as the Doppler width, under vapor phase conditions, is typically of the order of 1 GHz. A way of overcoming these difficulties is to use a pulsed dye laser, synchronized on a single mode c. w. dye laser[28,29]; this leads to a linewidth in the range of hundredths of MHz or better. This, up to now, has not been applied to experiments in external fields. It is a nice application of the concepts of stimulated emission.

Indeed several experiments[30,31,32,33] have recently combined atomic beam and c. w. dye laser excitation which shows that atomic densities are not in fact so small. As a consequence of the increase in the resolution and the well-controlled character of the irradiation, the detection of Rydberg states up to $n \cong 300$ turned out to be possible.[32] With pulsed dye laser excitation, it is hardly possible to go over $n \cong 100$, and $n = 30$ to 60 can be considered as typical values. Indeed, for $n = 100$, the Bohr frequency $2R/n^3$ is 6 GHz (cf. Table 1), hence comparable to pulsed laser linewidths.

Detection is performed through field ionization. A pulsed ionizing field is applied with a delay which allows one to explore the structure of the excited states. The ions are collected and amplified with electron multipliers or barrier surface detectors. This is an efficient and selective technique developed ten years ago.[34,35,36] However, it is worth noticing that the use of this technique requires a good knowledge of the physics of ionization for the system under consideration. For example, for the magnetic problem, the use of electric field ionization implies that one understands the physics of atoms in $(\mathbf{E}, \mathbf{B})$ fields. This is far from being known!

Ions are indeed formed. Knowing the states from which they originate is another question which is usually part of the problem to be solved. The same problem exists in conventional optical studies of atomic spectra, using spontaneous emission. But nobody ever deduces that the atom has only nP series from the fact that they are the only ones which radiate. There are firm arguments based on the isotropy of space in this case, which with field ionization are not so obvious. This means that confirmatory experimental evidence from other techniques is useful.

7.3.3 Vapor phase experiments

The drawbacks compared to an atomic beam situation come from collision processes, the excitation of discharges with electric fields and motional Stark effects in magnetic fields, due to the uncontrolled velocity of the atom.[37,38] Actually these turned out to be minor inconveniences in most of the situations, although atomic beam situations are potentially better.

It was long thought that Rydberg atoms studies were impossible under vapor phase conditions due to their sensitivity to the environment and stray fields. But this way of studying Rydberg physics proves to be an efficient one. One of the reasons is that the collisional perturbations are smaller than expected due to the smearing out of the electron's wavefunction.[40] Actually the collisional cross-section is far smaller than the geometrical one. Typically, collisional lineshifts are in the range 500 MHz. This is small compared to the external field perturbation of the atomic structure, and nearly unnoticeable in Doppler limited experiments, with 1 GHz resolution.

These experiments have been done usually using c. w. dye laser excitation and detection of the ions by means of a thermionic detector.[41,42,43] The use of c.w. dye laser excitation implies a perfectly controlled lineshape and a resolution in the range of a few MHz. This would seem to permit, combined with Doppler free two-photon techniques, the detection of atoms in states up to 1000, but there are other kinds of limitation (!), for example blackbody radiation around $n = 300$. In Doppler limited conditions (1 GHz), production of states up to 160 (for which $2R/n^3 \cong 1.5$ GHz) has been achieved this way.[21] Various exotic optical excitation schemes have been used for producing with ease highly hydrogenic states of alkali atoms. This can also be combined with stepwise excitation.

The basic detection system (optical detection is also possible) was devised by Hertz and Kingdom around 1920[41] and proved extremely efficient. This can be considered as a diode working in the space charge limited regime (see Fig. 7.3). An electronic space charge builds up around an 800 K heated tungsten wire. A thin mesh divides the surrounding electrode into two regions. The atoms are laser excited in one of the compartments which is electrostatically shielded from the other one. The stray electric field in the excitation region has been typically found in the range 20 mV/cm. Detection is based on the "spontaneous" ionization of Rydberg atoms which may result from a diversity of processes (collisions, laser induced, associative ionization or detachment, molecular complexes, field ionization). Upon drifting of the ions through the detection region, the perturbation of the electronic space charge induces an electron yield of 10^3 to 10^6 per ion.

The response time of the detector is quite long, in the range of 10 Hz which is the natural integration time in the experiments. Owing to its high sensitivity, pulsed laser excitation is to be avoided. A much wiser way of increasing the optical power in the interaction region is to combine the use of a Brewster-angle windowed detector with a servo-locked cavity of high Q. Effective c.w. optical powers in the range of 50 Watts can be obtained this way with a linewidth in the range of a few MHz. Further use of frequency modulation of the laser also leads to considerable enhancement of the performances as the dynamical response of the detector extends up to several kHz.[44,45] Several other variants have been successfully used, especially the one in which several wires (with indirect heating) are placed in an axially symmetrical configuration allowing the detection of n values up to 230.[40] This works at low pressures (10^{-3} torr) in glass cells or heat pipes.

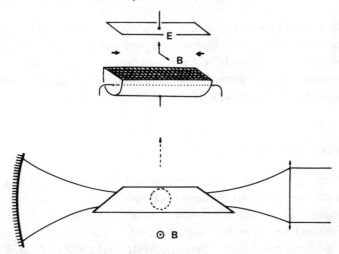

Fig. 7.3. Sketch of the experimental set-up in vapor phase experiments. C.w. dye laser excitation is combined with thermionic detection of the ions "spontaneously" produced from the Rydberg atoms. According to whether the optical excitation scheme is Doppler-limited or Doppler-free, the resolution is in the range 1 GHz or 10 MHz.

Experiments along this scheme have proved highly sensitive due both to the detector and to c.w. dye laser excitation. The positions of the lines are only slightly collisionally-perturbed. On large frequency scans, the intensities of the lines are not always reliable in the sense that the mechanism of "spontaneous" ionization is not controlled and may have changed. The more serious drawback is the lack of control of the atomic velocity. This gives rise to a motional Stark field $\mathbf{E} = \mathbf{v} \wedge \mathbf{B}$ when the vapor experiences a magnetic field.

But although motional field effects only weakly perturb the position of the lines in the magnetic problem, they may drastically affect the intensity distributions and the symmetries, as experimentally demonstrated.[46,47]. It makes clear that ultimate tests on these problems should be done using atomic beam arrangements.

7.3.4 Other experimental techniques

Garton and Tomkins, twenty years ago, performed their pioneering experiments — Rydberg states in strong magnetic fields seen by conventional absorption spectroscopy — with classical light sources. The quality of their data has hardly been superseded by laser spectroscopy, at least to the present time. Developments have shown that these experiments opened to a new field in atomic physics.

Another technique seems promising for atomic physics purposes as well as for plasma diagnostics. It is derived from optogalvanic spectroscopy[48-50] and has allowed recently a study of quasi-Landau resonances and the first study of the perturbations of natural auto-ionization in a magnetic field.[51,52,53] A discharge is thus not unduly aggressive to Rydberg atoms.

7.3.5   The production of fields

In both vapor phase and beam experiments, the production of magnetic fields has been effected by superconducting magnets (up to 8 Teslas),[21,54,55] conventional magnets (10 to 20 kG)[52] and even air coils (less than 1000 Gauss).[45] One experiment uses a pulsed magnetic field.[56] The ability to study the same magnetic phenomena at such different field strengths follows from the $n^3 \cdot B/B_c$ scaling law. The range on which η can be varied in vapor phase experiments is greater than in beam experiments, by at least one order of magnitude.

Electric fields are applied through a convenient arrangement of electrodes. While strengths of up to 100 kV/cm have been used in beam experiments, the limitations are about 20 V/cm in vapor phase conditions, in order to prevent the onset of discharges. Nevertheless, owing to the $n^4 E/E_c$ scaling law, it is possible to study all the regimes of Stark effect under absolute field strengths as low as these. In one experiment, on a relativistic ion beam, the use of relativistic principles allowed the transformation of a 7 T (pulsed) magnetic field into a MV/cm electric field in the atom rest frame.[57] Fields of such strengths were also used around 1930.[58]

7.3.6   About the future

Both types of experiments have fulfilled what was expected, validating and stimulating the development of theoretical models for the hydrogen atom, in the non-relativistic approximation. The next generation should be on hydrogen and helium, in atomic beams or traps, using c.w. dye laser excitation and state selective detection. In turn, the theory should be able to incorporate relativistic and Q.E.D. effects which seems to imply laser linewidths in the range of 10 to 100 kHz. In this respect, the two experiments based on hydrogen,[15,16] which are noteworthy accomplishments, are lacking resolution in their present stage. (But see the supplementary reference at (16): Ed.).

7.4　Electric Field Structure of One-Electron Rydberg Atoms

The electric field structure of the hydrogen atom was basically understood around 1930. The magnetic field structure began to be understood around 1982. Such differences in the status of problems which apparently are similar in nature is the result of a mathematical accident. The electric field problem was earlier found to be separable in parabolic coordinates of $R(3)$.[59,60] The magnetic field problem was shown around 1980 to be non-separable in the thirteen sets of natural coordinates of $R(3)$.[61]

Actually, the differences and similarities lie deeper, in the inner structure. While the Stark problem does possess a dynamical symmetry,[62] the magnetic field problem has none. The notion of dynamical symmetry does not refer to real space and geometrical properties but rather to some property of the motion in phase space. Stark effect is an "accident" in that the dynamical symmetry has a simple expression in parabolic coordinates, in real space.

7.4.1　Historical survey

As a consequence of this accident, the basic ideas for solving the electric field problem have been known for more than 60 years. Lo Surdo discovered the electric field action in 1913.[63] Around 1930, experiments on the hydrogen atom in fields in the MV/cm range gave the first experimental evidence of the so-called linear Stark effect. The theory was developed by Pauli, Epstein, Lanczos and many others [64,65,66] while the asymptotic nature of the solutions led to numerous developments in mathematical physics.[67,70]

7.4.2　The electric field structure of the hydrogen atom

The non-relativistic Hamiltonian takes the form (atomic units):

$$H = \frac{p^2}{2} - \frac{1}{r} + \boldsymbol{\beta} \cdot \mathbf{r} \; . \tag{7.6}$$

The non-compact character of H leads one to predict that there are no bound states, but only resonances.

7.4.2.1　The spectrum of resonances from a physical viewpoint

The width of the resonances is associated with the probability that the system ionizes. Obviously this is a process one would expect an electric field should lead to. Fortunately for experimentalists, the character of resonances of the states strongly depends on the energy of the electronic motion. This is clear from Fig. 7.4 which plots the effective potential in a one-dimensional approximation. On the anode side of the nucleus, the electric potential is binding, while of barrier-type on the other side. The energy W_c at the top of the barrier is the classical ionization energy. For states with energies $W \ll W_c$, the coupling with continuum states will be exponentially small. The states in the internal well are quasi-bound states and transform into resonances for $W \gtrsim W_c$.

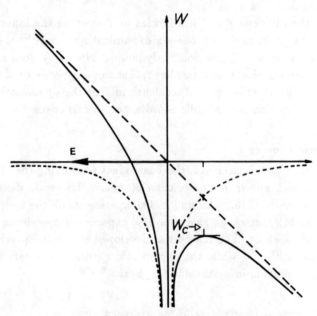

Fig. 7.4. One-dimensional approximation of the effective potential in the electric field problem. Dotted lines are respectively the electric field $(\mathbf{r} \cdot \mathbf{E})$ and Coulomb $(-1/r)$ contributions. The top of the barrier corresponds to the classical ionization energy W_c for which the Coulomb and Stark contributions are equal.

From Fig. 7.4, the classical ionization energy is negative and such that the Coulomb and Stark contributions are equal. W_c depends on the field value as $(E)^{1/2}$. The critical field (E_c) for which the n Coulomb state becomes a resonance is thus (a.u.):

$$E_c = W^2/4 = 1/16n^4 . \tag{7.7}$$

This obviously is in agreement with the phenomenological analysis of section 7.2.

The top (W_c, E_c) of the barrier corresponds to ξ values of the order of unity. This is the domain of Stark resonances[10,11] discovered in 1977.

7.4.2.2 *The hydrogen atom in an electric field — the standard treatment*

The standard treatment lies on the separability of Schrödinger's equation in parabolic or semi-parabolic coordinates.[70,71] A group theoretical approach will be indicated in sections 7.6 and 7.8. L_z being a constant of the motion, and $(\xi, \eta) = (r \pm z)$ the parabolic coordinates, with $\Psi = e^{im\varphi} g_1(\xi)g_2(\eta)/\sqrt{2\pi\xi\eta}$ one obtains:

$$\left\{ \frac{d^2}{d\xi^2} + (W/2 - 2V_1(\xi)) \right\} g_1(\xi) = 0$$

$$\left\{ \frac{d^2}{d\eta^2} + (W/2 - 2V_2(\eta)) \right\} g_2(\eta) = 0 . \tag{7.8}$$

The effective potentials V_i for the ξ and η coordinates are:

$$V_1(\xi) = (m^2 - 1)/8\xi^2 - Z_1/2\xi + E \cdot \xi/8$$
$$V_2(\eta) = (m^2 - 1)/8\eta^2 - Z_2/2\eta - E \cdot \eta/8 . \tag{7.9}$$

The three contributions are successively the rotational energy, the Coulomb energy and the Stark energy which is binding in ξ and not in η. The Z_i are the separation constants such that $Z_1 + Z_2 = 1$.

The spectrum and eigenfunctions being labeled with (W, Z_1, m), the problem amounts to finding the spectrum of the separation constants Z_i. Whether they are quantized or not depends on the shapes of the effective potentials which are schematically represented in Fig. 7.5.

The ξ motion is bounded leading to solutions of discrete type for a 1-dimensional potential. Bohr-Sommerfeld quantization can be used

$$\int_{\xi_1}^{\xi_2} [2(W/4 - V_1(\xi)]^{1/2}d\xi = \left(n_1 + \frac{1}{2} \right) \pi \tag{7.10}$$

where n_1 is the number of nodes of $g_1(\xi)$, as well as various other methods.[70,71,72,73]

The unbounded η motion (see Fig. 7.5) gives the whole spectrum the character of resonances. However this character depends on the importance of tunnelling through the barrier. For the low-lying states in the situation of Fig. 7.5, quantization can be performed through (7.10) for the η coordinate, introducing

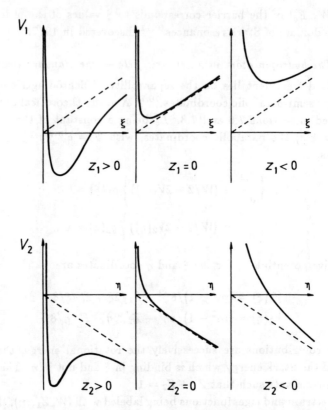

Fig. 7.5. Shapes of the $V_1(\xi)$ and $V_2(\eta)$ effective potentials in parabolic coordinates according to the separation constants Z_i. The ξ motion is always bounded while the η one gives the states the character of resonances. To the top of the barrier in the η motion is associated the ionization energy W_c (Z_2, m).[87] (Courtesy of C. Chardonnet, These 3eme Cycle (Paris, 1983), unpublished).

a new quantum number n_2. The spectrum is thus of quasi-discrete type (see subsection 7.4.2.1).

For higher values of the energy, the "resonance" character of the states is enhanced and the density of states can be schematically expressed through a Breit-Wigner formula $\Gamma^2(4(W - W_0)^2 + \Gamma^2)^{-1}$ where Γ is the ionization width. General solutions to this problem have been obtained only recently,[70,71,72] through various techniques.

An important point to notice from equation (7.9) and Fig. 7.5 is that the effective ionization field should be defined as:[74,75,76]

$$E_c = \frac{1}{16n^4} \cdot Z_2 \tag{7.11}$$

(as a first approximation[74]). It depends on the state (W, Z_1, m) under consideration. At constant energy, the higher the n_1 (or Z_1) values, the greater the stability of the states against ionization. This in turn is associated with some spatial localization of the wavefunctions, up-field or down-field, according as n_1 is maximum or minimum. From Fig. 7.6 it is clear that the electronic escape from "up-field" states should be more difficult as it occurs in a region closer to the nucleus where the binding action of the Coulomb field is enhanced.

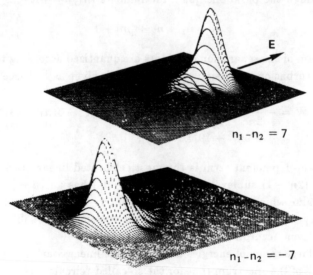

Fig. 7.6. Electronic density plots of the $(n = 8, m = 0)$ parabolic eigenstates (from Ref. 127). Only the two extreme cases $n_1 - n_2 = \pm 7$ are shown. They are associated with a spatial localization of the motion respectively up-field and down-field. Ionization from the up-field states at the top of the linear Stark manifold $(n_1 - n_2 = 7)$ is more difficult as it should take place in a region close to the nucleus where the effects of the Coulomb potential are enhanced (ionization occurs in the direction opposite to the field). (Reproduced with permission from Daniel Kleppner *et al.*, *Sc. Am.* **244** (1981) 108).

7.4.2.3 The crossing rule of the energy levels

The consequence of the separability of Schrödinger's equation is that the resonances labeled with (W, Z_1, m) will obey a general crossing rule, whatever the field and energy. This, being valid both for quasi-discrete and for ionizing channels, expresses the Stark dynamical symmetry for the orbital Coulomb problem. It is associated with the existence of an additional constant of the motion, built on the Lenz vector (see sections 7.6 and 7.8) which is field-dependent,[62] and expresses itself as $(Z_1 - Z_2)$.

7.4.2.4. The parabolic description of the quasi-bound states

The analytical description of the quasi-bound part of the spectrum is possible in the framework of perturbation theory.[70,71] In zero field, solutions to (7.8) and (7.9) are the well-known[59] parabolic eigenfunctions of the Coulomb problem which are introduced on more general grounds in section 7.6. The spectrum is labeled through the parabolic quantum numbers (n_1, n_2) with

$$n = n_1 + n_2 + |m| + 1 . \qquad (7.12)$$

The spectrum of the separation constants are quantized according to $Z_i = n_i/n$. The perturbation expansion of the energies is given as[70,71] (see section 7.6):

$$W = -\frac{1}{2n^2} + \frac{3}{2}n(n_1 - n_2)E - \frac{n^4}{16}(17n^2 - 3(n_1 - n_2)^2$$
$$- 9m^2 + 19)E^2 + \dots . \qquad (7.13)$$

The first field-dependent term is the most celebrated linear Stark effect. Each among the $(2n - 1)$ sublevels with $q = (n_1 - n_2)$ has an $(n - |q|)$ degeneracy on the m value which is removed at second order. The spectrum has a residual Kramers degeneracy on $|m|$. At fixed m, the spacing of the sublevels is $2\omega_E$ as q is odd or even according to (n, m).

Over a large range of energies and fields, the linear Stark effect is the dominant feature of the spectrum (but for the so-called "circular" states with $q = 0$). The quadratic term becomes comparable to the linear Stark term for fields of the order of the classical ionization field. Hence (7.13) applies to the whole quasi-discrete part of the spectrum. In spite of its asymptotic nature,[70,77] the series can also be used for describing the resonance spectra, through Padé resummations.[78,79,80]

Remark that, owing to the crossing rule, the states can be labeled through the (n_1, n_2, m) zero-field parabolic quantum numbers, whatever the field strength. Their energy curves will never anticross·· (but for the role of relativistic and Q.E.D. corrections not discussed here). The states with maximum n_1 values rise the faster with the E field, from (7.13). But as said in subsection 7.4.2.2, they are also the most stable against ionization. The electronic density plots on Fig. 7.6, of the parabolic eigenfunctions (the zeroth order approximation to the eigenfunctions in field), already show this property from their spatial localizations.

7.4.2.5 Distribution of oscillator strength and photoionization

Selection rules as well as oscillator strengths, in optical excitation, greatly depend on the symmetries of the initial and final states. For example, in the low field limit, they have a parabolic type symmetry which makes possible the excitation of several (n_1, n_2) subcomponents of the Stark manifold. The high selectivity on the l value of optical excitation techniques is lost, precisely because l is not a good quantum number. Indeed, the only selection rule is on the L_z value (the parity of the states being not defined).

In addition, the final states are likely to have very different properties against ionization. Hence the use of "spontaneous" (vapors) or "field forced" (beams) ionization should lead to very different appearances of the same spectra. Lacking some excitation - detection scheme really complying with the inner symmetries of the system and affording selective state detection, the interpretation of the experimental data may be challenging and requires numerical simulations. (See sub-section 7.4.4.3).

Analytical predictions made in the low field limit can be used as a guess in higher field conditions. An experimental spectrum in low field conditions (linear Stark effect) is shown in Fig. 7.7.[22] The spacing between the 39 components of the $m = 3$ manifold is $2\omega_E$. Such a distribution of oscillator strengths is typical and can be reproduced from the analysis of subsection 7.4.2.4.

7.4.3 One-electron Rydberg atoms in electric fields— the role of non-Coulombic corrections

This is an essential point to address in connection with experiments. The short range non-Coulombic correction to the potential breaks the Stark dynamical symmetry.

7.4.3.1 The problem to solve — Harmin's solution

For a core of closed shells, one can assume the non-Coulombic correction $v(r)$ to be centrally symmetrical, and short range. Equations 7.8 no longer hold. As the

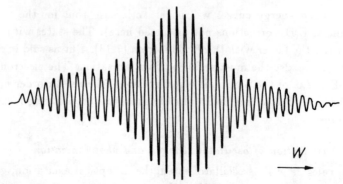

Fig. 7.7. Structure of the ($n = 42$, $m = 3$) linear Stark manifold of caesium using FM laser spectroscopy (resolution 1 GHz). The 39 Stark components spaced with $2\omega_E$ are exhibited (electric field value $E = 7.30$ V/cm). The signal is the derivative of the usual absorption profile (Gay, Steitz, Hartman, 1985, unpublished).

external field potential is long range, the asymptotic solution is still basically the Stark solution for the hydrogen atom. But near the core the symmetry should evolve from a parabolic type to a spherical one as $q r \cdot \mathbf{E} \ll v(r)$. The problem of the connection of the solutions across the transition region at r_0 has been solved by Fano[81] and Harmin.[82,83,84] This results in a coupling of the Stark parabolic channels compared to the pure Coulombic situation.

7.4.3.2　Energy diagram — the non-crossing rule

L_z is still a constant of the motion. At fixed m, all the energy levels should anticross as the Stark dynamical symmetry is broken. The energy spectrum being determined by the asymptotic behavior of Schrödinger's equation, at sufficiently high field strength, the diabatic behavior of the energy levels should look like that in the hydrogenic case. The argument supposes that the quantum defects are small (in terms of energy) compared to the external field perturbation.

7.4.3.3　Fano profiles and ionization

The coupling of the channels responsible for the anticrossings also takes place between quasi-bound states and continua. Hence, states which in the hydrogen situation would be stable are coupled with all the others, less stable. Huge modifications in the ionization mechanisms are thus to be expected.

Optical excitation of such states will lead to the formation of Beutler-Fano profiles of which this is the generic mechanism.

One should also remark another difference with the pure Coulomb situation. The absorption of radiation involves regions close to the core in which the symmetry is of spherical type rather than parabolic. Hence selection rules and oscillator strengths will be modified.

7.4.4 Experimental manifestations in Rydberg atoms

Indeed most of the advances in the area have an experimental origin. Their interpretations only involve basic physical mechanisms. These studies completely renewed the understanding of the electric field action on the atom and the concepts of ionization.

7.4.4.1 Linear Stark effect on quasi-hydrogenic species

As remarked in subsection 7.3.1, a convenient choice of the optical excitation process leads to situations as hydrogenic as desired. A criterion is that the quantum defects of the states involved in the Stark mixing be energetically small compared to the external field Hamiltonian. L_z being conserved in the Stark mixing, it proves sufficient to excite states with $m \geq 3$ (as the quantum defects for states with $l \geq 3$ in alkali atoms are negligibly small). For light alkali atoms, $m = 1$ states (lithium) and $m = 2$ states (sodium) are also a convenient choice.[85-87]

The linear Stark patterns of caesium $m = 3$ states are shown in Fig. 7.8 for n around 60[22] (the points diverging from P and D levels are an artefact of the excitation scheme, and should be ignored). This is a vapor type experiment using c.w. dye laser excitation and low field strengths $(E \leq 20 \text{ V/cm})$. The resolution is 1 GHz (Doppler-limited). The maximum quantum defect of the Stark mixed states is $\delta = 0.033$ (Fig. 7.7 refers to the same experimental conditions). For fields greater than 1 V/cm, this situation becomes highly hydrogenic. Only the Stark components at the center of the manifold (with $n_1 \simeq n_2$) are seen on the plot. Their spacings are $2\omega_E$. The linear Stark behavior is followed to within 1%. For the largest field values, the systematic red-shift of the components is due to the small effect of the quadratic hydrogenic Stark term in (7.13).

Such behavior does not preclude the existence of non-hydrogenic effects. For example, anticrossings exist in the inter n mixing regime, but in the range of 300 MHz at maximum they are hardly noticeable on the plot on Fig. 7.8. Using subdoppler techniques, the patterns would drastically change.

Such a dramatic alteration is shown on Fig. 7.9 for $m = 1$ caesium states. The P and D states with quantum defects of 3.57 and 2.57 are now involved in the Stark mixing. The patterns consequently become irregular, with strong

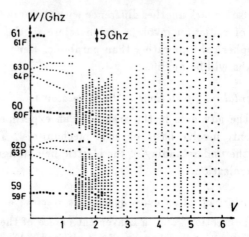

Fig. 7.8. Linear Stark map of atomic caesium around $n = 60$ ($m = 3$ quasi-hydrogenic states) at low field strengths ($E = 2.37V$ (V/cm) ≤ 15 V/cm). The resolution is about 1 GHz (Doppler-limited).[22] Series associated with other m values are seen owing to the pecularities of the excitation scheme.[87] (Courtesy of C. Chardonnet, These 3eme Cycle (Paris, 1983), unpublished).

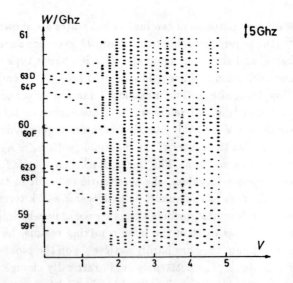

Fig. 7.9. Low field Stark map of $m = 1$ caesium states around n=60. As a result of the non-hydrogenic character of the situation there are strong anticrossings which make the patterns completely irregular even at low fields[87] (compare with Fig. 7.8).

anticrossings. This is the result of "tuning the quantum defects" through optical polarization.

7.4.4.2   Generation of pseudo profiles in the quasi-bound spectrum

Non-hydrogenic effects in the low field limit can be understood in terms of the interaction of one discrete "non-hydrogenic" channel with a set of discrete states acting as a quasi-continuum.[87] The latter are the components of the linear Stark manifold (see Fig. 7.10). The mechanisms of Stark re-distribution are thus analogous to those in the generation of Fano-profiles.[88,89,90] Actually, this is exactly the situation which Fano investigated in 1935.[89]

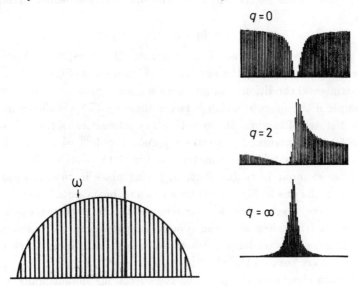

Fig. 7.10. The basic mechanisms of Stark re-distribution of non-hydrogenic states at low fields. Their interaction with the incomplete linear Stark manifold generates pseudo profiles in the intensity distributions of the discrete lines. Several examples are shown for various values of the Fano parameter q. This generalizes the anticrossing regime. (Courtesy C. Chardonnet.[87])

The basic physical parameters are the coupling V with the discrete channels, their spacing $\omega (\omega = 2\omega_E)$ and the number p of states (or the extension $p \cdot \omega$ of the equally spaced sublevels). The probability of re-distribution of the discrete state onto the q channel is thus

$$|a_q|^2 = \left[1 + \pi^2 \frac{|V|^2}{\omega^2} + (W_q - W_0 - \Delta)^2 / |V|^2 \right]^{-1} \qquad (7.14)$$

where Δ is the shift arising from the non-resonant part of the interaction. It follows that there are three types of regime in the Stark re-distribution. The weak coupling regime $(V/\omega \ll 1)$ is characterized with anticrossings as in Fig. 7.9. In the intermediate coupling regime $(V/\omega \cong 1)$ the discrete state is re-distributed onto several states of the manifold (the number of which being $\pi |\frac{V}{\omega}|^2$). Finally, the re-distribution is complete when $\frac{V}{\omega} \gg 1$ and $V^2 \gg p\omega^2$. This is for example the situation for the F state in the plots on Figs. 7.7 and 7.8.

Due to the mixing of the channels and according to the excitation-detection scheme, interference effects will affect the patterns and lead to Beutler-Fano profiles

$$I = (q + \varepsilon)^2/(1 + \varepsilon^2) \qquad (7.15)$$

where q measures the degree of interference. But here, this affects a set of discrete lines of which I is the envelope.[32] This is exemplified in Fig. 7.10.

Experimental conditions can give such a pseudo profile a continuous aspect, for example if the spectral width is larger than ω_E. This is shown in Fig. 7.11 for $m = 1$ states of caesium at low fields. The interaction of the $54P$ state with the $n = 50$ Stark manifold generates a pseudo profile[90] with a width 3.5 GHz and $q = 2$ (the situation is schematized in Fig. 7.11 — region I).

Another example of re-distribution, taking place in the intermediate coupling case is, the one in Fig. 7.12 from an atomic beam experiment.[91] ω_E being larger than the spectral width, this restores to the phenomenon its discrete character. It is manifested in the complete disappearance or enhancement of several discrete lines on large areas of the map. Both linear Stark effect and anticrossings are seen on the plot.[87]

It is quite clear that the previous considerations should apply also to the interaction of several discrete states with several manifolds. This leads to an important physical effect called stabilization or accidental decoupling of a state from the manifolds, at non-zero field values. This is shown in Fig. 7.13 and the situation is the one of region II in Fig. 7.11. At the anticrossing of two non-hydrogenic states coupled with several manifolds, one of them is decoupled as shown in Figs. 7.11 and 7.13 and leads to a sharp line, while the other one is more strongly coupled to the manifold states giving the large Fano profile. For $n = 59$ and 61, the decoupling is not perfect, and the line is broader.[87,90]

7.4.4.3 The Stark resonance spectrum

The previous phenomena which affect the quasi-bound spectrum for non-hydrogenic atoms are simple illustrations of the phenomena taking place close

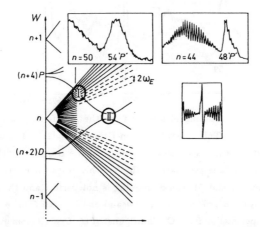

Fig. 7.11. Stark redistribution of nonhydrogenic states in the experimental situation of caesium (Doppler-limited spectrum).

–Region I leads to the generation of pseudo profiles. This is shown on the top insets. For $n = 50/54$ 'P' ($E = 4.5$ V/cm) the Stark components are not resolved while for $n = 44/48$ 'P' ($E = 8.4$ V/cm) the origin of the interference phenomena becomes clear. The situation is similar to the one theoretically investigated by Fao (1935).[89]

–Region II is associated with decoupling effects of non-hydrogenic states from the manifolds, at non-zero field values. This is shown in the inset for the $n = 54$ manifold (FM spectroscopy; $E = 7.1$ V/cm). Also see Fig. 7.13.

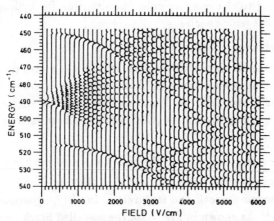

Fig. 7.12. Stark map of $m = 1$ lithium states from an atomic beam experiment.[91] The mechanisms previously described allow one to understand the disappearance of the discrete lines on large areas of the map.[87] The resolution is here far greater than w_E (the field strengths are three orders of magnitude greater than in the situation in Fig. 7.11.) (Reproduced with permission from Myron L. Zimmerman *et al.*, *Phys. Rev.* **A 20** (1979) 2251, (American Physical Society)).

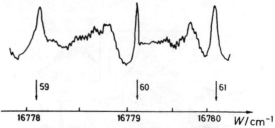

Fig. 7.13. Accidental decoupling of a state from the manifolds (caesium $m = 1$ states, $n \simeq 60$; $E = 4.25$ V/cm). This refers to region II in Fig. 7.11.[90] The sharp line associated with the decoupled state has a width of the order of the Doppler width (1 GHz) while the broad pseudo-Fano profile is associated with the other state (more strongly coupled to the manifolds). For $n = 59$ and 61, the decoupling is not perfect and the lines broader. Such a phenomenon obeys a n^5. $E = $ Const. law and can be seen from $n = 30$ to $n = 110$.[87] (Reproduced with permission from C. Chardonnet *et al.*, *Optics Commun.* **51** (1984) 249).

to the classical ionization threshold which involve now the mixing of resonances and continua.

(a)   Stark map across the classical ionization threshold

The dotted line in Fig. 7.14 represents the classical ionization field (7.11) associated with the saddle point in Fig. 7.5.[91] Close to this line, the tunnelling through the barrier becomes important and the resonance character of states is enhanced. Owing to the crossing rule, the situation for hydrogen differs from that for non-hydrogenic species.

(b)   The resonance character of states

Even below the classical ionization threshold, the states do present a width due to their coupling with ionizing channels. This is shown in Fig. 7.15 for rubidium atoms[92] and leads to Fano profiles. Similar effects for the $n = 4$ hydrogen states, at huge field strengths[93] have also been seen in the Los Alamos experiment on fast beams.

(c)   Stark resonances close to the classical ionization threshold

When crossing the saddle line, it was long thought that the response of the atom should become smooth and featureless. In 1977, experiments proved just the opposite.[10,11] As shown in Fig. 7.16, the so-called Stark resonances spectra extend up to positive energies.[94] The spacing of the resonances was found to scale approximately as $E^{3/4}$, for most alkali atoms.[95-98] The physical origin of the phenomena is the one described in section 7.2; it manifests the onset of the strong mixing regime in the electric field problem.

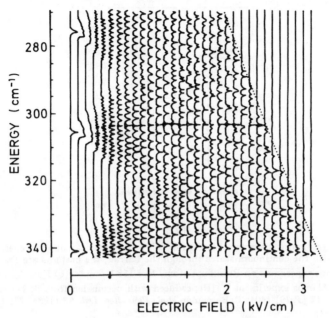

Fig. 7.14. Stark map across the ionization threshold on lithium $|m| = 1$ states.[34] The dotted line corresponds to the classical ionization field and top of the barrier. (Reproduced with permission from Michael G. Littman *et al.*, *Phys. Rev. Lett.* **41** (1978) 103 (American Physical Society)).

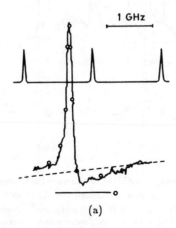

(a)

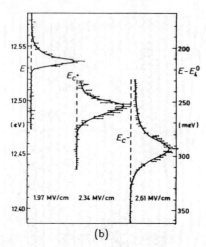

(b)

Fig. 7.15. The resonance character of the states below the classical ionization energy. Due to the coupling with the continuum states through the barrier, the patterns are Fano profiles. (a) Atomic beam experiment on Rubidium atoms $E = 158$ V/cm.[92] (b) Hydrogen fast beam from the Los Alamos experiment.[93] (Reproduced with permission from S. Feneuille *et al.*, *Phys. Rev. Lett.* **42** (1979) 1404; T. Bergeman *et al.*, *Phys. Rev. Lett.* **53** (1984) 775 (American Physical Society).

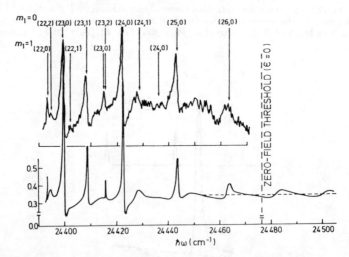

Fig. 7.16. The Stark-resonance spectra close to the zero-field threshold of sodium atoms ($E = 3.6$ kV/cm).[94] The assignment of lines is through the zero field parabolic quantum numbers (n_1, n_2). This compares fairly well with Harmin's theory.[83] Numerous Fano-shaped profiles can be seen which exemplify the role of non-Coulombic corrections in the sodium situation. (Reproduced with permission from Ting Shan Luk *et al.*, *Phys. Rev. Lett.* **47** (1981) 83, (American Physical Society)).

One of the most striking achievements of recent experiments on the hydrogen atom is shown in Fig. 7.17. These are Stark resonances exhibiting the $E^{3/4}$ law[99] (section 7.2.1; the spacing of the resonances should be proportional to $E^{3/4}$). Compared to non-hydrogenic conditions as in Fig. 7.16, there are almost no substructures. The quantum origin of the Stark resonances in the hydrogen atom situation lies in the excitation of long-lived resonances close to the threshold. From section 7.4.2 the most stable states are those with maximum n_1 (or Z_1) values with up-field spatial localization of the electronic density (upper states of the Stark multiplet in the low field limit). Moreover these states are not coupled to the others owing to the crossing rule (or the Stark dynamical symmetry).

The Stark resonances are the expression of the strong mixing regime for states (n_1, n_2, m) pertaining to the same exact "parabolic" type symmetry. As labeled through the zero-field parabolic quantum numbers, the components of a series are thus $(n_1, n_2, m), (n_1 + 1, n_2, m) \ldots$ and obey the $E^{3/4}$ law. In the lack of selection rules in optical excitation, several series with various n_2 values will be superimposed, with similar spacings but very different properties against ionization. They will not interact and their contributions to the signal are likely to decrease as n_2 increases. Moreover the sharp or broad aspect of the lines (Fig. 7.17) depends on the stability of the states, hence on the ionization field for each symmetry type.

These elements allow one to understand the seemingly more complicated patterns in Fig. 7.16. The basic difference is that the hydrogenic channels are now coupled by non-Coulombic corrections to the potential. In addition optical excitation of the various channels is usually possible. The situation, analogous to subsection 7.4.4.2 for the quasi bound spectrum leads to the formation of Fano-profiles on a large scale. The various series involved in the Stark resonance mechanism for hydrogen are now interacting. The most stable states (n_1 max., n_2 min.) lose their stability. Combined with interference effects in the optical process, this may lead to the local disappearance of lines from the most stable states while the others are enhanced. In that sense, the spectrum is richer but also far less regular than in the hydrogen case. Theoretical treatments allow a complete simulation of such spectra[83,84] as shown in the figure.

(d)   Stabilization of states through channel interactions

Another striking aspect of this problem, for non-hydrogenic atoms, is the creation of quasi-bound states isolated (without strong interactions) in a sea of resonances or continua. An elementary example for the quasi-bound spectra

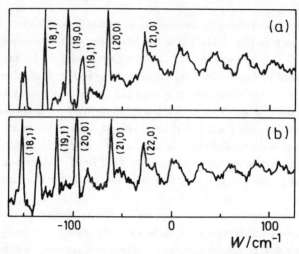

Fig. 7.17. The Stark-resonance spectrum for the hydrogen atom[99] close to the zero-field threshold. The appearance is much more regular than in Fig. 7.16. The series with $n_2 = 0$ is dominant and but for relativistic corrections the various series are not likely to interact. (m_1 = 0 states; (a) $E = 8$ kV/cm (b) $E = 6.5$ kV/cm). (Reproduced with permission from Wallace L. Glab and Munir H. Nayfeh, *Phys. Rev.* **A31** (1985) 530, (American Physical Society)).

has been given in subsection 7.4.4.2. Here it involves, along the same lines, resonances and continua. For a non-zero field value, as a result of interference effects, a resonance can be locally decoupled from the others, thus appearing as a sharp feature in the spectrum. Various mechanisms for such a process to occur have been discussed.[97,98]

7.4.4.4   The mechanisms of ionization

Obviously, the ionization process is very different according to the non-hydrogenic character of the species. From Fig. 7.14 the routes to ionization strongly depend on the non-crossing rule (hence on the symmetries). Diabatic ionization will occur as in the hydrogen case according to the crossing rule while adiabatic ionization will follow the non-crossing rule. Both the rates and the field strengths for ionizing the system strongly depend on the conditions,[74,76] except for the hydrogen atom for which the crossing rule holds. Numerous experimental demonstrations have been given.[74,76,100] The rates not only evolve according to the diabatic (hydrogen case) or adiabatic character of the ionizing pulse, but also can be tuned according to the m value. Usually several routes are possible.[100]

Consequently, when field ionization is used on a system, of which the structure is not known (atom in a **B** field, for example), the interpretation may not be simple.

7.4.5. Conclusions

The previous features have led to completely renewed conceptions on the mechanisms of ionization which by no means can be considered a single, invariable process. The notion of ionization limit has lost its sharp character, as a certain continuity of the phenomena exists across it. It also depends on the symmetries of the states.

Moreover, the electric field can be thought as exerting a certain binding action which manifests itself in the experimental data and in some sense (see section 7.8) makes the atom a tunable system, both as concerns the symmetries and structure of the energy levels. Such properties have already been applied to the study of collisions of Rydberg atoms[101,102] and to the study of natural autoionization[103,104] which is modified due to the discretization of the continuous channels. But this is only the very beginning of applications of these ideas.

7.5   Magnetic Field Structure of One-Electron Rydberg Atoms — the Diamagnetic Behavior

The theory starts around 1982 along non-traditional lines which are those described in the last sections. In contrast to the electric field problem, the use of higher concepts in which the full Coulombic dimension of the atom is incorporated is required here. As a matter of fact, the experimental data collected for ten years in the area will be interpreted only in sections 7.7 and 7.8.

7.5.1   Historical survey

The interest in using magnetic fields for studying atomic properties arose early with the works of Faraday,[5] Zeeman,[105] Van Vleck[106] among many others. The Zeeman effect, associated with the paramagnetic interaction, is still one of the finest tools for studying the properties of atoms, molecules and solids. Its role for establishing the foundations of quantum mechanics, in the discovery of the electron spin, is famous. It is also the dominant magnetic effect for low lying atomic states.

The diamagnetic interaction, our present concern, obeys different rules, and was considered in an atomic and molecular physics context in the most celebrated book of Van Vleck dealing with magnetic susceptibilities.[106] Landau in

1930[107] fully revealed the importance of diamagnetism in establishing the quantization law which charged particles obey in a magnetic field. The binding diamagnetic force leads to an oscillator-like structure with a spacing $\hbar\omega_c$. Around 1950, this became of great importance in solid state physics (semiconductors, excitons, impurities) and recently, some variations on the theme (including electronic correlations) led to the fractional quantized Hall effect.... [108,109]

The first theoretical and experimental attempts at the role of diamagnetism in atomic physics took place around 1939.[39,110] Jenkins and Segré took advantage of the B_c/n^3 scaling law obeyed in the Rydberg ladder, as well as of the high fields from the first Lawrence Cyclotron. Up to 1981, the accompanying theoretical approach by Schiff and Snyder was also the state of the art, though considerable efforts had been made in a solid state physics context in the meantime,[111] especially in the high field regime.

The experimental discovery, in 1969, of the quasi-Landau resonances, by Garton and Tomkins was the clue that something fundamental and unexpected was missing in our understanding of the problem. Close to the zero-field threshold of barium, in a magnetic field, the spectrum exhibited a system of resonances, with a spacing $\frac{3}{2}\hbar\omega_c$, extending far into the continuum regions. The phenomenon took the force of law when it was shown to exist whatever the atomic species.[112,113,114] But even a qualitative interpretation of the $\frac{3}{2}$ factor was missing for a long time.[115] This intriguing manifestation of the strong mixing regime triggered new experimental developments which in turn led to a theory. It turned out that nothing but the Zeeman effect was understood in the magnetic problem.

7.5.2   The magnetic structure of atoms — standard analysis of a non-separable problem

From (7.2) and (7.3) the Hamiltonian in the symmetrical gauge is written down as (atomic units):

$$H = \frac{p^2}{2} - \frac{1}{r} + \frac{\gamma}{2}L_z + \frac{\gamma^2}{8}(x^2 + y^2) . \qquad (7.16)$$

But this is already an approximation.[3] The two-body problem is not separable in the usual sense.[116,117,118] This leads in particular to the so-called motional Stark effect and alterations of the gyromagnetic factor. Equation (7.16) assumes the proton to be infinitely massive.

The only constants of the motion are L_z and parity, whence parity P_z along the **B** field is defined. L^2 does not commute with the diamagnetic interaction.

The Hamiltonian is not separable in any of the thirteen sets of natural coordinates in $R(3)$. There is no mathematical accident, in contrast to the electric field problem.

What can be said *a priori* hardly goes further than the phenomenological analysis of section 7.2. Scaling as $n^4 B^2$, the diamagnetic interaction can completely overwhelm the Coulomb one. The quantum spectrum should evolve between a Coulomb and Landau signature (oscillator-like), and from section 7.2 this is valid whatever the absolute strength of the field.

Lacking a third integral of the motion, the problem is a very difficult one which can hardly be tackled through standard theoretical methods or computations. The diamagnetic interaction couples any Coulomb state (n_0, l_0) to almost any other one (n, l) including continuum states. The only selection rule is on the l value: $\triangle l = 0, \pm 2$. There is no selection rule on the n value. The sizes of the basis set to consider in the strong mixing regime exceeds any practical possibilities. But experimentally, there are simple laws in this regime where the non- separable character of the Hamiltonian is at its most daunting.

There are two approximate treatments which are possible. In the high field limit, one is led to consider the perturbation of a Landau motion by the Coulomb interaction. The harmonic type Landau spectrum has an infinite degeneracy in the L_z value while the free motion along the **B** field gives rise to a continuous spectrum.[3] The perturbation of the motion due to the Coulomb force was studied early in solid state physics[119,120] in the framework of an adiabatic approximation. Some conclusions are discussed in Ref. 3.

The low field limit corresponds to the perturbed Coulomb spectrum. The diamagnetic interaction removes the n^2 degeneracy of the n Coulomb shell. This is basically the treatment of Schiff and Snyder[110] which can be extended through numerical work. But no insights into the structure can be obtained this way. One deduces only that there is an "inter-l-mixing" regime (when $\frac{1}{n^3} >> \gamma^2 n^4$) and next the inter-$n$-mixing regime. As the energies and matrix elements scale as $B^2 n^4$, this suggests that higher order terms should become at least as important (!) and in fact, they do.

For the sake of completeness, we remark that here, we are not concerned with other magnetic effects (or relativistic corrections) in the atom such as those associated with the spin of the electron, of the nucleus... These do not present any particular difficulty. We are here concerned with the orbital structure of the magnetic problem, which has required about 60 years for being understood. Far less time will be needed for building a fully relativistic version. Nevertheless, there are already experimental examples of crossed effects between diamagnetism and spin, nuclear motion...[73,121,122]

7.5.3 The diamagnetic behavior from experiments

The most important features were discovered experimentally or from numerical experiments. They are discussed below, prior to their interpretation (see section 7.7). They show there was a growing need for a change in the theoretical concepts in order to interpret simply, simple results.

7.5.3.1 The low field inter-l-mixing regime

Orbital Zeeman effect plays a role somewhat equivalent to linear Stark effect (see section 7.7). A difference is that the optical excitation of one given Zeeman sublevel m is usually possible and that the $\frac{\gamma m}{2}$ paramagnetic energy is constant throughout the spectrum (as L_z is an exact constant of the motion).

As the diamagnetic interaction H_D does not commute with L^2, this removes the $n - |m|$ degeneracy of the m Zeeman sublevel and results in the appearance of a "forbidden" line structure, with a spacing proportional to $n^4 B^2$. The validity is $\gamma^2 \cdot n^7 << 1$.

Several experimental aspects of the diamagnetic structure are shown in Fig. 7.18, for various atomic species and optical excitation conditions.[46] The differences in the patterns have their origins in non-hydrogenic effects which as in section 7.4 affects both the energies and oscillator strength distributions. According to the quantum defects of the diamagnetically mixed states, the whole ensemble of regimes described in section 7.4 should take place with, in particular, generation of Fano profiles.

Once again, the hydrogenic situation is the reference. It is nearly realized in caesium in Fig. 7.19.[21,123] The quantum defects of the $m = 3$ diamagnetically-mixed states are smaller than $\delta(F) = 0.033$ and field conditions are such that $\frac{2R}{n^3} \cdot \delta(F) << H_D$. While the positions of the lines are mostly unaffected by this quantum defect correction, the oscillator strengths still differ from the numerically-produced hydrogen pattern (Fig. 7.20).[123] The intensity of the highest energy line should be approximately twice the second one. To some extent all the patterns in Fig. 7.18 should tend to the hydrogenic one in Fig. 7.20 at sufficient field strengths, as the diamagnetic contribution is long ranged compared to the core.

The dramatic alteration of the patterns in Fig. 7.21 compared to the previous ones (Figs. 7.19 and 7.20) illustrates the role of exact symmetries (parity along the z axis) on the appearance of the diamagnetic band.[124] The lithium $m = 1$ states situation is highly hydrogenic. Notice the drastic change according to

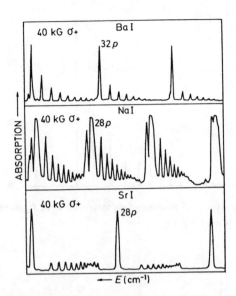

Fig. 7.18. Atomic diamagnetism in the low-field inter-l-mixing regime for various atomic species and even P_z parity.[46] (Reproduced with permission from K. T. Lu *et al.*, *Phys. Rev. Lett.* **41** (1978) 1034, (American Physical Society)).

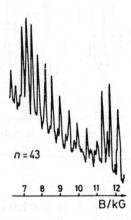

Fig. 7.19. Diamagnetic patterns of caesium $m = 3$ states ($n = 43$, even P_z parity) as a function of B field (fixed laser frequency). The situation is nearly hydrogenic but perturbations of the intensities of the lines are noticeable[21,123] (unpublished 1986).

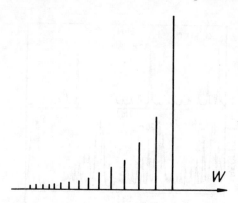

Fig. 7.20. Numerically-produced diamagnetic pattern for even P_z parity, assuming the quantum defects are zero. [Fixed B field].

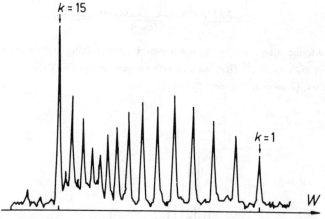

Fig. 7.21. The low field diamagnetic spectrum of lithium ($n = 31$, $m = 0$, odd P_z) for $B = 19.4$ kG. The situation is quasi-hydrogenic. The differences in the oscillator strength distributions, compared to Fig. 7.19, come from the odd character of the states in parity along the z axis.[124] (Courtesy of S. Liberman and J. Pinard from P. Cacciani, These 3eme Cycle (Orsay, 1984) unpublished).

whether the eigenfunctions have or do not have a node in the $z = 0$ plane; whether P_z is odd (Fig. 7.21) or even (Figs. 7.19 and 7.20).

A certain understanding of the features of the inter-l-mixing regime was possible through numerical diagonalizations of the diamagnetic Hamiltonian. For given m and parity, the m Zeeman manifold (with $n - |m|$ degeneracy) bursts out into about $(n - |m|)/2$ sublevels which can be labeled through (n, m, P_z) and an index k. It replaces l in some sense and can be given a physical meaning in the approach of section 7.7 only. Various conventions have been used.[125,126] We assume here the $k = 1$ states have the highest diamagnetic energy. The diamagnetic eigenvalues are $E_{nkm} = C_{nm}(k) \cdot B^2$ and it was shown[123] that the quantity $C_{nm}(k)/n^4$ presents a universal behavior. The (n, k, m, P_z) diamagnetic eigenfunctions are field independent. They can be considered as a class of eigenfunctions of the Coulomb problem which complies with the symmetries of the diamagnetic interaction at low fields. They are certainly not of a type separable in spatial coordinates. But plots of nodal surfaces as well as of electronic density[126,127] allows us to characterize their universal behavior. From Fig. 7.22, the nodal plots suggest that the symmetry in the diamagnetic band evolves from rotation to vibration. In addition, all the nodal lines anticross which confirms the lack of separability in spatial coordinates. But obviously the sizes of the anticrossings are fairly small compared to the extension of the eigenfunctions. In any case such behavior differs somewhat from that of the usual spherical eigenfunctions.[123] The electronic density plots allow confirmation of these views, showing that the spatial localization of the motion is different at the top and bottom of the band (Fig. 7.23). At the top, the motion tends to localize in the $z = 0$ plane perpendicular to the field. The extension of the $k = 1$ state along the B field scales as $n^{3/2}$ and thus, it is squeezed compared to spherical states.[128] Hence the diamagnetic energy is a maximum. In contrast, at the bottom of the band, the motion is localized along the B field and $\langle x^2 + y^2 \rangle$ is a minimum.

These were the first clues to the existence of a ro-vibrational structure in the diamagnetic band associated with a certain localization of the motion. This in turn does have some counterpart on the oscillator strength distribution. But one could not make further progress this way. New concepts were required.

7.5.3.2   The inter-n-mixing regime —
merging of the diamagnetic manifolds $(\gamma^2 \cdot n^7 \gtrsim 1)$

An experimental plot of this regime in which the diamagnetic manifolds merge is shown in Fig. 7.24. The spectrum becomes extremely complicated though remarkable features exist.[129]

An important question is to determine whether the manifolds anticross or not, in connection with the search for a dynamical symmetry. A puzzling answer to this question was the numerical discovery of the existence of exponentially small anticrossings, scaling as e^{-2n} between the states at the bottom of the

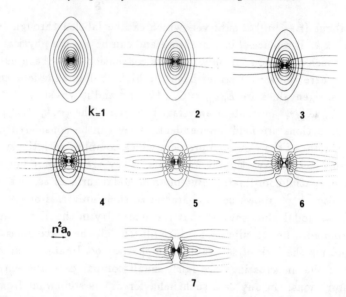

Fig. 7.22. Nodal surfaces plots of the diamagnetic eigenfunctions for $(n = 17, m = 3,$ odd parity) states. As parity P_z is even, there is no node in the $z = 0$ plane. The symmetry evolves from rotation to vibration from $k = 1$ to $k = 7$ (Reproduced with permission from D. Delande *et al.*, *Phys. Lett.* **A82** (1981) 393, (American Physical Society)).

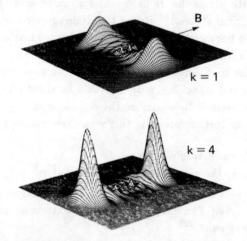

Fig. 7.23. Electronic density plots of the diamagnetic eigenfunctions $(n = 8, m = 0,$ even $P_z)$ for the $k = 1$ and $k = 4$ states at the top and bottom of the band respectively. While the states at the top of the band exhibit a certain localization in the $z = 0$ plane perpendicular to the field, the localization is along the B field at the bottom of the band (Reproduced with permission from Daniel Kleppner *et al.*, *Sc. Am.* **244** (1981) 108).

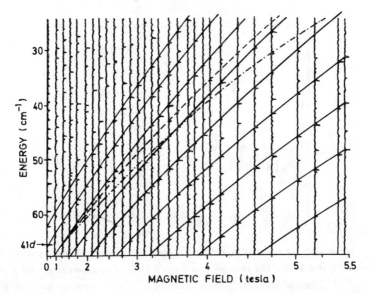

Fig. 7.24. The inter-n mixing regime (Na, $m = -2$ states) in diamagnetism (atomic beam experiment)[129]. Remarkable features still exist, especially a system of dominant lines branched to $k = 1$ states in the low field limit. (Reproduced with permission from Jarbas C. Castro *et al.*, *Phys. Rev. Lett.* **45** (1980) 1780, (American Physical Society)).

$(n + 1)$ band and those at the top of the n one.[125,126] From their nodal plots in Fig. 7.22 they do really present different symmetries. Various interpretations of this feature have been given.[130,131,132] But there is no exact dynamical symmetry, rather a transition to chaos as discussed in Ref. 133, at higher energy, of which the exponentially small anticrossings are the precursors.

In this regime, as a result of the interactions with the other manifolds and continua, the diabatic behavior of the energy levels should begin to depart from the B^2 law as seen in Fig. 7.24. Indeed it should scale as B in the Landau limit.

7.5.3.3 The quasi-Landau regime close to the zero field limit

Several aspects of this regime are shown in Fig. 7.25 and Fig. 7.26 for different experimental conditions.[16,17,21] In spite of the tremendous number of lines there are fortunately several distinguishable features, especially a system of dominant lines.

At constant field, the spacing of the dominant lines was experimentally shown to be about 1.5 $\hbar\omega_c$.[21,112,113,134] But according to conditions, especially the resolution, the phenomena may look like broad modulations which

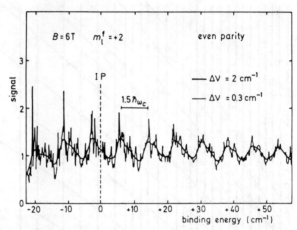

Fig. 7.25. Quasi-Landau resonances of the hydrogen atom ($m = +2$, even parity; $B = 6$ T) across the ionization limit. The resolution is about 6 GHz. The spacing between the clusters is 1.5 $\hbar\omega_c$.[16,17] (Reproduced with permission from A. Holle *et al.*, *Phys. Rev. Lett.* **56** (1986) 2594, (American Physical Society)).

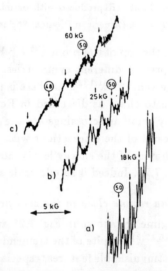

Fig. 7.26. Aspects of the quasi-Landau spectra in B field scans at nearly constant electronic energy W^{21} ((a) $W = -12.56$ cm^{-1} (b) $W = 0$ (c) $W = 99.53$ cm^{-1}). For zero energy, the quantization in field is $n^3 \cdot B/B_c = 1.56$ to within 5% (Reproduced with permission from J. C. Gay *et al.*, *J. Phys. B. Lett.* **13** (1980) L720).

are sharper below zero energy while smoother above, or entangled sets of sharp lines. In the latter case, the above periodicity still exists but the intensity distribution is irregular. Accidental coincidence with secondary lines may explain the perturbation of the spectra,[4] as well as general Fano-type mechanisms as discussed in section 7.4.[135] These intriguing features are actually a manifestation of chaos in a quantum system (see (7.9)).

At constant energy (or fixed laser frequency), typical aspects of the quasi-Landau spectra in B field scans are shown in Fig. 7.26. Again, there are numerous sharp substructures.[123] At zero energy, the experimentally determined quantization in the field is $n^3 B/B_c = 1.56$ (to within 5%).[21] For positive values of the energy of the order of 100 cm^{-1}, a Landau type $n \cdot B = $ const. quantization is fulfilled and the spacing is of the order of $1.1\hbar\omega_c$.[21,123,136]

These are the two-fold aspects in field and energy, of the quasi-Landau spectra. For a long time, the only understanding of the foregoing laws was through a semi-classical quantization[137,138,115] of the motion in the $z = 0$ plane perpendicular to the **B** field. This is a violent way of enforcing separability, suffering from various drawbacks.[4,21] But it certainly contains a part of the truth as the quantum states localized close to this plane play a major role in the formation of the spectra.

7.5.3.4 *The diabatic picture of the building of the magnetic spectra*

Around 1980, two series of experimental results[21,125] obtained on quasi-hydrogenic species, in very different experimental conditions, but on a large range of magnetic field strengths and energies, led to the formulation of a conjecture about the formation of the spectra. This was in agreement with a direct numerical simulation in the Rydberg region, through powerful techniques (indeed well adapted to the structure of the problem).[139–141]

The conjecture is that the states at the top of the diamagnetic band are the low field precursors of the ones giving rise to the quasi-Landau spectra. Such a diabatic picture agrees with the experimental behavior in field and energy for the positions of the dominant lines of the spectra (for even P_z parity) (see Fig. 7.24). Such a correlation is also satisfying in that both states do present some spatial localization close to the $z = 0$ plane, from subsections 7.5.3.1 and 7.5.3.3. The numerical simulation[140] proved also that the secondary lines were associated with the other states in the low field manifold which should give rise to the same quasi-Landau patterns for $W \cong 0$ (see Fig. 7.27).[141]

Finally, the whole picture for the magnetic spectra was that they represented the superposition of several series, weakly interacting, branched to well-defined

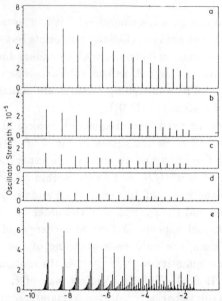

Fig. 7.27. Numerical sturmian simulations of the diamagnetic spectrum[141] (even P_z parity). The spectrum is the superposition of several series weakly interacting, each of which leads approximately to a $3/2\hbar\omega_c$ spacing at threshold. (Reprinted by permission from *Nature*, **292**, p. 437 ©1981 Macmillan Journals Limited).

states in the low field limit. Such a conception is very reminiscent of that for the Stark effect. But here there is no dynamical symmetry and in that sense, this is an oversimplified view, though proving extremely fruitful.

The first aspect to be clarified is thus that of the symmetries, especially in the low field limit: how does the diamagnetic interaction break the Coulomb symmetry? Addressing this question requires a change of viewpoint. But this leads to solving the problem, as well as some others.

7.6   Elementary Views on the Symmetries of the Coulomb Interaction

The orbital Coulomb problem does possess exceptional symmetry properties. Elementary consequences are well-known such as the symmetrical character of the classical trajectory and the n^2 degeneracy of the n Coulomb shell.

In analyzing the symmetry of the Coulomb field, one finds that there are two levels which are associated with simple physical pictures. The symmetry group associated with the first level of symmetry expresses the rotational invariance of the Coulomb interaction in a 4-dimensional space. It follows (see subsection 7.6.2.2) that all the properties of the n Coulomb shell can be expressed by means of two angular momenta, $\mathbf{j}_1$ and $\mathbf{j}_2$, acting in 3-dimensional spaces. The second level of symmetry expresses to some extent the equivalence of the dynamics with that of a 4-dimensional isotropic oscillator. This leads to the Coulomb dynamical group concept which allows all the Coulomb shells to be interconnected, whatever the energy. Both ideas can be introduced simply through consideration of the classical motion.

These concepts on the architecture of the Coulomb field were known in part around 1920.[64] The nature of the symmetry group was established by Fock[9] and Bargman[142] around 1935. Further developments around 1960 by Barut, Fronsdal among many others led to a formulation of the dynamical group concept. The analyses initiated a full branch of theoretical physics. It is amazing that these powerful and simple views on the Coulomb problem are mostly ignored in the field to which they rightly belong. They are found only with difficulty in standard textbooks in atomic physics. We will limit ourselves here to a non-relativistic approach and a naive exposition from first principles. The spin of the electron is not considered and the mention of "Lorentz spaces" throughout the text does not refer to the space-time manifold of relativity.

7.6.1   The overview from classical mechanics

The key ideas on the orbital structure can be deduced from a proper classical analysis. As a consequence of the uniformity of time and the isotropy of space respectively, the Hamiltonian H and the angular momentum $\mathbf{L}$ in space, are constants of the motion.

The symmetrical character of the conical trajectories implies the existence of an additional constant of the motion. This is the Laplace-Runge-Lenz vector $\mathbf{A}$, directed along the symmetry axis (cf. Fig. 7.28)) which is expressed:

$$\mathbf{A} = \frac{1}{2}(\mathbf{L} \times \mathbf{p} - \mathbf{p} \times \mathbf{L}) + me^2\mathbf{r}/r$$
$$\mathbf{A} \cdot \mathbf{L} = 0 . \tag{7.17}$$

The classical motion is degenerate as there are more constants of the motion than needed in a 3-dimensional problem. For given $(\mathbf{L}, \mathbf{A})$, the energy W is completely determined:

$$A^2 = 2mWL^2 + m^2e^4 . \tag{7.18}$$

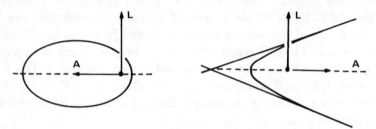

Fig. 7.28. The classical Coulomb trajectories showing the basic elements (**L**, **A**) of the Coulomb symmetry.

From (7.17) and writing φ the angle $(\mathbf{r}, \mathbf{A})$, the equation of the trajectory is:

$$\frac{1}{r} = \frac{me^2}{L^2}\left(1 + \alpha \cos \varphi\right) \tag{7.19}$$

where $\alpha = A/me^2$ measures the eccentricity. The existence of **A** is the first key feature in the Coulomb problem, associated with the symmetry group.

The second key feature involves less obvious considerations relating to time and energy. The time dependence of r and φ in (7.19) can be deduced from the conservation of $L = mr^2 \cdot \frac{d\varphi}{dt}$. But it is a complicated relation of the implicit type.[143] There is a well-known way of finding a better parametrization which amounts to the usual one for conical curves. For negative energies, the new parameter τ is chosen such that:

$$r = \frac{e^2}{2|W|}\left(1 - \alpha \cdot \cos \tau\right) \tag{7.20}$$

while the relationship with the time t in the physical problem is:

$$t = \left(me^4/8|W|^3\right)^{1/2}\left(\tau - \alpha \cdot \sin \tau\right)$$

$$\frac{dt}{d\tau} = \left(m/2|W|\right)^{1/2} \cdot r . \tag{7.21}$$

The parametric equations of the motion in the plane perpendicular to **L** are thus:

$$x = r \cdot \cos \varphi = \frac{e^2}{2|W|}\left(\cos \tau - \alpha\right)$$

$$y = r \cdot \sin \varphi = \frac{e^2}{2|W|}\left(1 - \alpha^2\right)^{1/2} \cdot \sin \tau . \tag{7.22}$$

This is the motion of a 2-dimensional harmonic oscillator in which the parameter τ plays the role of time. The second key feature in the Coulomb problem is thus its equivalence, whatever the energy, to harmonic motion, provided one abandons the uniformity of time. From (7.21) the "new time" τ does not flow uniformly on the Kepler ellipse. This is the essential idea for building the Coulomb dynamical group, a concept in which the energy no longer plays a privileged role but is treated as the other variables.

7.6.2   The symmetry group of the orbital Coulomb problem

Establishing the group structure from consideration of the Poisson brackets or commutators of $(\mathbf{L}, \mathbf{A})$ is straightforward. We limit the analysis to the bound spectrum $(W < 0)$. In order to give the Lenz vector the dimension of an angular momentum, we will from (7.17) introduce a scaled version $\mathbf{a}$:

$$\mathbf{a} = (-2mW)^{-1/2}\mathbf{A} \tag{7.23}$$

which is defined in the subspaces associated with the energy W of the Coulomb Hamiltonian. The commutation relations of $(\mathbf{L}, \mathbf{a})$ can be deduced from first principles. One obtains:

$$[a_i, a_j] = i \cdot \varepsilon_{ijk} \cdot L_k$$

$$[L_i, L_j] = i \cdot \varepsilon_{ijk} \cdot L_k$$

$$[L_i, a_j] = i \cdot \varepsilon_{ijk} \cdot a_k \tag{7.24}$$

Owing to the scaling (7.23), the algebra of the operators is closed. The set of equations (7.24) physically express that $\mathbf{L}$ is an angular momentum and $\mathbf{a}$, a vectorial operator, though not an angular momentum in 3 dimensions.

7.6.2.1   The SO(4) symmetry of the bound Coulomb spectrum

Actually, the meaning of (7.24) becomes clear in a 4-dimensional context. Writing 1 to 3 the spatial coordinates and 4 the extra-dimension, one introduces the operator $\mathcal{L}$, the components of which are:

$$\mathcal{L}_{23} = L_x \quad \mathcal{L}_{31} = L_y \quad \mathcal{L}_{12} = L_z$$
$$\mathcal{L}_{14} = a_x \quad \mathcal{L}_{24} = a_y \quad \mathcal{L}_{34} = a_z \ . \tag{7.25}$$

The commutation relations (7.24) express that $\mathcal{L}$ is an angular momentum in four-dimensions.[144,145] From (7.17) and (7.18) one deduces:

$$\mathbf{L \cdot a} = \mathbf{a \cdot L} = 0$$

$$\mathcal{L}^2 + 1 = L^2 + a^2 + 1 = -1/2H, \text{where } H \text{ is the Hamiltonian .}$$
$$(7.26)$$

For the bound spectra, the symmetry of the Coulomb field is one of rotation, but in a 4-dimensional space. This is labeled as SO(4) symmetry. A similar analysis can be made for the continuous spectrum upon which the symmetry can be shown to be of SO(3, 1) type, once again but in a 4-dimensional Lorentz space.[144,145] These properties which only refer to the orbital structure in the non-relativistic approximation were established by Fock[9] around 1935, using "Fock's" method.

The $\mathcal{L}_{ij}$ in (7.25) are the generators of the rotation group in four dimensions. The first three are the generators of an SO(3) subgroup associated with rotations in real space, while the last three from (7.24) do not build a 3-dimensional angular momentum.

7.6.2.2   The SO(4) vectorial model of the orbital problem

It is not so intuitive to deal with angular momenta in four dimensions. Fortunately, it turns out that a description of SO(4) in terms of two independent SO(3) subgroups is possible. From the structure of (7.26) it seems wise to consider the two operators:

$$\mathbf{j}_{\frac{1}{2}} = (\mathbf{L} \pm \mathbf{a})/2 \tag{7.27}$$

which from (7.24), satisfies ($\alpha = 1, 2$):

$$[j_{\alpha i}, j_{\beta j}] = i \cdot \delta_{\alpha\beta} \cdot \varepsilon_{ijk} \cdot j_{\alpha k} . \tag{7.28}$$

This proves that $\mathbf{j}_1$ and $\mathbf{j}_2$ are two commuting angular momenta in three dimensions. They are submitted to the constraints (7.26) which may be expressed as:

$$j_1^2 = j_2^2 = j(j+1) \quad \text{and} \quad j = \frac{n-1}{2} \tag{7.29}$$

where n is the principal quantum number ($W = -1/2n^2$). According to the value of n, j is integer or half-integer. Pauli made use of this description in

1926.[64] The symmetry properties of the Coulomb field can be described in terms of two independent angular momenta (j_1, j_2) acting in 3-dimensional spaces. This is a major simplification as the 3-dimensional angular momentum algebra O_3 is well-known. For example the spectrum and eigenfunctions of j_1 and j_2 are immediately known, though they are complicated differential operators from (7.17) and (7.27):

$$j_i^2 |j_i m_i\rangle = j(j+1)|j_i m_i\rangle$$
$$j_{iz}|j_i m_i\rangle = m_i|j_i m_i\rangle$$
$$-j \leq m_i \leq j .$$

(7.30)

All the properties of the n Coulomb shell follow from this description. As shown in Fig. 7.29, a rigorous vectorial model for the orbital structure[135] involves <u>two</u> vectors j_1 and j_2 (with $j = (n-1)/2$); this is a consequence of the high degree of symmetry of the situation and leads in particular to a justification of the Bohr model.

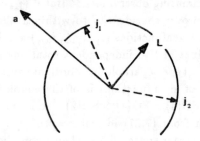

Fig. 7.29. The SO(4) vectorial model of the orbital Coulomb problem in the n shell.

7.6.2.3 The two classes of generalized eigenfunctions

The simplicity and strong predictive power of this description are well illustrated when looking at the Coulomb eigenfunctions. New basis sets can be deduced which nicely fit our purpose of understanding the physics of Rydberg atoms in external fields. There are actually two classes of Coulomb eigenfunctions according to the coupling schemes of two angular momenta in 3 dimensions.

The first class of eigenfunctions is of the decoupled type and associated with the subgroup chain reduction $SO(4) = SO(3) \otimes SO(3) \supset SO(2) \otimes SO(2)$. The two angular momenta are quantized along two different axes in space. The scheme is thus $(j_1^2, j_2^2 j_{1\omega_1} j_{2\omega_2})$ and the eigenfunctions are the direct products

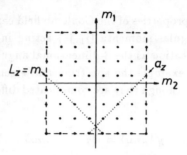

Fig. 7.30. The n^2 degeneracy of the $n = 8$ Coulomb shell as labeled through $(j_{1_{\omega_1}} = m_1;$ $j_{2_{\omega_2}} = m_2)$.

of those in (7.30) for each angular momentum. The $n^2 = (2j + 1)^2$ degeneracy of the n Coulomb shell can be labeled (Fig. 7.30) with (m_1, m_2). A special case is $\boldsymbol{\omega}_1 = \boldsymbol{\omega}_2$ for which the $(j_1^2 j_2^2 j_{1z} j_{2z})$ eigenfunctions coincide with those deduced from the separability of Schrödinger's equation in spatial parabolic coordinates (ξ, η). The latter can be shown to be associated with a description through the set (H, L_z, A_z) of commuting observables.[59] But if (j_{1z}, j_{2z}) are defined so also are (L_z, A_z), which proves the identity to within a phase-factor. Hence, the generalized parabolic eigenfunctions $(j_1^2 j_2^2 j_{1\omega_1} j_{2\omega_2})$ are deduced from the usual set through the product of two independent rotations $R^j(\boldsymbol{\omega}_1) \cdot R^j(\boldsymbol{\omega}_2)$. They are in general not of a type separable in coordinate systems in $R(3)$.

The second class of eigenfunctions is of the coupled type associated with the subgroup chain $SO(4) \supset SO(3) \supset SO(2)$. A natural choice is $(j_1^2 j_2^2 (j_1 + j_2)^2 (j_1 + j_2)_z)$ which from (7.27) identifies with $(j_1^2 j_2^2 L^2 L_z)$ where $\mathbf{L}$ is the angular momentum in real space. This scheme complies with the isotropy of space. The spatial representations in spherical coordinates coincide (to within a phase factor) with the usual spherical eigenfunctions. Notice that this describes both the radial and angular parts.

There are other ways of building a 3-dimensional angular momentum from the components of $\mathcal{L}$ (equation (7.25)). For example the quantity $\boldsymbol{\lambda}(a_x, a_y, L_z)$ which fulfills:

$$[\lambda_i, \lambda_j] = i\varepsilon_{ijk} \cdot \lambda_k \tag{7.31}$$

is such a 3-dimensional angular momentum associated with the non-standard $(j_{1x} - j_{2x}, j_{1y} - j_{2y}, j_{1z} + j_{2z})$ coupling scheme of (j_1, j_2). Hence the $(j_1^2 j_2^2 \lambda^2 \lambda_z)$ are eigenfunctions of the Coulomb problem such that:

$$\lambda^2 |j_1^2 j_2^2 \lambda^2 \lambda_z\rangle = \lambda(\lambda + 1)|j_1^2 j_2^2 \lambda^2 \lambda_z\rangle$$

$$0 \leq \lambda \leq n - 1, \quad -\lambda \leq \lambda_z \leq \lambda . \tag{7.32}$$

But they are of a completely unusual type. The 3-dimensional angular momentum λ turns out to play an important role in the theory of doubly excited states,[146] diamagnetism[147,148] and the crossed $(\mathbf{E}, \mathbf{B})$ fields spectrum. Obviously it is a pure product of the Coulomb symmetry. λ is the generator of rotations in three dimensions. But they are not of geometrical type in real space (the generator would be $\mathbf{L}$). Actually the rotation operator $e^{i\alpha\lambda_z}$ transforms a classical Keppler ellipse into another one, with the same energy but *different eccentricity*.

Actually, the λ and $\mathbf{L}$ operators are interconnected through a rotation in the 4-dimensional space as:

$$e^{i\frac{\pi}{2}(l_z-a_z)}\cdot\mathbf{L}\cdot e^{-i\frac{\pi}{2}(l_z-a_z)} = \lambda \tag{7.33}$$

Upon using the $(\mathbf{j}_1,\mathbf{j}_2)$ description this allows one to interconnect the eigenfunctions in the λ and $\mathbf{L}$ coupling schemes.[147] The λ type ones are not likely to separate in a coordinate system of $R(3)$.

7.6.2.4 Conclusions

Such an analysis takes full advantage of the specificity of the Coulomb field as it is based on the *inner symmetries* rather than on the *geometrical* ones in $R(3)$. New sets of Coulomb eigenfunctions have been deduced without referring to any particular coordinate system.

The strong predictive power is allied to simplicity once $(\mathbf{j}_1,\mathbf{j}_2)$ have been recognized as 3-dimensional angular momenta. The whole physics of the orbital problem can be picturesquely conceived in the framework of a double vectorial model. In addition, the spectrum and eigenfunctions of the operators are known immediately, though they are complicated differential operators from (7.27). Finally, most of calculations are simplified, being reduced to 3-dimensional angular algebra ones. For example, the parabolic and spherical bases differ by a recoupling of $(\mathbf{j}_1,\mathbf{j}_2)$. Hence (with $j = (n-1)/2$):

$$|j_1^2 j_2^2 Lm\rangle = \sum_{m_1 m_2} \langle jjm_1m_2|Lm\rangle |j_1^2 j_2^2 m_1 m_2\rangle \tag{7.34}$$

where the coefficients are the usual Clebsch-Gordan ones.

The physical relevance of this description, especially for understanding the physics of atoms in external fields, is made clear below. Most of the ideas can be deduced simply but the pictures are rigorous.

7.6.3   The Coulomb dynamical group

In order to interconnect all the Coulomb states, whatever the energy, one is led to building a more general structure which is the Coulomb dynamical group $SO(4,2)$.[149,150] The existence of such a structure lies in the possibility of thinking of the Coulomb problem as a pair of 2-dimensional harmonic oscillators. This was pointed out in subsection 7.6.1 in a special case. The symmetry group is embedded into such a structure.

The discussion of the full notion is outside the scope of this paper. But recent theoretical developments[151,152] have demonstrated its importance for describing the strong mixing regimes of atoms in external fields.

7.6.3.1   The general background

In order to get the flavor of what the structure is and of the applications, there are two key ideas to introduce. First the equivalence of the Coulomb problem to a pair of 2-dimensional oscillators. Second, the possibility of describing the dynamics whatever the energy in terms of two operators $(\mathbf{S}, \mathbf{T})$ acting in 3-dimensional Lorentz spaces, with the (u, v, w) non-Euclidean metric:

$$ds^2 = du^2 + dv^2 - dw^2 .$$

This seems similar to the approach with $(\mathbf{j}_1, \mathbf{j}_2)$ for the $SO(4)$ symmetry group. But the considerations are now valid whatever the energy, and provide us with a generalization of the symmetry group concept.

7.6.3.2   The orbital Coulomb problem as a pair of 2-dimensional oscillators

The equivalence can be established, in a special case, at constant $L_z = -i\hbar\partial/\partial\varphi$. Such an analysis applies to both the Stark and diamagnetic problems. Writing Schrödinger's equation in semi-parabolic coordinates $(\mu = (r + z)^{1/2}$; $\nu = (r - z)^{1/2})$ one obtains[151,153]:

$$F(\mu) = [\partial^2/\partial\mu^2 + \frac{1}{\mu} \cdot \partial/\partial\mu - m^2/\mu^2]/2$$

$$[F(\mu) + F(\nu) + W(\mu^2 + \nu^2) + 2]\Psi = 0 . \tag{7.35}$$

These are the equations of a pair of 2-dimensional isotropic oscillators with the frequency $\omega = \sqrt{-2W}$ (W is the Coulomb energy) written in polar coordinates.

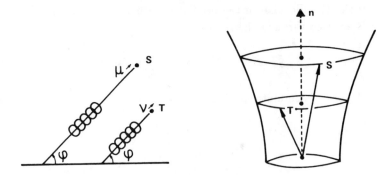

Fig. 7.31. The orbital Coulomb problem at constant $L_z = m$ as a pair of 2-dimensional isotropic oscillators. The dynamical properties of the system can be expressed in the framework of a SO(2, 2) vectorial model.[133,155] **S** and **T** are angular momenta in 3-dimensional Lorentz spaces which precess independently around **n** (cf. equation (7.40)) on the hyperboloid given by equation (7.37).

μ and ν are the radius vectors of the oscillators which are coupled in φ, due to the constraint $L_z = m = -i\hbar \partial/\partial\varphi$ (see Fig. 7.31).

In order to deduce equation (7.35), Schrödinger's equation has been multiplied by the factor $(\mu^2 + \nu^2)$ which coincides with "$2r$". From section 7.6.1 on the classical motion, this amounts to changing the definition of time, or performing a homogeneous canonical transformation on the Lagrangian. The dynamics then becomes oscillator-like. The Coulomb energy is the coupling constant of the harmonic potentials while the Coulomb potential leads to the constant "2" factor in (7.35). The solutions to (7.35) for the eigenvalue "2" map the solutions of the Coulomb problem for the energy W.

7.6.3.3   *The oscillator dynamics in a 3-dimensional Lorentz space*

The question is now how to describe the dynamics of the oscillators along the μ and ν axes, at constant L_z. This can be addressed in various ways[153,154] or from elementary considerations.[151]

The dynamics of the 2-dimensional oscillator *at constant* L_z can be described by introducing the 3-component operator **S** such that:

$$S_{w}^{u} = (\pm \partial^2/\partial\mu^2 \pm \frac{1}{\mu} \cdot \partial/\partial\mu \mp m^2/\mu^2 + \mu^2)/4$$

$$S_v = i(1 + \mu \cdot \partial/\partial\mu)/2 \tag{7.36}$$

S_w is half the Hamiltonian of the oscillator with unit frequency. The components of **S** are tied through the relations:

$$\tilde{S}^2 = S_u^2 + S_v^2 - S_w^2 = \frac{1}{4}(1 - m^2) \tag{7.37}$$

and the algebra of commutators is closed:

$$[S_u, S_v] = -iS_w$$
$$[S_v, S_w] = iS_u$$
$$[S_w, S_u] = iS_v$$
$$[\mathbf{S}, \tilde{S}^2] = 0 . \tag{7.38}$$

This is an $0_{2,1}$ Lie algebra.[153] It is very similar to the 0_3 one for the angular momentum in 3 dimensions. The quantity **S** can be thought of as an angular momentum in a 3-dimensional Lorentz space with the non-Euclidean metric ds^2. The associated group structure is labeled SO(2,1).

The properties of SO(2, 1) are as well-known as those of the rotation group SO(3). There are differences in that the representations are infinite dimensional. This is to be expected as **S** should allow the description of the infinite number of levels in the harmonic ladder whatever the energy, at constant L_z. Writing $|n, k\rangle$ the eigenstates, one has [144,151,153]:

$$\tilde{S}^2|n, k\rangle = k(1 - k)|n, k\rangle = \frac{1}{4}(1 - m^2)|n, k\rangle$$
$$S_w|n, k\rangle = (n + k)|n, k\rangle . \tag{7.39}$$

The latter equation from the fact that S_w is half the Hamiltonian of the oscillator is also to be expected.

7.6.3.4　The double vectorial model of the Coulomb dynamics in 3-dimensional Lorentz spaces

Upon introducing the sets **S** and **T** of operators for each oscillator along the μ and ν axes, one is led, once again, to think of all the dynamical properties of the Coulomb problem in terms of a double vectorial model. But this model now involves two operators (**S**, **T**) acting in 3-dimensional Lorentz spaces.[155] A major difference with the model in section 7.6.2.2 is that it describes the dynamics at constant $L_z = m$, whatever the energy, and thus allows for the interconnection of all the Coulomb states (at constant m in this special case).

This is a subgroup SO(2, 2) = SO(2, 1)⊗ SO(2, 1) of the complete SO(4, 2) dynamical group. With $\alpha = -2W$, equation (7.35) can be written

$$[(\mathbf{S} + \mathbf{T}) \cdot \mathbf{n} - 2]\Psi = 0 \qquad (7.40)$$

where $\mathbf{n}$ is the vector in the 3-dimensional Lorentz space with the components $((\alpha - 1), 0, -(1 + \alpha))$. Such an equation is similar to the equation $\mathbf{J} \cdot \mathbf{n} = k$ for the rotation group and can be diagonalized through a Lorentz rotation $e^{i\theta(S_v + T_v)}$.[133] As shown in Fig. 7.31, this describes the independent precession of $\mathbf{S}$ and $\mathbf{T}$ around $\mathbf{n}$ in a 3-dimensional Lorentz space. From (7.39) one immediately deduces the spectrum and eigenfunctions for the Coulomb problem.

The re-coupling rules of $\mathbf{S}$ and $\mathbf{T}$ are similar to those of the O_3 angular momentum algebra. As in section 6.2.3, one can obtain several types of subgroup chain reductions of SO(2, 2). The representation $(S^2 T^2 S_w T_w)$ is of decoupled type. It commutes with L_z. The standard coupling $\mathbf{V} = \mathbf{S} + \mathbf{T}$ leads to a representation $(S^2 T^2 V^2 V_w)$ which can be shown to commute with both L_z and L^2.[123,154] It is associated with the situation in section 7.6.1 where the Coulomb dynamics was shown to be equivalent to that of a 2-dimensional oscillator. As in section 7.6.2.3 a λ type description is possible with $\mathbf{W} = (S_u - T_u, S_v - T_v, S_w + T_w)$. The generators in this $(S^2 T^2 W^2 W_w)$ representation commute with both L_z and λ^2.[155,156]

This makes it clear that the dynamical group approach allows for a natural extension, valid whatever the Coulomb energy, of the symmetries described in section 7.6.2 for a given n shell.

7.7 The Low Field Regimes and Breaking the Coulomb Symmetry

In this section we apply the physical models previously established to understanding without calculations, all the major features of atoms in external fields, in the low field limit. The adequacy of the SO(4) vectorial model is certainly best proven in the crossed electric and magnetic fields situation. The theory was established by Pauli[64] in a way which can be considered as the first attempt at the symmetries of the orbital Coulomb interaction. The experimental evidence[156,157] was provided in 1984.

7.7.1 *Rydberg atoms in crossed* **E, B** *fields — the low field Pauli quantization*

In the general case, there are no longer any constants of the motion in the crossed $(\mathbf{E}, \mathbf{B})$ fields problem. In the low field limit, the Bohr frequency $2R/n^3$

is assumed to be large compared to the external field perturbation. We further assume that the diamagnetic term is small compared to the electric and paramagnetic ones.

7.7.1.1 *Pauli's prediction and the SO(4) analysis*

Even in the low field limit, a standard analysis of the problem is difficult. There are no obvious constants of the motion. Moreover, there is no longer any separability of Schrödinger's equation in some coordinate system. The only solution is to use the symmetry group formulation or to invent it!

It is indeed straightforward to solve the problem in the low field limit, using the results of section 7.6.3. The perturbation is expressed as:

$$W = \frac{q}{2m}\mathbf{B}\cdot\mathbf{L} + q\mathbf{E}\cdot\mathbf{r} \tag{7.41}$$

From either the classical picture, or symmetry considerations, $\mathbf{r}$ should be proportional to a *at first order* which is known as the "Pauli replacement":

$$\mathbf{r} \rightarrow \frac{3}{2}n\mathbf{a} . \tag{7.42}$$

From (7.41) this leads naturally to introducing the linear Stark frequency $\omega_E = \frac{3}{2}nE/E_c$ (equation (7.5)), which plays a role similar to the Larmor frequency ω_L (but is a pseudo-rotation vector as $\mathbf{E}$ is of polar type). Upon introducing $(\mathbf{j}_1, \mathbf{j}_2)$ from section 7.6.2, one immediately deduces:

$$W = \mathbf{\Omega}_1 \cdot \mathbf{j}_1 + \mathbf{\Omega}_2 \cdot \mathbf{j}_2$$
$$\mathbf{\Omega}_{\frac{1}{2}} = \boldsymbol{\omega}_L \pm \boldsymbol{\omega}_E . \tag{7.43}$$

This looks like a double Zeeman Hamiltonian and is interpreted as the independent precession of the two angular momenta $\mathbf{j}_1$ and $\mathbf{j}_2$ around respectively the $\mathbf{\Omega}_1$ and $\mathbf{\Omega}_2$ axes (Fig. 7.32). Their angular velocities are the same in the crossed field situation:

$$\Omega = \Omega_{1,2} = (\omega_E^2 + \omega_L^2)^{1/2} \tag{7.44}$$

and from the correspondence principle, this is the spacing of the sublevels (Fig. 7.32).[135,156–158]

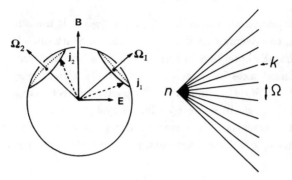

Fig. 7.32. The crossed $(\mathbf{E}, \mathbf{B})$ fields situation from the SO(4) vectorial model. $\mathbf{j}_1$ and $\mathbf{j}_2$ precess around $\mathbf{\Omega}_1$ and $\mathbf{\Omega}_2$ respectively at constant angular velocity $\Omega = \left(\omega_E^2 + \omega_L^2\right)^{1/2}$. This is the spacing of the sublevels which are labeled through the SO(4) quantum number k $= j_{1\Omega_1} + j_{2\Omega_2} = \lambda_{\Omega_1}$.[156]

The spectrum and eigenfunctions are immediately deduced from the fact that $(\mathbf{j}_1, \mathbf{j}_2)$ are two commuting 3-dimensional angular momenta. The eigenfunctions are thus of the generalized parabolic type $(j_1^2 j_2^2 j_{1\Omega_1} j_{2\Omega_2})$. The low field spectrum is:

$$W = (m_1 + m_2) \cdot \mathbf{\Omega} = (j_{1\Omega_1} + j_{2\Omega_2}) \cdot \mathbf{\Omega} = k \cdot \mathbf{\Omega}$$
$$-(n-1)/2 \le m_i \le (n-1)/2$$
$$-(n-1) \le k \le (n-1) \tag{7.45a}$$

where k is an integer. As shown in Fig. 7.32, the n^2 degenerate Coulomb manifold bursts out into $(2n-1)$ sublevels labeled through the quantum number k, with a spacing $\hbar\Omega$ in energy. Obviously, each sublevel presents an $(n - |k|)$ degeneracy which follows from the crossed character of the $(\mathbf{E}, \mathbf{B})$ fields leading to $\Omega_1 = \Omega_2$.

7.7.1.2 Physical meaning of the results

As shown in Fig. 7.32, the low field crossed $(\mathbf{E}, \mathbf{B})$ fields spectrum is analogous to the Zeeman and linear Stark ones which obviously are contained as limiting cases. In the Zeeman limit $(\boldsymbol{\omega}_E = 0)$, $\mathbf{j}_1$ and $\mathbf{j}_2$ precess around $\boldsymbol{\omega}_L$ in the same sense. In the linear Stark limit $(\boldsymbol{\omega}_L = 0)$, they precess around $\boldsymbol{\omega}_E$ in opposite senses (as $\boldsymbol{\omega}_E$ is a polar vector). To some extent, the previous results mean that one passes continuously from the Zeeman to the linear Stark regimes. This involves a rotation but in 4-dimensional space.

As to the quantum number $k = (j_1)_{\Omega_1} + (j_2)_{\Omega_2}$ it measures the quantized value of L_z in the Zeeman limit and of a_x in the Stark one. In general, it does not have a simple expression in terms of $\mathbf{L}$ or $\mathbf{a}$ but involves their projections onto two different axes in space. It only retains a clear interpretation in the framework of the SO(4) symmetry. Obviously, a direct search for (7.45) using the standard analysis in some coordinate system would be complicated and would miss the essential. It seems hardly possible to understand a process which amounts to breaking a symmetry without describing properly what the symmetry is.

7.7.1.3 The $\boldsymbol{\lambda}$ angular momentum and the low field crossed $(\mathbf{E}, \mathbf{B})$ fields spectrum

As pointed out previously, a degeneracy still remains in this regime at first order perturbation theory. This leads to assuming an independent description is possible. Introducing the angular momentum $\boldsymbol{\lambda}(a_x, a_y, L_z)$ one obtains $(0 \leq \lambda \leq n-1)$

$$W = \boldsymbol{\Omega}_1 \cdot \boldsymbol{\lambda} = \Omega \cdot \lambda_{\Omega_1}$$
$$-(n-1) \leq k = \lambda_{\Omega_1} \leq (n-1) \ . \tag{7.45b}$$

The physical process can be seen as a Zeeman effect, but it involves the angular momentum $\boldsymbol{\lambda}$ rather than the one $\mathbf{L}$ in real space. The associated set of eigenfunctions is $(j_1^2 j_2^2 \lambda^2 \lambda_{\Omega_1})$ deduced from $(j_1^2 j_2^2 \lambda^2 \lambda_z)$ through a rotation with the generator $\lambda_y = a_y$.

Hence the angular momentum $\boldsymbol{\lambda}$ which is a pure product of the SO(4) Coulomb symmetry is essential for understanding the previous basic situation of crossed $(\mathbf{E}, \mathbf{B})$ fields. Consequently, it is not an oddity arising from an academic game. As quantum mechanics puts the stress on the importance of measurements, the experimental demonstration of (7.45) allowing direct measurements of λ_{Ω_1} would lead to rooting $\boldsymbol{\lambda}$ in physical reality. This has recently been achieved using Doppler-free two-photon techniques and low external field conditions.[156,157]

7.7.1.4 An experiment in the geometry of the Coulomb field

Aside from the experimental difficulties, there is a reason for which such experiments have been undertaken only recently. This is because Pauli's prediction applies to a hydrogenic and non-relativistic situation. Alkali Rydberg atoms do not perfectly fit these requirements. For example, the nS, nP and nD series in alkali atoms usually have important quantum defects (see Fig. 7.1). The energy behavior at low fields will be smooth, with a quadratic Stark effect. This is precisely not the Pauli behavior but a non-hydrogenic one. Moreover these series are the only ones that can be excited through 1 or 2 photon transitions from the ground state. As discussed in the previous sections, the $(n^2 - 9)$ states with $l \geq 3$ behave as an incomplete hydrogenic manifold. Hence they are the ones which may be quantized according to Pauli's law, to within small non-hydrogenic corrections. Unfortunately owing to the lack of significant redistribution with fields of the nS, nP or nD series onto the incomplete manifolds, these states cannot be optically excited with efficiency at low fields.

The trick is indeed to use for example the nS states as a probe of the hydrogenic behavior in the incomplete manifold (see top part of Fig. 7.33). As the energy curves of the nS states are nearly field-independent, while the states in the incomplete manifold should present a linear-in-Ω field dependence, the two systems of curves should anticross. The tracking of the anticrossings positions on the nS energy curve, as a function of **E** and **B** fields, should provide us with a picture of the quantization *at nearly constant energy*. This is shown in Fig. 7.34 which displays the characteristic (E^2, B^2) dependence of (7.45).[156−158]

A direct record of these anticrossings with the $37S$ energy curve is shown in Fig. 7.33 (bottom part). This is a scan in **B** field (0–700 Gauss), for an electric field value $E_0 = 10.7$ V/cm, while the laser frequency is locked on the $37S$ two-photon line. To each line is associated a given value of $k = \lambda_{\Omega_1}$ which fulfills the relationship $k = \Delta/\Omega(B)$. The values of k measured this way (as well as from the plot on Fig. 7.34) are non-integer which is inessential and follows from the role of non-hydrogenic corrections[157] in Rb. The fact that λ_{Ω_1} varies by one unit from one sublevel to the other is much more important[158] and agrees with (7.45b).

A detailed analysis shows that (7.45) is fulfilled to within 1% by most of the states.[156,157] Actually, a more refined theory incorporating quantum defects (or the incomplete character of the manifold) has been established in[157] showing that the energy diagram is a bit more complicated than discussed here. But the key point is still that the experiment amounts to a direct test of the SO(4) structure.[158]

7.7.1.5 The extended tunability of the atom

Coming back to the eigenfunctions of generalized parabolic type $(j_1^2 j_2^2 j_{1\Omega_1} j_{2\Omega_2})$ or of $(j_1^2 j_2^2 \lambda^2 \lambda_{\Omega_1})$ type, they can be deduced from the usual ones (parabolic

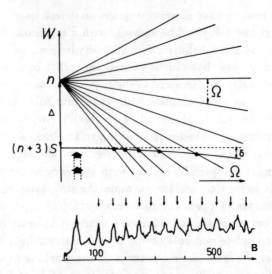

Fig. 7.33. Probing the quantization in the incomplete hydrogenic manifold through the anti-crossing with the $(n + 3)\,S$ "non-hydrogenic" state. (Rb atom - Doppler-free experiment[135]). A typical record of the $n = 34/37\,S$ anticrossings is shown for $E = 10.7$ V/cm, for B field scans from 0-700 G. (The frequency of the laser is locked on the 37 S two-photon line.[156]).

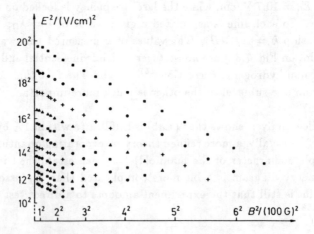

Fig. 7.34. The crossed $(\mathbf{E}, \mathbf{B})$ field quantization in Rubidium atom (anticrossing $n = 34/37S$) at nearly constant energy.[135,156] The positions of the anticrossings are plotted against (E^2, B^2).

or spherical) through a rotation with the generator $a_y = j_{1y} - j_{2y}$ and the angle $\beta = \tan^{-1}(\omega_E/\omega_L)$. This is a rotation in 4-dimensional space. The eigenfunctions express a continuous field tunability of the atom, which opens the way to a control of the atomic properties. Such eigenfunctions are not of a type separable in some coordinate system in real space. Their electronic density distributions are anisotropic and different from the usual ones. Recognition of the tunable character in $(\mathbf{E}, \mathbf{B})$ fields provides one with a way of building new classes of electronic states for various purposes. For example this provides with a simple way of building atomic circular states.[a]

It is striking to reconsider the plot in Fig. 7.34, knowing that to each point are associated eigenfunctions having completely different spatial representations. The fact that a nearly perfect sub-organization exists nevertheless is nothing but the manifestation of the existence of the Coulomb symmetry. It generates a localization of the electronic motion, not in space but rather in phase space, having a dynamical sense.

7.7.1.6. *Miscellaneous extensions*

The previous analysis provides us also with the adiabatic solution in more complicated cases involving inhomogeneous or time-dependent fields as in collisions of Rydberg atoms or hydrogenic ions with surfaces, with other atoms and so on. It also applies to atoms interacting with radiation or microwave fields.[133]

This is also a way for dealing with non-hydrogenic corrections in the low field limit and the field redistribution of non-hydrogenic states.[157] The "complete" manifold situation refers to the n^2 states with SO(4) symmetry. The incomplete manifold situation is deduced by removing a small number of states with a well-defined symmetry from this SO(4) manifold. The symmetry of the missing states is spherical. Such a picture extends to the treatment of Rydberg atoms in collisions in the Fermi approximation.

Such symmetry breaking processes can also be considered in the light of the recoupling of the angular momenta (j_1, j_2). The analogy with Hund's coupling cases in molecular physics then becomes obvious. For example, non-hydrogenic corrections cause the manifold to be characterized with two limiting coupling schemes at low fields. For penetrating states this is a $(j_1, j_2)\mathbf{L}$ coupling while the states of the incomplete manifold have a $(j_1, j_2)\lambda_{\Omega_1}$ coupling type in crossed $(\mathbf{E}, \mathbf{B})$ fields.

---

[a] The method is described in D. Delande and J. C. Gay *Laser Spectroscopy IX* – eds. Persson and Svanberg (Springer-Verlag, 1987) and *Europhys. Lett.* **5**, 303 (1988).

The molecular analogy is not an accident. Several striking examples will be given below. Indeed the rigid rotator in molecular physics and the orbital Coulomb problem share the same SO(4) symmetry group.[159]

7.7.2   The atom in a magnetic field — the low field diamagnetic behavior

The previous treatment assumes that the higher order terms in **E** and **B** are negligible. Among these are the diamagnetic Hamiltonian.

As said before, H_D commutes with L_z but not with L^2. It will remove the $(n - |m|)$ degeneracy of the Zeeman multiplet. The first theoretical description of the breaking of the Coulomb symmetry by the diamagnetic interaction was given in 1981 by Solovev.[160,161] He used classical perturbation theory to show that the secular evolution of the Kepler ellipse was such that the quantity $\Lambda = 4A^2 - 5A_z^2$ was invariant. In 1982, Herrick gave an SO(4) group theoretical description[147] using Fock's method[9] in momentum representation.

Indeed the result can be derived simply using the description of SO(4) symmetry in terms of two angular momenta $(\mathbf{j}_1, \mathbf{j}_2)$.[157,162,163]

7.7.2.1   The effective diamagnetic Hamiltonian in the n Coulomb shell

Conditions are such that $H_D = \gamma^2(x^2 + y^2)/8 << 1/n^3$ (or $\gamma^2 n^7 << 1$). The problem to solve amounts to applying the Wigner-Eckardt theorem to H_D, taking account of the SO(4) structure, which can be done in terms of $(\mathbf{j}_1, \mathbf{j}_2)$. The method developed in[162,163] is a two-step one. First the invariance of $(x^2 + y^2)$ under parity and plane reflections leads to an expression which is symmetric and quadratic in the two quasi spins. Secondly the evaluation of the coefficients is carried out using either one of the eigenbases in section 7.6.2 or averaging on four "degenerate" Kepler ellipses (two circles and two straight lines).[162] The exact quantum result, valid for all n and $L_z = m$ values, is

$$H_D = \frac{\gamma^2}{8}(x^2 + y^2) \rightarrow \frac{\gamma^2 n^2}{16}(3n^2 + 1 - 4m^2 + 20j_{1z}\cdot j_{2z} - 8\mathbf{j}_1 \cdot \mathbf{j}_2)$$

$$L_z = m = j_{1z} + j_{2z} \tag{7.46}$$

with $j_1 = j_2 = (n-1)/2$.

This can be cast under the equivalent form

$$H_D \rightarrow \frac{\gamma^2 n^2}{16}(n^2 + 3 + L_z^2 + 4a^2 - 5a_z^2) . \tag{7.47}$$

This gives a simple solution to the problem of the inter-l mixing regime. Within the n shell, the diamagnetic interaction is responsible for a coupling of the two quasi-spins (j_1, j_2), which is of non-standard type. Indeed (7.46) cannot be cast into diagonal form through some kind of recoupling or choice of a convenient basis. But all the structural information on the physics is now available.

7.7.2.2 The rovibrational structure of the diamagnetic band

We can readily deduce that there are two limiting symmetries in the manifold. This can be established in various ways. For example (7.46) or (7.47) can be recast into the form

$$H_D = \frac{\gamma^2 n^2}{16}(n^2 + 3 + 4\lambda^2 - 3L_z^2 - a_z^2) \qquad (7.48)$$

upon introduction of $\boldsymbol{\lambda}(a_x, a_y, L_z)$, with λ^2 replaced by its eigenvalue $\lambda(\lambda + 1)$. Hence the states at the top of the diamagnetic band are such that λ is maximum $(\lambda \cong (n-1)$ or $(n-2)$ according to parity) and a_z close to zero. This means that the Lenz vector $\mathbf{a}$ undergoes a secular motion close to the $z = 0$ plane with $a \cong n$; the two angular momenta (j_1, j_2) are correlated in opposite directions $j_1 + j_2 \cong 0$ in this $z = 0$ plane (though with the constraint $j_{1z} + j_{2z} = L_z = m$). The transverse part of the angular momentum $\mathbf{L}$ is nearly zero. Hence at the top of the band, the motion is of correlated type and the diamagnetic energy W_D expresses as

$$W_D \sim \frac{\gamma^2 n^2}{16}(n^2 + 3 + 4\lambda(\lambda + 1) - 3m^2) . \qquad (7.49)$$

The eigenfunctions tend to the $(j_1^2 j_2^2 \lambda^2 \lambda_z)$ coupling scheme. This expresses a symmetry of rotation of $SO(3)_\lambda$ type.

As to the states at the bottom of the band (for minimal though positive W_D values) a solution from (7.47) is to assume $a_x, a_y = 0$ and $a_z \cong \pm n$. The Lenz vector is directed along the $\mathbf{B}$ field axis. So also are j_1 and j_2, and thus coupled to the field rather than to each other. But this is subject to the constraint $j_{1z} + j_{2z} = m$. This indicates that this solution may or may not exist according to the m value. The motion at the bottom of the band is of uncorrelated type as:

$$H_D \sim \frac{\gamma^2 n^2}{16}(n^2 + L_z^2 - a_z^2) \qquad (7.50)$$

and the eigenfunctions are of the $(j_1^2 j_2^2 j_{1z} j_{2z})$ parabolic type associated with an $SO(2) \otimes SO(2)$ vibrational type structure as in the Stark effect.[131] There is a degeneracy according to a being up or down. Hence the states at the bottom of the band are parity degenerate as in a double well problem.[147] This is thus very similar to the behavior in molecules.

Hence from top to bottom, the structure evolves from rotation to vibration. This means that a crossover in the symmetries should occur in between. It corresponds to the condition on the Lenz vector, $4a^2 = 5a_z^2$ which indeed signals a fundamental change in the nature of the secular motion.[128,130,160]

The physics of diamagnetism in the low field limit thus becomes clear. "Inter-l-mixed" states are actually, for the most part, of approximate λ type, associated with a special class of Coulomb eigenfunctions. The remaining part has a vibrational type symmetry. As to the structure in energy, the splitting, which is proportional to B^2, evolves accordingly from a rotational type $\lambda(\lambda+1)B^2$ to a vibrational one $(\hbar\omega \cdot B^2)$, as shown in Fig. 7.35. The crossover in the rotation-vibration symmetries manifests itself as an irregularity in the energy level structures.

Fig. 7.35. The rovibrational structure of the diamagnetic band (inter-l mixing regime) as seen from the vectorial model. The two limiting symmetries are of SO(3) λ type at the top and SO(2)⊗SO(2) type at the bottom.[133] The crossover in the symmetries manifests itself as an irregularity in the energy level spacings.

7.7.2.3 Some experimental illustrations

The rovibrational structure of the diamagnetic manifold was shown in Fig. 7.19 and 7.21 and can be completely understood from the previous model. Especially the plot on Fig. 7.21 beautifully exhibits the crossover in the rovibrational symmetry. It is clear from the oscillator strength distribution that the appearance may drastically change according to conditions. This can be understood picturesquely from the previous model, especially the fact that for even P_z, the $k = 1$ state $(\lambda \cong n - 1)$ at the top of the diamagnetic band has the maximum oscillator strength in hydrogenic situations.[128] This was a major feature in connection with the interpretation of the experimental data.[21,129]

7.7.2.4　Dynamical correlations in the diamagnetic band

This structure is also a simple example of the role of dynamical correlations in atomic physics. Such features are likely to rule the behavior of doubly excited systems and double ionization processes, or the quantized Hall effect... among so many other effects to be discovered.[108,133]

The $SO(3)_\lambda$ type states at the top of the band are of such a correlated type. From the nodal surface plots in Fig. 7.22 they manifest both a rotational type symmetry and a spatial localization in the plane $z = 0$ perpendicular to the field. But this is only one aspect of localization, which indeed has to be thought of as localization in phase space rather than in real space. Indeed λ has no obvious geometrical meaning. In addition, the fact that these extremely stable states are localized close to a *maximum of the diamagnetic potential* should have more than a geometrical origin.

In contrast the vibrational type states of uncorrelated types, from Fig. 7.22, manifest a localization of the motion along the **B** field. It turns out that these states, localized close to a minimum of the diamagnetic potential, are *less stable* than the $SO(3)_\lambda$ type states. This is completely confirmed through studies of the classical motion[155] (see section 7.8).

7.7.2.5　Miscellaneous corrections

Obviously, the role of non-Coulombic corrections to the potential lead to effects which are similar to the ones described in section 7.4 for the Stark problem. The diamagnetic redistribution of non-hydrogenic states generates pseudo profiles and interferences. In the case of Fig. 7.19 this is the origin of the $k = 1$ line being not dominant.

7.7.2.6　Other classes of interactions

The analysis done previously can be applied to many other situations involving a perturbation to the Coulomb potential. Usually this will result in a band structure having a rovibrational symmetry, but its nature may differ from the previous one. For example, in the adiabatic approximation, the interaction of Rydberg atoms with surfaces leads to a rovibrational structure in which the rotational symmetry is of $SO(3)_L$ type (usual rotational symmetry in real space),[133,163] as the effective Hamiltonian is[133]:

$$W \sim n^2\{n^2 + 1 + m^2 - \mathbf{j}_1 \cdot \mathbf{j}_2 - 5\mathbf{j}_{1z} \cdot \mathbf{j}_{2z}\}$$

Note: A dynamical symmetry arising from perturbation of the hydrogen atom by the van der Waals interaction and others similarly classified has recently been reported[172].

7.7.3 Stark effect for quasi-hydrogenic species

The interpretation of the plots in section 7.4 can be obtained through the use of this formalism. The derivation of the linear Stark term has been given in section 7.7.1. The second order term is obtained immediately through elementary considerations analogous to the one in section 7.7.2. It may be written as (with $m = j_{1z} + j_{2z}$)

$$W_E^{(2)} = -\frac{E^2 \cdot n^4}{16}\left(17n^2 + 19 - 12j_{1z}^2 - 12j_{2z}^2 - 12j_{1z} \cdot j_{2z}\right)$$

which in contrast with (7.47) is diagonal in the $\left(j_1^2 j_2^2 j_{1z} j_{2z}\right)$ basis. The whole band is of vibrational $SO(2) \otimes SO(2)$ type. This feature is associated with the Stark dynamical symmetry.

7.8 The Strong Mixing Regimes and the Coulomb Dynamical Group

Section 7.7 has allowed us to show what was the actual meaning of experiments in the low field limit. From both the simplicity and the strong predictive power of this approach based on the Coulomb symmetry, it was clearly one of the right ways of understanding the matter.

The problem is now to extend this analysis into the strong mixing regimes, in which all the Coulomb shells are field-mixed, for example by the diamagnetic interaction. The use of the Coulomb dynamical group allows us to address this point, providing us with a generalization of the $SO(4)$ analysis, valid whatever the n value.

For dealing with problems at constant $L_z = m$, a restriction $SO(2, 2)$ of the $SO(4, 2)$ Coulomb dynamical group is sufficient. This is the approach outlined in section 7.6.3 in which the Coulomb problem is thought of as a pair of 2-dimensional oscillators. In turn, the dynamical properties of such a system are conveniently described through the introduction of two operators $(\mathbf{S}, \mathbf{T})$ acting in 3-dimensional Lorentz spaces.

7.8.1 The magnetic problem and coupled-oscillators dynamics

In this picture the diamagnetic interaction looks like an anharmonic coupling between the oscillators. The motions of the $\mathbf{S}$ and $\mathbf{T}$ generators become coupled through the non-linear term[151]:

$$H_D = \gamma^2 (S_w + S_u)(T_w + T_u)(S_w + T_w + S_u + T_u) . \tag{7.51}$$

Finally, the magnetic equation in the oscillator picture is written as

$$[(\mathbf{S} + \mathbf{T}) \cdot \mathbf{n} + H_D(\mathbf{S}, \mathbf{T}) - 2]\Psi = 0 \tag{7.52}$$

which can be solved in various ways, though not exactly.[133] Quantum predictions of the magnetic spectrum and eigenfunctions are now possible.[151,152]

7.8.2 The symmetries in the magnetic problem

Equation (7.52) makes possible a rigorous analysis of the symmetries including the dynamical ones, which, strictly speaking, do not exist, in this problem.[133,155] But the low field approximate symmetries of $SO(3)_\lambda$ and $SO(2) \otimes SO(2)$ types survive up to a certain point in the strong mixing regime. The dynamical group description in section 7.6.3.4 has allowed us to show how to extend the notion from the low field limit to the strong mixing regime. The limiting symmetries are of $(S^2 T^2 W^2 W_w)$ $SO(2, 1)_\lambda$ type and $(S^2 T^2 S_w T_w)$ type which notations allow for "inter n" mixing of the channels.

The states with λ type symmetries give rise to the quasi-Landau phenomena. They express a correlated behavior in the motion of the pair of oscillators which are strongly coupled and oscillate with a well specified relative phase. Moreover, the two operators $(\mathbf{S})$ and $(\mathbf{T})$ are coupled through a scheme which is approximately $(S_u - T_u, S_v - T_v, S_w + T_w)$. This expresses some localization of the motion in phase space rather than in real space. These states are ridge states in the approach of Fano.[164]

The other class of states are those with vibrational type symmetry in the low field limit. They are associated with the $(S^2 T^2 S_w T_w)$ coupling scheme and with uncorrelated motions of the oscillators. Their relative phase is not fixed. These states are valley states in the approach of Fano.[164] They are also less stable than the previous ones as shown below.

7.8.3. Classical chaos and its quantum analog in the magnetic problem

It is well known that the classical dynamics of coupled oscillators may turn out to be chaotic. From the previous pictures, it is not altogether unexpected that the classical magnetic problem also exhibits such a transition to chaos.[155,165,166,167] This is shown on the plot in Fig. 7.36 in the strong mixing regime, using the set $(\mathbf{S}, \mathbf{T})$ as the classical dynamical coordinates. The two types of quasi-periodic motion of rotational and vibrational types can be seen. One can further show that the correlated type motion is more stable than the other one. Above a critical energy,[155] the whole phase space motion becomes chaotic. This is a good indicator of the non-existence of a dynamical symmetry.

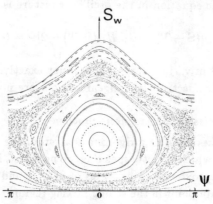

Fig. 7.36. Poincaré surface of section for the magnetic problem in the $(\mathbf{S},\ \mathbf{T})$ oscillator representation. The coordinates can be thought of as the energy of one oscillator and twice their relative phase. The closed curves are associated with the $SO(2,1)$ λ type (correlated) motion giving rise to quasi-Landau resonances. The curves near the boundaries express rather an uncorrelated behavior in the oscillator system as the relative phase is not bounded. The onset of chaos is seen close to the separatrix (crossover in the symmetries). (Courtesy D. Delande[155]).

Hence the complicated spectra in Figs. 7.24–7.26 are among the first examples of the quantum analog of classical chaos, in atomic spectra.

7.8.4 Conclusions

Equation (7.52) for the magnetic problem is similar (from the structure) to that which rules the dynamics of, for example, superfluidity (in which correlations play an important role); parametric oscillations (in which chaos is likely to occur); or simply the asymmetrical top, squeezing of the radiation field and so on. But, five years ago, it was somewhat unexpected that the atom in a **B** field can lead to such rich physical discoveries.

Perhaps it is worthy mentioning explicitly that the electric field problem for the hydrogen atom does not present any transition to chaos (in the non-relativistic approximation). The Stark Hamiltonian is

$$H_E = \beta\{(S_w + S_u)^2 - (T_w + T_u)^2\} \; . \tag{7.53}$$

There is no coupling between the oscillators and the dynamics of the **S** and **T** is always of the uncorrelated type. There is a dynamical symmetry in such a situation. It is associated with the conservation of the field-dependent quantity M_z (A_z being the component on the **E** field of the Lenz vector—equation (7.17)):

$$M_z = A_z + E(x^2 + y^2)/2 \; .$$

Finally, it was not unexpected that we should return, at the very end, to classical pictures, though they have been submitted to a complete rethinking in the meantime. The nature of non-separability in such problems is a matter of structure. Rydberg atoms in fields behave semi-classically. The question was to find the correct conceptual level for this to become obvious. It existed in the architecture of the Coulomb interaction which dominates the whole behavior of the atom.

7.9   Conclusions

Throughout these studies on Rydberg atoms in external fields, we have learned a lot about the properties of atoms in electric and magnetic fields. Direct applications to plasma, solid state and astrophysics follow. For example realistic evaluations of atomic ionization rates can be used in the equations of magneto-hydrodynamics. The field discretization of atomic continua also implies that the collision rates, absorption of radiation, electronic capture, formation of negative ions and so on should strongly depend on the field strengths. But these are basic aspects of the energy balance in plasmas. In magnetic fields, the absorption of radiation should take place not only at the cyclotron frequency ω_c, which is known as "cyclotron resonance", but also around $\frac{3}{2}\omega_c$ for which the concentration of atomic oscillator strengths is likely to be more important. Natural autoionization in external fields seems also promising and may lead to new "magnetically controlled" processes for lasers. Plasma diagnostics should also greatly benefit from advances made in the field. For example, the mechanisms leading to the generation of pseudo profiles may play a role in line formation, though the widths have no relationship with the local temperature. But the spectrum and eigenfunctions for such systems being known, it is now possible to go further in the study of the atomic properties.

But what have we learnt from the physics of these Coulombic systems? What makes Rydberg physics so fundamental? It is the importance of the symmetries which govern the whole dynamics of the atom. Indeed, Rydberg atoms in external fields provided historically the systems for which the Coulombic dimension of the atom became fully involved in the understanding of the physical properties. This led, in the strong mixing regimes, to laws as universal as the Rydberg formula, which indeed has the same profound origin. This is valid also for quasi-hydrogenic ions, in the non-relativistic approximation. These symmetries make the atom in external fields a tunable system both as concerns the spectrum and as concerns the eigenfunctions. The preparation of new classes of atomic states for various purposes becomes possible, by combining DC and pulsed external fields and optical excitation.

The last elementary problem in quantum mechanics, the atom in a magnetic field, has been analyzed making use of the concepts of Coulomb symmetry. Considered in a traditional way, the experimental results can hardly be understood. They present a puzzle with intriguing characteristics. Considered in the light of the Coulomb symmetry, the puzzle organizes itself and the quasi-Landau phenomenon expresses to some extent the survival of a symmetry which has more than a geometrical origin. But for dealing with symmetry breaking processes, is there a wiser thing to do than to describe properly what the symmetry is?

An important notion which arises from these studies is that of dynamical correlation in atomic systems, which expresses a certain localization of the motion in phase space. The magnetic problem provides us with a simple example. Although states giving rise to the quasi-Landau phenomenon are localized close to a maximum of the diamagnetic potential, they turn out to be the most stable ones in the problem. The origin of the phenomenon is clear in the $(\mathbf{S}, \mathbf{T})$ oscillator picture. It shows up as the existence of a collective motion of the two oscillators, with a well specified relative phase. This is certainly one of the most elementary manifestations of the "ridge" state behavior in the views of Fano.[164] Moreover, even in the lack of an exact dynamical symmetry, such a behavior is here unambiguously associated with the correlated or uncorrelated evolution of the set of quantum dynamical variables $(\mathbf{S}, \mathbf{T})$ for the problem.

There are other systems in atomic physics which are likely to exhibit such a correlated behavior and should be reconsidered to the light of the Coulomb symmetry: atoms in electromagnetic fields, the two-center Coulomb problem (molecules, collisions) and doubly-excited systems. In the latter case, new classes of atomic states are at present being sought which would express a correlated motion of the pair of electrons in the Coulomb field of the nucleus. Experiments have already proved that the double ionization process obeys to some extent the Wannier law,[168,169,170] which manifests the failure of the independent particle model.

Finally, it was not unexpected that we should come back to classical pictures for a deeper understanding of the dynamics of the atom. Classical notions have been considerably rethought in the meantime. The magnetic problem turns out to be a nice playground for studying the quantum analog of classical chaos in connection with the destruction of symmetries. How should such systems be quantized and what are the statistics of the quantum energy levels? Perhaps, in the near future, Rydberg atoms will allow us to address these fundamental questions.[171]

Acknowledgements

The author would like to thank those who contributed at various stages of this work: Christian Chardonnet, Francis Penent, Francois Biraben, Leslie Pendrill, and, especially, Dominique Delande.

REFERENCES

1. D. Kleppner, in Les Houches Summer School, session 28, eds. J. C. Adam and R. Balian (New York, Gordon and Breach, 1981).
2. H. Hasegawa, *Physics of Solids in Intense Magnetic Fields*, ed. E. D. Haidemenakis (Plenum, New York, 1969).
3. J. C. Gay, *Progress in Atomic Spectroscopy*, eds. M. J. Beyer and H. Kleinpoppen (Plenum, 1984) Vol. C.
4. H. A. Lorentz, *The Theory of Electrons* (Dover Pub., 1952).
5. H. Poincaré, *Acta Math.* **13** (1890) 1.
6. A. Einstein, *Ver. Deut. Phys. Ger.* **19** (1917) 82.
7. K. Van Leeuwen, G. V. Oppen, S. Renwick, J. B. Bowlin, P. M. Koch, R. V. Jensen, O. Rath, D. Richards and J. G. Leopold, *Phys. Rev. Lett.* **55** (1985) 2231.
8. W. R. S. Garton and F. S. Tomkins, *Ap. J.* **158** (1969) 83.
9. V. A. Fock, *Z. Phys.* **98** (1935) 145.
10. S. Feneuille, S. Liberman, J. Pinard and P. Jacquinot, *C. R. Heb. Acad. Sc.* **284** (1977) 291.

11. R. R. Freeman and N. P. Economou, *Phys. Rev. A* **20** (1979) 2350.

12. A. R. P. Rau, *Phys. Rev. A* **16** (1977) 613.

13. S. Chikazumi and N. Miura, *Physics in High Magnetic Fields*(Springer-Verlag, 1981).

14. J. C. Gay, in *Photophysics and Photochemistry in the V.U.V.*, eds. Mc Glynn *et al.* (Reidel, 1985).

15. M. H. Nayfeh, K. Ng and D. Yao, in *Atomic Excitation and Recombination in External Fields*, eds. M. H. Nayfeh and C. W. Clark (Gordon and Breach, New York, 1985).

16. A. Holle and K. H. Welge, in *Laser Spectroscopy VII*, eds. T. W. Hänsch and Y. R. Shen (Springer-Verlag, 1985); see also J. Main, G. Wiebusch, A. Holle and K. H. Welge, *Phys. Rev. Lett.* 1 December (1986).

17. H. Rottke, A. Holle and K. H. Welge, in *Atomic Excitation and Recombination in External Fields*, eds. M. H. Nayfeh and C. W. Clark (Gordon and Breach, New York, 1985).

18. R. M. Sternheimer, in *Progress in Atomic Spectroscopy*, eds. H. J. Beyer and H. Kleinpoppen (Plenum, New York, 1984) Part C.

19. K. B. Eriksson and I. Wenåker, *Phys. Scripta* **1** (1970) 21.

20. D. Popescu, C. D. Collins, B. W. Johnson and I. Popescu, *Phys. Rev. A* **9** (1974) 1182.

21. J. C. Gay, D. Delande and F. Biraben, *J. Phys. B, Letters B* **13** (1980) L720.

22. C. Chardonnet, F. Penent, D. Delande, F. Biraben and J. C. Gay, *J. Phys. Lett.* **44** (1983) L517.

23. C. J. Sansonetti and C. J. Lorenzen, *Phys. Rev. A* **30** (1984) 1805.

24. M. G. Littman, M. L. Zimmerman, T. W. Ducas, R. R. Freeman and D. Kleppner, *Phys. Rev. Lett.* **36** (1976) 788.

25. M. L. Zimmerman, J. C. Castro and D. Kleppner, *Phys. Rev. Lett.* **40** (1978) 1083.

26. E. Luc-Koenig, S. Liberman and J. Pinard, *Phys. Rev. A* **20** (1979) 519.

27. K. Liu and M. G. Littman, *Opt. Lett.* **6** (1981) 117.

28. J. Pinard and S. Liberman, *Opt. Commun.* **20** (1977) 344.

29. F. Trehin, Thesis (Paris, 1979); F. Trehin, F. Biraben, B. Cagnac and G. Grynberg, *Opt. Commun.* **31** (1979) 76.

30. F. Biraben and L. Julien, *Opt. Commun.* **53** (1985) 319.

31. E. R. Eliel, W. Hogervorst, K. A. Van Leeuwen and B. H. Post, *Opt. Commun.* **39** (1981) 41.

32. H. Rinneberg, J. Neukammer, G. Jönsson, H. Hieronymus, A. König and K. Vietzke, *Phys. Rev. Lett.* **55** (1985) 382.

33. S. Liberman and J. Pinard, *Phys. Rev. A* **20** (1979) 507.

34. M. G. Littman, M. M. Kash and D. Kleppner, *Phys. Rev. Lett.* **41** (1978) 103.

35. T. F. Gallagher, L. M. Humphrey, R. M. Hill and S. A. Edelstein, *Phys. Rev. Lett.* **37** (1976) 1465.

36. J. L. Vialle and H. T. Duong, *J. Phys. B* **12** (1979) 1407.

37. D. M. Larsen, *Phys. Rev. Lett.* **39** (1977) 878.

38. M. Rosenbluh, T. A. Miller, D. M. Larsen and B. Lax, *Phys. Rev. Lett.* **39** (1977) 874.

39. F. A. Jenkins and E. Segré, *Phys. Rev.* **55** (1939) 52.

40. R. Beigang, W. Makat and A. Timmermann, *Opt. Commun.* **49** (1984) 253.

41. G. Hertz, *Z. Phys.* **18** (1923) 307; K. H. Kingdom, *Phys. Rev.* **21** (1923) 408.

42. C. J. Lorenzen and K. Niemax, *J. Quant. Spectrosc. Radiat. Transfer* **22** (1979) 247.

43. K. C. Harvey and B. P. Stoicheff, *Phys. Rev. Lett.* **38** (1977) 537.

44. F. Penent, D. Delande, F. Biraben, C. Chardonnet and J. C. Gay, *Laser Spectroscopy VI*, eds. H. P. Weber and W. Luthy (Springer, 1983).

45. F. Penent, D. Delande, F. Biraben and J. C. Gay, *Proceedings of the 15th EGAS Conference*, ed. European Physical Society (1983).

46. K. T. Lu, F. S. Tomkins, H. M. Crosswhite and H. Crosswhite, *Phys. Rev. Lett.* **41** (1978) 1034.

47. S. Feneuille, S. Liberman, E. Luc-Koenig, J. Pinard and A. Taleb, *Phys. Rev. A* **25** (1982) 2853.

48. G. V. Marr and S. R. Wherrett, *J. Phys. B* **5** (1972) 1735.

49. D. R. Lyons, A. L. Schawlow and G. Y. Yan, *Opt. Commun.* **38** (1981) 35.

50. P. Hannaford and G. W. Series, *J. Phys. B* **14** (1981) L661.

51. J. P. Grandin and X. Husson, *J. Phys. B* **14** (1981) 433.

52. J. P. Lemoigne, J. P. Grandin, X. Husson and H. Kucal, *J. Phys.* **45** (1984) 249.

53. J. P. Grandin, X. Husson, H. Kucal and J. P. Lemoigne, in *Atomic Excitation and Recombination in External Fields*, eds. M. H. Nayfeh and C. W. Clark (Gordon and Breach, New York, 1985).

54. D. Delande and J. C. Gay, *Phys. Lett.* **82A** (1981) 399.

55. N. P. Economou, R. R. Freeman and P. F. Liao, *Phys. Rev. A* **18** (1978) 2506.

56. K. J. Drinkwater, J. Hormes, D. D. Burgess, J. P. Connerade and R. C. M. Learner, *J. Phys. B* **17** (1984) L439.

57. W. W. Smith *et al.*, in *Atomic Excitation and Recombination in External Fields*, eds. M. H. Nayfeh and C. W. Clark (Gordon and Breach, New York, 1985).

58. H. R. Traubenberg, R. Gebauer et G. Lewin, *Naturwissenschaften* **18** (1930) 417.

59. L. D. Landau and E. M. Lifschitz, *Mécanique Quantique* (MIR, Moscow, 1966).

60. A. Durand, *Mécanique Quantique* (Dunod, Paris, 1970).

61. P. M. Morse and H. Feshbach, *Methods of Theoretical Physics* (McGraw Hill, New York, 1953).

62. P. J. Redmonds, *Phys. Rev.* **133** (1964) 1352.

63. As pointed out in J. R. Oppenheimer, *Phys. Rev.* **31** (1928) 66.

64. W. Pauli, *Z. Phys.* **36** (1926) 336.

65. P. S. Epstein, *Phys. Rev.* **28** (1926) 695.

66. C. Lanczos, *Z. Phys.* **62** (1930) 518; *Z. Phys.* **65** (1930) 431; *Z. Phys.* **68** (1931) 204.

67. H. J. Silverstone, *Phys. Rev. A* **18** (1978) 1853.

68. P. Froelich and E. Brandas, *Phys. Rev. A* **12** (1975) 1.

69. C. Cerjan, R. Hedges, C. Holt, W. P. Reinhardt, K. Scheibner and J. J. Wendoloski, *Int. J. Quantum Chem.* **14** (1978) 393.

70. R. J. Damburg and V. V. Kolosov, in *Rydberg States of Atoms and Molecules*, eds. R. F. Stebbings and F. B. Dunning (Cambridge University Press, 1983).

71. S. P. Alliluyev and I. A. Malkin, *J.E.T.P.* **39** (1974) 627.

72. E. Luc and A. Bachelier, *J. Phys. B* **13** (1980) 1769.

73. C. W. Clark, K. T. Lu and A. F. Starace, *Progress in Atomic Spectroscopy*, eds. H. J. Beyer and H. Kleinpoppen (Plenum, New York, 1984) Part C.

74. D. Kleppner, M. G. Littman and M. L. Zimmerman, in *Rydberg States of Atoms and Molecules*, eds. R. F. Stebbings and F. B. Dunning (Cambridge University Press, 1983).

75. P. Jacquinot, S. Liberman and J. Pinard, in *Etats Atomiques et Moléculaires couplés à un continuum* - Colloque International CNRS n° 273 (Les Editions du CNRS, Paris, 1977).

76. W. Cooke and T. F. Gallagher, *Phys. Rev. A* **17** (1978) 1276.

77. H. J. Silverstone and P. M. Koch, *J. Phys. B* **12** (1978) L537.

78. G. A. Baker, *Essentials of Padé Approximants* (Academic Press, New York, 1975).

79. B. Simon, *Ann. Phys. (N.Y.)* **58** (1970) 76.

80. R. J. Damburg and V. V. Kolosov, *J. Phys. B* **12** (1979) 2637.

81. U. Fano, *Comments Atomic Mol. Phys.* **10** (1981) 223.

82. D. A. Harmin, *Phys. Rev. A* **24** (1981) 2491.

83. D. A. Harmin, *Phys.Rev. A* **26** (1982) 2656.

84. D. A. Harmin, *Comments Atomic Mol. Phys.* **15** (1985) 281.

85. M. G. Littman, M. L. Zimmerman and D. Kleppner, *Phys. Rev. Lett.* **37** (1976) 486.

86. P. Jacquinot, *Laser Spectroscopy IV*, eds. H. Walther and R. W. Rotke (Springer, New York, 1979).

87. C. Chardonnet, These 3é cycle, Paris, 1983.

88. C. Cohen-Tannoudji and P. Avan, *Colloque International CNRS n° 273* (Les Editions du C.N.R.S., Paris, 1977).

89. U. Fano, *Phys. Rev.* **124** (1961) 1866; U. Fano *Nuovo Cimento* **12** (1935) 156.

90. C. Chardonnet, D. Delande and J. C. Gay, *Opt. Commun.* **51** (1984) 249.

91. M. L. Zimmerman, M. G. Littman, M. M. Kash and D. Kleppner, *Phys. Rev. A* **20** (1979) 2251.

92. S. Feneuille, S. Liberman, J. Pinard and A. Taleb, *Phys. Rev. Lett.* **42** (1979) 1404.

93. T. Bergeman, C. Harvey, K. B. Butterfield, H. C. Bryant, D. A. Clark, P. A. M. Gram, D. Mac Arthur, M. Davis, J. B. Donahue, J. Dayton and W. W. Smith, *Phys. Rev. Lett.* **53** (1984) 779.

94. T. S. Luk, L. Dimauro, T. Bergeman and H. Metcalf, *Phys. Rev. Lett.* **47** (1982) 83.

95. R. R. Freeman, N. P. Economou, G. C. Bjorklund and K. T. Lu, *Phys. Rev. Lett.* **41** (1978) 1463.

96. A. R. P. Rau and K. T. Lu, *Phys. Rev. A* **21** (1980) 1057.

97. S. Liberman and Ch. Blondel, in *Proceedings of Les Houches Summer School — New Trends in Atomic Physics*, eds. R. Stora and G. Grynberg (North-Holland, 1985).

98. J. Y. Liu, P. Mc Nicholl, D. Harmin, J. Ivri, T. Bergeman and H. J. Metcalf, *Phys. Rev. Lett.* **55** (1985) 189.

99. W. L. Glab and M. H. Nayfeh, *Phys. Rev. A* **31** (1985) 530.

100. J. H. M. Neijzen and A. Dönszelmann, *J. Phys. B* **15** (1982) 1981.

101. M. Matsuzawa, *Phys. Rev. A* **20** (1979) 860.

102. A. P. Hickman, *Phys. Rev. A* **28** (1983) 111.

103. S. M. Jaffe, N. H. Tran, H. B. Van Linden Van Den Heuvell and T. F. Gallagher, *Phys. Rev. A* **30** (1984) 1828; W. Sandner, K. A. Safinya and R. F. Gallagher, *Phys. Rev. A* **24** (1981) 1647.

104. E. B. Saloman, J. W. Cooper and D. E. Kelleher, *Phys. Rev. Lett.* **55** (1985) 193.

105. P. Zeeman, *Philos. Mag.* **5** (1897) 43.

106. J. H. Van Vleck, *The Theory of Elastic and Magnetic Susceptibilities* (Oxford University Press, 1932).

107. L. D. Landau, *Z. Phys.* **64** (1930) 629.

108. R. B. Laughlin, *Physica Scripta* **1268** (1984) 254.

109. K. Von Klitzing, G. Dorda and M. Pepper, *Phys. Rev. Lett.* **45** (1980) 494.

110. L. I. Schiff and H. Snyder, *Phys. Rev.* **55** (1939) 59.

111. B. S. Monozon and A. G. Zhilich, *Sov. Phys. Semicond.* **1** (1967) 563.

112. W. R. S. Garton, F. S. Tomkins and H. M. Crosswhite, *Proc. R. Soc. London* **A373** (1980) 189.

113. R. J. Fonck, D. H. Tracy, D. C. Wright and F. S. Tomkins, *Phys. Rev. Lett.* **40** (1978) 1366.

114. R. J. Fonck, F. L. Roesler, D. H. Tracy, K. T. Lu, F. S. Tomkins and W. R. S. Garton, *Phys. Rev. Lett.* **39** (1977) 1513.

115. A. F. Starace, *J. Phys. B* **6** (1973) 585.

116. W. E. Lamb, *Phys. Rev. A* **85** (1952) 259.

117. L. P. Gorkov and I. E. Dzyaloshinskii, *J.E.T.P.* **26** (1968) 449.

118. B. P. Carter, *J. Math. Phys.* **10** (1968) 788.

119. H. Hasegawa and R. E. Howard, *Phys. Chem. Solids* **21** (1961) 179.

120. S. M. Kara and M. R. C. Mc Dowell, *J. Phys. B* **13** (1980) 1337.

121 H. Crosswhite, U. Fano, K. T. Lu and A. R. P. Rau, *Phys. Rev. Lett.* **42** (1979) 963.

122. A. R. P. Rau, *Comments Atomic Mol. Phys.* **10** (1980) 19.

123. D. Delande, Thése 3ᵉ, Paris, 1981.

124. P. Cacciani, Thése 3ᵉ cycle, Orsay, 1984.

125. M. L. Zimmerman, M. M. Kash and D. Kleppner, *Phys. Rev. Lett.* **45** (1980) 1092.

126. D. Delande and J. C. Gay, *Phys. Lett. A* **82** (1981) 393.

127. D. Kleppner, M. G. Littman and M. L. Zimmerman, *Sci. Am.* **244** (1981) 108.

128. D. Delande, C. Chardonnet, F. Biraben and J. C. Gay, *J. Phys.* **43-C2** (1982) 97.

129. J. C. Castro, M. L. Zimmerman, R. G. Hulet and D. Kleppner, *Phys. Rev. Lett.* **15** (1980) 1780.

130. E. A. Solovev, *J.E.T.P.* **82** (1982) 1762.

131. C. W. Clark, *Phys. Rev. A* **24** (1981) 605.

132. M. Robnik, *J. Phys. A* **14** (1981) 3195.

133. J. C. Gay, in *Atoms in Unusual Situations*, ed. J. P. Briand (Plenum, New York, 1986).

134. N. P. Economou, R. R. Freeman and P. F. Liao, *Phys. Rev. A* **18** (1979) 2506.

135. F. Penent, C. Chardonnet, D. Delande, F. Biraben and J. C. Gay, *Colloque International CNRS n° 352 J. Phys.* **44-C7** (1983) 193.

136. D. Delande, C. Chardonnet and J. C. Gay, *Opt. Commun.* **42** (1982) 25.

137. A. R. P. Rau, *J. Phys. B* **12** (1979) L193.

138. R. F. O'Connell, *Phys. Lett.* **60A** (1977) 481.

139. C. W. Clark and K. T. Taylor, *J. Phys. B* **13** (1980) L737.

140. C. W. Clark and K. T. Taylor, *J. Phys. B* **15** (1982) 1175.

141. C. W. Clark and K. T. Taylor, *Nature* **292** (1981) 437.

142. V. Bargmann, *Z. Phys.* **99** (1936) 576.

143. L. D. Landau and E. M. Lifschitz, *Mécanique* (MIR, Moscow, 1966).

144. A. M. Perelomov, *Sov. Phys. Usp.* **20**, 9 (1977) 703.

145. M. Bander and C. Itzykson, *Rev. Mod. Phys.* **38** (1966) 330.

146. M. E. Kellman and D. R. Herrick, *J. Phys. B* **11** (1978) L755.

147. D. R. Herrick, *Phys. Rev. A* **26** (1982) 323.

148. J. J. Labarthe, *J. Phys. B* **14** (1981) L467.

149. A. O. Barut, *Lectures in Theoretical Physics* (Gordon and Breach, New York, 1967).

150. A. O. Barut and C. Fronsdal, *Proc. Roy. Soc.* **A287** (1965) 532.

151. D. Delande and J. C. Gay, *J. Phys. B* **17** (1984) L335.

152. D. Delande and J. C. Gay, *J. Phys. B* **19** (1986) L173.

153. M. J. Englefield, *Group Theory and the Coulomb Problem* (Wiley, New York, 1971).

154. E. U. Condon and H. Odabasi, *Atomic Structure* (Cambridge University Press, 1980).

155. D. Delande, Thése d'Etat, Paris, 1988.

156. F. Penent, D. Delande, F. Biraben and J. C. Gay, *Opt. Commun.* **49** (1984) 184.

157. F. Penent, Thése 3e cycle, Paris, 1984.

158. J. C. Gay and D. Delande, in *Atomic Excitation and Recombination in External Fields*, eds. M. H. Nayfeh and C. W. Clark (Gordon and Breach, New York, 1985).

159. B. R. Judd, *Angular Momentum Theory for Diatomic Molecules* (Academic Press, New York, 1975).

160. E. A. Solovev, *J.E.T.P. Lett.* **34** (1981) 265.

161. T. P. Grozdanov and E. A. Solovev, *J. Phys. B* **15** (1982) 1195.

162. J. C. Gay, D. Delande, F. Biraben and F. Penent, *J. Phys. B Lett.* **16** (1983) L693.

163. J. C. Gay and D. Delande, *Comments Atomic Mol. Phys.* **13, 6** (1983) 275.

164. U. Fano, *Rep. Prog. Phys.* **46** (1983) 97.

165. A. R. Edmonds, *J. Phys.* **31-C4** (1970) 71.

166. A. R. Edmonds, preprints, Imperial College, 1981.

167. M. Robnik and E. Schrufer, *J. Phys. A* **18** (1985) L853.

168. S. J. Buckman, P. Hammond, G. C. King and F. H. Read, *J. Phys. B* **16** (1983) 4219.

169. P. Fournier-Lagarde, J. Mazeau and A. Huetz, *J. Phys. B* **17** (1984) L591.

170. G. H. Wannier, *Phys. Rev.* **90** (1953) 817.

171. Since the writing of this review, experimental developments, especially in the group of K. Welge, as well as theoretical analysis on classical chaos and its quantum counterpart, add a new chapter to the story... The hydrogen atom in a magnetic field is now the quantum prototype "par excellence" for both experimental and theoretical studies on this matter of chaos. Some of the recent advances are described e.g. in D. Delande and J. C. Gay Comments Atomic Mol. Phys. **19** (1986) 35, A.R.P. Rau, *Nature* **325** (1987) 577, and A. Holle *et al.* *Z. Phys. D* **5** (1987) 279.

172. Y. Alhassid, E. A. Hinds and D. Meschede, *Phys. Rev. Lett.* (1987) **5** October.

CHAPTER 8

SPECTROSCOPY OF POSITRONIUM

A. P. Mills, Jr.

8.1 Introduction

Positronium $(e^+ - e^-)$ is the lightest of the exotic hydrogen-like atoms made of a lepton and an anti-lepton. The other members of this set include muonium $(\mu^+ - e^-)$, anti-muonium $(\mu^- - e^+)$, muium $(\mu^+ - \mu^-)$, and all the analogous atoms containing a tauon. Since the leptons are considered to be structureless particles, careful measurements of the lifetimes and energy levels of these atoms should give us direct information about the lepton-anti-lepton interaction. The lighter atoms tell us mostly about the electromagnetic interactions; the heavier atoms can also probe weak interaction effects. To date only positronium (Ps) and muonium (Mu) have been produced in the laboratory. The present chapter will discuss only the positronium atom.

The intrinsic properties of positronium that can be measured are its mass, total charge, electric and magnetic moments, its annihilation rates and branching ratios, its energy levels and their radiative decay rates. Some of these properties are illustrated by the diagram in Fig. 8.1. The extrinsic properties

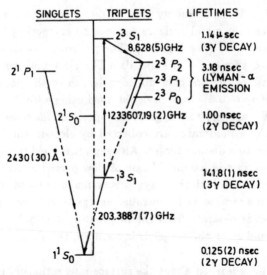

Fig. 8.1. Energy levels of the $n = 1$ and $n = 2$ states of positronium. The quantities with error estimates in parentheses are measured values.

would include the cross sections for excitation and ionization by photons and other projectiles, the electron affinity, the properties of the positronium ion, interactions with atoms, molecules, surfaces, and condensed matter. In the present context it is suitable that we discuss mostly the intrinsic properties, which we shall do after a brief digression on sources of positronium. There are many review articles and books that cover various aspects of positron and positronium physics; a few of these are listed in the references.[1-5]

8.2   Positronium Formation Methods

8.2.1   *Fast positrons*

In 1951, M. Deutsch[6] discovered that Ps can be formed in a gas that is being irradiated with energetic positrons from a radioactive source. The positronium formed in a gas is useful for measuring the decay rates and the hyperfine interval of the ground state, provided an extrapolation is made to zero gas pressure. While some progress has been made in choosing a good gas for these experiments, studies involving the excited states of Ps must be done in vacuum to eliminate the severe perturbations due to gas collisions.

8.2.2   *Slow positrons*

Positrons with a thermal energy spread and a total energy of a few eV are emitted from certain metal surfaces exposed to energetic positrons from a radioactive source.[7-14] (See Fig. 8.2). The positrons are emitted because they have a negative affinity for the metals. The slow positron yield is approximately the ratio of the positron diffusion length to the positron range in the solid. Yields of up to 0.2% have been obtained using a Co^{58} positron source and a well annealed single crystal tungsten slow-positron moderator.[15,16] The slow positrons from the moderator are collected by electric and/or magnetic fields and transported to a distant target. Although this yield is not very large, there are many advantages in having a beam of slow particles. At the distant target, the background from nuclear γ-rays is very much reduced; the positrons may be injected into a sample with controlled energy to achieve different conditions; since the target is separate from the moderator, it may be chosen to produce the most Ps; and since the target is in vacuum, the Ps so produced suffers no collisions.

Much has been learned about the surface interactions of positrons since the first experiments by Canter, Mills, and Berko.[17] For the present purposes it is sufficient to mention only that we can form thermal energy Ps by thermally desorbing positrons from the surface of a clean metal. (See for example Refs. 18 and 19.) The velocity spectrum of the slow Ps desorbed from an Al surface is illustrated in Fig. 8.3. In this way one may convert about half of the beam

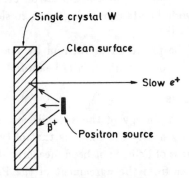

Fig. 8.2. Production of slow positrons. Energetic positrons from a radioactive source or from pair production are stopped in a clean single crystal moderator, typically made of W or Cu. Some of the positrons can diffuse to the surface where they are ejected into the vacuum because of their negative affinity for the metal.

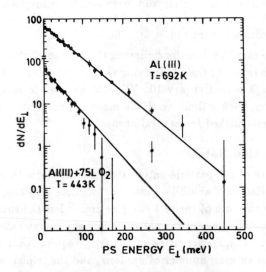

Fig. 8.3. Spectra of the perpendicular energy component of the positronium thermally desorbed from Al(111) surfaces at two different temperatures. The fitted exponential curves imply positronium temperatures of 464(51)K and 636(64)K for the two cases. (Reproduced with permission from A. P. Mills, Jr. and L. Pfeiffer, *Phys. Rev.* **B32** (1985) 53, (American Physical Society)).

positrons into slow Ps. A small laboratory positron beam can produce 10^6 Ps atoms per s or 5 nsec wide bursts containing some 20 slow Ps atoms each at a rate of 1000 bursts per s.[20]

Fast Ps atoms may be produced in vacuum by charge exchange at a thin foil[21] or a gas target,[22] or by ionization of beam-foil produced Ps$^-$ ions.[23]

8.3   Positronium Mass

Measurements of the total energy of the annihilation γ—rays imply that the total energy of Ps is very nearly the same as that of two electron masses.[24] However, the inertial mass of Ps has not been measured directly with very great precision. We may infer from the agreement of the Ps time-of-flight energy spectrum in Fig. 8.3 with a Maxwellian-beam that M_{Ps} is within 10% of being equal to two electron masses.[25] It presently might be feasible to measure the gravitational mass of Ps by cooling it with Lyman-α photons to 0.1 K,[26,27] exciting it to the $n = 100$ state,[28] and watching it fall for a fraction of one second. As emphasized by Leventhal,[29] such an experiment would be similar to the Witteborn-Fairbank[30] falling-electron experiment, but the problems associated with shielding electric and magnetic fields might not be so severe.

8.4   Positronium Charge

While there are many reasons for believing that the total charge of Ps is exactly zero, there is no reported test. Observing the energy of the annihilation photons from Ps at a high potential (say 10^7 V) could be used to establish that $Q_{Ps} <$ 10^{-8} e. However, such a limit would be much poorer than the analogous limit that has been established for neutral atoms.[31]

8.5   Annihilation Rates

A unique feature of the particle-antiparticle pair atoms is that they annihilate. Positrons and electrons annihilate into one or more photons, the total energy of which equals the sum of the rest energies $2m_e c^2$ less the total binding energy and plus the kinetic energy of the pair, and less any energy taken up by recoiling particles. In vacuum, conservation of C-parity requires that the singlet states of Ps decay into an even number of photons, and the triplet states into an odd number. Tests of C-parity non-conservation by searches for the violation of this rule[32,33] are hopelessly less sensitive than required to observe effects predicted by the standard model of the weak interaction.[34]

The decay rates of singlet and triplet Ps have been measured to a precision sufficient to test the first order corrections to the calculated rates by Gidley,

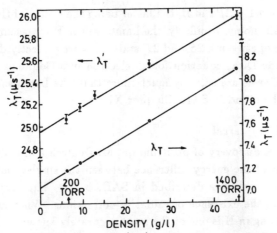

Fig. 8.4. Triplet $m = 0$ $\left(\lambda'_T\right)$ and $m = 1$ $\left(\lambda_T\right)$ decay rates measured at various isobutane gas densities in an applied magnetic field of 3.4432 kG. (Reproduced with permission from D. W. Gidley, A. Rich, E. Sweetman, and D. West, *Phys. Rev. Lett.* **49** (1982) 525 (American Physical Society)).

Rich, Sweetman, and West.[35] (See Fig. 8.4.) The singlet Ps decay rate is measured to be 7.994(11)ns^{-1} in agreement with the Harris and Brown calculation that includes a radiative correction of -0.07ns^{-1}.[36] The experiment is not accurate enough to be sensitive to the latest $\alpha^2 \ln \alpha$ corrections.[37] The triplet decay rate has been determined in the same experiment to be 7.0516(13)μs^{-1} in significant disagreement with the 7.0386μs^{-1} calculated in Ref. 37. The experiment is in agreement with several other determinations, and the reason for the discrepancy is not known at this time.

Singlet and triplet positronium can also decay into four and five photons respectively. The rates have not been measured, but the branching ratios of these higher order processes are calculated to be about 0.01α^2, or about one part per million.[38,39]

8.6 Positronium Energy Levels

The gross structure of the Ps energy levels is similar to H, except that the reduced mass of Ps is $\frac{1}{2}m_e$ instead of approximately m_e. The bound state energies are thus about half of the corresponding energies in H. (See Fig. 8.1.) The fine structure of the Ps bound states is qualitatively different from that of H because the magnetic moment of the positron is about 10^3 times greater than that of the proton. Thus the ground state hyperfine interval is 203 GHz for Ps, whereas it is only 1.4 GHz in H, the latter being the source of the famous 21 cm line of radio astronomy. Similarly, the Lamb shift in H represents a purely QED energy splitting between a $2S$ and $2P$ state that are perfectly degenerate when described by the Dirac equation for an electron in a Coulomb potential; in Ps the analogous levels are split by much more than the Lamb-shift terms. More details are to be found in SAH, Chapter XII.

8.6.1 *Hyperfine interval*

Shortly after his discovery of positronium, M. Deutsch and collaborators made a measurement of the energy difference between the singlet and triplet ground states.[40] This work was described in SAH, p. 81. The accuracy was sufficient to confirm the existence of radiative corrections. Whereas the analogous hyperfine splitting in H is one of the most accurately known quantities in all of physics, primarily because of the long lifetime of the hyperfine transition and the availability of atomic H, the Ps hyperfine interval is difficult to measure because of the short $^1S_0$ lifetime and the problems of producing the Ps atoms in quantity. Nevertheless, the Ps hyperfine interval has now been measured to a remarkable accuracy in a series of experiments done principally by Hughes and collaborators at Yale.[41,42]

The Zeeman levels of positronium are shown in Fig. 8.5. Since the hyperfine interval is at an inconvenient frequency, all measurements of this quantity have depended on the measurement of the triplet $m = 0$ to $m = 1$ splitting in a magnetic field, eg. the transition f_0 in Fig. 8.5. Positronium is formed in a microwave cavity in a uniform magnetic induction B. When the resonant condition of B and f_0 occurs, there is an increase in the 2γ annihilation yield. This is because the $m = 1$ states, ordinarily decaying only via 3γ's, are mixed with the triplet and singlet $m = 0$ states by the rf magnetic field. Since large rf fields are required, the experiments use a tuned microwave cavity, and one moves through the resonance condition by varying the magnetic field, as shown in Fig. 8.6. Using the Breit-Rabi formula, one may derive a value for the hyperfine interval from the position of the peak of the resonance. A substantial shift in the resonance due to Ps collisions with the molecules of the buffer gas is removed by extrapolating to zero gas density. Fig. 8.7 shows the latest result of the Yale group.[41]

The theoretical interpretation of experimental resonance curves like the one in Fig. 8.6 is complicated by a small bound state shift in the electron and positron g factors,[43,44] and by the resonance line shape being slightly asymmetrical.[45-47] The latest result, the weighted average of the Yale

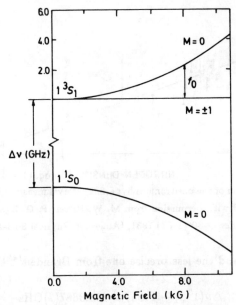

Fig. 8.5. Zeeman energy levels for positronium in the ground state. (Reproduced with permission from M. W. Ritter, P. O. Egan, V. W. Hughes and K. A. Woodle, *Phys. Rev.* **A30** (1984) 1331, (American Physical Society)).

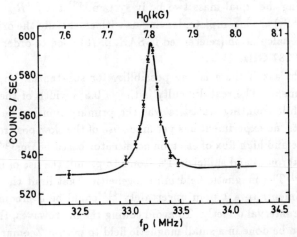

Fig. 8.6. Typical Zeeman resonance used to determine the positronium hyperfine interval (Reproduced with permission from M. W. Ritter, P. O. Egan, V. W. Hughes and K. A. Woodle, *Phys. Rev.* **A30** (1984) 1331, (American Physical Society)).

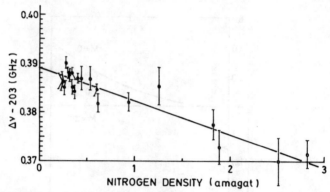

Fig. 8.7. Extrapolation of the positronium hyperfine interval measurements to zero buffer gas density. (Reproduced with permission from M. W. Ritter, P. O. Egan, V. W. Hughes and K. A. Woodle, *Phys. Rev.* **A30** (1984) 1331, (American Physical Society)).

measurement[41] and the less precise one from Brandeis[42] is

$$\Delta\nu(1^3S_1 - 1^1S_0) = 203.3887(7)\,\text{GHz} .$$

Unfortunately, the theory is very difficult, even though Ps is seemingly so simple. In spite of advances in the understanding of the Bethe-Salpeter equation describing the equal mass two body system,[48] the $\alpha^4 R_\infty$ corrections to $\Delta\nu(1^3S_1 - 1^1S_0)$ are incomplete, leaving us with essentially the original Karplus and Klein[49] calculation (referenced in SAH, p. 78) good to order $\alpha^3 R_\infty$. Their value was 203.37 GHz.

One might ask if there is any possibility for substantial improvements in the measurement. The real difficulty is the $\approx 0.5\%$ width of the resonance in Fig. 8.6. While counting statistics was the primary limiting factor in all the measurements, no experiment has yet made use of the slow positron beam. The pulsed nature and high flux of electron accelerator based beams[8,50] would give us a low background and sufficient Ps atoms to permit the use of line-narrowing techniques.[5] The magnetic field inhomogeneities that limit the Zeeman resonance experiments to a 1 ppm uncertainty[41] could be overcome by exciting the hyperfine interval directly. To avoid tuning the microwaves, the experiment would have to be done in a small magnetic field to permit Zeeman tuning; however, the precision of that field could be an order of magnitude poorer than in the usual Zeeman resonance experiment. Fig. 8.8 shows a calculation[51] of a typical hyperfine resonance at a microwave frequency of 207 GHz.

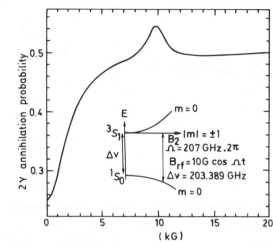

Fig. 8.8. Calculated 2γ annihilation probability for positronium in a variable magnetic field in the z-direction and excited by an rf magnetic field in the x-direction with constant frequency.

8.6.2   Ps Lamb shift

The production of excited states of positronium and the measurement of its fine structure was a long standing goal[52] finally reached in 1975 thanks to the introduction of the slow positron beam. The discovery that positronium can be formed in vacuum by slow positrons impinging on a metal surface[17] made it seem likely that a small fraction of the atoms would be formed in excited states. Whereas any Ps formed in excited states in the presence of a dense buffer gas would be rapidly quenched, the excited states of Ps in vacuum would have a relatively long lifetime in which to perform spectroscopic measurements. Fig. 8.9 shows the slow positron beam apparatus for observing Ps Lyman-α radiation.[53] Slow positrons from a MgO moderator[9] were guided by a bent solenoid to a target that is viewed by a uv-sensitive photomultiplier behind one of three interference filters centered on the expected Lyman-α wavelength of 2430Å and at 30Å on either side. Fig. 8.10 shows that there is a significant increase in the uv detector count rate detected in coincidence with an annihilation γ-ray when the filter is chosen to be at the Lyman-α wavelength. The signal goes away when the positron energy is increased from 25 eV to 400 eV. Ironically, when the positrons are implanted at higher energies, they diffuse back to the surface with a more nearly thermal energy distribution.[12] Evidently, formation of the $2S$ state requires positrons to reach the surface with several eV of kinetic energy.

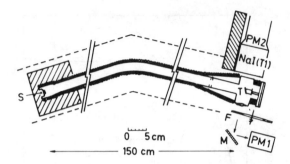

Fig. 8.9. Slow positron beam apparatus for observing positronium Lyman-α radiation. *S*, ⁵⁸Co source; *T*, target; *F*, ultraviolet filter wheel; *M*, mirror; PM1, Lyman-α photon detector; PM2-NaI(Tl), annihilation γ-ray detector. The slow positrons were produced by a MgO powder moderator and guided to the target in vacuum by a long bent solenoid (Reproduced with permission from K. F. Canter, A. P. Mills, Jr. and S. Berko, *Phys. Rev. Lett.* **34** (1975) 177, (American Physical Society)).

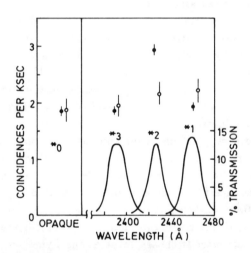

Fig. 8.10. Photon detector-annihilation γ-ray coincidence count rate versus uv filter. There is a Lyman-α signal at 2430Å with 25eV positrons (solid circles), but no signal with 400 eV positrons (open circles). (Reproduced with permission from K. F. Canter, A. P. Mills, Jr. and S. Berko, *Phys. Rev. Lett.* **34** (1975) 177, (American Physical Society)).

The yield of $2S$ positronium in this experiment was about 10^{-4} $2S$ atoms per incident positron, or about $10 s^{-1}$, sufficient to allow measurement of a fine structure interval in the $n = 2$ state.[54] The $2^3S_1 - 2^3P_2$ interval was chosen for the measurement because it was predicted to lie in the X band and involves the relatively long-lived triplet S state (see Fig. 8.1). The measurement was based on the observation of an enhanced Lyman-α emission when $2^3S_1 - 2^3P_2$ transitions were induced by an rf electric field at the resonance frequency. Fig. 8.11 shows the target of the slow positron beam apparatus used in the measurement. The positrons impinge on the back of a small microwave cavity. Lyman-α photons are detected in coincidence with $1S$ annihilation gammas, and the count rate is measured as a function of the microwave frequency. The resonance and its derivative shown in Fig. 8.12 lead to a value for the fine structure interval $\Delta\nu(2^3S_1 - 2^3P_2) = 8.628(5)$ GHz in agreement with the 8.625 GHz calculated by Fulton and Martin[55] (also referenced in SAH, p. 78).

The natural linewidth of the triplet $2S - 2P$ fine structure intervals in Ps is 50 MHz due to the 3 ns ^3P radiative lifetime. If the hyperfine interval measurements are any guide, we may expect that eventually one could improve the fine structure measurements by at least a factor of 100 given a sufficient count rate. There have been many improvements in positron beam techniques in the last ten years, but only small increases in the efficiency of $n = 2$ state production.[56] Recently Hatamian, Conti, and Rich[57] measured all three $2S - 2P$ intervals and obtained $\Delta\nu(2^3S_1 - 2^3P_2) = 8620(3)$ MHz.

8.6.3 1S-2S interval

In positronium, the largest shifts combined with the narrowest linewidths are to be found in the triplet S states. To measure the energy of one of these states, we could either weigh the atom, look at the annihilation photons, find the threshold for photoionization, or compare one bound state level with another. In the hyperfine and fine structure measurements, the comparison 1^1S_0 or 2^3P level has a large width. Using first order Doppler-free two photon spectroscopy,[58] we can measure the difference between two triplet S levels with a combined natural linewidth of only 1 MHz.

Figure 8.13 shows the transitions involved in the $1S - 2S$ resonant three-photon ionization of Ps. If the two photons that make the $1S - 2S$ transition are oppositely directed and of equal frequencies, the blue shift of one photon due to the motion of the Ps atom will be canceled by the red shift of the other in first order. This is very important in Ps because of its light mass and concomitant high velocity compared to H atoms at the same temperature.

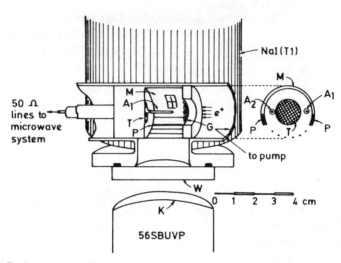

Fig. 8.11. Positron target chamber and microwave cavity for measuring a fine structure interval in the $n = 2$ state of positronium. G, grid; T, copper target; M, mirror; W, window; K, CsTe photocathode of the uv photon detector; A, antennae; NaI(Tl), annihilation γ-ray detector (Reproduced with permission from A. P. Mills, Jr., S. Berko and K. F. Canter, *Phys. Rev. Lett.* **34** (1975) 1541, (American Physical Society)).

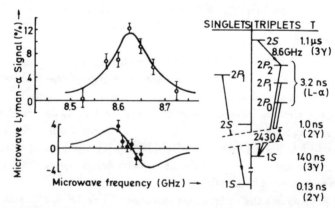

Fig. 8.12. Lyman-α signal and its logarithmic first difference as a function of microwave frequency. The inset is the schematic term diagram for the $n = 1$ and $n = 2$ Ps states indicating the relevant transitions and the lifetimes for each level (Reproduced with permission from A. P. Mills, Jr., S. Berko and K. F. Canter, *Phys. Rev. Lett.* **34** (1975) 1541, (American Physical Society)).

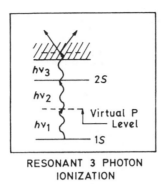

RESONANT 3 PHOTON
IONIZATION

Fig. 8.13. 1S-2S resonant three photon ionization of positronium.

Since the $1S - 2S$ transition is forbidden, a high power and therefore pulsed laser is required. The Ps atoms must be formed in vacuum to avoid collisions, and this dictates the use of a positron beam. The slow positrons must have a very high instantaneous intensity to be compatible with the time structure of the laser. The only successful experiments[59−62] were done using a small laboratory beam that was bunched using a suddenly switched on harmonic potential as shown in Fig. 8.14. It would of course be an improvement to use the intense positron pulses that can be produced by an electron accelerator.[8,50]

In the experiment of Chu, Mills, and Hall[62] Ps was formed by 10 ns bursts of about 100 positrons that struck a clean Al(111) target surface in ultrahigh vacuum. The target was kept at 300° C to desorb the surface state positrons as free thermal Ps in vacuum (see Fig. 8.3). The Ps was excited from the 1^3S_1 state to the 2^3S_1 state by two counter-propagating 486 nm laser pulses. As shown in Fig. 8.15, the light was narrowed in frequency by a Fabry-Perot interferometer in the vacuum system. Ps atoms from the Al target were ionized by the light and collected by an electron multiplier detector. Fig. 8.16 shows a single 5 min. scan of the Ps resonance along with a simultaneously recorded Te_2 reference line and the frequency marker signal. While the line center relative to the Te_2 reference line can be determined to 5 ppb in one scan, accounting for systematic effects limited the final accuracy of this experiment to about 12 ppb.

The major systematic effects are as follows:

(a) The ac Stark shift introduced by the intense laser beam. This was accounted for by extrapolating the Ps resonance line center measurement to zero laser power as shown in Fig. 8.17.

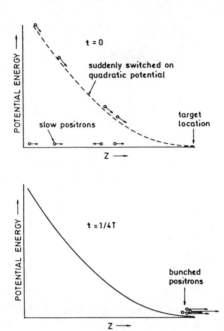

Fig. 8.14. Quadratic potential bunching of a slow positron beam.

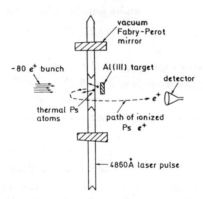

Fig. 8.15. Thermal positronium-laser beam interaction region. Positronium is formed by a bunch of positrons that is stopped by a clean Al surface in ultrahigh vacuum. Positronium atoms thermally desorbed form the surface are ionized by the laser and the e^+ fragments are collected by a single particle detector. The laser pulse is narrowed in frequency by the Fabry-Perot interferometer (Reproduced with permission from S. Chu, A. P. Mills, Jr., and J. L. Hall, *Phys. Rev. Lett.* **52** (1984) 1689, (American Physical Society)).

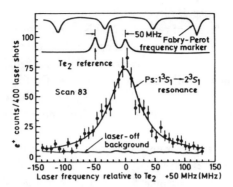

Fig. 8.16. Resonant three-photon ionization of positronium due to $1^3S_1 + 3h\nu \rightarrow 2^3S_1 + h\nu \rightarrow e^+ + e^-$. The Te_2 reference line has been split into three lines by acousto-optically modulating the cw dye laser at 50 MHz. For this scan, the line center was 25.9± 2.7 MHz above the Te_2 line (Reproduced with permission from S. Chu, A. P. Mills, Jr., and J. L. Hall, *Phys. Rev. Lett.* **52** (1984) 1689, (American Physical Society)).

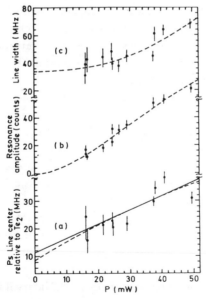

Fig. 8.17. Frequency shift, linewidth, and resonance amplitude plotted vs laser power transmitted through the Fabry-Perot cavity of Fig. 8.15. The simultaneous fit of three theoretical curves to the data is given by the dashed lines. The simple quadratic Stark shift is shown by the solid line (Reproduced with permission from S. Chu, A. P. Mills, Jr. and J. L. Hall, *Phys. Rev. Lett.* **52** (1984) 1689, (American Physical Society)).

(b) The second order Doppler shift (time dilation) due to the Ps motion relative to the laser frame of reference. This was removed by extrapolating the measurements to zero Ps velocity as shown in Fig. 8.18.

(c) The frequency offset between the cw dye laser sensing the Te_2 reference line and the high power pulsed laser causing the Ps $1S - 2S$ transition. This offset was measured directly using a scanning Fabry-Perot interferometer. The pulsed laser was typically 20 MHz lower than the cw frequency as shown in Fig. 8.19.

The measurements were reduced to an absolute value for the $1^3S_1 - 2^3S_1$ interval by measuring the frequency of the Te_2 reference line relative to the deuterium $2S_{1/2} - 4P_{3/2}$ Balmer β line. The absolute value of the deuterium line was obtained from the calculation by Erickson[63] updated using a newer value of the Rydberg constant.[64] The final result was $\Delta\nu(1^3S_1 - 2^3S_1) = 1\,233\,607\,185 \pm 16$ MHz. This differs from the calculation of Fulton[65] by $(13 \pm 10.5 \pm 10.6)$ MHz, where the first error is from the Ps measurement relative to the Te_2 line and the second error is from the absolute Te_2 calibration. The theory does not include $\alpha^4 R_\infty$ terms that might contribute about 10 MHz. The Te_2 reference line has been now re-calibrated by McIntyre and Hänsch.[66] According to these authors the new calibration leads to the value $1\,233\,607\,142.9\,(10.7)$ MHz for the $1^3\,S_1 - 2^3S_1$ interval, which significantly lessens the agreement between theory and experiment.

The precision of the $1S - 2S$ measurement can certainly be improved greatly. With some advances in laser metrology and positronium sources, one can expect an eventual measurement that splits the natural linewidth by a factor of 10^3. Given a significant advance in the theory of the relativistic two body system, we will then have a very good test of QED, and a very precise value for the positronium Rydberg constant. The present precision of 12 ppb implies that the masses of the electron and positron are equal to within 40 ppb, as expected from TCP conservation. An experiment capable of great precision might also allow us to observe E1-M1 vs E1-E1 interference effects due to the weak interaction.[34] As pointed out in Ref. 34, the measurement of such parity-violating interference terms offers the possibility of determining the weak angle Θ_W in a purely leptonic system at low energies.[34]

Note: A preliminary measurement of the $1S$-$2S$ interval in muonium, made by a similar method, has recently been reported.[67]

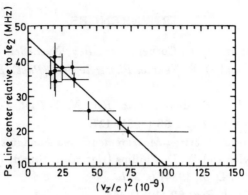

Fig. 8.18. Ps line center relative to the Te_2 reference line plotted vs ν_z^2/c^2 showing the second order Doppler shift (Reproduced with permission from S. Chu, A. P. Mills Jr. and J. L. Hall, *Phys. Rev. Lett.* **52** (1984) 1689, (American Physical Society)).

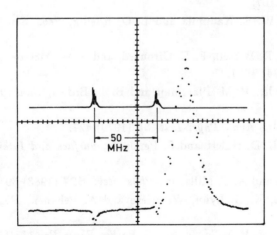

Fig. 8.19. Frequency offset of the pulsed laser (dotted curve) relative to the acousto-optically modulated cw laser beam (solid curve). The horizontal axis is the frequency scale reading from high to low frequency (Reproduced with permission from S. Chu, A. P. Mills, Jr. and J. L. Hall, *Phys. Rev. Lett.* **52** (1984) 1689, (American Physical Society)).

REFERENCES

1. S. deBenedetti and H. C. Corben, *Ann. Rev. Nucl. Sci.* **4** (1954) 191.
2. G. T. Bodwin and D. R. Yennie, *Phys. Reports (Phys. Lett. C)* **43** (1978) 267.
3. S. Berko and H. N. Pendleton, *Ann. Rev. Nucl. Part. Sci.* **30** (1980) 543.
4. A. Rich, *Rev. Mod. Phys.* **53** (1981) 127.
5. V. W. Hughes, in *Precision Measurements and Fundamental Constants II*, eds. B. N. Taylor and W. D. Phillips (Nat. Bur. Stand. (U.S), Spec. Publ. 617, Washington D.C., 1984) p. 237.
6. M. Deutsch, *Phys. Rev.* **82** (1951) 455.
7. W. Cherry, Ph.D. Thesis, Princeton University (1958) available from University Microfilms Inc., Ann Arbor, MI, USA.
8. D. G. Costello, D. E. Groce, D. F. Herring and J. W. McGowan, *Phys. Rev.* **B5** (1972) 1433.
9. K. F. Canter, P. G. Coleman, T. C. Griffith, and G. R. Heyland, *J. Phys.* **B 5** (1972) L167.
10. T. S. Stein, W. E. Kauppila and L. O. Roellig, *Rev. Sci. Instrum.* **45** (1974) 951.
11. S. Pendyala, D. Bartell, F. E. Girouard, and J. W. McGowan, *Phys. Rev. Lett.* **33** (1974) 1031.
12. A. P. Mills, Jr., P. M. Platzman and B. L. Brown, *Phys. Rev. Lett.* **41** (1978) 1076.
13. A. P. Mills, Jr., *Appl. Phys. Lett.* **35** (1979) 427.
14. J. M. Dale, L. D. Hulett and S. Pendyala, *Surface and Interface Analysis* **2**(1980) 199.
15. R. J. Wilson and A. P. Mills, Jr., *Phys. Rev.* **B27** (1983) 3949.
16. P. J. Schultz, K. G. Lynn, W. Frieze and A. Vehanen, *Phys. Rev.* **B27** (1983) 6626.
17. K. F. Canter, A. P. Mills, Jr. and S. Berko, *Phys. Rev. Lett.* **33** (1974) 7.
18. A. P. Mills, Jr., in *Positron Solid State Physics*, eds. W. Brandt and A. Dupasquier (North-Holland, Amsterdam, 1983) p. 432.
19. G. K. Lynn, in *Positron Solid State Physics*, eds. W. Brandt and A. Dupasquier (North-Holland, Amsterdam, 1983) p. 609.
20. A. P. Mills, Jrs. in *Positron Scattering in Gases*, eds. J. W. Humberston and M. R. C. McDowell (Plenum, New York, 1984) p. 121.
21. A. P. Mills, Jr. and W. S. Crane, *Phys. Rev.* **A31** (1985) 593.

22. B. L. Brown, in *Positron Studies of Solids, Surfaces and Atoms,* eds. A. P. Mills, Jr., W. S. Crane, and K. F. Canter (World Scientific, Singapore, 1986) p. 160.

23. A. P. Mills, Jr., *Comments on Solid State Phys.* **10** (1982) 173.

24. J. W. Knowles, *Can. J. Phys.* **40** (1962) 237, 257.

25. A. P. Mills, Jr. and L. Pfeiffer, *Phys.Rev.* **B32** (1985) 53.

26. W. Wing, *Phys. Rev. Lett.* **45** (1980) 631.

27. S. Chu, L. Hollberg, J. E. Bjorkholm, A. Cable, and A. Ashkin, *Phys. Rev. Lett.* **55** (1985) 48.

28. G. C. Bjorklund, R. R. Freeman, and R. H. Storz, *Opt. Commun.* **31** (1979) 47.

29. M. Leventhal, private communication.

30. W. M. Fairbank, F. C. Witteborn, J. M. J. Madey and J. M. Lockhart, *Experimental Gravitation,* Proc. Int. Sch. Phys. "Enrico Fermi" Course LVI, ed. B. Bertotti (Academic Press, New York, 1974) p. 310.

31. H. F. Dylla and J. G. King, *Phys. Rev* **A7** (1973) 1224 and references therein.

32. A. P. Mills, Jr. and S. Berko, *Phys. Rev. Lett.* **18** (1967) 420.

33. K. Marko and A. Rich, *Phys. Rev. Lett.* **33** (1974) 980.

34. W. Bernreuther and O. Nachtmann, *Z. Phys. C.* **11** (1981) 235. Refer also to Chapter 3 of this volume.

35. D. W. Gidley, A. Rich, E. Sweetman, and D. West, *Phys. Rev. Lett.* **49** (1982) 525. For more recent work see C. I. Westbrook, D. W. Gidley, R. S. Conti, and A. Rich, *Phys. Rev. Lett.* **58** (1987) 1328.

36. I. Harris and L. Brown, *Phys. Rev.* **105** (1957) 1656.

37. W. G. Caswell and G. P. Lapage, *Phys. Rev.* **A20** (1979) 36.

38. A. Billoire, R. Lacaze, A. Morel, and H. Navelet, *Phys. Lett.* **78B** (1978) 140.

39. G. S. Adkins and F. R. Brown, *Phys. Rev.* **A28** (1983) 1164.

40. M. Deutsch and S. C. Brown, *Phys. Rev.* **85** (1952) 1047.

41. M. W. Ritter, P. O. Egan, V. W. Hughes, and K. A. Woodle, *Phys. Rev.* **A30** (1984) 1331.

42. A. P. Mills, Jr. and G. H. Bearman, *Phys. Rev. Lett.* **34** (1975) 246.

43. H. Grotch and R. A. Hegstrom, *Phys. Rev.* **A4** (1971) 59; H. Grotch and R. Kashuba, *Phys. Rev.* **A7** (1973) 78.

44. E. R. Carlson, V. W. Hughes, M. L. Lewis, and I. Lindgren, *Phys. Rev. Lett.* **29** (1972) 1059; M. L. Lewis and V. W. Hughes, *Phys. Rev.* **A8** (1973) 625.

45. A. Rich, *Phys. Rev.* **A23** (1981) 2747.

46. A. P. Mills, Jr., *Phys. Rev.* **A27** (1983) 262.

47. F. H. M. Faisal and P. S. Ray, *Phys. Rev.* **A30** (1984) 2316.

48. G. P. Lepage, in *Atomic Physics 7*, eds. D. Kleppner and F. M. Pipkin (Plenum, New York, 1981) p. 297.

49. R. Karplus and A. Klein, *Phys. Rev.* **87** (1952) 848.

50. R. H. Howell, R. A. Alverez, and M. Stanek, *Appl. Phys. Lett.* **40** (1982) 751.

51. A. P. Mills, Jr., *J. Chem Phys.* **62** (1975) 2646.

52. H. W. Kendall, Ph.D. thesis, MIT (1954).

53. K. F. Canter, A. P. Mills, Jr. and S. Berko, *Phys. Rev. Lett.* **34** (1975) 177.

54. A. P. Mills, Jr., S. Berko and K. F. Canter, *Phys. Rev. Lett.* **34** (1975) 1541.

55. T. Fulton and P. C. Martin, *Phys. Rev.* **95** (1954) 811.

56. D. C. Schoepf, S. Berko, K. F. Canter, and A. H. Weiss, in *Positron Annihilation*, eds. P. G. Coleman, S. C. Sharma, and L. M. Diana (North-Holland, Amsterdam, 1982) p. 165.

57. S. Hatamian, R. S. Conti, and A. Rich, *Phys. Rev. Lett.* **58** (1987) 1833.

58. T. W. Hänsch, in *Atomic Physics 8*, eds. I. Lindgren, A. Rosen, and S. Svanberg (Plenum, New York, 1983) p. 55.

59. S. Chu and A. P. Mills, Jr., *Phys. Rev. Lett.* **48** (1982) 1333.

60. A. P. Mills, Jr. and S. Chu, in *Atomic Physis 8*, eds. I. Lindgren, A. Rosen, and S. Svanberg (Plenum, New York, 1983) p. 83.

61. S. Chu, A. P. Mills, Jr., and J. L. Hall, in *Laser Spectroscopy VI*, eds. H. P. Weber and W. Luthy, (Springer-Verlag, Heidelberg, 1983) p. 28.

62. S. Chu, A. P. Mills, Jr., and J. L. Hall, *Phys. Rev. Lett.* **52** (1984) 1689.

63. G. W. Erickson, *J. Phys. Chem. Ref. Data* **6** (1977) 831.

64. S. R. Amin, C. D. Caldwell, and W. Lichten, *Phys. Rev. Lett.* **47** (1981) 1234.

65. T. Fulton, *Phys. Rev.* **A26** (1982) 1794.

66. D. H. McIntyre and T. W. Hänsch, *Phys. Rev. A*, **34** (1986) 4504.

67. Steven Chu, A. P. Mills, Jr., A. G. Yodh, K. Nagamine, Y. Miyake and T. Kuga, *Phys. Rev. Lett.* (1988) 11 January.

CHAPTER 9

TEMPERATURE-DEPENDENT LEVEL SHIFTS

G. Barton

9.1   Introduction

The frequency shift of atomic transitions in electric fields is familiar, for static fields as the ordinary Stark effect, and somewhat more recently also for oscillatory fields as the "light-shift" or "dynamic Stark effect" (Townes and Schawlow, 1951; for recent references see for example Avan *et al.* 1976). The special case where the field is due to black-body radiation at some temperature T, though it is only just becoming accessible to experiment, has for long (if intermittently) fascinated theorists, perhaps because black-body radiation is so widely regarded as a quintessential quantum system.

Such thermal frequency shifts are given by differences between temperature-dependent level shifts $\Delta(T)$; correct expressions for these were first obtained, by two different methods, in 1972 (Barton 1972; Knight 1972). Thermodynamically, the $\Delta(T)$ are the coupling-induced shifts in the Helmholtz free energy of the coupled system atom plus electromagnetic field, in thermal equilibrium subject to the constraint that the atom be in a specified state. When the specified atomic state is changed by interaction with an external spectroscopic probe like a laser beam, the work done by the probe equals the increase in the (constrained) free energy; hence it is this free energy change that is determined spectroscopically from the frequency of the probe (Barton 1986; this and the two 1972 papers cited above also identify some of the errors in other work, many of which continue to recur in the literature with embarrassing frequency).

Knight's (1972) approach exploits the standard formalism of the dynamic Stark effect, setting out to capture directly the influence of the Planck black-body field on the atomic frequencies. Here we follow the other approach (Barton 1972, 1986), which is perhaps the more heuristic: it starts by focussing on the influence of the atom on the field. The end-results of the two approaches are identically the same.

The field is characterized by the photon occupation numbers n_λ, where λ symbolises both the two-valued polarization index s, and the wave number $\mathbf{K}$. Presently we shall specialize to the unperturbed Planck distribution $\langle n_\lambda \rangle = 1/(e^{\beta \omega_\lambda^{(0)}} - 1)$, where $\beta \equiv 1/kT$, and $\omega_\lambda^{(0)} = cK_\lambda$ is the mode frequency in absence of the atom.[a] The allowed values of $\mathbf{K}$ are fixed conventionally by imposing fictitious boundary conditions on the surface of a large quantiza-

---

[a] We use atomic units, where $\hbar = m_e = e = 1$, and Gaussian units for the electromagnetic field. Then the speed of light is $c = 1/\alpha$, where $\alpha \equiv e^2/\hbar \, c \approx 1/137$ is the fine-structure constant; the Bohr magneton is $\mu_e = e\hbar/2m_e \, c = \alpha/2$. The unit of length is the Bohr radius $a = \hbar^2/m_e \, e^2$ (recall that scattering amplitudes have the dimension of length, and polarizabilities of length cubed). The unit of energy is $e^2/a = 6.578 \times 10^{15}$ Hz, and Boltzmann's constant in a.u. is $k = 3.167 \times 10^{-6}/{}^\circ$K.

tion volume V: then one has $V^{-1}\sum_\lambda = \sum_s V^{-1}\sum_K = \sum_s (2\pi)^{-3}\int d^3K = 2(2\pi)^{-3}4\pi\int_0^\infty dK K^2 = \pi^{-2}\int_0^\infty dK K^2 = \frac{1}{\pi^2 c^3}\int_0^\infty d\omega\omega^2$. The third equality applies because the black-body radiation is unpolarized and isotropic.

We need the energy-level shifts of the system consisting of the field and the atom, due to the interaction between them, and calculate these to leading order in the fine structure constant α. The idea is that the atom shifts the frequency of each field mode by $\delta\omega_\lambda$, and hence the energy of that mode by $n_\lambda\delta\omega_\lambda$; the ordinary Lamb shift can also be approached in this way (Feynman 1961: Power 1966). To determine $\delta\omega_\lambda$, consider first a sufficiently dilute medium consisting of N atoms in the volume V; the refractive index, close to unity, is $\mu(\omega) = 1 + (N/V)2\pi\mathrm{Re}\,f(\omega)c^2/\omega^2$, where $\mathrm{Re}\,f(\omega)$ is the real part of the forward scattering amplitude of the atom for light of circular frequency ω (Ditchburn 1976; Newton 1982). (A common approximation to $f(\omega)$ appears in equation (9.5b) below.) The allowed values of K are the same as before, because they are determined by the boundary conditions independently of μ: but, because the phase-velocity of electromagnetic waves in the medium is c/μ rather than c, the vacuum frequency is replaced by

$$\omega = \omega^{(0)} + \delta\omega = \omega^{(0)}/\mu$$
$$\approx \omega^{(0)}\{1 - (N/V)2\pi\mathrm{Re}\,f(\omega)c^2/\omega^2\}\;.$$

Therefore the frequency shift induced by just one atom is

$$\delta\omega = -2\pi\mathrm{Re}\,f(\omega)c^2/\omega V\;, \tag{9.1}$$

where we no longer need to distinguish, on the right, between $\omega^{(0)}$ and ω.

Thus, the energy-level shift induced by the coupling in the state with photon occupation numbers $\{n_\lambda\}$ is $\sum_\lambda n_\lambda\delta\omega_\lambda$. But according to standard thermodynamic perturbation theory (Landau and Lifshitz 1958; Peierls 1979), the (constrained) free energy shift $\Delta(T)$ that we require is just the average of this level shift over the unperturbed canonical distribution for the field. Hence we obtain $\Delta(T)$ simply by replacing the n_λ by their Planckian averages $\langle n_\lambda\rangle$:

$$\Delta(T) = \sum_\lambda\langle n_\lambda\rangle\delta\omega_\lambda = -\frac{2}{\pi c}\int_0^\infty d\omega\frac{\omega}{e^{\beta\omega}-1}\mathrm{Re}\,f(\omega)\;. \tag{9.2}$$

Equation (9.2) applies equally to any particle, whether atom, ion or electron.

It is instructive to represent f by means of the dispersion relation (see for example Davydov 1976)

$$\mathrm{Re}\,f(\omega) = -Q/Mc^2 + (\omega^2/2\pi^2 c)\int_0^\infty d\omega'\sigma(\omega')/(\omega'^2 - \omega^2)\;, \tag{9.3}$$

where the integral is understood as a Cauchy principal value; Q, M are the particle's total charge and mass (e.g. $Q = 0$ or e and $M = m_H$ or m_e for a hydrogen atom or an electron), and σ its total cross section for photons, accurately to order α (to this order this is the same as the total absorption cross section, since the scattering cross section is of order α^2). Notice that (9.2) and (9.3) allow one to express Δ entirely in terms of the directly measurable quantities Q, M, σ. This ensures in particular that the result is gauge invariant, a point that sometimes causes trouble in other approaches.

9.2   Nonrelativistic Approximation

9.2.1   Free electrons

From now on we consider only temperatures and energies low enough that the particles can be treated non-relativistically.[b]   Then for free electrons $f = -e^2/m_e c^2 = -1/c^2 = -\alpha^2$; we substitute this into (9.2), change the integration variable to $y = \beta\omega$, and use $\int_0^\infty dy\, y/(e^y - 1) = \pi^2/6$ to obtain

$$\Delta(T, \text{free electron}) = \frac{2\alpha^3}{\pi\beta^2} \int_0^\infty dy\, \frac{y}{e^y - 1}$$
$$= \alpha^3 \pi (kT)^2/3 \approx 2.42 \times 10^3 (T/300)^2 \text{Hz} . \tag{9.4}$$

As far as the writer can tell, this result was first given correctly by Englert (1959).

9.2.2   Atoms

For an atom, $f(\omega)$ is just the forward Rayleigh scattering amplitude, averaged over directions relative to the atom, related to the similarly averaged (i.e. scalar) dynamic polarizability $\Pi_0(\omega)$ by $f(\omega) = (\omega/c)^2 \Pi_0(\omega) = \alpha^2\omega^2\Pi_0(\omega)$. ($\Pi_0$ is the sum of the electric and magnetic polarizabilities, but in most cases one can neglect the latter.) In a one-optical-electron atom in state 0, and in the standard electric-dipole approximation, one writes

$$\text{Re}\, f_{E1}(\omega) = \alpha^2\omega^2\Pi_0(\omega) = \frac{2\alpha^2}{3}\omega^2 \sum_i E_{i0} \frac{|\mathbf{r}_{i0}|^2}{E_{i0}^2 - \omega^2} , \tag{9.5a,b}$$

---

[b] In true thermal equilibrium, black-body radiation naturally contains electrons and positrons as well as photons. Their effects are of relative order $\exp(-\beta\, m_e\, c^2)$ when this factor is small (Landau and Lifshitz 1958, pp. 325-326); hence they are totally negligible under all laboratory conditions, though in principle they must be included in any fully relativistic field theory.

where the sum is over all states $i \neq 0$, $E_{i0} \equiv E_i - E_0$, and $\mathbf{r}_{i0} \equiv \langle i|\mathbf{r}|0\rangle$. (The frequency shift $\delta\omega$ for a neutral system, given in terms of Π_0 by (9.5a) and (9.1), is derived by an alternative method by Boyer (1969), following Casimir (1951)). From (9.2) and (9.5b) we obtain

$$\Delta(T, \text{atom}) = -\frac{4\alpha^3}{3\pi\beta^2}\sum_i E_{i0}|\mathbf{r}_{i0}|^2 \int_0^\infty dy\frac{y^3}{(e^y - 1)}, \quad \frac{1}{(\beta^2 E_{i0}^2 - y^2)}. \quad (9.6)$$

9.3 Approximations for High and Low Temperatures

Simple expressions result only if, in all the important terms of $\sum_i$, one has either $(\beta E_{i0})^2 \ll 1$ (high T approximation), or $(\beta E_{i0})^2 \gg 1$ (low T approximation). The important states i are those with principal quantum numbers n_i close to n_0, so that in a hydrogenic atom (one electron, nuclear charge Z, and effectively infinite nuclear mass) one has, very roughly, $\left|E_{i0}\right| \sim \left|\frac{d}{dn}\frac{1}{2}\frac{Z^2}{n^2}\right| = Z^2/n^3$; this is confirmed by the detailed calculations of Farley and Wing (1981). Thus $|\beta E_{i0}| < 1$ for those states 0 with principal quantum number $n > n_{\text{crit}} = (Z^2/kT)^{\frac{1}{3}} \approx 10.2Z^{\frac{2}{3}}(T/300)^{-\frac{1}{3}}$. If all the important $(\beta E_{i0})^2 \ll 1$, one simply drops this term from the denominator in (9.6): then the integral reduces to $\pi^2/6$ (as above) independently of i, and the sum is given by the oscillator sum rule $\sum_i E_{i0}|\mathbf{r}_{i0}|^2 = 3/2$ (Landau and Lifshitz 1977, p. 632). Accordingly, to this accuracy Δ becomes independent of the state 0, and reduces to the free electron shift (9.4). This could have been foreseen because, when $kT \gg |E_{i0}|$, the dominant frequencies $\omega \sim kT$ are also much larger than $|E_{i0}|$, and the scattering amplitude (9.5) then reduces to the free electron amplitude $-\alpha^2$, again by virtue of the sum rule.[c]

Conversely, if all the important $(\beta E_{i0})^2 \gg 1$, one drops the term y^2 from the denominator in (9.6); since $\int_0^\infty dy y^3/(e^y - 1) = \pi^4/15$, this yields

$$\Delta(T) = -\frac{4\pi^3}{45}\alpha^3(kT)^4\sum_i |\mathbf{r}_{i0}|^2/E_{i0}$$

$$= -\frac{2\pi^3}{15}\alpha^3(kT)^4\Pi_0(0)$$

$$= -\frac{1}{2}\Pi_0(0)\langle\mathcal{E}^2\rangle_T, \quad (9.7)$$

---

[c]This is an artefact of the non-relativistic approximation rather than a universal truth. When relativistic effects are taken into account, the high-frequency limit of the bound-state scattering amplitude differs from $-\alpha^2$; correspondingly, the high T limit of $\Delta(T)$ differs from (9.4). (cf. Barton 1986, and references given there.)

where the factor $\Pi_0(0) = (2/3)\sum_i |\mathbf{r}_{i0}|^2/E_{i0}$ is just the electrostatic polarizability of the atom in state 0, and where we recognize $\langle \mathcal{E}^2 \rangle_T = (4\pi^3/15)\alpha^3(kT)^4$ as the mean-square electric field of the black-body radiation. The rightmost expression in (9.7) is due to Knight (1972). Equation (9.7) applies for instance to the ground state of hydrogen at room temperature.[d] Then one has (Landau and Lifshitz 1977, p. 286) $\Pi_{1s}(0) = 9/2Z^4$, and (9.7) yields

$$\Delta_{1s} = -3\pi^3\alpha^3(kT)^4/5Z^4 \approx -0.0388(T/300)^4 Z^{-4} \text{ Hz}. \tag{9.8}$$

Notice that the low T and the high T (or free electron) approximations (9.7) and (9.4) have opposite signs. When neither is appropriate, one must resort to numerical calculations, like those of Farley and Wing (1981).

It is worth stressing that one could be badly misled if one tried to determine the effects of the radiation in terms of some shift in the effective (inertial) mass of the electron, which one then attempted to incorporate into an ordinary atomic Hamiltonian. A correct but necessarily rather involved interpretation in terms of an effecO7tive Hamiltonian has however been given by Avan *et al.* (1976); see also Barton (1986).

9.4 The $n = 2$ Lamb Shift

A delicate complication besets the room-temperature correction to the Lamb shift, i.e. to the difference $\Delta_{2s_{1/2}}(T) - \Delta_{2p_{1/2}}(T)$ (Barton 1972). Here the sums in (9.6) must be subdivided into contributions Δ_{opt} from intermediate states i with $n_i \neq 2$, and contributions Δ_{rf} from states with $n_i = 2$. (The labels opt and rf allude to spectroscopic intervals in the optical and radio frequency ranges respectively; they correspond also to non-degeneracy and degeneracy when relativistic and ordinary (zero temperature) radiative corrections are ignored.) The Δ_{opt} are spin-independent, and are given by the low-T expressions (9.7): these are readily calculable in closed form by adapting standard methods (Landau and Lifshitz 1977, p. 286; Ford and Weber 1984) to the exclusion $n_i \neq 2$. One finds $\Pi_{2s}(0) = 120/Z^4, \Pi_{2p}(0) = 530/3Z^4$, whence

---

[d]Total energy $U = F + TS$, entropy S and free energy F are related by $U = (1 + \beta\partial/\partial\beta$ For a free electron, or in the high T regime, where $\Delta \equiv \Delta F \propto \beta^{-2}$, this entails $\Delta U = -\Delta$; in the low T regime, where $\Delta \propto \beta^{-3}$, it entails $\Delta U = -3\Delta$. The energy balance $\Delta U -\Delta = T\Delta S$ is supplied to the coupled system atom plus field not as work done on the atom by the spectroscopic probe, but by the emission or absorption of thermal photons by the heat reservoir responsible for the black-body radiation. The changes in ΔS are governed by the changes in the average photon occupation numbers: the perturbed values $1/\{\exp[-\beta(\omega_\lambda^{(0)} + \delta\omega_\lambda)]-1\}$ depend on the state of the atom, and differ from the perturbed averages $\langle n_\lambda \rangle$ used in section 9.1 to calculate the free-energy shift $\Delta(T)$.

$\Pi_L(0) \equiv \Pi_{2s}(0) - \Pi_{2p}(0) = -170/3Z^4$. Rescaling Δ_{1s} by the ratio of the appropriate polarizabilities, namely by -340/27, immediately leads to

$$\Delta_{\text{opt}}(T) = (68/9)\pi^3\alpha^3(kT)^4/Z^4 \approx 0.489(T/300)^4 Z^{-4}\text{Hz} . \tag{9.9}$$

By contrast, the high T approximation is appropriate to the Δ_{rf}, for which the effects are dominated by the shift of the $2s_{1/2}$ state, due to the $i = 2p_{3/2}$ intermediate state. This turns out to be

$$\Delta_{\text{rf}} = \pi\alpha^5 Z^2(kT)^2/8 \approx 0.0483(T/300)^2 Z^2 \text{ Hz} . \tag{9.10}$$

9.5 Hyperfine Structure

The hyperfine structure is also subject to corrections. There are separate contributions due to the electric-dipole and magnetic-dipole parts of the Rayleigh amplitude f in (9.2), or in other words, to the electric and magnetic fields of the black-body radiation. In hydrogen the hyperfine interaction raises the $F = 1$ above the $F = 0$ component of the ground state by $\omega_{hfs} \approx 1.42 \times 10^9$ Hz (measured to one part in 10^{13}) $\approx 32\mu_e\mu_p/3$ (where $\mu_e = \alpha/2$ and μ_p are the spin magnetic moments of the electron and proton). It also alters their electric-dipole scattering amplitudes, inducing a difference $\Pi_{hfs} = \Pi_{F=1} - \Pi_{F=0} = (31/4)\omega_{hfs}$ between their static scalar polarizabilities (Sandars 1967; Stuart *et al.* 1980). Rescaling Δ_{1s} by the ratio of the appropriate polarizabilities yields the electric-dipole correction

$$\frac{(\Delta\omega_{hfs})_{E1}}{\omega_{hfs}} = \frac{31}{4}\cdot\frac{2}{9}\cdot\Delta_{1s}(\text{in a.u.}) \approx -1.02 \times 10^{-17}(T/300)^4 . \tag{9.11}$$

The other correction stems from the averaged forward magnetic-dipole scattering amplitude

$$\text{Re } f_{M1}(\omega) = \frac{2\alpha^2}{3}\omega^2\sum_i E_{i0}\frac{|\mathbf{M}_{i0}|^2}{E_{i0}^2 - \omega^2} , \tag{9.12}$$

where $\mathbf{M} = 2\mu_e\mathbf{S}_e + 2\mu_p\mathbf{S}_p$ is the relevant magnetic moment operator. In this approximation, when 0 is the $F = 0$ state, the only intermediate states in the sum are the components of its $F = 1$ partner, and vice-versa. A brief calculation yields

$$\text{Re }\{f_{M1}(\omega, F = 1) - f_{M1}(\omega, F = 0)\} = -\frac{4}{3}2\alpha^2\omega^2(\mu_e - \mu_p)^2\frac{\omega_{hfs}}{\omega_{hfs}^2 - \omega^2} . \tag{9.13}$$

We need the shift (9.2) with Re f replaced by (9.13). At room temperatures the important frequencies far exceed ω_{hfs}, so that ω_{hfs}^2 is dropped from the denominator in (9.13), giving

$$\frac{(\Delta\omega_{hfs})_{M1}}{\omega_{hfs}} = -\frac{8\pi\alpha^3}{9}(\mu_e - \mu_p)^2(kT)^2 \approx -1.31 \times 10^{-17}(T/300)^2 . \quad (9.14)$$

This expression was first reported by Itano *et al.* (1982).

The correction (9.14) can be understood in terms of the temperature-dependent change $\Delta\mu_e(T)$ in μ_e, which (as quoted above) is a factor of ω_{hfs}. For a free electron, $\Delta\mu_e(T)/\mu_e = -(5\pi\alpha^5/9)(kT)^2$; but for a bound electron, $\Delta\mu_e(T)/\mu_e = -(2\pi\alpha^5/9)(kT)^2$ in the regime where, with B the applied field, $\mu_e B << kT$, yet kT is still well below the main orbital excitation energies. (See Barton 1985 for details and for further references.) The latter ratio is the same as (9.14), if μ_p is disregarded relative to μ_e.

9.6 Experiment

The corrections (9.7) to low-lying states and (9.9) and (9.10) to the Lamb shift have not yet been observed. Though in hydrogen both the hyperfine corrections (9.11) and (9.14) are also very small, in alkali atoms the shift $(\Delta\omega_{hfs})_{E1}$ can be larger by a factor of over 100, essentially because the polarizabilities are much larger. Itano *et al.* (1982) have suggested that $\Delta\omega_{hfs}$ should be observable in Cs, and indeed that such corrections already need to be taken into account in current work on frequency standards.

A temperature-dependent frequency shift has been measured by Hollberg and Hall (1984), when low-lying states in Rb are excited to high n Rydberg states, using two-photon Doppler-free absorption spectroscopy. The T dependent correction to the low-lying state, given by (9.7), is negligible: the observed frequency shift is predicted to be that of the Rydberg state, which in this respect behaves exactly like a high n state in hydrogen. Hence it is shifted according to (9.4), exactly as a free electron would be, and independently of n once n is high enough. In other words, as regards its physical interpretation this is the same shift as one would find in measuring the ionization threshold. Hollberg and Hall confirm the T^2 proportionality and the magnitude of the predicted shift (9.4), within their error of $\pm^{10}_{30}\%$, using effective temperatures up to the order of $1000°K$, and $n = 36$.

Though, as we have seen, the temperature-dependent energy shifts are only just accessible to present techniques, emission induced by black-body radiation is competitive with spontaneous emission in governing the relative populations of high Rydberg levels: such effects are reviewed by Gallagher (1983).

It is a pleasure to acknowledge the advice of Peter Knight.

REFERENCES

1. Avan, P., Cohen-Tannoudji, C., Dupont-Roc, J. and Fabre, C., *J. Phys.* **37** (1976) 993.
2. Barton, G., *Phys. Rev.* **A5** (1972) 468; *Phys. Lett.* **162B** (1985) 185; *J. Phy. B. At. Mol. Phys.* **20** (1987) 879.
3. Boyer, T. H., *Phys. Rev.* **180** (1969) 19.
4. Casimir, H. B. G., *Philips Res. Rep.* **6** (1951) 162.
5. Davydov, A. S., *Quantum Mechanics*, second ed. (Pergamon, Oxford, 1976) Sec. 123.
6. Ditchburn, R. W., *Light*, third ed. (Academic, London, 1976) Sec. 15.45. (Ditchburn's β is our $f(\omega)$.)
7. Englert, F., *Bull. Cl. Sci. Acad. R. Belg.* **45** (1959) 782.
8. Farley, J. W. and Wing, W. H., *Phys. Rev.* **A23** (1981) 2397.
9. Feynman, R. P., in *La Théorie Quantique des Champs:* (Interscience, New York, 1961) p. 61.
10. Ford, G. W. and Weber, W. H., *Phys. Reports.* **113** (1984) 195, Sec. 4.2 (Beware of misplaced factors Z.)
11. Gallagher, T. F., in *Rydberg States of Atoms and Molecules*, eds. R. F. Stebbing and F. B. Dunning (Cambridge University Press, Cambridge, 1983).
12. Hollberg, L. and Hall, J. L., *Phys. Rev. Lett.* **53** (1984) 230.
13. Itano, W. M., Lewis, L. L. and Wineland, D. J., *Phys. Rev.* **A25** (1982) 1233.
14. Knight, P. L., *J. Phys.* **A5** (1972) 417.
15. Landau, L. D. and Lifshitz, E. M., *Statistical Physics* (Pergamon, London, 1958).
16. Landau, L. D. and Lifshitz, E. M., *Quantum Mechanics*, third ed. (Pergamon, Oxford, 1977).
17. Newton, R. G., *Scattering Theory of Waves and Particles*, second ed. (Springer, New York, 1982) Sec. 1.5.
18. Peierls, R., *Surprises in Theoretical Physics* (Princeton University Press, 1979) Sec. 3.3.
19. Power, E. A., *Am. J. Phys.* **34** (1966) 516.

20. Sandars, P. G. H., *Proc. Phys. Soc.* **92** (1967) 857. (Apparently through a misprint, equation (3.6) here lacks a factor $1/3$. Sandars' α_{10} is our $\Pi_{hfs}(0)$.)
21. Stuart, J. G., Larson, D. J. and Ramsey, N. F., *Phys. Rev.* **A22** (1980) 2092.
22. Townes, C. H. and Schawlow, A. L., *Microwave Spectroscopy* (McGraw-Hill, New York, 1955) Secs. 10.8 and 10.9.

CHAPTER 10

HYDROGEN AND THE FUNDAMENTAL ATOMIC CONSTANTS

G. W. Series

10.1   The Atomic Constants, Physical Theory, and the Base Units of Measurement

There is a three-fold interconnection between the fundamental atomic constants, physical theory, and the base units of measurement. The connection between the first two is obvious — for how are the values of fundamental constants to be ascertained except by measurements interpreted in the light of theory, and is it not an important constraint on theories that their quantitative predictions should be consistent with values of the constants derived from other sources? And in the business of obtaining a self-consistent and accurate set of atomic constants the spectrum of hydrogen and the theories which have been used to interpret it have long been to the fore.

It is obvious, too, that the determination of physical quantities requires a set of base units of measurement. But it is generally not realized that agreement on what shall constitute the base units, as, for example, the units of the Système Internationale, requires more than agreement on definitions: beyond the definitions are the problems of realization, and the need for agreement on how the base units shall, in practice, be maintained in standards laboratories.

Consider for a moment the dominant interaction in hydrogen: electrical. It follows that any quantitative test of a theory which relates hydrogen to other physical entities must at some stage refer back to measurement of the electronic charge. Within the SI the base electrical unit is the ampère whose definition — based on the force between currents travelling along 'infinitely long wires' of 'negligible, circular cross section' presents the standards laboratories with an extremely difficult technical problem: even as late as 1985 the definition simply could not be implemented at a level as relatively crude as one part per million, and in 1986 the best that could be achieved was agreement to within a few tenths of a microamp. In practical terms, if you wanted your ammeter calibrated correctly to 1 in 10^7, no standards laboratory anywhere in the world could do it for you, simply because different methods of going from the definition to the standardization could not agree to within 0.3 μamp. A few years ago the disagreement was about 6 parts in 10^6, the uncertainty in this 6 being 1.4 [COH 83].

So what is to be done in this situation? The solution adopted is to bring together into one comprehensive analysis as large as possible a set of measurements of physical quantities where

(a) the theory relating the quantity measured to the atomic constants is believed to be reliable, and

(b) the measurements have been carried out at a level of accuracy, in relation to the base units, which is at the frontier of physical measurement at the time the measurement is made.

These measurements are then subjected to close scrutiny to assess the quality of the work, and to a least squares analysis in which the quantities ultimately to be derived are not only values of the atomic constants such as e, h, m_e, and so on, but also the *ratios of maintained units to units defined in the Système Internationale*.

It should be understood that, in this analysis, the number of physical quantities where measurements qualifying for inclusion have been made is larger than the number of constants to be ascertained: consequently the set of equations for solution is over-determined, and what is sought in the least-squares analysis is a set of mutually consistent values of the constants. Furthermore, the 'input' values are not of equal worth. For some quantities the techniques of measurement lend themselves to much higher accuracy than for others, and such quantities, called 'auxiliary quantities', are treated as constants in the analysis. For these the least squares analysis will leave their values unaltered, whereas for the less precisely measured quantities the 'output' value may differ from the 'input' value, in the interests of consistency.

A great deal of experience is required in carrying out this work, for personal judgements are called for as to the choice of material to be included in the assessment, and as to which quantities should be treated as 'auxiliary' and which subject to variation; and among these, what weight should be attached to each measurement. Pre-eminent in this work have been Birge [BIR 41], DuMond and Cohen [COH 65], Cohen and Taylor [COH 73]. A further, thorough-going assessment by Cohen and Taylor has been in preparation for several years and the results have very recently been published [COD 86].

The consistency checks imposed by this least-squares analysis lead to the recognition that maintained units of measurement may differ from defined units by significant amounts. This has been recognized for several years and is clearly a most important finding for those interested in precision measurements. It illustrates an important function of these analyses: shortcomings are revealed, either in experimental technique, or in the theoretical understanding of phenomena, or in both. This testing of one measurement against another, at the frontier of measurement, is a continuing story. Its fascination, its achievements, and the fundamental questions which it raises are skilfully exposed by Petley [PET 85] in a recently published book.

10.2 The Rydberg Constant

From the earliest days the Rydberg constant R has been one of the most important quantities to be included in these least-squares analyses for the atomic constants. Defined originally as the constant relating term values in Rydberg series in simple spectra to the effective principal quantum number n_e

$$T = -R/n_e^2 = -R/(n - \sigma)^2 \qquad (10.1)$$

(n integral, σ: the quantum defect), and recognized as applying to hydrogen by putting $\sigma = 0$, it received a theoretical interpretation with Bohr's theory:

$$R_\infty = [m_e e^4/4\pi\hbar^3 c][\mu_0 c^2/4\pi]^2 , \qquad (10.2)$$

(here, according to normal practice, R_∞ is expressed as a reciprocal length). But Bohr's theory takes no account of fine structure, and it is a mistake to suppose that the R of equation (10.1) can be strictly identified with the R_∞ of equation (10.2). And while, strictly speaking, it is the quantity appearing in (10.1) which is Rydberg's constant, the quantity generally known as the Rydberg constant is the R_∞ defined by (10.2).

Experimental determinations of R_∞ based on the spectrum of atomic hydrogen therefore require

(a) a precision measurement of the wavelength (or frequency) of some resolved component of a line in the spectrum relative to some standard of wavelength or frequency, corrected for systematic effects inherent in the particular experiment,

(b) correction of this measurement by the reduced mass factor $(1 + m_e/M)$ to bring it to a value appropriate to a nucleus of infinite mass, and

(c) correction for hyperfine and fine structures, for radiative effects (which may include additional reduced mass effects) and for nuclear volume effects.

Details of these reductions are to be found in section 5.4.

The importance of R_∞ in the history of the fundamental constants stems from the facts that

(a) spectroscopic measurements of wavelength (or frequency) can normally be made with very high accuracy,

(b) it has been possible to make these measurements in conditions where the perturbations of the environment on the radiating atoms are very small,

(c) the corrections needed to relate the measured quantity to R_∞ are small in relation to R_∞ itself, and are understood to a very high level of accuracy —

and in these circumstances it has nearly always been the case that the accuracy of the experimentally-determined value of R_∞ has been sufficient to place it among the auxiliary constants: it has been a cornerstone in the history of the evaluation of the fundamental constants. But statements such as these are always to be read in the context of the contemporary scene. As improvements in the measurement of other quantities have been reported, so also have the demands for improved values for R_∞.

Most determinations have been made with reference to the Balmer-α or Balmer-β lines, both of which lie conveniently in the visible spectrum and are the obvious lines to study with a new technique. Balmer-α was more convenient for interferometry before dielectric films were generally available. Balmer-β has a more complicated fine structure than Balmer-α. In the higher members of the Balmer series the fine structure is dominated by the structure of the $n = 2$ levels. The He$^+$ line 468 nm $(n = 4 - 3)$ is again conveniently placed in the spectrum, and although its fine structure is more complicated than that of the Balmer lines, it is more widely spread. Deuterium offered a useful reduction in Doppler broadening in the pre-laser era. Measurements within the laser era were first made on Balmer-α, but Balmer-β is now receiving attention, and transitions to higher levels $(n = 8$ and $10)$ have been studied by two-photon laser spectroscopy.

Values obtained over a period of nearly sixty years are given in Table 1. The importance of making measurements on different transitions, and on different isotopes, to provide a check on theory, need hardly be stressed. What is perhaps not so obvious as a check on theory is the usefulness of measurements on one transition relative to another in the same system: an external frequency standard can be regarded as an intermediary which allows measurements on one physical system to be related to measurements on another.

In recent years particular attention has been given to the way in which 'errors' are reported: it is now customary to quote one standard deviation, but it was not always so, and the estimated errors quoted in parentheses are strictly not comparable quantities. Nevertheless, it is clear that the value current 60 years ago would command assent today at the level of a few parts in 10^7. Notable also is the dramatic improvement in accuracy brought about by the laser revolution. The contemporary value, say 109 737.315 7cm^{-1}, is reliable to a few parts in 10^9.

Table 1. Values obtained for the Rydberg Constant, R_∞

Date	Workers	Lines studied	R_∞ - 109 737 cm^{-1}
1927	Houston	H_α, H_β, He$^+$ (468)	.336 (16)
1939	Chu	He$^+$ (468)	.314 (20)
1940	Drinkwater *et al.*	H_α, D_α	.314 (9)
1968	Csillag	D_α, ... D_η	.3060 (60)
1971	Masui	H_α	.3188 (45)
1973	Kibble *et al.*	H_α, D_α	.3253 (77)
1977	Kessler	He$^+$ (468, 656, 1012)	.3208 (85)
1974	Hänsch *et al.*	H_α, D_α	.31410 (100)[1]
1978	Goldsmith *et al.*	H_α	.31490 (32)[1]
1979	Petley *et al.*	H_α	.31513 (85)[1]
1986	Zhao *et al.*	H_α, D_α	.31569 (7)[1]
1987	Zhao *et al.*	H_β, D_β	.31573 (3)[1]
1987	Biraben *et al.*	H, D: $2S$ - 8, 10D	.31569 (6)[1]
1986	CODATA		.31534 (13)[1]

[1] These values were obtained by laser spectroscopy. The pre-laser values are as tabulated by Petley [PET 85], p. 72: references to the original work will be found in SAH for the first three entries and in Chapter 5 for the others. The figures in the last column reflect the wavelength standards and the values of m_e/M used by the authors; for the last three entries the value $m_p/m_e = 1836.152701$ [VAN 86] was used.

The CODATA 86 value is based on values available before the three most recent determinations had been published.

10.3　The Fine Structure Constant

Whether or not $\alpha = (e^2/\hbar c)[\mu_0 c^2/4\pi]$ be called a fundamental constant or an atomic constant is irrelevant, what is important is that it has proved to be a key constant in relation to fundamental theory in general and in relation to the spectrum of hydrogen in particular.

It entered physics in Bohr's theory as the speed (relative to the speed of light) of an electron in Bohr's first orbit, v_1,

$$v_1/c = (e^2/\hbar c)[\mu_0 c^2/4\pi] ,$$

and explicitly in Sommerfeld's theory where it appears in the expression for the relativistic energy:

$$E_{n,k} = -\frac{R}{n^2}\left[1 + \frac{\alpha^2}{n}\left(\frac{1}{k} - \frac{3}{4n}\right) + \dots\right] .$$

10.3.1   *Experimental study of fine structure in hydrogen*

At this point α became capable of experimental determination by study of the fine structure in the spectrum, though hydrogen itself proved less attractive (on account of Doppler broadening) than He^+, where the scaling factor $Z^2(RZ^2/n^2; \alpha^2 Z^2/n$, above) enlarges the fine structure in relation to the gross structure. In fact, the spin doublets in X-ray spectra were better resolved and provided the basis for an experimental evaluation which was not superseded until the nineteen fifties. It was, of course, the new quantum theory rather than the old by which the later measurements were interpreted. The Dirac theory, without QED, was adequate. For the X-ray measurements electron-shell screening effects were important, and nuclear volume effects were considered as a refinement.

These older measurements had in common that the spectroscopic intervals which were the center of interest — intervals between fine structure levels having the same value of n — were determined as differences between fine structure components of lines in the gross structure. The experimental advances which improved the accuracy in the determination of α a thousandfold were based on

(a) measurement of radiofrequency resonances between the fine structure levels within a given n, and

(b) level-crossing experiments between fine structure levels.

The resonance experiments were usually of the electric dipole type, as was the Lamb-Retherford experiment. These give intervals such as $S_{1/2} - P_{1/2}$ and $S_{1/2} - P_{3/2}$, and the sum of these intervals (since $S_{1/2}$ lies between $P_{1/2}$ and $P_{3/2}$) gives directly $P_{1/2} - P_{3/2}$. This is the interval whose physical origin is the spin-orbit interaction, directly proportional to the magnetic moment of the electron, and whose value for $n = 2$ is given by the Dirac theory as $R\alpha^2/16$.

Level-crossing experiments give this interval directly in terms of a magnetic field, and knowledge of the Landé g-factor of the levels gives the interval as a frequency. The technique is described in Chapter 1.

The value of α derived from the level-crossing experiments was

$$\alpha^{-1} = 137.0354(6) \quad [\text{BAI 71}] \,.$$

Values derived from the resonance experiments were

$$\alpha^{-1} = 137.0356(7) \quad [\text{SHY 71}]$$
$$\text{and } \alpha^{-1} = 137.0358(5) \quad [\text{VOR 71}] \,.$$

With either of these techniques the line width of the resonances is determined principally by the 'natural width' (against radiative decay) of the levels involved, though in point of detail the two techniques differ in the difficulties encountered. While it is undoubtedly true that the refinements of QED are necessary to give a proper interpretation of the measurements, it should not be overlooked that these intervals $P_{1/2} - P_{3/2}$ are highly sensitive to the value of α, but not at all sensitive as tests of QED. Further, it should be noticed that, although the Rydberg constant enters directly into the expression for these intervals, the uncertainty in its value is so small as to be quite irrelevant in the evaluations of α. Conversely, the uncertainty in α enters into the evaluation of R_∞ only as an uncertainty in small correction terms where its influence is minimal.

10.3.2 *Fine structure in helium*

α dominates the fine structure in elements other than hydrogen. Helium — the neutral atom — is appealing to the experimenter in that the 2^3P levels are more widely spaced than the 2^2P levels in hydrogen, and furthermore, the lifetime of the helium P levels is longer than that of the hydrogen levels, so the resonances can be made narrower. Against this, the theory is more complicated. Nevertheless, by the mid-seventies the theory had progressed to the point where a comparison with experiment was meaningful. A brief account is given in section 2.12. A review by Lewis [LEW 74] gives an account of the theory in a historical setting and derives the value

$$\alpha^{-1} = 137.03598(27)$$

from experimental values of the interval $2^3P_1 - 2^3P_0$ in ^{4}He. A refinement of the theory by Lewis and Serafino [LEW 78] allowed the value

$$\alpha^{-1} = 137.036113(110)$$

to be deduced. This work refers to the even isotope ^{4}He.

The 2^3P states of ^{3}He have also been explored by radio frequency [PRE 85] and by level-crossing spectroscopy [GER 67] and the results interpreted, on the basis of the analysis for ^{4}He, to give self-consistent values of the hyperfine interaction constants for those states [HIN 85, HIN 86]. Though these analyses affirm the reliability of the QED treatment by which the fine structure is calculated, it would appear that work on helium now stands at the limit imposed by uncertainty in the theory. More accurate values of α may be derived from other systems [COH 83].

10.3.3   *Hyperfine structure*

The hyperfine interaction *a*-factor (magnetic dipole interaction between the electron and the nucleus) for hydrogen in the ground state is given, it will be recalled (sections 2.10 and 4.1.1) by an expression which can be written

$$ a = \frac{4}{3} R_\infty c \alpha^2 g_e g_I (m_e/m_p) \times C_{\text{rel}} \times C_{\text{red. mass}} \times C_{\text{QED}} \times C_{\text{nucl}} , $$

in which the Landé *g*-factor of the electron is shown explicitly, as are also the relativistic, reduced mass and radiative corrections. C_{nucl} depends on nuclear structure, polarizability and recoil, and has played an enigmatic role in the history of the subject. Notwithstanding the difficulty in evaluating the QED and nuclear corrections, it is to be acknowledged that these are small. The main point to be noticed is that an experimental determination of the hyperfine intervals gives a direct route to α.

As is spelt out in detail in Chapter 4, the experimental determinations of these intervals are almost as accurate as the primary standard itself can be realized. For the intervals are those on which the hydrogen maser is based, and inter-comparisons between hydrogen masers and caesium clocks represent metrology at its highest level of accuracy. So the accuracy of experimental determinations — a level better than 10 parts in 10^{12} — far exceeds the possibility of evaluating the theoretical expressions to comparable accuracy.

Since the early measurements of this structure, when discrepancies at the level of parts in 10^4 between the experimental values and the Fermi formula (in which g_e had its Dirac value of 2 and all the *C*-factors were unity) disclosed the anomaly in g_e, the focus of interest in the hyperfine formula has swung between

(a) taking a value of α from other sources and using the equation to test the QED and nuclear corrections, and

(b) assuming the validity of the corrections (or placing close limits on the various terms) and using the equation to obtain a value for α.

Either way, the uncertainty in the experimental value is irrelevant.

In the late nineteen sixties it became apparent that the values of α derived from fine and hyperfine structures were uncomfortably discordant. The QED calculations had been carried far enough to direct the finger of suspicion towards the nuclear corrections, particularly proton structure. But nuclear theorists could not reconcile the 'hyperfine' α with the 'fine structure' α (at that time based on resonance methods) without postulating an unlikely model for the proton, and the discrepancy remained, for some years, at the level of a few parts in 10^5 for α.

As we shall see below, the discrepancy was resolved in favor of the hyperfine value. Nevertheless, the existing theory is not capable of treating proton structure and proton recoil to yield a value of α as good as that from the sources we mention below. The hyperfine value stands at

$$\alpha^{-1} = 137.03597(22) \qquad [\text{COH 83}] .$$

10.3.4   The Josephson effect

The new light in the nineteen sixties came from an entirely unexpected source — superconductivity. The AC Josephson effect — quantized jumps in the voltage as the flow of electrons is increased across a junction between two semiconductors, the junction being irradiated by microwaves — provides a new path to an old constant, e/h, (whose determination by studies of the photoelectric effect is quite obsolete). In point of fact it is not e/h but $2e/h$ which the Josephson 'plateaux' yield through the relation

$$2eV = h\nu ,$$

and, though the exactness of that equation has been doubted, confidence in it has been built up through years of testing with different types of junction and with different combinations of superconductor until now the phenomenon itself is used to provide a maintained unit of voltage, instead of standard cells, whose deficiencies had long been recognized. The ratio

$$2e/h = 483\ 594\ \text{GHz per volt}$$

is now used to relate the measured frequency (of the electromagnetic radiation falling on a Josephson junction) to the voltage difference between adjacent plateaux, and so to establish a voltage standard in terms of a frequency.

The relevance of this to the problem of α is that it came to be realized [TAY 69] that the expression

$$\alpha^{-1} = \left[\frac{1}{4R_\infty} c \left(\frac{\Omega}{\Omega_M} \right) \left(\frac{\mu_p'}{\mu_B} \right) \frac{(2e/h)}{\gamma_p'} \right]^{\frac{1}{2}}$$

gives α in operational terms, that is, in terms of quantities measured against a particular set of *maintained* units. The suffix M indicates the maintained unit:

thus, (Ω/Ω_M) is the ratio of the SI defined to the maintained ohm. (μ_p'/μ_B) is the ratio of the magnetic moment of the proton to the Bohr magneton — the prime on μ_p' being a conventional symbol to indicate that the protons are not free, but in a spherical water sample. γ_p' is the gyromagnetic ratio of the proton (water sample) measured by the method of precession in a magnetic field proportional to a current (whose value, taken with the maintained resistance, relates to the voltage entering the $2e/h$ experiment). Moreover — and this point is of great significance — no QED corrections are required to evaluate α in this way. Indeed, with confidence in the Josephson equation and with the realization that the least accurately known quantity in the expression for α is γ_p', we are led to

$$\alpha^{-1} = C(\gamma_p')^{-1/2} \, ,$$

where C does not depend on QED, and whose value is known to a degree of accuracy such that the uncertainty in α derives almost entirely from the uncertainty in the measurement of γ_p'. This is the situation at the time of writing (1986). But even in 1969 it was evident that the 'Josephson' value of α supported the hyperfine value (with an acceptable range of values for the nuclear corrections) and the spotlights were trained upon the fine structure experiments and theory, with the conviction that the source of the fine/hyperfine discrepancy was to be found there.

It was indeed. On the experimental side some overlooked systematic effects were discovered, leading to a revision of the value of α, and on the theoretical side, it was admitted that one of the many terms in the evaluation of the QED corrections had been entered with the wrong sign. When this was put right (in the early nineteen seventies) the values of α obtained from the two spectroscopic routes in hydrogen stood in satisfactory agreement with the Josephson value, and the QED theory was, to that extent, validated.[a]

The Josephson value, quoted in a recent publication [TAY 85], is

$$\alpha^{-1} = 137.035981(12).$$

The fine structure values quoted at the end of section 10.3.1 are free of the errors mentioned above.

It appears now that hydrogen has yielded all the accuracy it is capable of in the determination of α. It is not without some pangs of regret that we are

---

[a] A new route to α via γ_p' has recently been proposed by B. W. Petley and B. P. Kibble (Metrologia **23**, (1986/7) 167). The method depends on relating the Josephson constant, $2e/h$, to the dimensions and velocity of a rectangular coil moving across a magnetic field.

obliged to look elsewhere to find a more accurate value. In the next sections we shall enter these fields: the hyperfine structure of muonium, the g-factor anomaly of the electron, and the quantized Hall effect.

10.3.5　Muonium hyperfine structure

Muonium $(\mu^+ e^-)$ is, in principle, a simpler structure than hydrogen but is possessed of a similar pattern of energy levels: in particular, the hyperfine structure in the ground state offers a path to the determination of α. From the early nineteen sixties it became possible to measure this structure with ever increasing precision. Current measurements stand at the level of about 3 parts in $10.^8$ The same apparatus used for the hyperfine structure measurements can be used also to obtain values for the ratios of magnetic moments and ratios of masses which enter into the theoretical expression. An account of these measurements and of the theory by which they are interpreted is given in section 4.1.6. At the time of writing the experimental result is an order of magnitude more reliable than the theory.

The value of α derived from a comparison of theory with experiment is

$$\alpha^{-1} = 137.035988\,(20)\ ,\quad [\text{SAP 83}]$$

where the uncertainty derives mainly from theory.

10.3.6　The electron and the positron: the g-2 anomaly

Free particles, strictly speaking, are outside the scope of this book. But the g-2 anomaly of the electron is so closely bound up with the hydrogen spectrum that little excuse is needed for devoting a section to it, more particularly since it offers, at the time of writing, a value of α of higher accuracy than is to be found elsewhere. And again, the particular set of experiments which are the source of these exceedingly accurate measurements of $g - 2$ present the electron as a particle bound in its apparatus, and hence to the earth. The name given is geonium. It is stretching a point to call this a variety of hydrogen (the binding is non-Coulombic), but if excuse be needed, this will serve.

It was in the late nineteen fifties that it began to be possible to measure the g-factor of free electrons. The technique, introduced by Dehmelt [DEH 58] and followed up by Franken, was optical pumping combined with 'spin-exchange' (section 1.5). If you have a mixture of two paramagnetic species in a vapor and orient one of them by optical pumping, the other becomes oriented by collisions with the first. In this case the two species were sodium atoms (which are easy to orient by optical pumping) and 'free' electrons produced by a short pulse of gas discharge. A magnetic resonance experiment carried out on the electrons — in the course of which the orientation of the electrons is changed — is monitored by detecting the collisionally-induced re-orientation of the sodium atoms through changes in their opacity to polarized resonance radiation. In the same experiment the resonance of the sodium atoms themselves is easily displayed, so the *g*-value of the electrons is measured in relation to the *g*-value of the ground state sodium atoms.

But the techniques now in vogue measure $g-2$ in one way or another directly. It is usual to work with a quantity a defined by

$$a = \frac{1}{2}g - 1 ,$$

with g defined by

$$\hat{\mu} = g(\mu_B/\hbar)\hat{s} ,$$

in which μ_B is the Bohr magneton, $e\hbar/2m_e c$.

Now, the angular frequency of precession, ω_s, of the electron spin in a magnetic field B is

$$\omega_s = g(\mu_B/\hbar)B = g(e/2m_e c)B ,$$

whereas the cyclotron frequency of the electron in the same field is

$$\omega_c = (e/m_e c)B ,$$

so

$$\omega_s = \frac{1}{2}g\omega_c ,$$

whence

$$a = (\omega_s - \omega_c)/\omega_c = \omega_D/\omega_c .$$

A measurement of the difference ω_D between the two precessional frequencies in the same magnetic field, as a fraction of the cyclotron frequency itself in that field, is therefore a direct determination of a.

Relativistic effects are taken into account in work at the highest level of accuracy, and the foregoing analysis is usually expressed in quantum terms. The energy level diagram is shown in Fig. 10.1

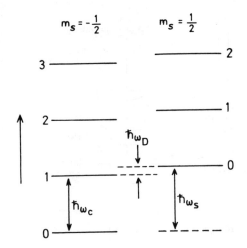

Fig. 10.1. Cyclotron (Landau) levels for the two spin-states of the electron. ω_c, ω_s and ω_D are defined in the text.

10.3.6.1   Resonance experiments

The experiments to which we have referred have been developed over a period of many years at the University of Washington (Seattle) by Dehmelt, Van Dyck, Schwinberg, and others [VAN 84], and their characteristic is that the electrons (or other charged particles — for similar experiments have been carried out on positrons and atomic ions) are bound in a configuration of electric and magnetic fields called a Penning trap (Fig. 10.2). In its simplest form this is a quadrupolar electric field with potential $V(x, y, z) = A_0(2z^2 - x^2 - y^2)$, generated by means of electrodes as indicated in the figure, on which is superimposed a uniform magnetic field in the z-direction. The charged particles undergo cyclotron motion in the $x - y$ plane, combined with a harmonic oscillation in the z-direction. The cyclotron motion will generally be centred away from the centre of the trap and will drift round it at a slower frequency — magnetron motion. The magnetron frequency also is one of the characteristic frequencies of the device.

Within the trap single particles can be isolated, held for periods of time measured in months, 'cooled' so that their motion is restricted to a very small region of space, irradiated with oscillatory fields, and interrogated from time

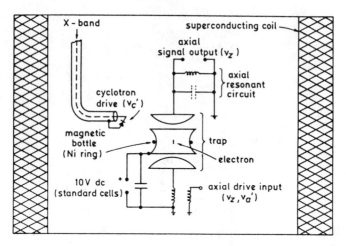

Fig. 10.2. Penning trap, after Van Dyck *et al.* [Van 84].

to time by external electrical pulses applied to the $\pm z$ electrodes to determine what is their quantum state. Hence, it can be discovered whether or not they have undergone some transition between the levels indicated in Fig. 10.1, and, by resonance experiments based on subjecting the particles to radio frequency fields, ω_D and ω_c can be determined.

The very high accuracy in the determination of the resonance frequencies is a result of the long time that the particles can be held in a controlled way, their phase evolution being effectively locked to the oscillators.

An interesting feature of the experiments is that the transitions whose frequencies are to be compared, ω_c and ω_s, are, respectively, electric dipole and magnetic dipole transitions. The cyclotron (or, rather, cyclotron-magnetron) transitions are induced by direct irradiation of the particles with an rf electric field at the appropriate frequency; the spin-flip transitions by introducing a non-uniformity into the magnetic field. To electrons undergoing cyclotron motion this non-uniformity of the magnetic field will appear to rotate at the cyclotron frequency: a modulation at the small difference frequency ω_D will provide the necessary stimulus at the frequency ω_s to bring about the spin-flip transitions.

A result reported in 1984 [VAN 84] for the value of the anomaly was

$$a_e = 1\,159\,652\,193\,(4) \times 10^{-12} \,,$$

an astounding claimed accuracy of about 4 parts in 10^9, which corresponds to a determination of g_e to an accuracy of 4 parts in 10^{12}.

On the theoretical side, the evaluation of the QED coefficients in the expansion of g_e as a function of α has proceeded sufficiently far that the largest source of uncertainty in its evaluation lies in the value of α. Hence the experimental value may be used with the theoretical relationship to determine α. The result is

$$\alpha^{-1} = 137.035\,994\,2 \binom{5}{\text{exp}} \binom{89}{\text{theor}}, \qquad \text{[VAN 84]}$$

the combined uncertainty now being less than 1 part in 10^7.

We return to these experiments in section 10.4 in connection with the electron/proton mass ratio.

10.3.6.2 *Free precession experiments*

The names of Crane and Rich, among others, are associated with these experiments which have been carried out with successive refinements over many years at the University of Michigan. Polarized electrons are injected into a long solenoid where the uniform field is slightly modified to trap them while they undergo many cycles of orbital motion and spin precession. These two motions get progressively out of step. The electrons are ejected after a time T, their spin orientation tested, whereby the precession of spin relative to cyclotron motion is determined for a calculable number of cyclotron orbits. This is precisely the information needed to determine a.

The result quoted by Wesley and Rich in 1971 [WES 71] was

$$a_e = 1\,159\,657.7\,(35) \times 10^{-9}$$

which, although it has now been superseded, was at the level of accuracy required for comparison with the theoretical value in the early nineteen seventies.

10.3.7 *Muons: the g − 2 anomaly*

Just as measurements of muonium hyperfine structure have contributed to knowledge of α (with reliance on QED) or can be used as checks of QED (with reliance on a value of α obtained elsewhere), so also have measurements of the $g - 2$ anomaly for μ^- and μ^+ featured in the same debates.

As for electrons, the technique is to determine the differential precession of the spin with respect to the orbital motion. In an experiment at CERN [BAI 79] muons, produced by the decay of pions, were injected into an annular storage ring of radius about 7m, where, on entering, their spin direction was in the direction of motion, and therefore perpendicular to the constraining magnetic field. The differential precession over the time interval until the muons decayed was monitored by observing the flux of decay electrons in a particular direction — these are preferentially emitted in the spin direction.

The results obtained were

$$a(\mu^+) = 1\,165\,912\,(11) \times 10^{-9}$$
$$a(\mu^-) = 1\,165\,938\,(12) \times 10^{-9}$$

both of which are in agreement with the theoretical prediction of

$$a(\mu) = 1\,165\,920(10) \times 10^{-9} \; .$$

10.3.8 The quantized Hall effect

This most unexpected phenomenon entered the field of fundamental constants from semiconductor physics in 1980 [KLI 80]. From the many articles which have since appeared on this subject we select for reference a review by the discoverer of the effect, von Klitzing [KLI 83] and one which emphasizes the importance of the effect in metrology as a possible realization of the ohm, [BLI 83].

Measurements of the transverse resistance to the flow of electrons in a two-dimensional layer between two semi-conductors, the layer being at low temperature and in a high magnetic field (configuration for measuring the Hall coefficient for semi-conductors) reveal that this resistance is quantized as follows:

$$R_\perp = h/e^2 i, \qquad \text{where } i \text{ is an integer.}$$

The exactness of this equation, not at first fully appreciated, has been much investigated and is not yet (1986) fully understood. Nevertheless its validity is gradually gaining acceptance. Its importance lies in its relation to the fine structure constant: to within the dimensions of velocity and certain fixed numbers, it *is* the fine structure constant, for

$$\alpha = \left(\frac{\mu_0 c}{2}\right) \frac{e^2}{h} = \left(\frac{\mu_0 c}{2i}\right) \frac{1}{R_\perp}$$

with $\mu_0 = 4\pi \times 10^{-7}$ Hm^{-1} (definition) and $c = 299\,792\,458$ ms^{-1} (definition).

An early determination of $R_\perp$ [TSU 82] led to the value

$$\alpha^{-1} = 137.035\,968(23) \ .$$

With due caution, measurements of the quantized Hall resistance are now being included in assessments of the values of fundamental atomic constants.

The mean of six recent measurements (COD 86) leads to

$$\alpha^{-1}(\Omega/\Omega_{\mathrm{BI85}}) = 137.036\,204\,4(85) \ ,$$

the recommended value of $(\Omega/\Omega_{\mathrm{BI85}})$ being $1 + 1.563(50) \times 10^{-6}$.

10.3.9   The values of α

We conclude this section by bringing together the values of α^{-1} which we have quoted or implied in the text:

Method	Values of $(\alpha^{-1} - 137)$	Source
H fine structure (resonance)	.0356(7)	SHY 71
	.0358(5)	VOR 71
H fine structure (level-crossing)	.0354(6)	BAI 71
He fine structure	.036 113(110)	LEW 78
H hyperfine structure	.035 97(22)	COH 83
Josephson effect and γ_p	.035 981(12)	TAY 85
Muonium hfs	.035 988(20)	SAP 83
$R_\perp$	.035 990 2(85)	COD 86
g_e	.035 994 22 $\binom{5}{\exp}\binom{89}{\mathrm{theor}}$	VAN 84

10.4   The Electron-to-Proton Mass Ratio

This ratio entered our discussion of the Rydberg constant (section 5.4.1) and of spectroscopic isotope shifts (section 5.4.3). In the latter section it was recalled that, historically, the measurement of displacements between the spectral lines of hydrogen and its isotopes had provided a basis for the determination of

m_e/m_p, but that the accuracy of measurement at the present time was such that nuclear size effects, and the uncertainty in their evaluation, were beginning to provide an obstacle to the determination of the mass ratio by this route.

Meanwhile, the determination of mass ratios by the method of precession in Penning traps (section 10.3.6.1) is not subject to these uncertainties, and the accuracy of the method is now well beyond that of the spectroscopic methods. The technique of the Penning trap method is, essentially, to compare the cyclotron frequencies $\omega_c = (e/mc)B$ of the particles in the same magnetic field, B. For the electron and the proton the ratio

$$m_p/m_e = 1836.152\,701(37)$$

was obtained [VAN 86]. In a different approach Wineland *et al.* [WIN 83] have made a variety of measurements on characteristic frequencies of the ion $^9\text{Be}^+$ in its ground state. They measure Paschen-Back (magnetically-decoupled) hyperfine transitions and the cyclotron frequency. By use of auxiliary constants they obtain from these measurements the mass ratio of the Be ion to the electron. Again, relying on a mass spectroscopic determination of the mass of the ion to that of the proton, they ultimately arrive at m_p/m_e. By virtue of the control they are able to exercise over the Be ions in the trap they are able to claim very high accuracy for their result:

$$m_p/m_e = 1836.152\,38(62) \ ,$$

which is consistent with the result of Van Dyck *et al.*

10.5   CODATA

There exists an international body, the Committee on Data for Science and Technology (CODATA), a committee of the International Council of Scientific Unions. CODATA has a Task Group on the Fundamental Physical Constants, for many years chaired by E. R. Cohen and latterly by T. J. Quinn of the Bureau International des Poids et Mesures, Pavillon be Breteuil, 92310 Sèvres, France. The chairman of CODATA receives information on precision measurements of fundamental interest, and from time to time circulates a Newsletter and a Bulletin to collaborating persons and institutions. The Bulletin is an archival journal available in the libraries of many institutions.

From time to time the personnel of CODATA evaluate the information that is before them and produce a consistent set of constants which they recommend for general adoption. The CODATA Newsletter No. 38 (1986) contained, among others, the following recommended values:

$$R_\infty = 1.097\,373\,153\,4(13) \times 10^7 \text{m}^{-1}$$
$$\alpha^{-1} = 137.035\,989\,5(61)$$
$$m_p/m_e = 1\,836.152\,701(37)\;,$$

and these values, together with the recommended values of a very large number of other fundamental physical constants, the measurements from which they were derived, and the arguments leading to their adoption, appear also in the CODATA Bulletin No. 63 (November, 1986).

REFERENCES

BAI 71   J. C. Baird, J. Brandenberger, K.-T. Gondaira and H. Metcalf, in *Precision Measurement and Fundamental Constants I*. eds. D. N. Langenberg and B. N. Taylor. NBS Special Publication 343 (1971) p. 345

BAI 79   J. Bailey and 18 others, *Nucl. Phys.* **B150** (1979) 1.

BIR 41   R. T. Birge, *Rep. Prog. Phys.* **8** (1941) 90.

BLI 83   L. Bliek and V. Kose, in *Quantum Metrology and Fundamental Physical Constants*, eds. P. H. Cutler and A. A. Lucas, (Plenum, 1983) p. 249.

COD 86   CODATA Bulletin, No. 63, Pergamon Press (November, 1986). Reprinted in *Rev. Mod. Phys.* (1987) October.

COH 65   E. R. Cohen and J. W. M. DuMond, *Rev. Mod. Phys* **37** (1965) 537.

COH 73   E. R. Cohen and B. N. Taylor, *J. Chem. Ref. Data* **2** (1973) 663.

COH 83   E. R. Cohen, same publication as BLI 83, p. 523 (1983).

DEH 58   H. G. Dehmelt, *Phys. Rev.* **109** (1958) 381.

GER 67   K. R. German, Ph.D. Thesis, University of Michigan (1967).

HIN 85   E. A. Hinds, J. D. Prestage and F. M. J. Pichanick, *Phys. Rev.* **A32** (1985) 2615.

HIN 86   E. A. Hinds, J. D. Prestage and F. M. J. Pichanick, *Phys. Rev.* **A33** (1986) 68.

KLI 80   K. von Klitzing, G. Dorda and M. Pepper, *Phys. Rev. Lett.* **45** (1980) 494.

KLI 83   K. von Klitzing, in *Atomic Physics 8*, eds. I. Lindgren, A. Rosen and S. Svanberg (Plenum, 1983) p. 43.

KPO 71   A. Kponou, V. W. Hughes, C. E. Johnson, S. A. Lewis and F. M. J. Pichanick, *Phys. Rev. Lett.* **26** (1971) 1613.

LEW 74   M. L. Lewis, in *Atomic Physics 4*, eds. G. zu Putlitz, E. W. Weber and A. Winnaker (Plenum, 1974).

LEW 78   M. L. Lewis and P. H. Serafino, *Phys. Rev.* **A18** (1978) 867.

PET 85   B. W. Petley, *The Fundamental Physical Constants and the Frontier of Measurement* (Adam Hilger Ltd, 1985).

PRE 85   J. D. Prestage, C. E. Johnson and E. A. Hinds, *Phys. Rev.* **A32** (1985) 2712.

SAP 83   J. Sapirstein, *Phys. Rev.* **A51** (1983) 985.

SHY 71   Tong Wha Shyn, R. T. Robiscoe and W. L. Williams, same publication as BAI 71, p. 355 (1971).

TAY 69   B. N. Taylor, W. H. Parker and D. N. Langenberg, *Rev. Mod. Phys.* **41** (1969) 375.

TAY 85   B. N. Taylor, *J. Res. Nat. Bur. Stand.* **90** (1985) 91.

TSU 82   D. C. Tsui, A. C. Gossard, B. F. Field, M. E. Cage and R. F. Dziuba, *Phys. Rev. Lett.* **48** (1982) 3.

VAN 84   R. S. Van Dyck, Jr., P. B. Schwinberg and H. G. Dehmelt, in *Atomic Physics 9*, eds. R. S. Van Dyck, Jr. and E. N. Fortson (World Scientific, 1984) p. 53 and earlier refs. there given.

VAN 86   R. S. Van Dyck, Jr., F. L. Moore, D. L. Farnham and P. B. Schwinberg, *Bull Am. Phys. Soc.* **31** (1986) 244.

VOR 71   T. V. Vorburger and B. L. Cosens, same publication as BAI 71, p. 361 (1971).

WES 71   J. C. Wesley and A. Rich, *Phys. Rev.* **A4** (1971) 1341.

WIN 83   D. J. Wineland, J. J. Bollinger and W. M. Itano, *Phys. Rev. Lett.* **50** (1983) 628; *erratum* **50** (1983) 1333.

SUPPLEMENTARY REFERENCES

Atoms in cavities
Radiative properties of Rydberg states in resonant cavities.
S. Haroche and J. M. Raimond, *Adv. Atom Mol. Phys.* **20** (1985) 347.

Quantum-electrodynamic level shifts between parallel mirrors.
G. Barton, *Proc. Roy. Soc.* **A410** (1987) 141 and 175.

Atoms interacting with 'squeezed states' of light
Atomic level shifts in a squeezed vacuum.
G. J. Milburn, *Phys. Rev.* **A34** (1986) 4882.

Atoms in non-Euclidean spaces
Atomic fine structure calculations in spaces of constant curvature.
N. Bessis, G. Bessis and R. Shamseddine, *J. Phys.* **A15** (1982) 3131;
N. Bessis, G. Bessis and D. Roux, *Phys. Rev.* **A30** (1984) 1094.

Diagnostics of plasmas
H. R. Griem, *Spectral Line Broadening by Plasmas* (Academic Press, 1974).
Doppler-free spectroscopy in a dense plasma.
K. Danzmann, K. Grützmacher and B. Wende, *Phys. Rev. Lett.* **57** (1986) 2151.

Masers in the sky
A. H. Cook, *Celestial Masers* (Cambridge, 1977).

Quasars
M. J. Rees, *Observatory* **98** (1978) 210.

SUBJECT INDEX